Robotic Engineering

An Integrated Approach

Richard D. Klafter

Professor
Department of Electrical Engineering
Temple University
Philadelphia, Pennsylvania 19122

Thomas A. Chmielewski

Unit Manager
General Electric, Advanced Technology Laboratories
Moorestown, New Jersey
and
Drexel University
Electrical and Computer Engineering Department

Michael Negin

President
Mnemonics, Inc.
Mount Laurel, New Jersey 08054

Prentice Hall, Englewood Cliffs, New Jersey 07632

Library of Congress Cataloging-in-Publication Data

Klafter, Richard David.
    Robotic engineering : an integrated approach / Richard D. Klafter,
Thomas A. Chmielewski, Michael Negin.
      p.   cm.
    Includes bibliographies and index.
    ISBN 0-13-468752-3
    1. Robotics. I. Chmielewski, Thomas A. II. Negin, Michael.
III. Title.
TJ211.K555 1989
629.8'92—dc19   88-31899
                                          CIP

Editorial/production supervision
  and interior design: *Denise Gannon*
Cover design: *20/20 Services, Inc.*
Manufacturing buyer: *Mary Noonan*

 © 1989 by Prentice-Hall, Inc.
A Division of Simon & Schuster
Englewood Cliffs, New Jersey 07632

Printed in the United States of America

10 9 8 7 6 5 4 3 2 1

ISBN 0-13-468752-3

PRENTICE-HALL INTERNATIONAL (UK) LIMITED, *London*
PRENTICE-HALL OF AUSTRALIA PTY. LIMITED, *Sydney*
PRENTICE-HALL CANADA INC., *Toronto*
PRENTICE-HALL HISPANOAMERICANA, S.A., *Mexico*
PRENTICE-HALL OF INDIA PRIVATE LIMITED, *New Delhi*
PRENTICE-HALL OF JAPAN, INC., *Tokyo*
SIMON & SCHUSTER ASIA PTE. LTD., *Singapore*
EDITORA PRENTICE-HALL DO BRASIL, LTDA., *Rio de Janeiro*

Contents

3   Mechanical Systems: Components, Dynamics, and Modeling   **101**

B Motor Selection in the Design of a Robotic Joint

C Digital Control of a Single Axis   729

Index   741

Preface

Although industrial robots have been available for a number of years, it is only since the early 1970s that research efforts into these sophisticated computer-controlled devices has begun to accelerate. The primary reasons for this are the advent and availability of the microprocessor and, in this country, the realization by industry that robots must be used to meet the increased competition from foreign manufacturers.

As a result of the industrial experience gained during a leave of absence, the principal author organized a senior/graduate course in robotics in the early 1980s. In assemblying the material for this course, the author found that very little was written on the specific subject of robotics. The reason for this is that, quite simply, robotics is not a single discipline. Rather, it is a highly multidisciplinary field that combines the areas of controls, computers (both the hardware and software aspects), measurement technology (i.e., sensors), pattern-recognition techniques and hardware (e.g., vision systems), and various aspects of mechanical engineering, including statics, dynamics, kinematics, and mechanical design. A complete study of the subject should also involve some discussion of applications as well as the economics of robots and the sociological consequences of placing them in the workplace. Although it was certainly possible, at the time, to find material on many of these individual subjects, there was no single compilation of the topics that existed which would permit a comprehensive course to be taught. Moreover, many of the papers written were extremely low level and were often nothing more than glorified sales pitches.

A number of years later, the situation has changed somewhat, with a relatively large number of books on the subject having come out in the interim. However, these are, for the most part, descriptive, rather low-level texts that are aimed primarily at the two-year technology student and are therefore inappropriate for engineering courses at any level. Of the few that are written at a higher level, some are rather sketchy and others are extremely detailed in only a few areas. Thus neither group is really applicable for comprehensive "core" (or first-level) courses that seniors and/or graduate students would (and it is our feeling should) want to take.

It is quite apparent that robotics is a "hot" area and that there will be a definite need for a book that will permit an engineering core course (or courses) to be taught. Although there are still relatively few of these being offered at universities throughout the country, it seems to us that this is a result of there being *no appropriate text available* rather than there being a lack of interest in teaching such a course. Clearly, people will always want to teach their "specialties" (e.g., robotic controls, machine vision, etc.). It is our belief, however, that the more specialized courses that cover only a few topics in depth will have a greater impact on the student, and therefore, should be taught only after the relationship among the various disciplines that go into producing a working robot are clearly understood. Thus we feel, for example, that it is not appropriate to begin talking about optimal or adaptive control of a robot until one fully appreciates the advantages and disadvantages of the type of control currently utilized and how the large swings in inertia (inevitably occurring as the manipulator moves in its work volume) affects the particular control strategy selected. Having said this, it is our judgment that a comprehensive text such as this one should provide the reader with the "why" and "how to" aspects of robotics. Theorems and proofs are better left to follow-up specialty courses. This *does not* mean, however, that we utilize the anecdotal, often pseudotechnical approach that characterizes many of the currently available texts and papers on the subject. Rather, we have utilized our extensive pedagogical and practical experience (with robots) to present to the reader many of the theoretical and practical concepts and ideas that are essential to understanding how a robot is designed and how it works. In doing this, it is our hope that the book will be extremely useful in the (engineering) academic sector and in the engineering workplace. With these ideas in mind, we have organized the book in the following manner:

In Chapter 1, a fairly detailed introduction is presented where the terminology and various robot types, as well as the history, sociological, and economic implications of these forms of automation, are discussed. In addition, current and future applications are given. The chapter goals are for the reader to be able to understand what an industrial robot is and what it is not, where it is applicable and where it is not, and finally, how such devices have evolved and how they well may cause another industrial revolution to occur.

Chapter 2 deals with the robot's various component parts as well as how these devices are normally utilized in an *automated system*. At the conclusion of the chapter, it is expected that the reader will be able to identify the major system components of a robot from a high-level, black-box point of view and will also be able to understand the considerations that go into both the development of robotic systems specifications and the selection of system components.

The next chapter presents the mechanical structure and discusses a variety of devices and components as they relate to robots. Various methods of converting rotary to linear motion are given from both the ideal and "real-world" points of view. It is the purpose of this chapter to provide an understanding of how certain mechanical components behave and how power is transmitted from an actuator to a load. The reader will also learn about how many of these devices are used in a practical manner to produce a working robotic manipulator.

The typical control structure of modern industrial robots is presented along with a fairly detailed discussion of various types of actuators and power amplifiers in Chapter 4. The reader will not only gain an understanding of how classical servo theory is applied to a robotic system to produce the desired robotic joint performance but will also learn about the various actuators and amplifiers available to the robot designer and which are preferred in a given application. Many practical considerations that affect the proper operation of a robotic joint are included here.

In the following chapter, the topic of nonvision-based robotic sensors is presented in great detail. A large number of internal sensors are discussed, with special emphasis being given to the practical aspects of several, including the optical encoder. External sensors are also introduced, with the topics of proximity, welding, and tactile sensors being discussed. The purpose of the chapter is to demonstrate clearly the role played by internal sensors in the control of individual robotic joints and also by external sensors in providing the robot with knowledge about its external environment. Also, the practical aspects of the presentation should assist the reader in understanding why certain sensors are to be preferred over others in a given application.

Robotic (or machine) vision is discussed in Chapter 6. Various components of a vision system, as well as a number of image recognition techniques are presented. The reader will be able to understand the similarities and dissimilarities of computer vision relative to other types of sensors and will appreciate the magnitude of the information-processing problem associated with using computer vision in a robotics application. The material in the chapter covers various vision sensors and systems, and discusses the capabilities (e.g., object detection versus inspection) of currently available, practical cost-effective vision technology.

In Chapter 7 the architectural and hardware considerations related to the computers utilized in a robotic system are discussed. In addition, the role played by the computational elements in robotic applications is given and a summary of various robotic programming languages is presented. Various trade-offs that are required when using different computer architecture implementations for robotic systems are discussed. The reader will also learn about the practical considerations that go into the selection of a robot computer system, including the hardware, software, and task programming aspects.

The important topics of coordinate transformations, along with how to obtain the forward and inverse or back solutions, are presented in Chapter 8. Homogeneous transformations are introduced and how they are applied to a robot's kinematic structure. Additional discussion involves the method used in a robot to represent points in space and then how to utilize this information to produce continuous-path, straight-line, and other types of coordinated motions.

Chapter 9, the concluding chapter, brings together many of the important technological ideas presented in the preceding chapters. This is accomplished by designing various aspects of a robot required to perform a specific task (e.g., sorting eggs). From the material in this chapter it is expected that the reader will be able to take a set of given specifications and actually come up with a potential robot design. This should include the mechanical configuration, the control and computer structures, and the choice of actuators that will meet all of these specs.

Three appendices are included and should be of interest. The first is a compilation of existing commercial robots and their specifications/attributes. The second presents an orderly method of selecting a servomotor for a specific task. The last one discusses the digital control of a single robotic joint.

As a text, the book is ideal for courses at the senior/graduate level in electrical engineering since it places a good deal of emphasis on subjects that are traditionally considered to be "electrical in nature." However, many modern mechanical engineering curricula now require their students to take courses in controls (and systems), computers, and mathematics beyond the standard calculus, analytic geometry, and differential equations sequence. For such departments, this book could be utilized in a robotics course with the assurance that much of the material would be within the abilities of their students. In fact, over the years that the authors have used the manuscript in a classroom environment, there have always been a number of mechanical engineers who successfully completed the course. The same is also true for the few computer science students, although, admittedly, they had a much more difficult time because of their lack of specific engineering knowledge. Although this book is definitely not an engineering technology text, since it assumes a fairly extensive analytical background, there are a *small number* of four-year technology programs (primarily, electrical) that could use some of the material in a robotics course at the senior level.

As mentioned above, we have utilized the book in graduate courses that had both graduate and selected undergraduate students enrolled. There is more than enough material provided to cover a two-semester course. Clearly, the instructor may wish to elaborate in some areas and gloss over others. This would obviously depend on the backgrounds of the students and their needs. In our case, Chapters 1 and 2 were covered in about three classes with the remaining part of the semester devoted to (sometimes expanded versions of) the third, fourth, and parts of the fifth and seventh chapters. Also included was much of the material contained in Appendices B and C. The second semester was then devoted to sensors (vision and nonvision based), kinematics, and computer systems and robotic languages. Also, Chapter 9 was discussed in great detail, with the students encouraged to submit other designs for the same task.

As a final word, it is our belief that the practical engineering approach that is utilized throughout the text will most certainly interest engineers who are working in the fields of controls and automation (e.g., those with backgrounds in electrical engineering, mechanical engineering, and computer science/engineering). In addition, engineers working in industries that may be *users* of robots may find this book helpful in providing them with the background needed to select the correct type of robot (and the various options) to perform a specific task at their company's plant.

ACKNOWLEDGEMENT

The authors wish to thank a number of people who have made this book a better work with their constructive criticisms, helpful suggestions, and overall support. In particular, we wish to extend our appreciation to our former colleagues at United States Robots, Inc. with special thanks going to John Stetson. Also, our colleagues and friends at Drexel University's ECE Department and Temple University's EE Department were extremely supportive. Special mention is due Dr. Paul Kalata of Drexel University who assisted in the preparation of Appendix C.

As is usually the case with a project that is carried out within a university environment, our undergraduate and graduate students have greatly influenced the final version of the text. Thus, we would like to thank all those students who gave us comments and suggestions concerning various aspects of the book. In this respect, the following individuals (at Temple University) are worthy of special mention: J. Coyle-Byrne, D. Knoll, B. Maber, K. Reed, and K. Roy.

In addition, Mrs. Rachel Balaban, Mrs. Oksana Bilyk, Mrs. Carol Dahlberg, and Ms. Pat Taddei assisted in the (intelligent) proofreading of the manuscript and in performing other important clerical activities associated with the project.

1

Introduction

1.0 OBJECTIVES

As the reader will begin to appreciate, the study of robotics involves understanding a number of diverse subjects. For example, several engineering disciplines as well as those relating to physics, economics, and sociology must be mastered before one can truly acquire more than a nodding acquaintance with the field. This book is intended to be primarily an *engineering* text. However, before beginning a discussion of the technical aspects of robotics, it is necessary for the reader to become conversant with the language of the subject. Thus the overall objective of this chapter is to provide an overview of robotics, presenting the material in a descriptive, fairly nontechnical manner.

Specifically, the topics that are covered are as follows:

- Historical perspective of robots
- Classification of robots
- Description of the major robot components
- Discussion of fixed versus flexible automation
- Economic considerations used for the selection and justification of robots
- Sociological consequences of automation/robots
- Robot state-of-the-art survey
- Current and future applications of industrial robots

1.1 MOTIVATION

When one first hears the word *robot*, the image that probably comes to mind is that of a mobile biped that is both humanoid in structure and capable of independent actions (e.g., thinking). The George Lucas (*Star Wars*) creation C3-P0 is an example. Such devices are, unfortunately, still relegated to science fiction novels and motion pictures. The truth is that we simply do not, at this time, know how to create machines with this degree of intelligence and mobility. In fact, it is likely that unless there is a significant breakthrough in a number of areas such as artificial intelligence, computers, and power storage devices, most of the readers of this book will never see a robot that has anywhere near the capability of C3-P0. For this reason, in this book we choose to discuss industrial robots almost exclusively.

As we will see in subsequent chapters, an industrial robot is a complex electromechanical device that brings together a large number of disciplines in what could be termed a "polygamous relationship." Despite the fact that this chapter will be a relatively descriptive, nontechnical introduction to the subject, a variety of important questions must still be answered. For example, what types of mechanisms can be classified as industrial robots and what types cannot? From where did robots evolve, and what was the nature of the first devices produced? Also, since the robots about which we will be speaking will almost always be used in a manufacturing environment, what are the economic justifications for utilizing such devices for a given task? Moreover, what applications are appropriately handled by robots both now and in the future? Finally, what type of robots are available in the marketplace today? Besides these important questions that we will attempt to answer later in this chapter, there are other considerations worthy of our attention.

No study of robots would be complete without some discussion of their sociological consequences. It is clear that they will have a significant impact on the manufacturing environment. But will workers in industries utilizing robots be displaced from their jobs, or will more jobs actually be created? How will these workers accept this new form of technology, and what can management and/or the government do to ease any problems that result from the introduction of this type of automation into the workplace? We will attempt to answer these extremely difficult questions, but the reader is warned not to expect easy answers: the authors unfortunately do not have magical solutions to the inevitable social problems. Regardless, we believe that if the United States is to reestablish itself as a world leader in manufacturing, robots will have to be utilized in ever-increasing numbers.

We now trace the origin of the modern industrial robot and indicate how these devices have evolved in relationship to other technological developments.

1.2 A HISTORICAL PERSPECTIVE OF ROBOTS

The word *robot* was first used in 1921 by the Czech playwright, novelist, and essayist Karel Capek in his satirical drama entitled *R.U.R.* (*Rossum's Universal Robots*)

[1]. It is derived from the Czech word *robota*, which literally means "forced laborer" or "slave laborer." In his play, Capek pictured robots as machines that resembled people but worked twice as hard. These devices had arms and legs and no doubt were similar in many ways to C3-P0 in the 1977 film *Star Wars*. The industrial robot of today does not look the least bit human and therefore has little in common with Capek's robots.

Although Capek introduced the word "robot" to the world, the term "robotics" was coined by Isaac Asimov in his short story "Runaround," first published in 1942. This work is also notable because the so-called "Three Rules (or Laws) of Robotics" are presented for the first time:

1. A robot may not injure a human being, or, through inaction, allow one to come to harm.

2. A robot must obey the orders given it by human beings except where such orders would conflict with the First Law.

3. A robot must protect its own existence as long as such protection does not conflict with the First or Second Laws. [2]

Asimov has stated that workers in the field of artificial intelligence indicated to him that these three laws should serve as a good guide as the field progresses.

Before proceeding with the history of robots themselves, it is interesting to trace briefly the antecedents of these devices. Surprisingly (perhaps), the concept of a programmable machine dates back to eighteenth-century France and includes inventors such as Bouchon, Vacaunson, Basile, and Falcon. Possibly the best known of the group is Joseph Jacquard who developed the mechanical loom controlled by punched cards. Its mass production occurred around 1801. In the third decade of the nineteenth century, an American, Christopher Spencer, produced a programmable lathe called the *automat* that was capable of turning out screws, nuts, and gears. Its "programming," and hence the pattern that was to be cut, was modified through the use of a set of interchangeable cam guides that were fitted on the end of a rotating drum.

The problem of removing hot ingots from a furnace was solved by Seward Babbit in 1892. He developed and patented a rotary crane equipped with a motorized gripper. In 1938–1939, Willard Pollard invented a *jointed* mechanical arm that was utilized primarily in paint spraying. A similar device was developed by an employee of the DeVilbiss Co. (a current manufacturer of robots), Harold Roselund.

A "relative" of the robotic manipulator, the *teleoperator* or *telecheric,* was developed during World War II to permit an operator to handle radioactive materials at a safe distance. Just after the conclusion of this war, George Devol, the acknowledged "father of the robot," developed a magnetic process controller that could be used as a general-purpose playback device for controlling machines. In the same year (1946), Eckert and Mauchly built the ENIAC, the first large-scale electronic computer, and at the Massachusetts Institute of Technology (MIT) a general-purpose digital computer (*Whirlwind*) solved its first problem. One year later in 1947, a servoed* electric-powered teleoperator was introduced by Raymond Goertz. It permitted the servo-controlled slave to follow the position command of the master (i.e., the operator). However, no force control was incorporated into the design until the following year. By permitting the load to back-drive the mechanical interface to the master, the sense of touch was restored to the operator. In 1952 the first numerically controlled machine tool was developed at the MIT Servomechanism Laboratory.

It is generally acknowledged that the "robot age" began in 1954 when Devol patented the first manipulator with a playback memory. This device was capable of performing a controlled motion from one point to another (i.e., point-to-point motion). In addition, Devol also coined the phrase *universal automation*. (This was to be shortened later to *unimation*.) Five years after this, the first commercial robot was sold by the Planet Corporation. However, in 1960 Devol chose to sell his original robot patents (approximately 40 in all) to Consolidated Diesel Corporation (Condec), which actually developed the Unimate robot at its newly formed subsidiary, Unimation, Inc.† The design of the Unimate combined the playback features of numerically controlled devices (e.g., milling machines) with the servo-controlled capabilities of the telecherics developed by Goertz. Two years later, in 1962, General Motors installed the first Unimate on one of its assembly lines in a die-casting application.

By the mid 1960s, the new field of robotics sparked the formation of several centers of research into this area and the related topic of artificial intelligence (AI) at such institutions as MIT, Stanford University, Stanford Research Institute (SRI) International, and the University of Edinburgh in Scotland. In 1967, General Electric Corporation produced a four-legged vehicle (under a Department of Defense contract) that required simultaneous control of the appendages by a human operator. This proved to be extremely difficult to achieve and the project was scrapped. A year later, SRI demonstrated an "intelligent" mobile robot that had some vision capability (using a TV camera), an optical range finder, and touch sensors (see Figure 1.2.1). The device also had the ability to understand and react to verbal commands in English. Because it moved in a highly irregular and jerky manner, it was given the name "Shakey."

One of the early innovators in the field of robotics, Victor Scheinman, while working at Stanford University in 1970 demonstrated a computer-controlled manipulator that was powered by servomotors rather than by hydraulics. This six-axis device, shown in Figure 1.2.2 and variously referred to as the *Scheinman* or *Stanford arm,* was extremely sophisticated and technically complex and, in fact, is

---

*A servomechanism (or more commonly, a servo) is a feedback control system in which the variable being controlled is a mechanical quantity such as velocity or position.

†This company became the largest robot manufacturer in the world (although it does not currently enjoy this position) and gained its independence from Condec in 1981. In the latter part of 1982, Unimation was acquired by the Westinghouse Corporation.

Figure 1.2.1. Shakey, a wheeled, nonautonomous robot that was developed in 1968 by Stanford Research Institute. (Courtesy of SRI International, Menlo Park, CA.)

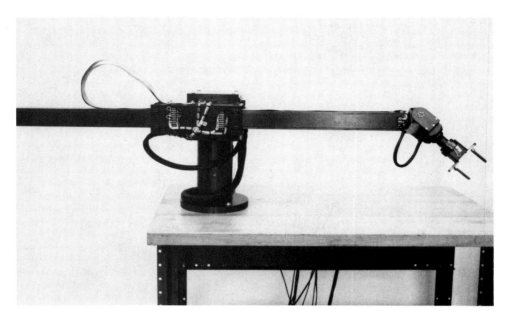

Figure 1.2.2.   Jet Propulsion Laboratory—Stanford arm.   This 6-axis, electrically actuated robot utilizes the Scheinmann design with JPL modifications.   (Courtesy of Dr. A. K. Bejczy and Jet Propulsion Laboratories, Pasadena, CA.)

still used today by a number of research centers.   Three years later, in 1973, Richard Hohn of the Cincinnati Milacron Corporation produced the first minicomputer-controlled commercial robot, the T^3, which was a hydraulically actuated machine capable of lifting payloads of up to 100 lb.* Scheinman, recognizing the growth potential of industrial robots, founded his own company (Vicarm Inc.) and in 1974 introduced the first servomotor-actuated, minicomputer-controlled manipulator. In the same year (1976) as the NASA *Viking 1* and *2* landers used their manipulators to collect samples from the surface of Mars, Vicarm developed the first micro-processor-controlled robot under Navy contract.   This year also saw a significant industrial development at the MIT Draper Laboratory, the invention of a compliant robot wrist (called a remote-centered compliance or RCC).   Such a device per-mitted certain assembly operations to be performed by a rigid robot manipulator even when there was a significant misalignment between the robot tool and the part being worked on.

A workable robotic vision system was developed by SRI in 1977 and resulted in a system commercialized by Machine Intelligence Corporation.   In 1978, Uni-mation, working with a set of specifications provided by General Motors, developed the programmable universal machine for assembly (PUMA).   This five- (or six-) axis robot was servomotor driven and controlled by a number of microprocessors.

---

*This is true for the T^3-566.   However, the more robust T^3-586 will now carry up to 225 lb.

Unimation (which by now had acquired Vicarm) used as their model an arm that Scheinman had developed while at MIT.

In the mid-1970s, the importance of robots to the future of American industry was recognized when the National Bureau of Standards was directed to develop the Automated Manufacturing Research Facility (or AMRF) by the latter part of the 1980s. Well along in their work, a group at NBS under the leadership of James Albus has already made significant and practical contributions in the fields of robotic vision and controls. 1980 saw the establishment of the largest university-related robotics laboratory, at Carnegie–Mellon University. The Westinghouse Corporation was instrumental in getting this facility started by providing the initial funding. This year also saw the University of Rhode Island demonstrate a prototype robotic vision system that could handle the "bin picking" problem. Utilizing the techniques developed at URI, a robot was able to pick up randomly stacked rods. A modification of the system was marketed by Object Recognition Systems, Inc. and demonstrated in 1982.

The area of mobile robots saw a substantial development when in 1983 a company called Odetics, Inc. introduced a unique experimental six-legged device

Figure 1.2.3.   ODEX II, a second generation functionoid developed by Odetics, Inc. for use in nuclear power plants. Called "Robin," this unique hexapod is capable of lifting loads up to 5.6 times its own weight and changing its geometry so as to facilitate moving through doorways or passageways. It is also equipped with a six-axis manipulator. (Courtesy of ODETICS, Inc., Anaheim, CA. Owned by Dept. of Energy and sponsored by the Savannah River Laboratory operated by DuPont.)

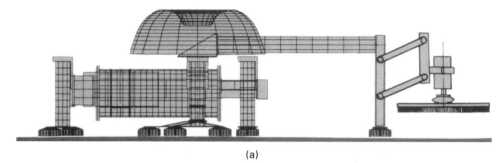

(a)

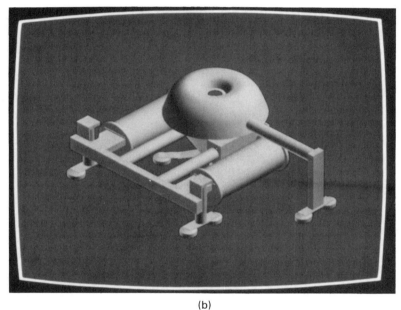

(b)

Figure 1.2.4. The RM3 Marine robot, an amphibious mobile device consisting of 3 legs and one arm that can walk at speeds up to 150 meters/hour. The unit is used to clean the hulls of ships both above and below the water: (a) CAD-CAM generated line drawing; (b) CAD-CAM generated 3D solid version; (c) actual device. (Courtesy of P. Kroczinski, International Robotic Technologies, Inc., Marina del Rey, CA.)

that was designed by studying the gait of both human beings and certain insects. Originally called a *functionoid*, it demonstrated the ability to walk over obstacles and to lift loads up to 5.6 times its own weight while stationary, and 2.3 times its weight while moving. Although the experimental unit was teleoperated, the company is currently working on producing a fully autonomous unit that can be used in nuclear power plant installations. A second generation of such a device, called

(c)

Figure 1.2.4.   Continued.

ODEX II, is equipped with an industrial type manipulator and gripper (see Figure 1.2.3).   The French shipbuilding company of Chantiers du Nord et de la Méditerranée successfully tested a marine mobile robot in early 1984 (see Figure 1.2.4). Intended for cleaning the sides and bottoms of large ships, this remarkable device has already been used by Renault to paint the walls of a large gas tank.

Although it would appear from this brief chronolog that the Japanese are "Johnny-come-latelies," such is not the case.   In fact, as early as 1968, Kawasaki Heavy Industries was granted a license from Unimation to manufacture their robots. The robot industry grew so rapidly that in 1971, the Japan Industrial Robot Association (JIRA) was founded.   It is interesting to note that despite all of the research activity in the United States, the Robotic Institute of America (RIA), now called the Robotic Industries Association, an organization primarily for manufacturers and users of robots, was begun only in 1975.   An even more revealing statistic is found in Joseph F. Engelberger's excellent book entitled *Robots In Practice* [3].   At the time of its publication (September 1980), the author listed nine Japanese, nine European, and only four American companies manufacturing robots.   Significant industrial effort in the United States has occurred since then, with the RIA (in its 1982 World Wide Robotics Survey and Directory) listing approximately 28 American firms now involved in the manufacture of robots. [The same publication indicates that the numbers for both Japan and Europe have increased also (to 16).] Nevertheless, this does demonstrate that the Japanese have been exceedingly active in the industrial application of robots for quite a long time.

Now that the reader has been given a brief historical perspective, it is time to become more specific and to describe what is and what is not a robot, and also what forms industrial robotic manipulators can take.

1.3 CLASSIFICATION OF ROBOTS

What exactly *is* a robot? As mentioned in the preceding section, it is not a C3-P0-like humanoid that has multiple appendages and is capable of decisions and actions based solely on moment-to-moment processing of information acquired through its external sensors. Webster defines a robot as

> An automatic apparatus or device that performs functions ordinarily ascribed to humans or operates with what appears to be almost human intelligence. [4]

Although this definition may be adequate for a dictionary, it does not tell the entire story as far as an industrial device is concerned. Recognizing this, the RIA developed the following, more complete definition:

> A robot is a reprogrammable, multifunctional manipulator designed to move material, parts, tools, or specialized devices through variable programmed motions for the performance of a variety of tasks.

Based on this definition, it is apparent that a robot must be able to operate automatically, which implies that it must have some sort of programmable memory.

In this section we follow the approach suggested by Engelberger (in his book) to classify industrial robotic manipulators in two different ways, one based on the *mechanical configuration* of the device and the other based on the general method used to *control* its individual members (i.e., the "joints" or "axes"). Before doing this, however, we wish to consider several devices that are not truly robots but are often called by this name in the media.

1.3.1 Robotic-Like Devices

There are a number of devices that utilize certain facets of robot technology and are therefore often mistakenly called robots. In fact, Engelberger has referred to them as "near relations." There are at least four such classes of mechanisms, two of which we have already briefly encountered in the preceding section. Each of these is now briefly described in turn.

1. Prostheses.   These are often referred to as "robot arms" or "robot legs." Even though they can make use of either hydraulic or servomotor actuators, utilize servo control, and have mechanical linkages, they do not have their own "brains" and are not truly programmable.   The impetus to produce an action (called the "command signal") in such a device originates in the brain of the human being. It is then transmitted via nerves to the appropriate appendage, where electrodes sense the nerve impulses.   These are processed electronically by a special-purpose computer (on board the prosthesis), which, in turn, controls the motion of the substitute limb (or hand).   Although there are a number of serious research efforts into producing such an appendage, we are probably many years away from realizing the concepts described in the popular television series of the 1970s, "The Six Million Dollar Man."   Among the many difficult problems that remain to be solved is the ability to reliably extract (from all the electrical activity produced by neighboring muscles) and process the low-level nerve impulse signals appropriate to controlling the missing limb.

2. Exoskeletons.   As shown in Figure 1.3.1, these are a collection of mechanical linkages that are made to surround either human limbs or the entire human frame.   They have the ability to amplify a human's power.   However, it is clear

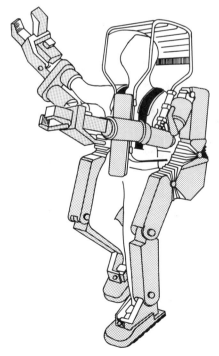

Figure 1.3.1.   Artist's conception of a General Electric Hardiman, an exoskeletal device developed in the 1970s.   It allowed a human operator to lift loads up to 1500 lbs. and utilized hydraulically actuated servos.

that they cannot act independently and, as such, are not robots. In fact, when an exoskeletal device is used, the operator must exercise extreme caution, due to the increased forces and/or speeds that are possible. An example of such a device is the General Electric Hardiman, developed in the 1970s, which utilized hydraulically actuated servos. Loads of up to 1500 lb could be lifted by a worker "wearing" this piece of equipment (see Figure 1.3.1).

3. Telecherics. As mentioned previously, these devices permit manipulation or movement of materials and/or tools that are located many feet away from an operator. Even though telecheric mechanisms use either hydraulic or servomotor actuators, which are usually controlled in a closed-loop manner, they are not robots because they require a human being to close the entire loop and to make the appropriate decisions about position and speed. Such devices are especially useful in dealing with hazardous substances such as radioactive waste. It has also been

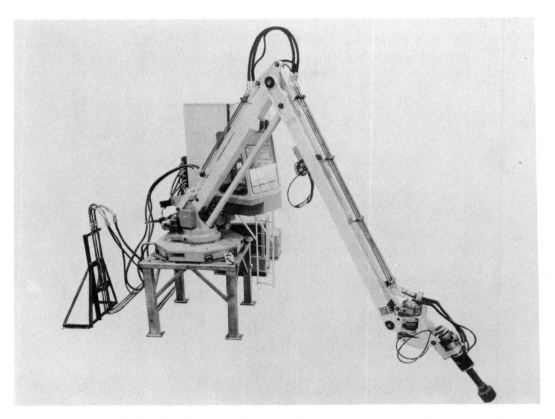

Figure 1.3.2. The G.E. Manmate Industrial Manipulator is an articulating arm boom that can be used for material handling. The device utilizes servo controlled hydraulic actuators and force feedback to the master control. (Courtesy of Peter Offierski, Canadian General Electric Company, Peterborough, Ontario, Canada.)

proposed that they be used in undersea exploration.   An example of an existing telecheric mechanism is the arm that is installed on the NASA *Space Shuttle* (mistakenly referred to by the press as a "robotic arm").   Another example is the General Electric Corporation's *Manmate*, a device developed in 1967 (See Figure 1.3.2).

   4. Locomotive Mechanisms.   These are devices that imitate human beings or animals by having the ability to walk on two or four legs.   Although the multiple appendages can be highly sophisticated collections of linkages that are hydraulically or electrically actuated under closed-loop control, a human operator is still required to execute the locomotive process (i.e., make decisions concerning the desired direction of the device and to coordinate limb motion to achieve this goal).   An artist's rendering of the previously mentioned and ill-fated General Electric four-legged vehicle is shown in Figure 1.3.3.

   Having described what is not a robot, we now devote the remainder of this section to classifying the various types of robotic devices.   As mentioned above, the approach taken will be similar to that suggested by Engelberger in his book. Classification will be performed in two different ways, based on:

- The particular coordinate system utilized in designing the mechanical structure
- The method of controlling the various robotic axes

   We consider the coordinate system approach first.

1.3.2 Classification by Coordinate System

   Although the mechanics of a robotic manipulator can vary considerably, all robots must be able to move a part (or another type of "load") to some point in space.   The major axes of the device, normally consisting of the two or three joints or degrees of freedom that are the most mechanically robust (and often located closest to the base), are used for this purpose.   The majority of robots, therefore,

Figure 1.3.3.   The General Electric four-legged walking machine.   Because of severe stability problems, the project was never completed.

fall into one of four categories with respect to the coordinate system employed in the design of these axes. That is, they can be described as being either cylindrical, spherical, jointed, or Cartesian devices. Each of these categories is now discussed briefly.

1.3.2.1 Cylindrical coordinate robots

When a horizontal arm (or "boom") is mounted on a vertical column and this column is then mounted on a rotating base, the configuration is referred to as a cylindrical coordinate robot. This is shown in Figure 1.3.4. As can be seen,

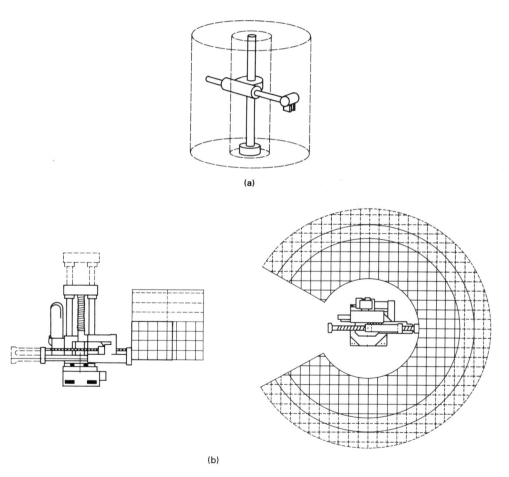

(a)

(b)

Figure 1.3.4. A cylindrical coordinate robot: (a) a general view of the geometry of the robot's major axes; (b) vertical and top views of the workspace of such a robot. (Courtesy of J. Coshnitzke, Cincinnati Milacron, Cincinnati, OH).

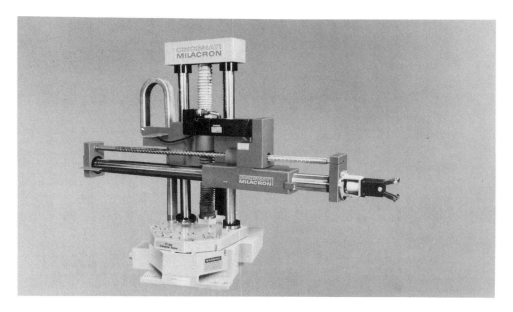

Figure 1.3.5.   An actual cylindrical coordinate robot manufactured by Cincinnati Milacron.   Called a T3300, this unit has 3 or 4 axes, is electrically servo controlled, and can handle loads up to 50 kg.   (Courtesy of J. Coshnitzke, Cincinnati Milacron, Cincinnati, OH.)

the arm has the ability to move in and out (in the r direction), the carriage can move up and down on the column (in the z direction), and the arm and carriage assembly can *rotate* as a unit on the base (in the θ direction).   Usually, a full 360° rotation in θ is not permitted, due to restrictions imposed by hydraulic, electrical, or pneumatic connections or lines.   Also, there is a minimum, as well as a maximum extension (i.e., R), due to mechanical requirements.   Consequently, the overall volume or *work envelope* is a portion of a cylinder.   Commercial robots having this configuration are (or have been) manufactured by companies such as Prab, Versatran, Autoplace, General Numeric, Seiko, and Cincinnati Milacron.   One such unit is shown in Figure 1.3.5.

1.3.2.2 Spherical coordinate robots

When a robotic manipulator bears a resemblance to a tank turret, it is classified as a spherical coordinate device (see Figure 1.3.6).   The reader should observe that the arm can move in and out (in the r direction) and is characterized as being a *telescoping boom*, can pivot in a vertical plane (in the ϕ direction), and can rotate in a horizontal plane about the base (in the θ direction).   Because of mechanical and/or actuator connection limitations, the work envelope of such a robot is a portion of a sphere.   Commercially available spherical coordinate robots

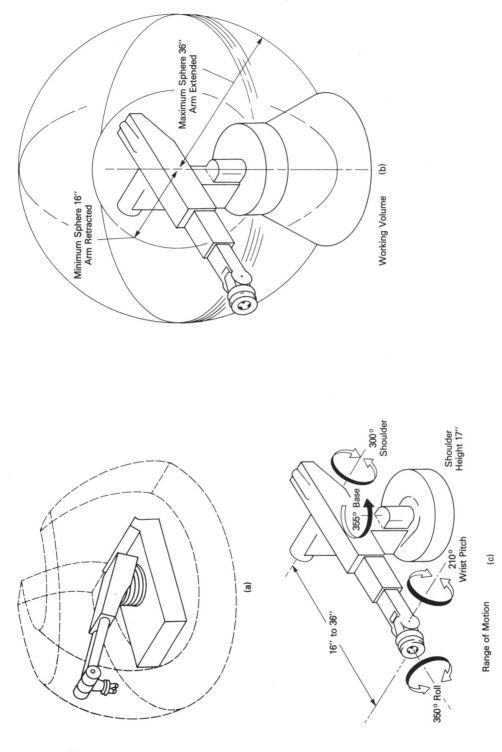

Figure 1.3.6. A spherical coordinate robot: (a) general view of the geometry of the robot's major axes; (b) working volume (workspace) of such a robot; (c) range of motion for each of the five axes of a typical spherical coordinate robot. (Courtesy of G. Heatherston and U.S. Robots, a Square D Company.)

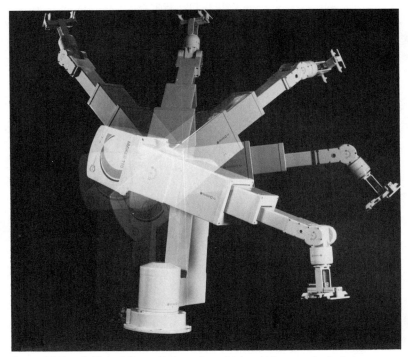

Figure 1.3.7.   A United States Robots, Inc., Maker 110 pure spherical geometry robot.   The unit has 5 electrically servo controlled axes and can lift loads up to 10 lbs.   (Courtesy of G. Heatherston and U.S. Robots, Inc., a Square D Company.)

are being built by Prab, Unimation, and United States Robots.   A United States Robots' Maker 110 is shown in Figure 1.3.7.

1.3.2.3 Jointed arm robots

There are actually three different types of jointed arm robots: (1) pure spherical, (2) parallelogram spherical, and (3) cylindrical.   We briefly describe each of these in turn.

1. Pure Spherical.   In this, the most common of the jointed configurations, all of the links of the robot are pivoted and hence can move in a rotary or "revolute" manner.   The major advantage of this design is that it is possible to reach close to the base of the robot and over any obstacles that are within its workspace.   As shown in Figure 1.3.8, the upper portion of the arm is connected to the lower portion (or forearm).   The pivot point is often referred to as an "elbow" joint and permits rotation of the forearm (in the α direction).   The upper arm is connected to a base (or sometimes a *trunk*).   Motion in a plane perpendicular to the base is possible at this *shoulder* joint (in the β direction).   The base or trunk is also free

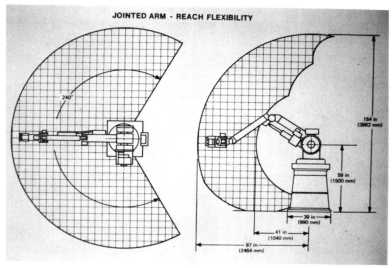

Figure 1.3.8.   Geometry of a pure spherical jointed robot. (Courtesy of J. Coshnitzke, Cincinnati Milacron, Cincinnati, OH.)

to rotate, thereby permitting the entire assembly to move in a plane parallel to the base (in the γ direction). The work envelope of a robot having this arrangement is approximately spherical. Examples of commercial manipulators having this geometry are the Puma (Unimation), the Cincinnati Milacron T^3, and those made by ASEA, Niko, and GCA. Three different sizes of Pumas are shown in Figure 1.3.9.

    2. Parallelogram Jointed.   Here the single rigid-member upper arm is replaced by a multiple closed-linkage arrangement in the form of a parallelogram (see Figure 1.3.10). The major advantage of this configuration is that it permits the joint actuators to be placed close to, or on the base of, the robot itself. This means that they are not carried *in* or *on* the forearm or upper arm itself, so that the arm inertia and weight are considerably reduced. The result is a larger load capacity than is possible in a jointed spherical device for the same-size actuators. Another advantage of the configuration is that it produces a manipulator that is mechanically stiffer than most others. The major disadvantage of the parallelogram arrangement is that the robot has a limited workspace compared to a comparable jointed spherical robot. Examples of such commercial units are those manufactured by ASEA, Hitachi, Cincinnati Milacron, Yaskawa, and Toshiba. The latter manipulator is shown in Figure 1.3.11.

    3. Jointed Cylindrical.   In this configuration, the single *r*-axis member in a pure cylindrical device is replaced by a multiple-linked open kinematic chain, as

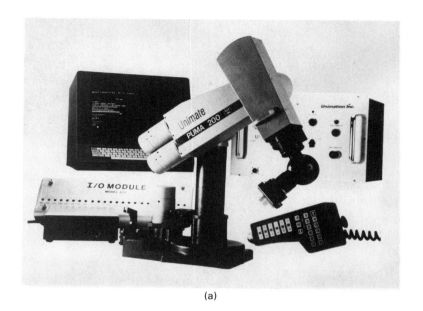

(a)

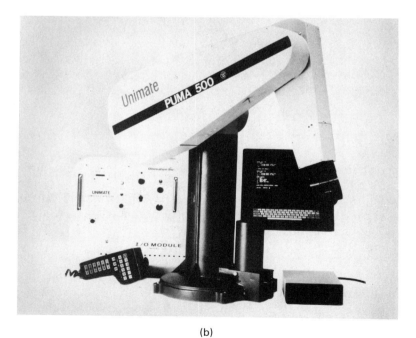

(b)

Figure 1.3.9. The Unimation PUMA is an example of a pure spherical jointed robot. Shown are 3 different sizes, the 200, 500, and 700 series. (Courtesy of Unimation, Inc., a Westinghouse Company, Danbury, CT.)

(c)

Figure 1.3.9. Continued.

shown in Figure 1.3.12. Such robots tend to be precise and fast but will generally have a limited vertical (z direction) reach. Often the z-axis motion is controlled using simple (open-loop) air cylinders or stepper motors, whereas the other axes make use of more elaborate electrical actuation (e.g., servomotors and feedback). Robots having this configuration are made by Hirata, Reis, GCA, and United States Robots.

A subclass of the jointed cylindrical manipulator is the *selective compliance assembly robot arm* (or SCARA) type of robot [23]. Typically, these devices are relatively inexpensive and are used in applications that require rapid and smooth motions. One particularly attractive feature, selective compliance, is extremely useful in assembly operations requiring insertions of objects into holes (e.g., pegs or screws). Because of its construction, the SCARA is extremely stiff in the vertical direction but has some lateral "give" (i.e., compliance), thereby facilitating the

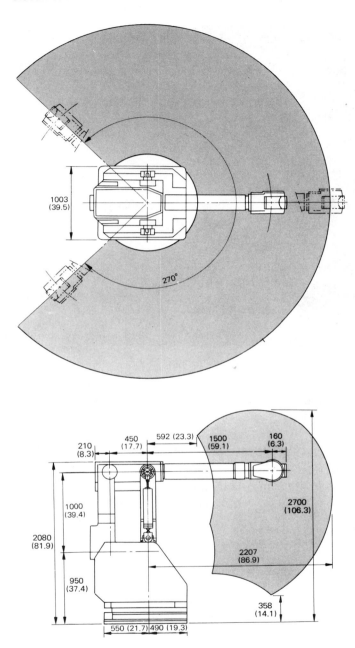

Figure 1.3.10.   Workspace and geometry of a parallelogram spherical jointed robot.   (Courtesy of Toshiba/Houston International Corp., Houston, TX.)

Figure 1.3.11. Example of an actual parallelogram spherical jointed robot. This unit, manufactured by Toshiba Corp., has 6 servo controlled axes, can lift loads up to 120 kg, and has a repeatability of ±1 mm. (Courtesy of Toshiba/Houston International Corp., Houston, TX.)

insertion process. Some SCARAs even permit the lateral compliance to be increased during an operation by reducing appropriate electronic amplifier gains. Such a device is shown in Figure 1.3.13.

1.3.2.4 Cartesian coordinate robots

In this, the simplest of configurations, the links of the manipulator are constrained to move in a linear manner. Axes of a robotic device that behave in this way are referred to as "prismatic." Let us now consider the two types of Cartesian devices.

1. Cantilevered Cartesian. As shown in Figure 1.3.14, the arm is connected to a trunk, which in turn is attached to a base. It is seen that the members of the robot manipulator are constrained to move in directions parallel to the Cartesian *x, y,* and *z* axes. Such a robot is shown in Figure 1.3.15. Devices like these tend to have a limited extension from the support frame, are less rigid, but have a less restricted workspace than other robots. In addition, they have good repeatability and accuracy (even better than the SCARA types) and are easier to program because of the "more natural" coordinate system. Certain types of motions may

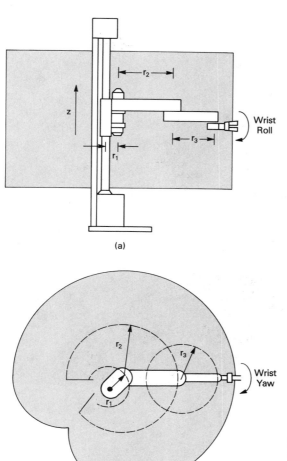

(a)

(b)

Figure 1.3.12. Jointed cylindrical workspace and geometry robot: (a) vertical cross-section; (b) top view. In some SCARA robots, $r_1 = 0$ and the z axis is located at the wrist. Also, wrist could have a pitch axis.

be more difficult to achieve with this configuration, due to the significant amount of computation required (e.g., straight line in a direction not parallel to any axis). In this respect, Control Automation did manufacture a robot that was capable of unrestricted straight-line paths. However, since 1985, the company has stopped marketing such a manipulator.

   2. Gantry-Style Cartesian.   Normally used when extremely heavy loads must be precisely moved, such robots are often mounted on the ceiling. They are

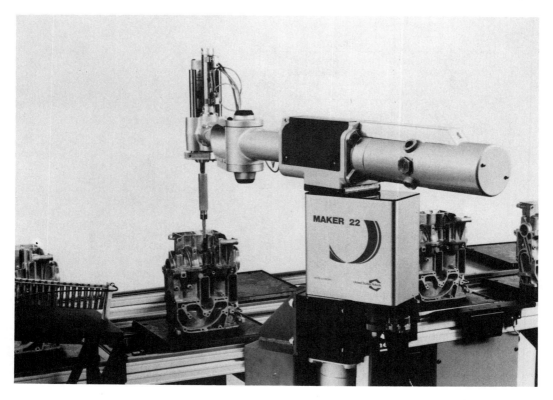

Figure 1.3.13.   An example of an actual SCARA robot, the U.S. Robot's Maker 22.   (Courtesy of G. Heatherston and U.S. Robots, Inc., a Square D Company.)

generally more rigid but may provide less access to the workspace.   In the last few years, a number of smaller devices in this class have emerged.   In this instance, a framed structure is used to support the robot, thereby making it unnecessary to mount the device on the ceiling.   The geometry of a gantry Cartesian device is shown in Figure 1.3.16.   A unit made by Cincinnati Milacron is shown in Figure 1.3.17.

It is important to understand that the classifications above take into account only the *major* axes (defined at the beginning of Section 1.3.2) of the manipulator. However, a robot is not limited to only three degrees of freedom.   Normally, a wrist is affixed to the end of the forearm.   This appendage is, itself, capable of several additional motions.   For example, as shown in Figure 1.3.18, axes that permit *roll* (i.e., motion in a plane perpendicular to the end of the arm), *pitch* (i.e., motion in vertical plane passing through the arm), and *yaw* (i.e., motion in

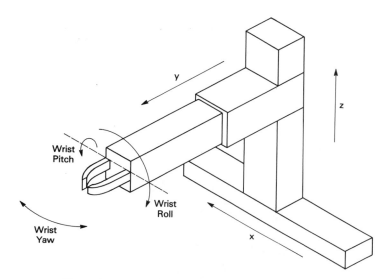

Figure 1.3.14.   Cantilevered Cartesian robot geometry.

a horizontal plane that also passes through the arm) are possible. Moreover, the entire base of the robot can be mounted on a device that permits motion in a plane (e.g., an *x-y* table or a track located in either the ceiling or floor). From this discussion it should be clear to the reader that robots with as many as eight (and as few as two) axes can be constructed.

Having classified robots according to the geometry of their major axes, we now look at another way of organizing robot types.

1.3.3 Classification by Control Method

As mentioned above, the second method of classification looks at the technique used to control the various axes of the robot. The two general classes are (1) non-servo controlled, and (2) servo controlled. We now consider each one separately.

1.3.3.1 Non-servo-controlled robots

From a control standpoint, the non-servo-controlled or *limited-sequence robot* is the simplest type. Other names often used to described such a manipulator are *end point robot, pick-and-place robot*, or *bang-bang robot*. Regardless of mechanical configuration or use, the major characteristic of such devices is that their axes remain in motion until the limits of travel (or "end stops") for each are reached. Thus only two positions for the individual axes are assumed. The non-servo nature

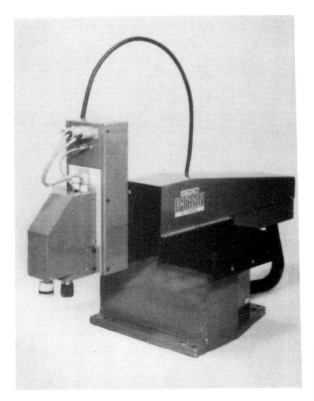

Figure 1.3.15. An actual cantilevered Cartesian robot. (Courtesy of Seiko Instruments, USA, Inc., Torrance, CA.)

of the control implies that once the manipulator has begun to move, it will continue to do so until the appropriate end stop is reached. There will be no monitoring (via external sensors) of the motion at any intermediate points. As such, one refers to this class of robot as being controlled in an *open-loop* manner.

"Programming" a limited-sequence robot is accomplished by setting a desired *sequence* of moves and adjusting the end stops for each axis accordingly. The manipulator "brain" consists of a controller/sequencer. The "sequencer" portion is generally a motor-driven rotary device (similar to the "timer motor" found in certain home appliances) with a number of electrical contacts. Unlike the timer motor on a washing machine, for example, a series of jumper plugs is used and permits the appropriate contacts to be *enabled* by the sequencer in the desired order. Each such enabled contact will cause power to be switched to an axis actuator (e.g., pneumatic or hydraulic valve/piston arrangement) by the controller portion. The energized axis will continue to move until the "programmed" end stop is reached. This information is then used to cause the sequencer to index to the next step in its "program." It is important for the reader to understand that this is the only time that information is "fed back" to the sequencer.

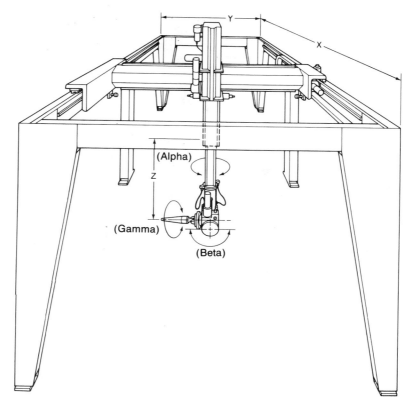

Figure 1.3.16.   Geometry of a Cartesian gantry style robot.   (Courtesy of CIM-CORP, Inc., St. Paul, MN.)

A typical operating sequence for a hydraulic or pneumatic non-servo-controlled robot is as follows:

1. A program "start" causes the controller/sequencer to signal control valves on the manipulator's actuators.

2. This causes the appropriate valves to open, thereby permitting air or oil to flow into the corresponding pistons (actuators) and the member(s) of the manipulator begin to move.

3. These valves *remain open* and the members continue to move until they are physically restrained from doing so by coming into contact with appropriately placed end stops.

4. Limit switches, generally located on the end stop assemblies, signal the end of travel to the controller/sequencer, which commands the open valves to close.

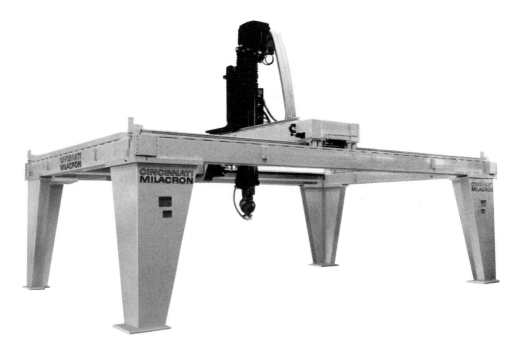

Figure 1.3.17. An actual Cartesian gantry robot, a Cincinnati Milacron T3886. This unit can lift loads up to 90 kg and has 6 electrically actuated servo controlled axes. (Courtesy of J. Coshnitzke, Cincinnati Milacron, Cincinnati, OH.)

5. The sequencer now indexes to the next step and the controller again outputs signals to actuator valves, thereby causing other members of the manipulator to move. Alternatively, signals can be sent to an external device such as a "gripper," causing it to open or close as desired.

6. The process is repeated until all steps in the sequence are executed.

Other attributes and/or capabilities worthy of mention for this class of robot are as follows:

- Conditional modification of the programmed sequence is possible if some type of external sensor is employed. For example, if a simple optical interrupter (see Chapter 5) is used, it may be possible to have the manipulator pause in its sequence until a peripheral tool (e.g., a punch press) has cleared the work envelope. Robots having this ability normally can perform one program.

- Open-loop or non-servo control is often used in smaller robots because of its

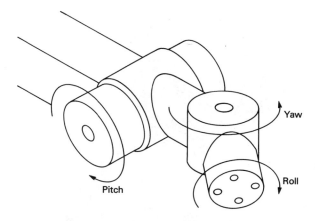

Figure 1.3.18. Example of a three degree-of-freedom wrist showing the roll, pitch, and yaw axes. These robotic joints are used for orienting objects in space.

low cost and simplicity.   An example of such a device is the Seiko PN-100, shown in Figure 1.3.19.

- It is possible to have a number of "intermediate" stops for each of the axes. This allows the manipulator to be programmed for more complex paths and permits a limited degree of *path control.*

- Although the controller normally applies full power to an axis that is selected by the sequencer and turns this power off only when the limit stop is reached, it is possible to achieve a degree of deceleration into the stop by using shock absorbers or appropriate valving at the end stops.   This results in less stress on the components of the manipulator and on the part being moved.

Even though limited sequence robots can be configured in a variety of ways (e.g., Cartesian, cylindrical, etc.), a number of characteristics are common to all such devices.   In particular:

- They are relatively high speed machines because of the small size of the arm and the full power applied to the axis actuators.

- They are low cost and easy to maintain and operate.   Also, they are *extremely* reliable devices.

- They have a repeatability of about ± 0.01 inch.   That is, they have the ability to return to the same point within ± 10 mils.   (A few small pneumatically actuated robots such as the Seiko PN-100 advertise repeatabilities of about ± 0.5 mil).

- This class of robot has limited flexibility with respect to positioning and programming.   Thus although more than one axis can be moved at a time, it is generally not possible to cause a tool held at the end of the manipulator to move in a straight line (except if the desired line happens to coincide with one of the robot axes).   Also, *coordinated motion* cannot be produced whereby the axes reach the endpoint of the desired motion at the same instant.

Figure 1.3.19. Seiko PN-100 pneumatically actuated 2-4 axis non-servo (point-to-point) robot. The unit shown has 2 axes, can handle a payload of up to 1.5 kg, and uses adjustable limit stops for "programming." (Courtesy of Seiko Instruments, USA, Inc., Torrance, CA.)

1.3.3.2 Servo-controlled robots

Servo-controlled robots are normally subdivided into either *continuous-path* or *point-to-point* devices. In either case, however, information about the position and velocity (and perhaps other physical quantities) is continuously monitored and fed back to the control system associated with each of the joints of the robot. Consequently, each axis loop is "closed." Use of closed-loop control permits the manipulator's members to be commanded to move and stop *anywhere* within the limits of travel for the individual axes. (The reader should contrast this with the non-servo-controlled machines described above, where only axis extremes could be programmed.) In addition, it is possible to control the velocity, acceleration, deceleration, and jerk (i.e., the time derivative of acceleration) for the various axes between the endpoints. Manipulator vibration can, as a consequence, be reduced significantly. Besides the above, servo-controlled robots also have the following additional features and/or attributes:

- A larger memory capacity than in non-servo-controlled devices. This implies that they are able to store more positions (or points in space) and hence that the motions can be significantly more complex and smoother. It also means that more than one program can be created and stored, thereby permitting the robot to be used in a variety of applications with a minimum of downtime required for the changeover.

- The end of the manipulator can be moved in any one of three different classes of motion: *point-to-point* (where the endpoints of the motion are important but the path connecting them is not), *straight line* [where it is important to cause a specified location on the manipulator, often referred to as the *tool point*, to move from the initial point to the final one in a *linear* fashion (in three-dimensional space)], or *continuous path* [where points along the path are connected so that the instantaneous position *and* either its spatial or time (i.e., velocity) derivative are continuous]. Note that not every servo-controlled robot is capable of performing straight-line and/or continuous-path motion. Also, it may not always be possible to maintain a constant path velocity if all points along a desired path have been taught at the same speed. (This is due to the often complex geometry that relates the tool point to the individual joints. See Chapter 8.)

- Within the limits imposed by the mechanical components, positional accuracy can be varied by adjusting the gains of appropriate amplifiers in the servo loops.

- Joint actuators are usually either hydraulic valve/piston arrangements or servomotors, although until about 1985 there did exist at least one commercially available robot that used pneumatic servos (see Figure 1.3.20).

- Programming is generally done in what is referred to as *teach mode*. The manipulator is manually moved to a sequence of desired points. The coordinates of each of these are stored in the robot's (semiconductor) memory. Some of the more sophisticated systems actually have a specialized computer language that permits these stored points to be utilized in a variety of motions, paths, orientations, and so on. An example of such a language is Unimation's VAL (or VAL II [24]).

- It is possible to program each axis to move to almost any point along its entire range of travel. Consequently, this affords the user with a great deal of flexibility in the type of motions that are possible. Moreover, "coordinated motion" can be achieved whereby two or more joints move simultaneously so that the end of the manipulator is capable of tracing out an extremely complex path. It is important to understand that such coordination among the robot axes is normally done "automatically" under mini- or microcomputer control.

- It is possible to permit branching operations whereby alternative actions are taken by the manipulator based on data obtained from external sensors. For example, it might be possible for the robot to repeat a particular set of moves if a part did not appear at a workstation because of a faulty feed mechanism.

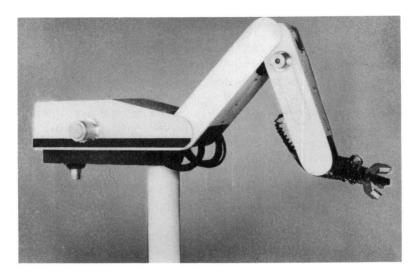

Figure 1.3.20. Until approximately 1985, IRI (International Robomation Intelligence) produced an air servo robot that looked very similar to its newer M 50 E AC servo-controlled robot pictured here. The air version had severe vibration problems caused, in part, by the compressibility of air. (Courtesy of C. Gordanella, International Robomation Intelligence, Carlsbad, CA.)

This capability arises from the extensive use of microprocessors in the robot controller.

- Because servo-controlled robots generally have considerably more complex control, computer, and mechanical structures than non-servo-controlled devices, they may be more expensive and somewhat less reliable. Nevertheless, their great flexibility makes them extremely attractive and cost-effective in a large number of applications.

With these features in mind, the following represents a typical operating sequence for a general servo-controlled robot (it is assumed that the desired points have been taught and stored in memory prior to running the program):

1. At the beginning of the program, the actual position of all of the manipulator joints is obtained from appropriately mounted sensors. The desired (or command) position information is sent out to the individual axes from a master computer.

2. For each joint, the actual and desired positions are compared and an "error" signal is formed. This is used to drive the individual joint actuators.

3. As a result, the members of the robotic manipulator move. Position, velocity, and any other physical parameter of the motion are monitored or estimated (again utilizing appropriate sensors), and this information is used to automatically modify the error signals accordingly.

4. When the error signals for all the individual axes are zero, the members stop moving and the manipulator is "home" [i.e., at the desired (or taught) final point in space].

5. The master computer then sends out the next taught point, and steps 1 through 4 are repeated. This process continues until all of the desired points (or actions, e.g., opening or closing of a gripper) have been reached (or performed).

Although most servo-controlled robots behave in the general manner described above, there are certain features that are specific to the point-to-point and the continuous-path robots. We next consider these briefly.

1.3.3.3 Point-to-point servo-controlled robots

Point-to-point robots are widely used for moving parts from one location to another and also for handling various types of tools. Although they can perform all of the tasks of the pick-and-place robot, they are far more versatile because of their ability to be multiply programmed and also because of their program storage capability. A typical point-to-point application might be the unloading or loading of a pallet of parts. In the former case, the robot would be taught (i.e., programmed) each of the n locations on the pallet. (Alternatively, the first point and the x and y offsets for each of the other pallet locations would be taught.) It would then move to the first of these taught points, pick up the part, move to a position above the conveyor, and place the part onto the conveyor. The manipulator would repeat the action for each of the remaining $(n - 1)$ locations on the pallet. Such an application, while possible with a simple, nonservo pick-and-place device, would probably require a servo-driven x-y table that would actually move the pallet relative to the fixed pickup point. An example of loading a pallet is shown in Figure 1.3.21.

For the class of closed-loop control robot being considered here, only the initial and final points are taught. The path used to connect the two points is unimportant and is, therefore, not programmed by the user. (The computer calculates the actual path of the manipulator.) More sophisticated point-to-point robots permit straight-line or piecewise-linear motions. Others also permit the velocity of the individual joints to be a continuous function of time and also to be changed by the user, that is, the speed with which the device performs a desired task is user selectable. If no changes in what the robot will do are expected, the initially taught points can be stored in a permanent or read-only memory (ROM). Alternatively, a combination of temporary or random access memory (RAM)* and ROM can be used for teaching new points and storing the old ones.

In general, these robots have a working range and load capacity that is quite

---

*With the advent of inexpensive, low-power, high-speed CMOS RAM, many robots now have battery-backup memory, which effectively makes such read/write memory "permanent."

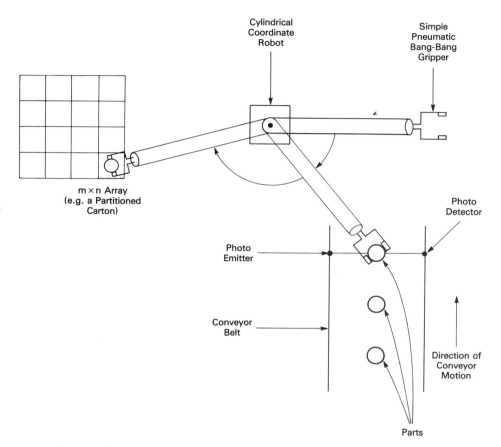

Figure 1.3.21. Use of a robot in a palletizing operation. When a part moving on a conveyor interrupts the light beam from a photo emitter, the controller commands the robot to acquire the part. This part is then moved to and placed in one of the (empty) locations in a partitioned carton. This process is repeated until all such locations are filled, at which time the carton is removed from the loading station, an empty one replaces it, and the operation is repeated.

high, that is, loads of up to 500 lb and reaches of 10 to 11 ft are possible. They most often use hydraulic actuators, although recently, the trend has been toward servomotor-actuated systems. Examples of this type of robot are those made by ASEA, Cincinnati Milacron (the T³), Unimation (the Unimate 2000), and Versatran (see Figure 1.3.22).

1.3.3.4 Continuous-path servo-controlled robots

Many applications do not require that the manipulator have a long reach or be able to carry a large load. In particular, there is an entire class of applications

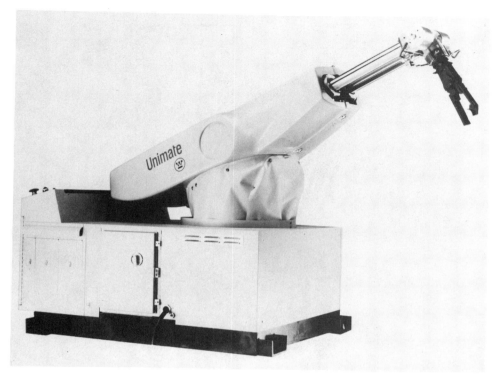

Figure 1.3.22.   A UNIMATE 2000, hydraulically actuated point-to-point servo controlled robot.   (Courtesy of Unimation, Inc., a Westinghouse Co., Danbury, CT.)

where it is most important to follow a complex path through space and possibly to have the end of the arm move at high speeds.   Examples of these applications include spray painting, polishing, grinding, and arc welding.   In all instances, the tool carried by the manipulator is fairly light but the required motion to perform the task may be quite complex.   A continuous-path (CP) robot is usually called for in these cases.

Although points must still be taught prior to executing a program, the method of teaching is usually quite different from that used for the point-to-point servo-controlled robot.   Unlike the procedure described above, points are not *recorded* manually in the CP robot.   What happens is that in the teach mode, an automatic sampling routine is activated which can record points (and/or velocity information) at a rate of 60 to 80 times a second for approximately 2 minutes.   An operator simply moves the tool over the desired path with the sampler running.   The sampling rate is usually high enough so that when the recorded points are "played back" (i.e., the program is run), extremely smooth motion results.   It is clear that a large memory is required since as many as 9600 points may be recorded in the

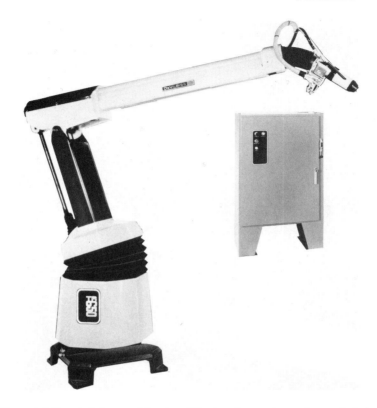

Figure 1.3.23. A 6-degree-of-freedom CP robot that can be used to apply industrial coatings including paint. (Courtesy of The DeVilbiss Co., Toledo, OH.)

2-minute period. To facilitate the accurate recording of complex paths (e.g., in arc welding applications), the tool can be moved over the desired path during the teaching phase at a slow speed. Playback, however, will be independent of the recorded speed, so rapid and accurate curve tracing is possible. An example of a CP robot is that produced by DeVilbiss and shown in Figure 1.3.23.

It is important to understand that, in general, CP robots can be used for only a limited number of tasks and are often single-task devices (e.g., spray painting and welding).* On the other hand, point-to-point robots sometimes have the ability to perform CP motion, although the method of teaching the large number of points is not nearly as convenient since each point must still be recorded manually. Examples of such devices are the PUMA and the Maker 110 (see Figure 1.3.24).

It should be apparent from the above that even though there are two general

---

*As of this writing, the lack of flexibility is more than offset by the increase in productivity achieved with CP robots when used to perform these tasks.

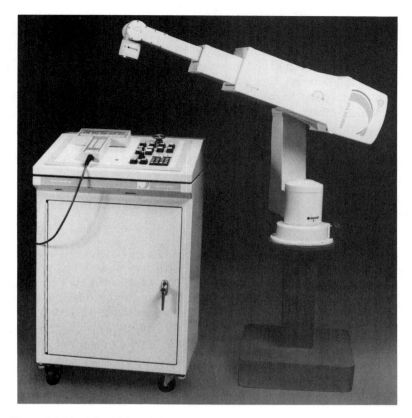

Figure 1.3.24. The Maker 110 robot can execute continuous path motion by teaching a large number of discrete points. (Courtesy of G. Heatherston and U.S. Robots, a Square D Company.)

methods of robot classification, there are still a large number of different robot types. Despite this fact, many points of commonality among these diverse mechanisms exist. This is the subject of the next section.

1.4 MAJOR COMPONENTS OF A ROBOT

Although the mechanical, electrical, and computational structure of robots can vary considerably, most have the following four major components in common: (1) a manipulator or arm (the "mechanical unit"), (2) one or more sensors, (3) a controller (the "brain"), and (4) a power supply. Let us briefly describe each of these in turn.

1. The Manipulator. This is a collection of mechanical linkages connected by joints to form an *open-loop kinematic chain* (see Chapter 3). Also included are gears, coupling devices, and so on. The manipulator is capable of movement in various directions and is said to do "the work" of the robot. In fact, the terms "robot" and "manipulator" are often used interchangeably, although, strictly speaking, this is not correct.

Generally, joints of a manipulator fall into one of two classes. The first, *revolute,* produces pure rotary motion. Consequently, the term *rotary joint* is often used to describe it. The second, *prismatic,* produces pure linear or trans-lational motion and as a result, is often referred to as a linear joint. Each of the joints of a robot defines a *joint axis* about or along which the particular link either rotates or slides (translates). Every joint axis defines a *degree of freedom* (DOF), so that the total number of DOFs is equal to the number of joints. Many robots have six DOFs, three for *positioning* (in space) and three for *orientation,* although, as discussed in Section 1.3.2, it is possible to have as few as two and as many as eight degrees of freedom.

Regardless of its mechanical configuration, the manipulator defined by the joint–link structure generally contains three main structural elements: the *arm*, the *wrist*, and the *hand* (or *end effector*).* Besides the mechanical components, most manipulators also contain the devices for producing the movement of the various mechanical members. These devices are referred to as *actuators* and may be pneumatic, hydraulic, or electrical in nature (see Chapter 4). They are invar-iably coupled to the various mechanical links or joints (axes) of the arm either directly or indirectly. In the latter case, gears, belts, chains, harmonic drives, or lead screws can be used (see Chapter 3).

2. Sensory Devices.   These elements inform the robot controller about the status of the manipulator. This can be done continuously or only at the end of a desired motion. For example, in some robots, the sensors provide instantaneous position, velocity, and possibly acceleration information about the individual links that can be fed back to the control unit to produce the proper control of the mechanical system. More simply, the controller can be informed only when the individual links of the manipulator have reached their preprogrammed final or end positions. Regardless of how it is used, the information provided by the sensors can be either analog, digital, or a combination.

Sensors used in modern robots can be divided into two general classes:

- Nonvisual
- Visual

The first group includes limit switches (e.g., proximity, photoelectric, or mechanical), position sensors (e.g., optical encoders, potentiometers, or resolvers),

---

*The term *arm* is sometimes used in place of *manipulator*.

velocity sensors (e.g., tachometers), or force and tactile sensors (for overload protection, path following, calibration, part recognition, or assembly work). These are discussed more fully in Chapter 5. The second group consists of vidicon, charge-coupled device (CCD), or charge injection device (CID) TV cameras coupled to appropriate image-detection hardware. They are used for tracking, object recognition, or object grasping and are discussed in Chapter 6.

3. The Controller.   Robot controllers generally perform three functions:

- They initiate and terminate the motion of the individual components of the manipulator in a desired sequence and at specified points.
- They store position and sequence data in their memory.
- They permit the robot to be interfaced to the "outside" world via sensors mounted in the area where work is being performed (i.e., the workstation).

To carry out these tasks, controllers must perform the necessary arithmetic computations for determining the correct manipulator path, speed, and position. They must also send signals to the joint-actuating devices (via interfaces) and utilize the information provided by the robot's sensors. Finally, they must permit communication between peripheral devices and the manipulator.

Robot controllers usually fall into one of the following classes:

- Simple step sequencer
- Pneumatic logic system
- Electronic sequencer
- Microcomputer
- Minicomputer

The first three are generally used in less expensive, open-loop-control robots, discussed in Section 1.3.3.1. The microcomputer-based robotic controller is the most commonly used device in the servo-controlled robots described in Section 1.3.3.2. Minicomputer controllers are not common because they are currently not as cost-effective as microcomputers.

4. The Power Conversion Unit.   The purpose of this part of the robot is to provide the necessary energy to the manipulator's actuators. It can take the form of a power amplifier in the case of servomotor-actuated systems, or it can be a remote compressor when pneumatic or hydraulic devices are used.

Up to this point, we have been concerned primarily with the classification of robots according to their geometry or control scheme (see Section 1.3). In addition, we have briefly described in the current section the major components that one expects to find in any industrial robotic device. The remaining portions of the chapter are devoted to the reasons and justifications for using robots, the potential consequences of placing robots in the workplace, and finally, some current and possible future applications of these devices.

1.5 FIXED VERSUS FLEXIBLE AUTOMATION

The age of automation started in the eighteenth century when machines began to take over jobs that had previously been performed by human beings. Since that time, new machines have been finding their way into factories as more and more new products have been conceived. Up to the time of the first robot (i.e., the early 1960s), these machines have had one major thing in common: They have been designed to perform essentially one task with little capability for changing. For example, whereas the devices that produce bottles can be adjusted to produce bottles of different sizes, they cannot produce light bulbs. Generally, machines of this type are referred to as *fixed automated devices* and the process that incorporates them is called *fixed* (or *hard*) *automation*.

With the advent of the industrial robot, a new method of automating products became possible. Called *flexible automation*, a single complex machine was now able to perform a multitude of jobs with relatively minor modifications and little "downtime" needed when changing from one task to another. It is believed by some economists and sociologists that the introduction of the robot into the manufacturing process in the early 1960s signaled the arrival of the "Second Industrial Revolution." This rather remarkable statement gives some idea of the impact that flexible automation has had in the industrial environment. To see why this should be so, let us look at the three major advantages (as identified by Engelberger in his book) that this new approach to automation has over the more traditional one.

1. Reaction Time. In general, when a fixed automated device is to be used in a process for the first time, it must be *designed, built*, and *tested* before it can be used. As an example, let us suppose that a plant manager decides to introduce a new product into an existing facility. To do this, an assembly process requiring new machinery is necessary. The "traditional" approach is to have the plant's manufacturing engineering staff study the problem and then generate a set of specifications for the device that will perform the required tasks. After evaluating competitive bids, a manufacturer of this special-purpose device will be selected. A period of time will then go by while the machine is fabricated. It is not unlikely that during this time, the original specifications will have to be modified, thus postponing the actual delivery date of the equipment. Eventually, however, the fixed automated device will be installed at the manufacturing facility. At this point, it probably cannot be used to produce anything yet, because it must first be tested and adjusted. Such a process may take months. When it is finally ready to go, many months and even years may have elapsed since the idea to produce the new product was conceived. The long lead time may be acceptable in some instances, but it may also mean that in certain highly competitive industries, the "edge" has been lost.

How could flexible automation help solve the problem?   First, a robot is an *off-the-shelf* device.   Once the appropriate type of unit is selected, a rather short period will elapse before the robot is delivered to the factory.   Once it is uncrated, it is essentially "ready to go" (this is what is meant by "off the shelf").   In reality, a period of time must be allocated for personnel to become acclimated and for programming.   Also, techniques and devices that permit the appropriate parts in the particular process to be properly presented to the robot must be developed, although this will often be done during the planning stage and while the robot is being built.   These devices are referred to as the robotic *tooling* and might consist of a specialized gripper and various parts presentation mechanisms, such as sorters and shakers.   In point of fact, it is the (possibly) unique gripper that permits the off-the-shelf robot to be customized to a particular task.   (Note that a similar sort of tooling would also be necessary in the case of the fixed automated device, except of course, for the end effector.)   In any case, it is most likely that the robot will be able to do the job after a relatively short period.   Moreover, if any variability develops in the process, it will usually be quite easy to compensate for this with the robot.   For example, if small size changes in metal castings occurred with time due to mold wear, it might be possible to handle any misalignment problems by modifying the robot's program (or reteaching a few points).   Such might not be true with the fixed automated device.

It is clear that use of a robot may significantly reduce the lead time required to start producing a new product and will facilitate changes necessitated by process variability.   Thus even though a robot may cost significantly more than the fixed automated equipment initially,* the robot will actually be less costly when time is factored in.

2. Debugging.   As mentioned above, once a fixed automated device is delivered to the plant, it must be placed into operation.   Due to the fact that it is a special-purpose electromechanical device for which there is little or no past history of operation, this will often require a good deal of "fine tuning."   For example, limit switches and perhaps other sensors will have to be correctly positioned, solenoids properly adjusted, and so on.   In some instances, it may even be necessary to redesign and rebuild entire portions of the machine before satisfactory operation is achieved.   All of this will, no doubt, make the debugging or shakedown part of the procedure a time-consuming affair.

On the other hand, if a robot is to be used to perform the same task (or tasks), the debugging operation will take a significantly shorter time.   Since the robot is an off-the-shelf piece of automation, power connections, perhaps compressed air lines, and proper positioning (on a stand and near the workstation) will be required.   (Note that the fixed automated device may need power and actuator feeds too.)   Also, the appropriate gripper (or grippers) will have to be available, although such devices were probably ordered at the same time as the robot.

---

*This is certainly not always true since robots may actually be less costly.

As noted above, the robot itself will be operational almost immediately. However, additional time will be required for the programming of the device. Generally, this will mean teaching the points in the workspace that the manipulator must move to in order to perform the desired tasks. In addition, it may also be necessary to install simple sensors (e.g., optical switches) that permit the robot to interact with other equipment in its workcell. Nevertheless, it should be fairly evident to the reader that debugging time for the robot is likely to be considerably shorter than for the special-purpose, hard automated device.

3. Resistance to Obsolescence. Engelberger has said that resistance to obsolescence is the "very essence of a robot." Unlike a piece of fixed automation which is capable of performing only a single, specific task, the robot is not limited by the nature of the product, the type of operations to be performed, or the particular industry. In fact, many of the robots that were purchased in the early 1960s are still operational despite the fact that they are considerably less sophisticated than modern-day units. It is this aspect that makes flexible automation such an attractive alternative to companies that regularly require model changes that necessitate retooling (e.g., the automobile industry). These industries can now retool, in part, by reprogramming their robots and also by utilizing different types of grippers (although this may not always be necessary). Consequently, downtime and costs can be reduced considerably.

The "conventional wisdom" concerning the use of robots is that they should be considered in operations that require periodic modifications to the process or where it is expected that a variety of items will be produced over a specific time. Although this wisdom seems to be reasonable, manufacturers who normally would use hard automation in making their limited variety of products are now considering or actually utilizing robots also. The reason for this apparent "misuse" of flexible automation is simply stated—*cost*!

These manufacturers have discovered that because a robot can be placed into operation in a much shorter time when the design, building, and debugging of a fixed automated device are taken into account, they can probably begin to produce their product much faster. Also, even though a robot is a complex device, its capital cost may actually be *lower* than that of a comparable hard automated machine. For although the cost of developing a robot may be great, it can be amortized over a large number of units and many different customers, whereas all of the development costs for special-purpose devices must usually be borne by a single user. Consequently, on a per unit basis, these costs will be relatively small when a large number is to be purchased.

Another reason for selecting a robot in "traditional" fixed automated applications is *ease of operation*. A robot tends to be quite "user friendly." To become proficient in its use generally requires only a few days, although additional time will inevitably be needed to program and debug the input/output devices, feeder mechanisms, and any other peripherals. Consequently, it is possible to use the robot in the manufacturing process in ways that were not imagined initially. For example, as experience is gained in making a new product, the robot can readily adapt to efficiency-increasing changes in the process that become apparent only after some time. With a fixed automated device, any unanticipated problems or manufacturing shortcuts generally require a major redesign. This is likely to be both costly and time consuming.

The final reason for choosing a robot to perform a limited range of tasks is that, besides the time and cost factors discussed above, the device *can* always handle other manufacturing tasks, if necessary. The manufacturer recognizes that even though modification or complete change over of a process cannot be (or is not) anticipated at the time of purchase, the robot *will* be able to adapt to the new situation if the time ever arises when change is necessary or desirable. Although it does not cost any more to get the ability to change, it is the knowledge that it is there which is comforting. Of course, this is one of the major advantages of flexible automation.

We have shown in this section that the use of robots in the manufacturing environment certainly seems to be justified from a qualitative point of view. However, in business, it is often the "bottom line" that dictates whether or not a certain policy will or will not be acceptable. In the next section, we explore briefly the economic justifications of using robots.

1.6 ECONOMIC CONSIDERATIONS

Although it is certainly true that robots can relieve humans of the need to perform what has been called "3D jobs" (i.e., a very dirty, very dangerous, or very difficult jobs), the fact remains that manufacturing plant managers are extremely concerned with the "bottom line." A survey of robot users and potential users conducted in 1981 by the Carnegie–Mellon University Robotics Institute indicates that the primary reason for selecting a robot is to reduce labor costs. See Table 1.7.3 in the following section. Thus regardless of how potentially beneficial robots may appear to be with respect to humans, if they cannot be justified economically, they will not be purchased. The purpose of this section is to present briefly some simple techniques that have been used to demonstrate that, indeed, robots can rather easily be shown to be an economically justifiable capital expenditure. It is not our intention, however, to develop sophisticated economic theories, as that is beyond the scope of this book.

Today, the price of a single industrial robot ranges from about $10,000 to well over $100,000. To this must be added the cost of the associated tooling and fixturing that are to be used within the robot work cell and also the cost of the installation itself. It has been found that approximately 55% of the overall system cost is for the robot, 30% is for the additional tooling, and about 15% is for

installation. Let us consider a system with a total cost of $100,000 broken down
as follows:

- Materials-handling robot $55,000
- Tooling and fixturing $30,000
- Installation $15,000

It should be noted that the figure used for this type of robot is about the
current average in the United States. Also, the tooling and fixturing figure includes
engineering development costs.

To determine the economics of such a robot, we need to know the cost of
labor and of the operation of the robot itself. It was estimated that in 1982, an
automobile worker earned about $17 per hour, including fringe benefits. In ad-
dition, the Draper Laboratory at MIT has estimated that it costs about $6 per hour
to run a robot based on operating 16 hours per day (i.e., two shifts per day) and
a useful life of about 8 years. (Although other sources suggest a figure of $2/
hour,* many robot manufacturers use the more conservative number.) Since a
worker will normally put in about 2000 hours per year (40 hours/week $\times$ 50 weeks),
it can be seen that the $11/hour differential in labor costs ($17 $-$ $6) produced by
the robot results in a yearly "saving" of about $22,000. Thus it will take about
2.8 years to pay back the original cost of the robot ($55,000/$22,000). After this
time, the user will be "making" $22,000 per year or, more correctly, will be ex-
periencing a positive cash flow. If we assume a two-shift-per-day activity, the
payback period will be only 1.4 years, after which time a cash flow of $44,000/year
will occur.

Even if we take into account the entire system cost of $100,000, the payback
period for single- and double-shift operations will require only about 4.5 and 2.3
years, respectively. Positive cash flow will still be realized before the 8-year "tax
life" of the robot is over. However, it should be recalled that one of the advantages
of robots is that they can be used for a very long time since they do not become
obsolete. Thus it is to be expected that the user will realize a considerable profit
over the robot's *useful* lifetime.

Obviously, this analysis is an oversimplification, since it does not look at all
economic factors, such as the cost of money and the escalation of labor costs.
Nevertheless, it does provide one with the idea that robots *can* be justified economically, and rather easily at that. Now let us refine the analysis somewhat by
including such factors as corporate tax rates, depreciation, and the savings resulting
from using less material in a particular process.

---

*Richard C. Dorf, *Robotics,* Addison-Wesley Publishing Company, Inc., Reading, Mass., 1985,
p. 149.

It can be shown that the payback period Y can be calculated from the following equation:

$$Y = \frac{(P + A + I) - C}{(L + M - O) \times H \times (1 - \text{TR}) + D \times \text{TR}}$$

where Y = number of years required to break even
 P = price of the robot = \$55,000
 A = cost of the tooling and fixturing = \$30,000
 I = installation cost = \$15,000
 C = investment tax credit* (assumed to be 10%) = \$10,000
 L = hourly cost of labor, including fringe benefits = \$17
 M = hourly savings in the cost of materials = \$1
 O = cost of running and maintaining the robot system = \$6
 H = number of hours per year per shift = 2000
 D = annual depreciation assuming an 8-year "tax life,"* the straight-line method, and a salvage value of \$10,000: = (\$100,000 − \$10,000)/8 = \$11,250
 TR = corporate tax rate,* assumed to be 40% (= 0.4).

Substituting these values into the foregoing equation gives a payback period of 4.8 years for a single-shift operation and 2.7 years for a double-shift operation. It should be noted that this result does not take into account the time value of money, which could be done by using discounted cash flows. We will, however, not do such an analysis.

Another economic yardstick that is often used in determining whether a particular capital expenditure is warranted or not is the *return on investment* (ROI). Defined as the ratio of the total annual savings realized from the equipment divided by the total investment (and expressed as a percentage), if the ROI is larger than the current percentage rate of borrowing money (e.g., assumed here to be about 20% but obviously subject to change, depending on the economic and political climate existing at the time the analysis is performed), the purchase is usually justified. We may write

$$\text{ROI} = \frac{\text{total annual savings}}{\text{total investment}} \times 100\%$$

In terms of the quantities defined above, this can be expressed as

$$\text{ROI} = \frac{(L + M - O) \times H - D}{P + A + I - C} \times 100$$

---

*The Tax Law of 1986 modifies the actual value of this quantity.

Using our example values in this equation indicates that the ROI is only 14.2% for a single-shift operation. However, this figure increases to an impressive 40.8% when the robot is used two shifts per day. When compared with the 20% cost of borrowing money, it appears that a robot used in a multiple-shift application is clearly a good investment whereas it is marginal in the case of a single shift. It is important to realize that a more or less favorable result will be obtained if different assumptions are made concerning labor and/or robot costs. For example, if we use $2/hour for the running and maintenance of a system, the one- and two-shift ROIs become 23.1% and 58.6%, respectively.

Other economic measures can also be used to determine if the purchase of a robot is valid. These include the *internal rate of return,* which permits the time value of money to be included, and the total cost of labor analysis, which allows the labor costs over an extended period to be compared with the cost of running the robot for the same period of time. We will not discuss these here and the reader wishing to learn more about them or other techniques is referred to the References at the end of this chapter [25, 26, 27, 28, 29, 30].

One final point is worthy of mention. The quantitative measures described above do not take into consideration the economic benefits that can be derived from using a robot to produce a product that is of a consistently high quality. In addition, they do not permit an estimate of what the savings will be if a robot takes appropriate emergency action, thereby preventing an expensive process from being ruined. [One user actually estimated that because of frequent power outages in

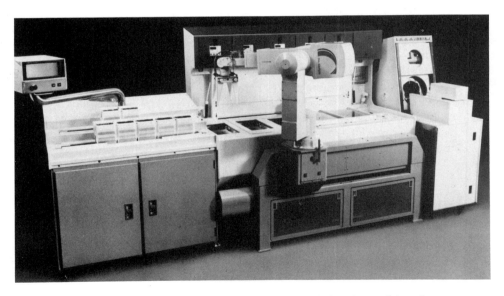

Figure 1.6.1. A Maker 110 Robot from U.S. Robots, designed to work in a clean room environment, is pictured in a semiconductor wafer-etching application. (Courtesy of G. Heatherston and U.S. Robots, Inc., a Square D Company.)

his plant's area, the payback period for his robot would be *one day*! (See Figure 1.6.1.) Also, based on the total amount of product that the robot could salvage over the period of one year, he felt that he could realize a potential savings of about $1 million.] Obviously, each user would have to make such a calculation using the set of circumstances and assumptions that are appropriate to the operation in question. It should be clear to the reader, however, that regardless of the application, such considerations are extremely important when trying to justify the purchase of a robot. In some instances, in fact, they may even be more important than a favorable ROI.

The important result of this section has been to demonstrate clearly that from an economic point of view, robots seem to make a great deal of sense. However, what about the human element? What will be the impact on the workers themselves of introducing these devices into the workplace? We present possible answers to these and other problems in the next section.

1.7 SOCIOLOGICAL CONSEQUENCES OF ROBOTS

The problems created by the introduction of machines into the workplace are not new. For example, the early part of the Industrial Revolution saw Adam Smith, in *An Inquiry into the Nature and Causes of the Wealth of Nations,* published in 1776, expressing his concern that workers who sought to improve their economic circumstances might drive themselves so hard that their health would be adversely affected, even to the point of shortening their lives. Twenty-three years later, David Ricardo perceived that mechanization might be a double-edged sword, in that some workers who were displaced by machines might not be able to find new jobs. Thus what was good for the employer might not be good for the worker. By the middle of the nineteenth-century, Karl Marx wrote about the extensive job displacement suffered by millions of workers caused by the introduction of machines into the workplace, but incorrectly predicted the subsequent self-destruction of the capitalist system as a consequence. His error was in not recognizing that increased productivity due to mechanization would actually produce a substantial improvement in the economic and social well-being of future worker generations.

With the introduction of the robot, the twentieth-century worker may well face many of the same problems as those of his eighteenth-century counterpart and, in addition, a host of others. If, as has been said, the robot will be the catalyst for initiating the second industrial revolution, an important question that must be asked is: What will be the effect on society as a whole and the individual worker in particular? Clearly, there are no pat answers to such a question, nor are there easy solutions to the problems that will inevitably arise, and in certain instances have already arisen, when robots and other high-level intelligent automation devices are introduced into the manufacturing environment. In this section we wish to make the reader aware of the difficulties that American society will face as this new form of technology becomes a "way of life."

In a paper presented at the Robots VI Conference held in Detroit in March 1984, Sandra Pfister of Prab Robots, Inc. wrote that "it is increasingly apparent that robotics, automation, and related high technology are the keys to national [economic] survival."[5]   If the reader doubts the validity of this statement, the following facts should be considered:

- In the past six years, U.S. productivity (defined as the ratio of total production to the units of labor) has grown at an annual rate of 0.01%, whereas in 1979 alone, Germany had a 4.4% increase.[5]   It should be noted, however, that in congressional hearings held in the latter part of 1982, it was pointed out that the Japanese worker is actually about 60% *less* productive than his American counterpart.   In the United States the problem has been that productivity has remained at the *same* level for a number of years, whereas countries such as Japan and Germany have had large *increases* in their productivity [6].

- In 1981, a large Japanese company, Hitachi Ltd., assembled a force of 500 technical people with the express purpose of developing by 1985 a universal assembly robot with visual and tactile sensing.   In contrast to this, the largest manufacturer of robots in the United States, Unimation, Inc., a Division of Westinghouse, has only about 90 robotics engineers [7].

- Renault, the French automobile giant, developed a "vertical robot" equipped with TV and sonar sensors that is used to locate, lift, turn, and move a 26-lb crankshaft from a pallet to an assembly line.

- Fujitsu Fanuc Ltd. has opened a plant that employs robots to *produce* about 100 robots per month.   Human beings are still used for final assembly, but they represent only 20% of the normal work force that would be required in a plant of the same size [8].

These and numerous other examples of foreign developments in the field of robots and associated automation have placed great pressure on American industry. In Table 1.7.1 it is observed that since 1980, Japan has been producing almost six times the number of robots per year as the United States.   However, it is estimated that by 1990, this production gap will narrow to about three times.

Although some of these devices have been exported, many have been used in Japanese industries.   As shown in Table 1.7.2, it can be seen that the Japanese have enjoyed a significant numerical superiority over the rest of the world in the actual number of robots in use since 1980.   It is important to note that the numbers for the Japanese units represent only those that satisfy the RIA definition of a robot.   As of 1983, there were about 96,000 additional nonprogrammable, fixed-sequence, and/or manual manipulators in use in Japanese industry.

Obviously, a company cannot continue to ignore the fact that its foreign competitors are making use of advanced technology if it is to remain financially healthy.   Not surprisingly, many U.S. firms *are* introducing robot systems into their production plants.   For example, McDonnell-Douglas is employing a million-dollar robotic manipulator to control a laser beam that cuts out sheets of graphite

TABLE 1.7.1   U.S. AND JAPANESE ROBOT PRODUCTION

Year	Units/year		Value(millions)	
	U.S.	Japan	U.S.	Japan
1979	614	2,763	$ 62.5	$ 81.1
1980	1,118	4,493	101.0	205.3
1981	1,993	8,182	155.0	310.7
1982	2,585	14,937	190.0	471.0
1983	3,060	18,599	240.0	612.9
1984	5,137[a]	23,249(E)	332.6[a]	766.1(E)
1985	6,209[a]	31,900(E)	442.7[a]	2,150.0(E)
1990	21,575(E)	57,450(E)	1,884.0(E)	4,450.0(E)

[a]Robot Industries Association, from *Robotics Today*, December 1986, p. 9.

(E) Estimate by Paul Aron

Source except as noted: Paul Aron Report (#28), "The Robot Scene in Japan: The Second Update," Daiwa Securities America Inc., April 15, 1985.

used in aircraft brakes. This device also forms tail and wing section parts that are then welded together by an arc-welding robot. Two workers and the robots now do work that previously took 30 people to do [9]. Also, Texas Instruments uses three robots to test its hand calculators. One brings a particular unit into the test area, another one presses the keys in a prescribed sequence, and the third looks for the correct numbers on the calculator's display. In addition, Cheesebrough-Pond has a robot that places jars of skin cream into cartons from an assembly line.

TABLE 1.7.2   INSTALLED ROBOT POPULATION 1980—1983 (U.S. DEFINITION)

Country	1980	1981	1982	1983
Japan	14,246	21,684	33,961	48,825
Soviet Union	NA	6,650	12,050	25,000(E)
United States	4,100	4,700	6,301	9,361
West Germany	NA	1,420	4,300	4,800
France	NA	620	993	3,600
Sweden	600	700	1,450	1,900(E)
Great Britain	500	713	977	1,753
Canada	NA	NA	273	700
Belgium	NA	44	305	514
Poland	NA	NA	285	245

(E) Estimate

NA = Information not available

Source: Paul Aron Report (#28), "The Robot Scene in Japan: The Second Update," Daiwa Securities America Inc., April 15, 1985.

In the highly competitive and labor-intensive automobile manufacturing industry, U.S. companies have had robots working in many sections of the assembly line alongside their human employees for a number of years. Most of these units have come from American companies, probably due, in large part, to union restrictions on using foreign devices and parts. However, the pressure to meet foreign competition has recently produced a significant change. General Motors, a traditional stronghold of the "buy American" philosophy, in 1982 joined with Japan's Fujitsu Fanuc Ltd. to form the GMF corporation, for the express purpose of designing, building, and marketing robots in the United States. The plant is located in Michigan, and GM currently buys about 75% of the units produced there for its own use. (The first robot delivered to GM actually came from Japan [10].)

It is clear from the small number of examples cited above that robots are here to stay and will be used by American industry. With this as a given, what will be the effect on the American worker? Writing in the *New Republic*, Mark Miller of the University of Pennsylvania suggested that "the fear of robots is an apprehension of gradual displacement . . . a foreboding of our own annihilation" [11]. It appears that this "fear" is, to a certain extent, well founded. For example, General Electric projects that eventually it will replace almost half its 37,000 assembly workers with robots. The company feels that in some instances robots will increase productivity in their plants by 50% [12]. Also, it has been predicted that in the auto industry alone, as many as 100,000 jobs will be lost to robots by 1990 [13]. Moreover, in a study conducted at Carnegie–Mellon University in 1980, Ayres and Miller predicted that by the year 2000 the current class of non-sensor-based robot would replace as many as 1 million manufacturing production workers. They also predicted that robots possessing some rudimentary tactile and vision sensing capabilities would replace about 3 million of these workers. Finally, they projected that by the year 2025, all the current 8 million manufacturing production workers could be expected to be replaced by highly sophisticated robots. To be sure, this would represent "only" about 8% of the total workforce today. Nevertheless, when combined with unemployment in other areas, it could have a significant effect on the overall state of the nation's economy [14]. Another study performed by International Resource Development, Inc. in 1982 suggests that of the 32.1 million blue-collar workers now in the United States, approximately 13 million could be replaced by robots. Of these, about 25% would be retrained to program and maintain the robots. Interestingly enough, however, it is estimated that no more than about 18,000 workers have actually been displaced by robots despite the intense economic pressure from abroad. (The rule of thumb used by industry is that one robot displaces three workers.) Engelberger has stated that the reason for this is that few middle- and high-level managers in American industry have recognized that they need robots in their plants in order to meet foreign competition successfully [13].

The fear of displacement is apparently not restricted to American workers. In Japan, for example, the impact of robots on the production worker is just beginning to be felt. Several of that country's largest labor unions have forced the government to undertake a study on this subject. These unions are frightened at the prospect of potential wide-scale job elimination within the next five to ten years. The reason for this concern is that companies that have ordinarily moved people displaced by robots to other areas of the same plant are now moving these people to new locations, sometimes in other cities. Moreover, Japanese plant managers indicate that they are slowly running out of jobs for the displaced and for new workers who are just coming into the work force, although unemployment in Japan is still a low 2.3% [15]. It is fairly clear that unemployment must someday begin to rise even in Japan if automation continues to replace people without a corresponding increase in jobs. Note that this runs contrary to the popular notion in the United States that all Japanese workers enjoy "cradle to grave job security" (in reality, only about 30% have this benefit).

The key question, then, seems to be: Will the introduction of robots produce a net gain in the number of jobs, and if so, what will be the time frame to realize such an increase? History reveals that when machines were introduced at the beginning of the first industrial revolution, many jobs were lost for a significant period of time. No one would deny, however, that in the "long run," the machines created many more jobs than were lost. What happened was that industries that had not existed before now began to develop. While workers in the traditional "cottage" industries lost their jobs, new jobs were created and people were trained for these. But robots may not be just another machine. George Brosseau of the National Science Foundation has been quoted as saying: "In the past, whenever a new technology has been introduced, it has always generated more jobs than it displaced. But we don't know whether that's true of robot technology. There's no question that new jobs will be created, but will there be enough to offset the loss?" [10]

Others are more optimistic in their outlook. For example, James Albus of the National Bureau of Standards has said that "robots create profits, profits create expansion in industry and expanding industries hire more people" [16]. Ayres and Miller [14], [22], [31] have suggested that these new jobs will require workers who are significantly more skilled than before: for example, those who are capable of building, repairing, and maintaining the robots. They also feel that a large number of jobs could be created if the robot revolution follows a course that is parallel to that of the computer, that is, if a market for the "home" (or "personal") robot develops.

Even if Albus, Ayres and Miller, and the robot manufacturers are correct, it is possible that history may repeat itself and initially there will be a significant loss of jobs directly traceable to the introduction of this new technology. Hazardous, repetitious, and boring jobs that *human beings really should not be doing* will be the first to go. In fact, this is already occurring in some industries. This may be applauded by some, but what of the people who lose these jobs? The United Auto Workers (UAW) union has estimated that by the end of this decade, assembly-line labor could be reduced by as much as 50% because of robots and

TABLE 1.7.3 MOTIVATIONS FOR USING ROBOTS

Ranking	Users	Prospective users
1	Reduced labor costs	Reduced labor costs
2	Elimination of dangerous jobs	Improved product quality
3	Increased output rate	Elimination of dangerous jobs
4	Improved product quality	Increased output rate
5	Increased product flexibility	Increased product flexibility
6	Reduced materials waste	Reduced materials waste
7	Compliance with OSHA regulations	Compliance with OSHA regulations
8	Reduced labor turnover	Reduced labor turnover
9	Reduced capital cost	Reduced capital cost

Source: CMU Robotics Survey, April 1981.

other automated devices. This union has often encouraged the introduction of technological advances in the automobile industry. But with robots it has had to look for novel ways to counteract the potential impact on its members. As a consequence, the union has negotiated a contract with the automobile companies that provides for extra personal paid holidays. The idea is that existing jobs can then be distributed to more workers without a significant loss in pay. The problem with such a solution is that productivity may actually decrease.

Other industries may not want or be able to do the same thing. For example, as shown in Table 1.7.3, when Carnegie–Mellon University questioned 38 members of the RIA (19 users and 19 potential users of robots), they discovered that the number one consideration for placing robots into factories was *Reduction in Labor Costs* [17]. The same conclusion was arrived at in a 1980 study conducted by MIT's Draper Laboratory. These results are not that startling when one understands that between 1968 and 1978, labor costs in the United States rose by 250%.

Besides the potential for creating unemployment in the near term, the introduction of the robot into the work place will create some rather unique problems for human workers who remain employed. McVeigh points out that in the future, such workers may be forced to interact with machines, not other human beings [18]. He predicts that this situation may cause worker alienation since communication with others in a work situation is important in relieving monotony and making time pass more quickly on the job. As an example of this, a large chemical corporation decided to replace its human mailmen with a mobile robotic device. After it was put into service, the office manager found that it did not stop and talk to people or take coffee breaks, so it was more efficient. However, "news" about the firm, rumors, and office gossip, which had previously been supplied by the human being delivering the mail, no longer came from this link. Instead, the job of delivering the mail was performed silently much to the annoyance of the other office workers.   It turned out that the human mailman provided an important communications link which was broken by the robot [19].

McVeigh also states that new and old workers will have to compete in a "new system of labor" that is "functionally equivalent to slavery."   Moreover, where human beings and robots work together, the human being will be forced to compete with an entity that:

- Does not tire
- Does not seek or obtain wage increases and fringe benefits, such as paid holidays and vacations
- Does not go out on strike or slow down due to disagreements
- Does not argue or debate the supervisor's ideas

What can be done to soften the inevitable impact that robots will have on many American industrial workers?   Ayres and Miller suggest the following possibilities:

- Industry should identify categories of jobs and workers that will be adversely affected by robots.   For example, it is already quite clear that the need for human welders and machinists will diminish significantly in the near future. Also, Table 1.7.4, which compares the current distribution of jobs performed by robots with that expected by the year 1990, indicates that human beings will be competing with robots for jobs in several additional categories in the near future (e.g., certain types of assembly work).   The important thing is that identification of the potentially affected job categories should be done well in advance of the reduction and/or elimination.   For example, Table 1.7.4 clearly shows that assembly jobs will be severely impacted by robots

TABLE 1.7.4 JOBS DONE BY ROBOTS NOW AND IN THE FUTURE[a]

Task	Through 1981	By 1990
Spot welding	35–45	3–5
Arc welding	5–8	15–20
Materials handling including machine loading and unloading	25–30	30–35
Paint spraying	8–12	5
Assembly	5–10	35–40
Other	8–10	7–10

[a]Numbers represent the percentage of total jobs performed by robots in the given year.

Source: Bache Halsey Stuart Shields, Inc.

since by 1990, it is projected that 35–40% of the total jobs performed by
robots will be in this category (as compared to only 5–10% in 1981). Note
also that the reduction in the percentage of spot welding and spray painting
jobs handled by robots is due to the large increases in other areas and does
not imply that the actual number of jobs in these categories will decrease.

- Industry, government, and unions should cooperate in long-range planning
  of employment needs.
- These groups should identify and publicize the new job skills that will be
  required in the future. This will alert young people to the changing patterns
  in the job market so that they can acquire *marketable* skills.
- Education and job training facilities should be established to help retrain
  workers whose jobs will be lost.
- Industry and government should create facilities to locate suitable jobs for
  displaced workers and to help pay for their relocation. Of course, this assumes the availability of additional jobs.
- Workers and employers should finance an actuarily sound national job security fund which would be used to pay for the transportation, maintenance,
  and retraining of displaced workers.

Clearly, some of these suggestions would be difficult to implement and may
not solve the problems of all displaced workers. For example, it will be difficult
(or perhaps even impossible) to retrain a worker who has performed the same task
for over 20 years and has suddenly been replaced by a robot.

From the discussion above, it seems apparent that robots will displace workers
but may also create jobs. However, these jobs will be highly technical in nature
and will therefore require a significant amount of training. Another consequence
of the robot revolution is that workers may find themselves with more leisure time.
It has been suggested that this would stimulate growth in industries related to
recreation. In fact, a prediction made in 1956 "that the importance of work for
people as a 'central life interest' may decline as robots replace them" may become
a reality as the result of the use of this technology [20].

Having discussed some of the possible effects that robots may have on human
workers and having presented some writers' ideas of how society can soften the
impact of these devices, we now return to more concrete ideas and look very briefly
at the characteristics of several commercial robots that are currently available or
have been available in recent years.

1.8 STATE-OF-THE-ART SURVEY

One of the most comprehensive surveys of robots that are commercially available in
the United States was performed in 1982 by Stock Drive Products of New Hyde Park,
New York. In this work, the specifications provided by 74 manufacturers for a total

of 152 different models were tabulated. As an indication of the rapid changes that are expected to take place in the robot industry in the next few years, the people at Stock Drive Products caution the reader to keep in mind the following when using the complete table (which is reproduced in its entirety in Appendix A).

- The list is not complete and additional models (from these and other companies) may be available.
- Some manufacturers may not carry a large inventory of spare parts. Others may actually be out of business now or in the near future.
- Some of the manipulators have undergone little if any field testing and development. Some may be shipped without adequate (i.e., at least 50 hours) life testing in the factory.
- The performance characteristics quoted in the table may not actually be achievable.
- Not all manipulators listed in the table may currently be available.
- There is a wide variety of capability between manufacturers in regard to both current and potential future customer service.
- Some of the manipulators listed are for educational, not commercial use.

To assist the reader in using the complete table, however, we have excerpted some of the voluminous information included in this compendium for a few select robots. This is shown in Table 1.8.1.

In using this table (and the one in Appendix A), the reader should understand that the data contained therein must be carefully interpreted. For example, while the *maximum* tip speed may be 50 in./s, it is probably *not* possible to do this with the stated maximum load. Also, the price of the robot may, in certain cases, be for the minimum system and in other cases for the "Rolls-Royce" or fully accessoried system. Finally, memory-capacity comparisons may be difficult since there is no single standard for this specification. For example, some manufacturers specify the number of program steps, others the number of programs, and still others the number of memory bytes. A more complete discussion of these and other robot features is given in Section 2.5.

In addition to the data entries in Table 1.8.1, the more complete compilation found in Appendix A also gives the company's telephone number, the extent of the major joints, the dexterity of the wrist joints, the type of memory devices used (e.g., semiconductor, magnetic tape, air logic, or mechanical stop sequencer), and whether or not the robot is considered suitable for "educational purposes."

It is important to understand that the robotics industry is an extremely dynamic one, with companies going into and out of business quite routinely. For example, as of this writing, Bendix, Copperweld, and Nordson are no longer in the field and Unimation has been acquired by Westinghouse Corporation. On the plus side, as mentioned previously, in 1983, General Motors joined with Fujitsu Fanuc of Japan to form the GMF Corporation. It is expected that this type of

TABLE 1.8.1 SELECTED CHARACTERISTICS OF SEVERAL COMMERCIAL ROBOTS

Name and model	Price	Load cap. (lb)	Repeat. (in.)	Max. tip speed (no load) (in./s)	Coordinate system	Drive type	Type of control	Memory capacity	Programming method[a]	Applications[b]
Cincinnati Milacron T³-566	$80,000	100	0.050	50	Spherical	Hydraulic	Point-to-point Servo	450 points	Keyboard, pendant	1, 3, 4, 5, 6, 9, 10, 11, 14
Control Automation	82,000	10	0.001		Cartesian	Electrical	Servo	16k bytes	Keyboard, pendant	12, 13
IBM 7535	28,500	13.2	0.002		SCARA	Electrical	Servo	5 programs	Keyboard, pendant	1, 10, 11, 12, 13
Seiko Instruments 200	5,500	1.7	0.0004	180°	Cylindrical	Pneumatic	Point-to-point nonservo	Varies	Mechanical setup	1, 5, 8, 12, 13
Unimation Puma 760	60,000	22	0.008	40	Jointed spherical	Electrical	Servo	Varies	Keyboard, pendant	1, 5, 9, 11, 12, 13
U.S. Robots Maker 100	36,000	5	0.004	65	Spherical	Electrical	Servo	350 steps	Pendant	1, 2, 5, 11, 12, 13
Yaskawa Electric Motoman L3	50,200	6.6	0.004	70	Parallel-ogram	Electrical	Servo	2200 points	Keyboard, pendant	1, 4, 5, 7, 9, 11, 12, 13

[a] Pendant means either a teach pendant or box.

[b] Application code: 1, materials handling; 2, die casting; 3, forging; 4, plastic molding; 5, machine tool loading/unloading; 6, investment casting; 7, general machine loading/unloading; 8, spray painting; 9, welding; 10, machining; 11, other tool applications; 12, assembly; 13, inspection.

Source: Stock Drive Products, New Hyde Park, N.Y.

situation will continue to occur into the foreseeable future as the inevitable "shake-out" (perhaps accelerated by the worldwide economic weakness of the late 1970s and early 1980s) takes place.

The robots listed in this section (and in Appendix A) can perform a variety of manufacturing tasks. The next section describes some of these in more detail and also indicates possible uses for robots in the future.

1.9 ROBOTIC APPLICATIONS: CURRENT AND FUTURE

In its relative infancy, the state of the art of robotic applications is, in some ways, paralleling the development of digital computers. When they were first introduced, computers were used for tasks that had previously been performed by people (with perhaps the assistance of some type of manual aid, such as a slide rule or mechanical calculator). This was a natural application, for it was obvious that the new device would be able to perform such jobs much faster and even more reliably than people could perform them. However, as time progressed, it was recognized that tasks that had heretofore been rejected as being impossible to undertake because of excessive manpower and/or time requirements were now possible to attempt. Thus problems that were "not practical" to solve were handled with relative ease. Besides being able to solve such problems, it became apparent that there were many applications for the computer that had never been thought of before its development. In a sense, what happened was that people took off their "blinders" and allowed their imaginations free reign. The result of this has been that computers are now applied in many areas other than the more traditional "number crunching" that was initially envisioned as the major use. The fields of control (of large-scale systems), learning and teaching devices, handling of large data bases, and artificial (or perhaps more descriptive, "autonomous") intelligence come to mind, to name but a few nontraditional applications. But where do we stand with robots?

As already mentioned in earlier sections of this chapter, the first applications of the robot have been in areas where human beings have traditionally been working. Although there have been some significant technological advances in the design of robots (i.e., the hardware) since the first one was developed more than 20 years ago, the manipulators currently being manufactured are, as a general rule, rather simple (e.g., most lack the ability to sense their external or working environment). As a result, the state of the art in robot applications is probably where the computer was when it was used primarily for "computing." It has taken a much longer time for the blinders to be taken off when talking about robots than it did with computers. One can cite a number of possible reasons for this, including the problems of recessions, fear of people losing their jobs, and the lack of a major scientific breakthrough comparable to the development of the transistor and later, the integrated circuit. Also, some of the first big users and/or developers of computers were in government, the military, and the universities. These three entities, which were responsible for developing many of the unique computer applications, have only recently entered the robot field in a large way. (The program at the National Bureau of Standards, having been started in the 1970s, is a notable exception.) The industrial sector has been the major user, and as might be expected, the need to produce a "good bottom-line result" has prevented or at least significantly reduced the risk taking required to produce new ideas (i.e., applications) and developmental research by manufacturers. The recent emergence of robot programs supported by both the military and state and federal government may indicate that this situation is beginning to change, however. As a consequence, it is to be assumed that over the next few years, nontraditional robotic applications will begin to appear which will, in part, contribute to the development of the fully automated factory or *factory of the future.*

In the first part of this section we briefly summarize some of the more traditional uses for robots, some of which have already been mentioned in earlier sections of this chapter. In the concluding portion of the section we indicate some of the more futuristic applications that have been proposed by some workers in the field.

1.9.1. Current Robotic Applications

In the preceding two sections we encountered a number of applications of today's industrial robots. For example, in Section 1.7 it was indicated that welding, grinding, and spray painting account for the majority of applications of the current generation of robots. In addition, Section 1.8 listed a total of 13 apppplications for robots that were available in 1982.* We now briefly describe a number of these.

1.9.1.1 Welding

Welding is one of the major uses for an industrial robot. Actually, two distinct types of welding operations are readily and economically performed by robots: spot and arc welding. In the former case, the robot is taught a series of distinct points. Since the metal parts that are to be joined may be quite irregular (in three dimensions), a wrist with good dexterity is often required (e.g., three degrees of freedom). This permits the welding tool to be aligned properly at the desired weld point without the gun coming into contact with other portions of the part. Typically, the welding tools carried by these robots are large and reasonably heavy. Also, it is usually necessary for the manipulator to have a long reach. As a consequence, large point-to-point servo-controlled robots (either hydraulically or electrically actuated) such as those produced by Cincinnati Milacron (i.e., the T^3-566), Yaskawa (i.e., the Motoman L3), or General Motors Fanuck (GMF) are normally used for this purpose. The automobile industry is a heavy user of this type of robot (see Figure 1.9.1). Since the weld points are pretaught, sensory

---

*There are, in fact, many more than 13 applications for robots. However, these are currently the major ones.

Figure 1.9.1. A PUMA 700 series robot performing a spot welding operation on an automobile part. (Courtesy of Unimation, Inc., a Westinghouse Company, Danbury, CT.)

information is generally not required in order to energize the welding gun. It is, however, possible to utilize the increased motor current that results when the tool makes contact with the part to initiate the welding operation.

The second type of welding application, arc welding, is also utilized extensively by the auto industry. Here, an often irregularly shaped seam or a wide joint must be made. In this case, a continuous-path servo-controlled robot that is often specifically designed for this single application is most usually the choice (e.g., the Unimation Apprentice robot). If the parts to be welded can be accurately positioned and held in place, the complex three-dimensional path can be pretaught and no external sensors may be necessary. At present, a number of manufacturers include a position sensor that is placed in front of the welding tool and can therefore provide information concerning irregularities in the weld path. Several manufacturers provide additional sensory feedback, among them Automatix and GE. Where a wide joint is to be handled, the robot can be programmed to produce a weave type of motion. This ensures that the weld covers the entire gap. A major advantage of a robotic welder is that the arc time (a critical parameter in determining the weld's strength) can be carefully controlled.

1.9.1.2 Spray painting

The spray-painting operation is one that human beings should not perform, both because of the potential fire hazard and the fact that a fine mist of paint (both lead and modern plastic based) is carcinogenic. As such, this task is a natural

Figure 1.9.2. Spray painting application at a GM plant in Baltimore, MD. (Used with permission from General Motors Corporation, Detroit, MI.)

application for a robot and so it is not surprising that there are a large number of manipulators that perform only this particular job. Another advantage in using a robot for spray painting is that the resultant coating will be far more uniform than a human being could ever produce. This results in a higher-quality product, less reworking of parts, and considerably less paint being used (reductions of 40% are often achieved). Robots employed for this purpose are usually capable of performing both straight-line and continuous-path motions (see Figure 1.9.2).

Programming a spray-painting robot is usually performed by the best human operator. His actions are then mimicked by one or more robots. The spray-painting application generally does not require the use of external sensors. However, it is necessary that the part to be painted be accurately presented to the manipulator.

1.9.1.3 Grinding

As a result of arc welding two pieces of metal, a bead is formed at the seam. Where a smooth surface is required for appearance sake (such as on auto bodies) or for functionality (e.g., to maintain necessary tolerance of parts), it is usually necessary to perform a grinding operation. This is also a natural task for a robot since the manipulator can use the same program that was employed in the arc welding operation. All that must be done is to remove the welding tool and replace it with a rotary grinder (see Figure 1.9.3).

Another important grinding task is on metal castings. Here the robot is taught the correct shape of the casting using continuous-path programming. The

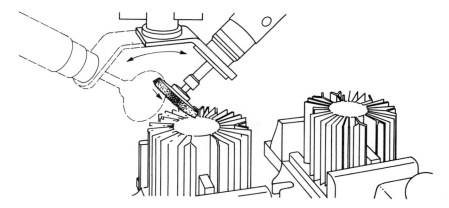

Figure 1.9.3.   Robot used to perform a grinding operation.   Depicted here is the smoothing of the top part of large heat sinks.   (Courtesy of Unimation, Inc., a Westinghouse Company, Danbury, CT.)

grinder then removes any undesired high spots and corrects areas of the casting that are too large.   A third robotic grinding application is that of deburring.   Here the unwanted material that remains around the back side of a drilled hole is ground away to leave a smooth surface.   For increased productivity, it is especially important to be able to perform this task automatically after the holes have been drilled automatically (perhaps by a robot).

In these grinding applications, there is always some uncertainty in the dimensions of the part being worked on.   As a result, sensory information is often needed to permit the robot to more accurately "feel" the actual contour of the part.   This is especially important in the case of smoothing of the arc weld bead.   Relatively simple touch sensors that provide this information are currently available.   For example, the Swedish company ASEA uses such a sensor with its IRB-60 robot.

1.9.1.4 Other applications involving a rotary tool

In addition to the rotary grinding or deburring applications, robots are also currently used for drilling holes, routing, polishing, nut running, and driving of screws.   In the first two cases, preprogramming of either points or paths can be performed when extreme accuracy is not required.   However, where exact placement of drilled holes is needed (e.g., in the structural components of aircraft), it may be necessary to utilize a template (see Figure 1.9.4).   The difficulty with doing this is that unless the robot wrist has some "give" (i.e., compliance), any misalignment of either the part or the robot itself will result in a damaged template and/or an inaccurately placed hole.   This problem is overcome by means of a compliant wrist which permits the drill bit to be aligned in the template hole even if there is a positional error.   The remote-centered compliance (RCC) has been used for this purpose and is discussed more fully in Chapter 3 (see Figure 1.9.5).

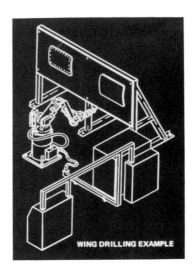

Figure 1.9.4. A Cincinnati Milacron T³ robot drilling holes in an aircraft wing. (Courtesy J. Coshnitzke, of Cincinnati Milacron, Cincinnati, OH.)

1.9.1.5 Parts handling/transfer

The simple task of moving a part or object from one location to another within the work area is one of the most common applications for robots today. Often, it is necessary to acquire a part from a remote location and to place it in a compartmentalized (e.g., a rectangular array) box or carton (see Figure 1.9.6). Once all the compartments are filled, the box is moved (by either the same robot or a larger one) to another location within the work cell, where it is sealed and stacked for future use. Such an operation is referred to as *palletization*. The inverse operation of unloading an array of objects and placing them in another place within the work space (e.g., on a conveyor belt) is called *depalletization*. Since cartons of parts are often stacked one on top of the other, it is necessary to teach a vertical offset to the pallet points so that the robot can unload objects in boxes that are under the topmost one. Some robots have languages that make this a relatively easy programming task (e.g., Unimation's Puma series utilizes VAL II, which permits such an offset to be accomplished with a single program statement). Others, such as the U.S. Robots' Maker 110/2, allow the user to program the offsets via the teach pendant.

Other important parts-handling applications involve the acquiring of blank or unfinished parts and feeding them into some type of machine tool for finishing (e.g., a punch press). This application is often a dangerous one for human beings and so is ideally suited for a robot.

Similarly, in the metalworking industries, a common task is to produce finished castings or extrusions. Such work can be dangerous since it is necessary for

(a)

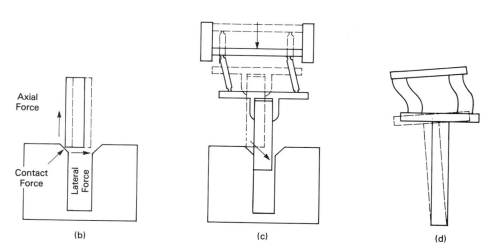

Figure 1.9.5.   A Remote Centered Compliance (RCC): (a) variety of actual units; (b) problem misalignment in a peg insertion operation; (c) with an RCC, the lateral force generated by such a misalignment causes the peg to translate horizontally thereby producing a successful insertion; (d) actual motion of an RCC under such a misalignment condition.   (Courtesy J. Rebman and Lord Corp., Cary, NC.)

a human being to work in the vicinity of hot furnaces, punch presses, lathes, and/ or drill presses.   Robots are ideally suited for this type of job because they can resist the high-temperature environment and can be programmed to avoid collision with the various other machine tools present in the work cell.   As an example, consider a portion of the automated workstation for forging airfoils shown in Figure

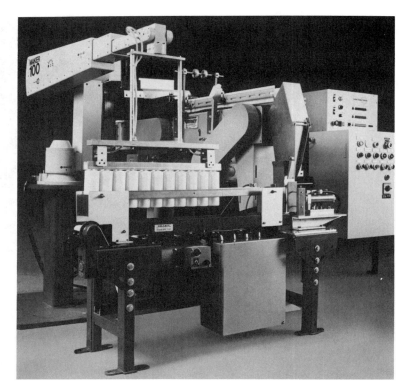

Figure 1.9.6. A United States Robots, Inc., Maker 100 is used to perform a palletization operation. The shampoo bottles are first filled and then the robot places eleven of them into a compartmentalized carton. (Courtesy of G. Heatherston and U.S. Robots, Inc., a Square D Company.)

1.9.7. It is seen that two simple and small cylindrical pick-and-place robots R1 and R2 (e.g., both Auto-Place Series 50 manipulators) and one larger spherical coordinate robot (a Unimate 1005, for example) are used in this application. A vibratory feeder mechanism deposits an unfinished (i.e., uncoated) steel slug at point A. R1 acquires the part and places it at location B where a coat of lubricant is applied automatically. This treated part is then indexed to C, where R1 is again used this time to place it on a drying turntable at D. When the slug reaches E, it is dry and the Unimate uses its long linear reach to grasp the part and place it on the furnace turntable at F. The softening process is completed at G and the Unimate again reaches into the hot furnace to pick up the heated slug and place it into the extrusion die on the press table at H. The work cell controller senses this and causes the press to be activated. This part of the forging process is completed when R2 first positions the die for lubrication and then places the finished extrusions in a bin or on a conveyor.

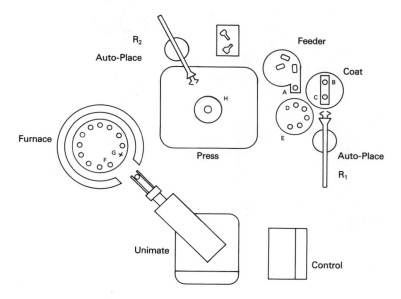

Figure 1.9.7.   Several robots being employed in an automated airfoil forging work-station. The large one, a Unimate, is a 1005 and is used to move the parts into and out of the furnace. The two smaller manipulators, Auto-Place series 50 pick-and-place units, move the parts variously from the feeder to a coating station and finally to a conveyor or finished parts bin. (From J. M. Perkins, "Three Robot Extrusion Workstation," Robots VI Conference, Paper MS82-132, Detroit, MI, March 24, 1982. Redrawn with permission of the author.)

1.9.1.6 Assembly operations

Human beings are capable of assembling a group of diverse parts to produce either a finished product or a subassembly because of their ability to utilize good eye–hand coordination in conjunction with the important sense of touch. However, these jobs may be extremely tedious because of their repetitive nature. As such, assembly operations represent an attractive application of robots. For example, consider the assembly of smoke detectors shown in Figure 1.9.8. Here, although not shown, a group of servo-controlled robots (e.g., U.S. Robots' Maker 100) is actually used. First, the finished printed circuit board is acquired and then is loaded into the bottom portion of the plastic case. Next, a 9-volt battery (with its terminals reversed to increase shelf life) is inserted into the battery compartment. Finally, the top portion of the plastic case is placed onto the finished bottom assembly. It should be noted that this last operation also requires that the robot exert a downward pressure so as to ensure proper locking of the two parts of the case. The finished detector is then stacked in a carton utilizing a palletizing program.

Other assembly applications performed by robots include putting together

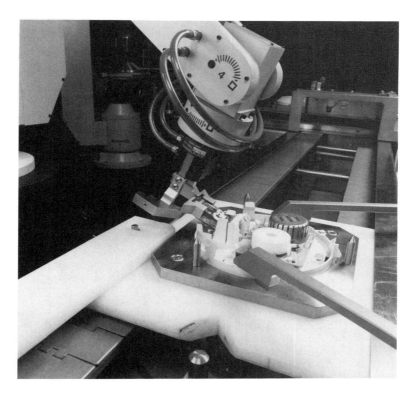

Figure 1.9.8. The assembling of smoke detectors is accomplished using several United States Robots, Inc., Maker 100, five-axis, servo-controlled robots. (Courtesy of G. Heatherston and U.S. Robots. Inc., a Square D. Company.)

scissors, pliers, and other simple hand tools, the fabrication of small electric motors, and the assembling of electrical plugs and switches. In most of these examples, the robot is taught the desired points and the sequence of operations. The only external sensory information that is normally utilized is whether or not a part or subassembly is at a particular location within the work cell. (Such an indication can be obtained using simple optical interrupters or mechanical switches, as discussed in Chapter 5.)

As mentioned above, some applications depend on the robot wrist being compliant. This is especially important in certain assembly operations, for example, insertion of shafts or rods into small clearance holes or the screwing of a screw into a threaded hole. To prevent the binding and/or bending of the rods or cross threading of the screws, an RCC is often used between the end effector and the robot's wrist flange. Alternatively, force and/or tactile feedback can be utilized to provide better external sensing capability, thereby permitting the robot to adapt better to any positional errors caused by either the devices which hold and/or position (i.e., "present") parts or by the robot itself. However, such sensing is, for the most part, not well developed, so most assembly applications are currently geared toward those that either do not require external sensing or else can be performed with an RCC device.

A number of assembly applications do not require the use of a compliant wrist (e.g., electronics assembly). In this case it is necessary to insert a variety of electronic components (e.g., resistors, capacitors, etc.) into a printed circuit (PC) board. As the leads on these components are easily bent, extremely accurate placement before insertion is usually required. Although human beings can perform these operations, the work is tedious and repetitive, with the result that mistakes are often made. Thus a robot is a good choice for this task. However, the high degree of accuracy demands that the manipulator be equipped with an external sensor (e.g., a vision system, see Figure 1.9.9). Although vision peripherals tend to reduce system throughput, it is expected that such applications will become more common as the cost of vision hardware and software drops and the systems themselves become faster. In fact, this is already happening.

1.9.1.7 Parts sorting

Often, groups of parts are produced in an unsorted manner either to reduce costs or because of tolerance variations inherent in the manufacturing process. Robots have been used to perform the extremely boring task of sorting such objects (e.g., washers or O-rings). As seen in Figure 1.9.10 the robot, equipped with an appropriately designed gripper, acquires the part from a gravity-feed magazine and brings it to a workstation that is equipped with an electronic gaging device. The manipulator places the object over a conically shaped anvil. The center hole (or inner) diameter of the object is a function of how far down it travels on this anvil. The gaging device informs the robot controller as to the correct inner dimension value, which causes an appropriate branch in the program to be taken. The part is then reacquired and placed in the correct bin. It has been found that because of the tedious nature of such gaging operations, human beings often make mistakes. In contrast, the robot can perform such tasks faster, for longer periods of time, and far more accurately.

Another less common sorting application involves the use of a vision system. A group of parts is randomly placed on a conveyor belt. The vision system is placed upstream of the robot and determines the type and orientation of the part passing within the field of view of the camera. This information is passed to the robot controller, which then directs the manipulator to move to the correct location with the appropriate gripper orientation. The robot is then able to place the acquired object into the correct bin (or onto another conveyor). This type of application is found in factories where many different castings must be handled (e.g., in the automobile industry).

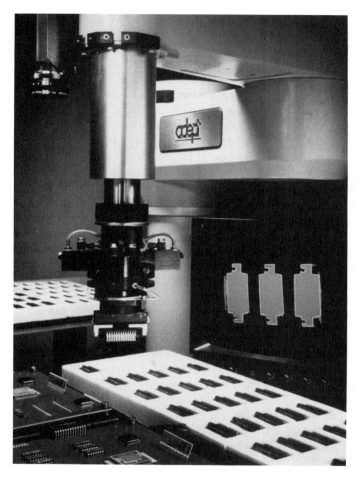

Figure 1.9.9. A vision system is used with an Adept robot to perform printed circuit board assembly. (Courtesy of Adept Technology Inc., San Jose, CA.)

1.9.1.8 Parts inspection

Robots have been used to inspect finished parts or subassemblies in order to increase product quality. The automobile industry is an example of a group of companies that is striving to upgrade its product by automating the inspection process. Here, for example, portions of auto bodies can be checked for dimensional accuracy by placing a special-purpose tool having a large number of movable spring-loaded probes against the part. The distance moved by each probe point (as given by a voltage, for example) can be compared to a predetermined value (stored in a computer data base) for a "good" part. This system not only permits the rejection of out-of-tolerance parts but also gives an indication of potential

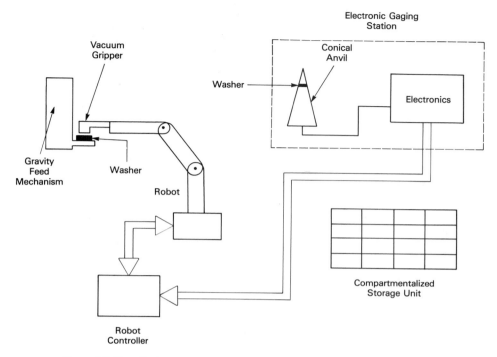

Figure 1.9.10.   Gaging and sorting of washers.   A special vacuum gripper on the robot is used to acquire a washer from the gravity-feed mechanism.   The washer is then placed on the conical anvil at the electronic gaging station where the diameter of the washer's hole is determined.   This information is sent to the robot's controller which causes the washer to be reacquired and then placed in the proper bin of the compartmentalized storage unit.

problems before they become serious.   Vision systems have also been used for such inspections, but because of their current high cost, this is not a common occurrence.

Inspection of electronic devices has also been performed by robots.   For example, a printed circuit (PC) board often must be checked for missing or improperly drilled holes before the components are placed on the board.   Two techniques can be used to accomplish this.   In the first, the robot picks up the board and puts it onto a special-purpose jig.   Probes extend through properly drilled holes and make contact with electrodes on the other side of the board.   If all contacts are made, the part is good and the robot can place it in the appropriate bin. If there is even one missing contact, the robot's program branches and causes the bad PC board to be placed in the reject bin. The second technique utilizes a vision system that consists of one or more video cameras.   The robot places the PC board on a light table.   The pattern of light passing through the drilled holes is sensed by the vision system and compared to a stored pattern for a good board. The vision system then commands the robot to place the part in the correct bin.

We next consider how robots will (or may) be used in the future.

1.9.2 Robot Applications in the Future

Having gone through the preceding section, the reader was probably struck by at least two things. The first was that the current applications of industrial robots are far from esoteric. In fact, they are, for the most part, relatively simple. The second observation was that most of these applications involved using robots for tasks previously performed by human beings. As mentioned earlier, it is expected that this will not always be so. The "blinders" will undoubtedly be taken off, with the result that non-human-oriented applications will become more the rule than the exception. What can we expect along these lines?

A number of studies have been made over the last few years which try to anticipate some of the new applications of robots. One common thread in these studies is that advanced applications will almost always depend on the development of sophisticated, dependable, and low-cost external sensors. The "deaf, dumb, blind, and tactileless" robot of the mid-1980s is limited in its ability to perform complex tasks. Thus, unless sensor-based robots are developed, applications will remain fairly simple.

As an example, Harmon of Case Western Reserve University identified 25 separate tasks that either currently require or will require one or more of four different types of external sensing capabilities (e.g., simple touch, higher-resolution tactile, vision, and simple proximity sensing). As can be seen from Table 1.9.1, some of these applications have already been realized and have been described in the preceding section. In some instances, the desired sensory needs have also been identified. However, to a lesser or greater degree, items 8, 10, 11, 12, 13, and 19 all would fall into the category of "futuristic" applications. In Table 1.9.2, the technological developments required to fully realize many of these applications are given. The reader should observe that very few of the needed developments can be achieved by integrating existing technology. In fact, most applications will require major technological breakthroughs in order to be sucessfully realized. The areas of vision and tactile sensing are two examples of this.

The medical applications of robots (e.g., routine examinations, surgery, or prosthetics) are certainly many years away from reality. For example, the "six-million-dollar man" will be possible only with the development of real-time signal-processing techniques that permit the desired signals emanating from the brain and transmitted over nerves to be separated from muscle noise so as to control the prosthesis reliably. In addition, the power pack for such appendages will have to be small, light, and powerful, and tactile-sensing elements with resolutions approximating those of the human hand (e.g., about 1 mm) will be necessary.

The ability of a robot to carry out surgical procedures or examinations will depend on the development of a variety of external sensors and real-time computer

TABLE 1.9.1   MAJOR SENSORY NEEDS FOR ROBOTIC TASKS

Function	Simple touch	Taction	Vision	Proximity
1. Assembly	—	X	X	X
2. Gauging (quantitative)	X	X	X	X
3. Grinding/Deburring	—	X	X	X
4. Harvesting	X	X	X	—
5. Inspection (qualitative)	—	X	X	X
6. Medical Exam	—	X	X	X
7. Pick and Place (manipulation)	—	X	X	X
8. Prosthetics/Orthotics				
sensory	—	X	X	X
orthopaedic	—	X	—	—
9. Sorting	—	X	X	X
10. Space (assembly, prospecting, repair)	X	X	X	X
11. Surgery	—	X	X	X
12. Teleoperators (biological, nuclear, space, etc.)	—	X	X	X
13. Underwater (prospecting, repair, salvage)	X	X	·	X
14. Machining	X	—	X	X
15. Stacking	X	—	X	X
16. Welding	X	—	X	X
17. Casting	X	·	—	X
18. Forging	X	·	—	X
19. Mining	X	·	—	X
20. Molding	X	·	—	X
21. Painting	X	·	X	X
22. Polishing	X	·	X	X
23. Pouring	X	·	—	X
24. Stamping	X	·	—	X
25. Transporting (gross movement)	X	·	X	X

X   = Definitely useful

—   = Somewhat useful

·   = Useless

Source: Harmon, Leon, D., "Automated Tactile Sensing," Robot 6 Conference, March 1982, Detroit, paper MSR82-02, MAP I, p. 26.

processing techniques.   More important, the robot will have to be intelligent in order to make rapid decisions based on current sensory information.   It is certain that a set of preprogrammed actions will be totally unsatisfactory for such applications.   This implies that significant advances in artificial (or autonomous) intelligence (AI) will be required.   It should be noted, however, that a Unimation PUMA Mark II series 200 robot has already been used during stereotactic neurosurgery [21].   The robot's controller was interfaced to a computerized tomography (CAT) system which determined and outputted the desired points in space to which the robot was required to move.

TABLE 1.9.2 ESTIMATES OF TECHNOLOGICAL DEVELOPMENTS REQUIRED TO ACHIEVE FUNCTIONS
(Rank ordered for overall estimated difficulty, D*, of achieving, 1980–1990)

Function	Transducers	Articulators	Pattern recognition	Adaptive software	Overall system control	D
1. Teleoperators	X	·	N.A.	N.A.	—	16
2. Gauging	—	—	—	·	·	17
3. Grinding	—	—	—	—	·	21
4. Space	—	—	—	—	—	25
5. Underwater	—	—	—	—	—	25
6. Sorting	X	·	X	—	—	31
7. Inspection	X	—	X	—	—	35
8. Pick and Place	X	—	X	—	—	35
9. Assembly	X	X	X	—	X	45
10. Harvesting	—	X	X	X	X	45
11. Prosthetics	X	X	X	X	X	50
12. Medical Exam	X	X	X	X	X	50
13. Surgery	X	X	X	X	X	50

* Arbitrary sum of · = 1, — = 5, X = 10

· Integrate existing technology

— Moderate new development

X Major developments required

Source: Harmon, Leon D., op. cit., Map III, p. 31.

Underwater applications of robots will involve prospecting for minerals on the floor of the ocean (e.g., nodules of manganese), salvaging of sunken vessels, and the repair of ships either at sea or in dry dock. In the latter case, a prototype version of a mobile robot that is used to clean barnacles from the sides of ships has been built and tested in Dunkerque, France, by Chantiers du Nord et de la Méditerranée, a ship-building company (see Figure 1.9.11). This rather remarkable tripod is capable of moving in either air (i.e., above the waterline) or in water. It grips the ship's sides with both vacuum and magnetic feet, a technique that has proven to be reliable. The scrubbing action is produced by a rotating brush mounted on the end of a rotary axis arm. Currently, no sensory information is available, so the device acts more like a telecheric rather than an autonomous mobile robot. In addition to an underwater application, the Renault Company has used the device to clean the inside of a large fuel storage tank. Since such a task is extremely hazardous, this is potentially an excellent application. However, a considerable amount of work remains to be done before the robot is feasible commercially. A modification of the device (using rectangular, rather than cylindrical leg geometry) is currently being tested for a window-washing application by International Robotic Technologies, Inc. (see Figure 1.9.12).

The military is currently looking at robots for use in a variety of areas. For

Figure 1.9.11.   RM3 Marine robot shown climbing a vertical wall in air.   The device is also capable of walking under water.   It adheres to the side of a ship using magnetic and/or vacuum feet.   Its single arm is equipped with a scrubbing brush.   (Courtesy of P. Kroczinski, International Robotic Technologies, Inc., Marina del Rey, CA.)

example, the air force and navy are both interested in mobile firefighters.   These devices would be equipped with infrared sensors and could react more quickly than people in an emergency and in extremely hazardous situations.   Moreover, they would be expendable.   Other military applications of robots will be on the battlefield itself.   Although it is not inconceivable that robots might someday be used to fight other robots, more realistic short-term applications from the military's point of view would be in the areas of surveillance (e.g., guard and sentry duty), mine sweeping, and artillery-loading devices.

In the latter instance, the concept of two coordinated robots being used on a large mobile gun platform has been developed.   A large manipulator (e.g., a Unimate 4000) would be used to acquire the 200-lb round from a magazine and to load it into the field piece.   After firing, a smaller robot would unload the shell casing and the process would be repeated.   The impetus for such an application comes from the fact that modern electronic tracking techniques make it relatively easy to destroy fixed gun emplacements, thereby putting the lives of soldiers operating nearby in jeopardy.   Also, the work of loading and unloading the shells

Figure 1.9.12. The International Robotic Technologies, Inc. Building Washing Robot. The device can climb up and down the sides of a modern building. Vacuum suction-cup feet permit the robot to adhere to either glass or metal. (Courtesy of P. Kroczinski, International Robotic Technologies, Inc., Marina del Rey, CA.)

is itself physically demanding and difficult, especially where the ground is wet and muddy. It is clear that this type of application would place severe environmental constraints on the robots being used. Manipulators capable of operating in such hostile environments remain to be developed.

The application of robots for surveillance and guard duty is not restricted to the military, however. For example, power generating plants, oil refineries, and other large civilian facilities that are potential targets of terrorist groups are being considered as potential users. The robots for these applications would probably be mobile (running on wheels, treads, or tracks), equipped with some form of vision system and other types of sensors (e.g., infrared), and even have defensive and/or offensive capability. In fact, several police forces (e.g., those in New York City and in London) have already employed prototypes of this class of robot for bomb disposal and for entering a dwelling suspected of harboring armed criminals.

Having mentioned the electric power industry above, it is interesting to note that a group called the Utilities/Manufacturers-Robotic Users Group (U/M-RUG) recently conducted a survey of its members to determine potential uses for robots within both fossil fuel and nuclear power plants. The excerpted results of this survey are shown in Table 1.9.3 and indicate the extremely wide range of potential uses for robots in this single industry.

As mentioned in an earlier section of this chapter, a potential futuristic application of robots would be in the home. Such devices would need to be small, mobile, sensor based, easy to program (or better yet "instruct"), and autonomous.

TABLE 1.9.3 POTENTIAL APPLICATIONS OF ROBOTS IN NUCLEAR AND FOSSIL FUEL POWER PLANTS

- Assisting in test and repair of meters
- Surveillance of uninhabitable areas of plants during operation
- Lubricating fittings on travelling screens
- Inspection of travelling screens and dams to determine need for maintenance
- Weld tube repairing (both girth and longitudinal)
- Loading $MgSo_3$ trucks
- Inspection of 4 KV bus and compartments in plants
- Valve packing inspection and repair
- Testing of fossil plant asbestos levels in work areas
- Location of air, gas, fuel leaks in unit compartments
- Checking for and capping leaking tubes in tubular air heaters
- Cleaning of condensors
- Inspection of condensors for tube leaks
- Plugging of leaking tubes and recording actions taken
- Hydro station inspection and cleaning of guide rails and leak detection
- Performing corona probe tests in air gap of generators
- Control rod drive removal and installation
- Control rod drive exchange
- Pipe welding
- Inspection of the inner diameter surface of moisture separator inlet line
- Pipe inspection and removal of blocking objects (e.g., tools)
- Sorting of laundry
- Decontamination of respirators
- Sorting and compacting of trash
- Tile scrubbing
- Firefighting

Source: 1987 U/M-RUG survey.

Clearly, a significant amount of research in AI, mechanisms, and computers remains to be done before a device such as the Heath Hero 1 or the more advanced Hero 2000 robot can be made to do windows or similar tasks (see Figure 1.9.13).   The goal of robots taking over most mundane household chores is probably worthy of reaching for purely sociological reasons.   In addition, it would cause the robot industry to "take off" and would probably create a host of new jobs.

Although we have already discussed the use of robots in the area of electronic assembly, there is a definite trend toward manufacturing higher-density PC boards with increased utilization of surface-mounted discrete devices (as opposed to radial and axial), together with chips and leadless chip carriers. As shown in Table 1.9.4, the robots of the mid-1980s must be improved in a number of ways before such assembly can be performed economically.   For example, Cartesian and SCARA-type robots must be developed with accuracy and repeatability that are 50% better than those currently available.   In addition, it is necessary for these robots to be equipped with intelligent grippers that provide reliable position and tactile feedback information.

TABLE 1.9.4 CURRENT AND FUTURE ELECTRONIC ASSEMBLY ROBOT
SPECIFICATIONS

Specifications	Today	1990
Work area	24 inches × 24 inches	24 inches × 24 inches
Insertion rate[a]	1200/hr	2000/hr
Number of axes	4	4
Accuracy	.002 inch	.001 inch
Repeatability	.001 inch	.0005 inch
Minimum component spacing	.02 inch	.01 inch
Payload	5 lbs with tooling	5 lbs with tooling
End-effector	Hard-tooled, chuck, turret and some intelligent grippers	Intelligent grippers with sensing feedback
Controller	Autonomous decentralized microprocessor and teach box	Supervisory computer download to microprocessor machine control
Programming languages	Numerous low and high-level robot programming languages	Hopefully standard high level factory automation operating systems
Price (robot and controller)	$20K to $150K	$20K

[a]varies with the number of pins per component
Source: CEERIS International, Inc., "Flexible Automated Assembly Systems Configuration and Implementation," March 1984.

Another extremely important future use of robots will be as a component in a fully automated factory or machine shop. Along these lines, the work at the National Bureau of Standards has been mentioned previously. It is expected that the flow of raw materials using robots and/or automated mobile carts from the warehouse to factory floor, the selection of the proper program to accomplish the desired manufacturing task, and the carrying out of this task by the appropriate numerically controlled machine tool or more sophisticated sensor-based industrial robot will all be handled by a hierarchical control system. A robot design that could be used for facilitating parts and/or materials acquisition from a storehouse in such a control scheme is shown in Figure 1.9.14. Currently under development at Temple University in Philadelphia, this novel double-wrist, four-axis mobile manipulator (called MOBI, for "*mo*bile *bi*ped") is capable of performing a type of walk from one array point to another in a manner similar to a (rigid) "Slinky" or a football player running through a course of rubber tires during training.

Today, most robots must be programmed by a human being to perform a desired task. However, robot controllers that permit interfacing with data base or CAD/CAM systems are just now becoming available. This development will

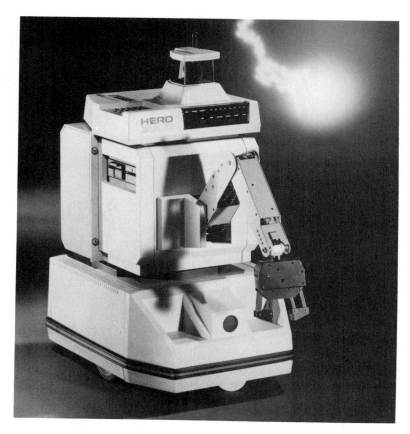

Figure 1.9.13.   All axes of the Hero 2000 mobile robot are servo-controlled. Autonomous operation is not possible.   (Courtesy of Health/Zenith Educational Systems.)

permit the user to simulate the work cell, robot, and other machine tools and/or parts feeding mechanisms "off-line."   The appropriate robot commands will then be easily coordinated with the actions of the other devices operating within the work cell itself.   In this way, the manufacturing process can be optimized with respect to cycle time and throughput.

Our discussion of future robotic applications ends by noting that the Committee on Science, Engineering, and Public Policy (COSEPUP), composed of experts from industry, academia, and government, gave a series of briefings to a number of federal agency directors in the early 1980s.   The Committee predicted that with the development of sophisticated vision systems, tactile sensors, and programming techniques/languages, it would be possible to use robots in industries that handle soft materials such as leather, foam rubber, and vinyl furniture coverings.   The ability to handle soft materials might also lead to the development

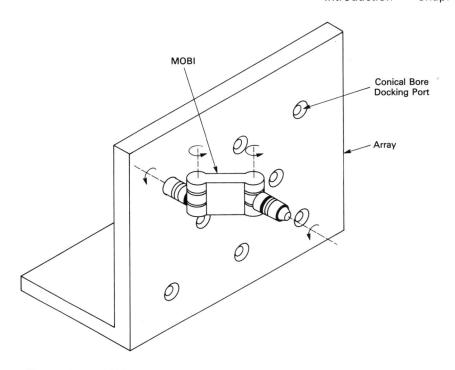

Figure 1.9.14. MOBI, a mobile biped robot. The four-axis device is capable of performing a "walk" maneuver by first off-loading its gripper or tool (not shown) and inserting its free end into an adjacent docking port. The other end of the manipulator is then removed from the original docking location. In this manner, it can move over an array of docking ports like a rigid slinky. At any location on the array, the device can perform operations associated with nonmobile industrial robots.

of robot tailors. Other potential applications would be in the area of package delivery services, trash collection, and automobile maintenance and repair. Most of these predicted applications still remain to be developed.

1.10 SUMMARY

In this fairly detailed, nontechnical introduction, we have attempted to give the reader an understanding of what an industrial robot is and what it is not, where it is applicable and where it is not, and finally, how such devices have evolved and how they may cause another industrial revolution to occur. In particular, the reader has been introduced to most of the terminology associated with these devices and has been shown how to categorize them either by geometry of their major axes or by the type of control utilized.   In addition to tracing the development of robots historically, the economic and sociological consequences of these forms of automation have been discussed.   Finally, the current and possible future applications of robots have been presented.

It should be apparent from the material contained in this chapter that there exist a wide variety of manipulators and that they can perform a large number of tasks.   Moreover, as vision and tactile sensors are incorporated and the controllers become "smarter," the complexity of these tasks will no doubt increase.   Applications that were not originally envisioned and involve more than just replacing a human worker with a robot will then be feasible.   To be sure, there will be an impact on some workers, who, unfortunately will be displaced by these machines. However, it is expected that in the longer term, more jobs will be created as new and expanded industries are developed as a direct consequence of this new, more flexible form of automation, the robot.

1.11 REVIEW QUESTIONS

Due to the fairly descriptive nature of this chapter, the problems that will be included at the end of subsequent chapters are replaced here by review questions. The purpose of these is to help the reader clarify the mostly nontechnical ideas presented in this introduction to robots.

1.1 Discuss the major differences between servo-controlled and non-servo-controlled robots.

1.2 Define the following terms:
  a. Work envelope
  b. Work cell
  c. Tip speed
  d. Coordinated motion
  e. Return on investment

1.3 Discuss the roles that the major and minor axes of a robot play in positioning a part in space.

1.4 Discuss the differences between fixed and flexible automation.

1.5 Discuss at least five robotic applications in terms of the type of robot that is best suited for the job, the level of external sensory information required, and the repeatability of the manipulator demanded by the task.

1.6 Discuss three methods of "teaching" a robot.

1.7 The end effector is the single component that "personalizes" the robot to a particular task.   Explain this statement.

1.8 Describe the function of the four basic components of a robot.

1.9 Why is the NASA *Space Shuttle* robotic arm not a true robot?

1.10 Discuss the need for certain robots to perform straight-line motion and provide several applications where this feature is absolutely necessary.

1.11 Discuss at least three reasons for using a robot instead of a human being to perform a specific task.

1.12 Discuss several reasons why robots should and possibly must be used in the workplace even though human beings may initially lose some jobs to this advanced form of automation.

1.12 REFERENCES AND FURTHER READING

1. Capek, Karel, *Rossum's Universal Robot*, English version by P. Selver and N. Playfair. New York: Doubleday, Page & Company, 1923.

2. Asimov, Isaac, *The Complete Robot*. Garden City, N.Y.: Doubleday & Company, 1982, pp. 209–220. A collection of robot science fiction short stories by the author, including "Runaround," where the "three laws of robotics" were first expounded.

3. Engelberger, Joseph F., *Robotics in Practice: Management and Applications of Industrial Robots*. New York: AMACOM, division of the American Management Association, Inc., 1980. The first book devoted entirely to robots. It is descriptive in nature but contains excellent material on applications and economic justification of robots.

4. *Webster's New Collegiate Dictionary*. Springfield, Mass.: G. & C. Merriam Co., 1974, p. 1001.

5. Pfister, Sandra L., "Robotics—The Development of the Second Industrial Revolution," *Proceedings of the Robots 6 Conference*, Detroit, Mich., March 2–4, 1982. Dearborn, Mich.: Society of Manufacturing Engineers, 1982, pp. 3–15.

6. "The Robot Market Explosion," Section 8, Publishers International Resource Development, Inc., Norwalk, Conn.

7. "The Push for Dominance in Robotics Gains Momentum," *Business Week*, March 29, 1981, pp. 108–109.

8. "Fanuc Edges Closer to a Robot-Run Plant," *Business Week*, Vol. 56, November 24, 1980.

9. "When Robots Take Over People's Jobs," *US News and World Report*, February 16, 1981, pp. 75–77.

10. Rosenblatt, Jean, "The Robot Revolution," *Editorial Research Reports*, 1982, pp. 347–364.

11. Miller, Mark C., "Tools and Monsters," *The New Republic*, May 16, 1981, pp. 26–32.

12. "Robotics and the Economy: A Staff Study," *Joint Economic Committee of Congress, Subcom. on Monetary and Fiscal Policy*, 97th Congress, 2nd Session, Item 1000-B, March 26, 1982.

13. Shaiken, Harley, "A Robot is after your job," *The New York Times*, September 30, 1980, p. A-19.

14. Ayres, R. U., and Miller, S. M., "The Impact of Robotics on the Workforce and Workplace," *Carnegie-Mellon University Report*, 1981.

15. "The Robot Invasion Begins to Worry Labor," *Business Week*, March 29, 1982, pp. 46–47.

16. Albus, James S., *Brains, Behavior, and Robotics*, Peterborough, New Hampshire: BYTE Publications, 1981.

17. Robotics Survey, Carnegie-Mellon University, April 1981.

18. McVeigh, Frank J., "The Human Implications of Robotics," presented at the annual meeting of the Association for Humanistic Sociology, Hartford, CT., October 27–30, 1983.

19. Kraft, R., "He's Geared for Service . . . All 700 lbs," *Morning Call*, January 14, 1977, pp. 21–22.

20. Dubin, Richard, "Industrial Worker's Worlds: A Study of Central Life Interests of Industrial Workers," *Social Problems*, Vol. 3, No. 1, 1956, pp. 131–142.

21. Kwoh, Y. S., Reed, I. S., et al., "A New Computerized Tomographic-Aided Robotic Stereotaxis System," *Robotics Age*, June 1985, pp. 17–22.

22. Ayres, R. U., and Miller, S. M., "Industrial Robots on the Line," *Technology Review*, May/June 1982, pp. 34–47.

23. Makino, H., Furuya, N., Soma, K., and Chin, E., "Research and Development of the SCARA Robot," *Proceedings of the 4th International Conference on Production Engineering*, Tokyo, 1980, pp. 885–890.

24. Unimation, Inc., *Programming Manual—User's Guide to VAL II*, (398T1), Version 1.1, August 1984.

25. Van Blois, John P., "Robotic Justification Considerations," *Proceedings of the Robots 6 Conference*, Detroit, Mich., March 2–4, 1982. Dearborn, Mich.: Society of Manufacturing Engineers, 1982, pp. 51–83. An excellent paper on the economics of robots.

26. Naidish, Norman L., "Return on Robots," Technical Paper MS82-136. Society of Manufacturing Engineers, Dearborn, Mich., 1982.

27. Estes, Vernon, "Robot Justification—A Lot More Than Dollars and Cents," *Proceedings of the Robots 8 Conference—Applications for Today*, Detroit, Mich., June 4–7, 1984. Dearborn, Mich.: Society of Manufacturing Engineers, 1984, Vol. 1, pp. 2-1 to 2-11.

28. Newman, Dennis J., and Harder, Michael J., "Comprehensive Calculation of Life Cycle Costs for Robotic Systems," *Proceedings of the Robots 8 Conference—Applications for Today*, Detroit, Mich., June 4–7, 1984. Dearborn, Mich.: Society of Manufacturing Engineers, 1984, Vol. 1, pp. 2-83 to 2-96.

29. Naidish, Norman L., "Realistic Robot Justification," *Proceedings of the Robots 10 Conference*, Chicago, April 20–24, 1986. Dearborn, Mich.: Society of Manufacturing Engineers, 1986, pp. 2-83 to 2-96.

30. Dorf, Richard C., *Robotics and Automated Manufacturing*, Reston, Va.: Reston Publishing Co., Inc., 1983. A descriptive work containing good material on applications and economics of robots.

31. Ayres, R. U., and Miller, S. M., *Robotics, Applications and Social Implications*. Cambridge, Mass.: Ballinger Publishing Co., 1983. An excellent book on the social and economic consequences of placing robots in the workplace.

The following is a list of books that are primarily descriptive in nature and contain good information on applications.

32. Heath, Larry, *Fundamentals of Robotics*. Reston, Va.: Reston Publishing Co., Inc., 1985.

33. Kafrissen, Edward, and Stephans, Mark, *Industrial Robots and Robotics*. Reston, Va.: Reston Pubishing Co., Inc., 1984.

34. Malcolm, Douglas R., Jr., *Robotics, An Introduction*. Boston, Mass.: Breton Publishers, 1985.

The following books contain more technical information but are also good sources of robot applications:

35. Critchlow, Arthur J., *Introduction to Robotics*. New York: Macmillan Publishing Company, 1985.

36. Groover, Mikell P., Weiss, Mitchell, Nagel, Roger N., and Odrey, Nicholas G., *Industrial Robotics: Technology, Programming, and Applications*. New York: McGraw-Hill Book Company, 1986.

2

Systems Overview of a Robot

2.0 OBJECTIVES

In this chapter we define the components of a robot from a systems approach, expanding on some of the ideas presented in Chapter 1. Some of the functions (and/or features) that a robot should be capable of performing whether alone or as part of a more extensive system are described. Upon completion, the reader should be able to identify the major system components of a robot and should understand the required functionality of a robot and its controller necessary for the unit to be properly integrated and utilized in a real-world environment. In addition, the reader should have an appreciation for interpreting robot system specifications. Specifically, the topics that will be covered are:

- The basic components of a robot system
- The robot as part of a workcell
- The functions required of a robot system
- The specifications of robot systems

2.1 MOTIVATION

The field of robotics draws on a multitude of engineering disciplines. Obviously, there are mechanical, electrical, and software considerations. However, the interaction among these and other disciplines is quite complex. Consider, for example, some of the design criteria required for a multijointed arm so that it is capable of moving along a straight path.

- The geometry of the manipulator must be such that it can position its tool along the path.
- The required positions (*set points**) for the servos that drive each joint must be generated in real time, usually by a computer.
- The servo system must be capable of responding to the set points and of driving each joint so that the tool traces out a straight trajectory (note that this is related to bandwidth and linear operation of the servos).
- The joint actuators (motors) must be sized properly to provide the torques needed as the arm moves (note that the inertias reflected into each joint may be a function of position).
- The feedback transducers must have the proper resolution so that the servos can control the joint positions within some defined error.
- The mechanical system itself must meet some predefined degree of stiffness, accuracy, and repeatability. (Thus the proper materials must be chosen for its construction to meet these requirements. Also, consideration must be given to the thermal properties of these materials.)

This brief discussion demonstrates the interdisciplinary nature of a robot and points out the need for good communication among the various engineers involved in the design of a robot system.

The following sections use a "top-down" description of a relatively sophisticated computer-controlled robot system so as to acquaint the reader with its many subsystems or "functional blocks" and give some indication of where and why interactions among them occur.

2.2 BASIC COMPONENTS OF A ROBOT SYSTEM

Recall from Chapter 1 that the four basic components of a robot system are:

- Manipulator
- Sensory devices
- Controller
- Power conversion unit

Figure 2.2.1 shows these components connected as a system. It is important to note that the sensory devices are spread throughout the system. For example, in

---

*Set points are commands given to a servo system. The system attempts to adjust its output so that it coincides with a given set point.

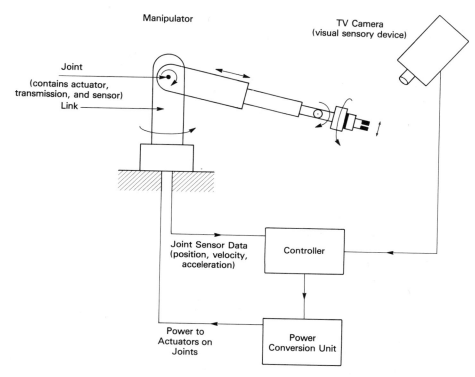

Figure 2.2.1.   Components of a robot system.

addition to the TV camera (a visual sensor), each joint contains sensors for position, velocity, and/or acceleration.   In addition, some of the power conversion hardware may be located inside the manipulator.   The following discussion defines the functions and subdivisions of these major components.

2.2.1 Manipulator

The manipulator consists of a series of rigid members, called *links*, connected by *joints*.   Motion of a particular joint causes subsequent links attached to it to move. The motion of the joint is accomplished by an actuator mechanism.   The actuator can be connected directly to the next link or through some mechanical transmission (in order to produce a torque or speed advantage or "gain").   The manipulator ends with a link on which a tool can be mounted.   The interface between the last link and the tool or end effector is called the *tool mounting plate* or *tool flange*.

   The manipulator itself may be thought of as being composed of three divisions:

- The major linkages
- The minor linkages (wrist components)
- The end effector (gripper or tool)

The major linkages are the set of joint–link pairs that grossly position the manipulator in space. Usually, they consist of the first three sets (counting from the base of the robot). The minor linkages are those joints and links associated with the fine positioning of the end effector. They provide the ability to orient the tool mounting plate and subsequently the end effector once the major linkages get it close to the desired position. The end effector, which is mounted on the tool plate, consists of the particular mechanism needed at the end of the robotic arm to perform a particular task. The end effector may be a tool that does a function such as welding or drilling, or it may be some type of gripper if the robot's task is to pick up parts and transfer them to another location. A gripper may be a simple pneumatically controlled device which opens and closes or a more complex servo-controlled unit capable of exerting specified forces, or measuring the part within its grasp (i.e., gaging).

2.2.2 Sensory Devices

For proper control of the manipulator we must know the state of each joint, that is, its position, velocity, and acceleration. To achieve this, a sensory element must be incorporated into the joint–link pair. Sensory devices may monitor position, speed, acceleration or torque. Typically, the sensor is connected to the actuator's shaft. However, it could also be coupled to the output of the transmission (so that monitoring of each joint's actual position with respect to the two surrounding links is possible).

Other types of sensors may also be included in a robot system. Figure 2.2.1 shows a TV camera which is part of a vision system. For the purpose of our discussion, this sensor, along with its associated electronics and control, is used to locate a particular object in its field of view. Once found, it relays the coordinates of the object to the robot's controller so that the robot can position its gripper over the object in order to pick it up.

Not to be excluded are numerous other types of sensors, such as those associated with touch (tactile sensors) and ranging (sonic or optical-type devices). These sensors can also be used by the robot system to gain information about itself or its environment.

2.2.3 Controller

The controller provides the "intelligence" to cause the manipulator to perform in the manner described by its trainer (i.e., the user). Essentially, the controller consists of:

- A memory to store data defining the positions (i.e., such as the angles and lengths associated with the joints) of where the arm is to move and other information related to the proper sequencing of the system (i.e., a program).

- A sequencer that interprets the data stored in memory and then utilizes the data to interface with the other components of the controller.

- A computational unit that provides the necessary computations to aid the sequencer.

- An interface to obtain the sensory data (such as the position of each joint, or information from the vision system) into the sequencer.

- An interface to transfer sequencer information to the power conversion unit so that actuators can eventually cause the joints to move in the desired manner.

- An interface to ancillary equipment. The robot's controller can be synchronized with other external units or control devices (e.g., motors and electrically activated valves) and/or determine the state of sensors such as limit switches located in these devices.

- Some sort of control unit for the trainer (or operator) to use in order to demonstrate positions or points, define the sequence of operations, and control the robot. These can take on the form of a dedicated control panel with fixed function controls, a terminal and programming language, and/or a "teach pendant" or similar device containing "menu"-driven instructions with which the operator can train the robot.

2.2.4 Power Conversion Unit

The power conversion unit contains the components necessary to take a signal from the sequencer (either digital or low-level analog) and convert it into a meaningful power level so that the actuators can move. As an example, this element would consist of electronic power amplifiers and power supplies for electric robots, while in the case of hydraulic drives, it would consist of a compressor and control valves.

2.2.5 An Implementation of a Robot Controller

Figure 2.2.2 shows the details of the four major components of a robot system discussed above and their interconnections. Based on this figure, we can propose a number of possible implementations for the robot controller. Figure 2.2.3 shows one such configuration. Here a single microprocessor is used as both the sequencer and the computational element. The common bus is the link that connects the microprocessor, its memory, the vision system, the binary I/O interface, and the servo loops. By partitioning the system as shown, only the servo loops have to interface to the sensory data from the joints and provide drive signals to the power amplifiers. Also in this implementation, the vision system is self-contained and incorporates all the necessary hardware and software to perform its function. By distributing the system, we have removed some of the burden from the sequencer.

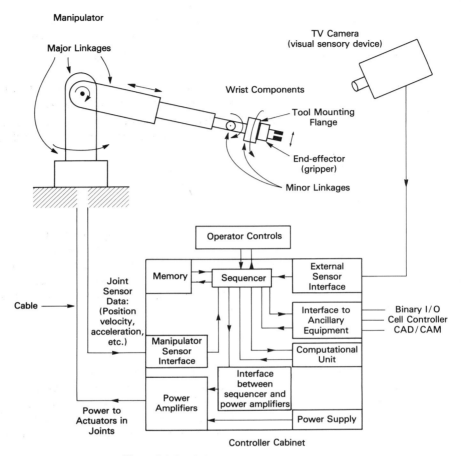

Figure 2.2.2. Subsystems of robot components.

The real-time clock is used to implement delays and to synchronize information transfer among the various devices connected to the bus. It may generate interrupts so that the servo controllers always sample the joint positions and generate new set points at the same instant, thus ensuring a uniform sample rate. These concepts are discussed further in Chapter 7.

From Figures 2.2.1 and 2.2.2 we can infer another way to organize and describe the components of a robot system. That is:

- Manipulator
- Connecting cable
- Controller cabinet
- Operator controls
- External sensors

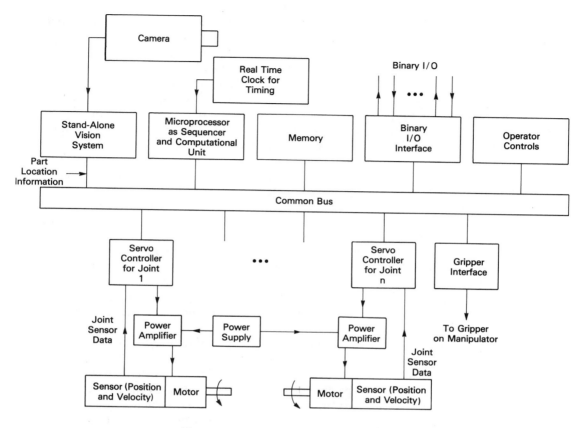

Figure 2.2.3. Possible implementation of a robot controller.

As opposed to the functionality approach just described, this organization is based on the physical packaging of the components and as a matter of fact, most industrial robots are packaged this way.   Clearly, such a description is not as meaningful to the user in terms of the functionality of each subunit.   However, it has the advantage of corresponding directly to the actual pieces of hardware.

2.3 THE ROBOT SYSTEM IN AN APPLICATION

By itself, a robot system has limited utility.   Normally, it must be integrated with other components so that it can be programmed or trained to do some useful task. The term "workcell" is used to describe a collection of automated equipment and controls dedicated to performing one or more specific tasks.   The workcell may contain several robots in addition to fixed automation devices (e.g., part feeders and conveyors), control devices (e.g., computers or programmable controllers), or other devices, such as machine tools.   In a typical manufacturing process, a part or subassembly may pass through many different workcells before it is finished.

In the case of a workcell, once its function is defined and all the components of the system are identified, installed, and operational, the process is essentially fixed.   Even though we define robots as "flexible automation" devices, once they become part of a system, they act as "fixed automation" devices.   That is, the operations the robots perform will not change until there is a revision of the original process or they are used in a new application.   In either case, the applications engineer may easily alter the part of the process performed by the robot by retraining it and if necessary, changing its end-of-arm tooling.

Based on the flexible automation concept, using a robot for a given application reduces the time needed to procure equipment and the setup time required for a revision or change in the process.   Recall that the robot is an off-the-shelf component, and one does not have to wait for the design and fabrication of a specific machine.   Additionally, robots may be considered very versatile since they can easily be retrained.   This idealized versatility may make the use of a robot attractive for short runs since one piece of equipment (i.e., the robot) can be used in place of many fixed-cycle machines.   Of course, we must recognize that this "utopian machine" also has its limitations and, as in most applications today, the majority of the "cost" (whether dollars, design time, or setup time) is in the end-effector tooling and parts-handling devices.

The "intelligence" and flexibility of a robot may reduce the need for other equipment or controllers in the workcell.   For example, by using a robot to transfer a part, the need to have that part oriented for pickup in a specific way may be eliminated.   The robot may be able to adapt to this situation by the use of a vision system or by moving the part around until it is properly seated in the gripper. These approaches might eliminate the cost of a part orientation station (implemented in fixed automation).   Of course, there may be a time penalty by not having oriented parts, but this trade-off must be weighed against the cost of the parts orientation station and quick implementation.

Regardless of the specifics of the application, one may view the robot in one of two ways:

- As the cell controller
- As a peripheral device

By examining a robot in a workcell (either as the controller or as a peripheral) we shall begin to see the required functionality and requirements of a general robot system.

2.3.1 The Robot as a Cell Controller

Figure 2.3.1 shows an application in which the robot is the cell controller.   In this case, the robot controls the activity and senses the states of the three conveyor

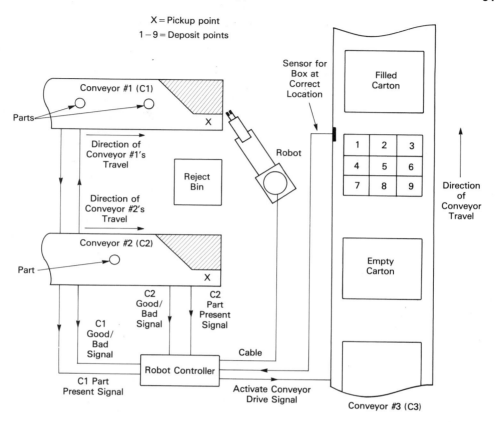

Figure 2.3.1. The robot as a cell controller.

belts and also keeps track of any exceptions (or errors) that occur in the process. Here the robot takes a part from either of the input conveyors (C1 or C2) and places it into one of the nine compartments in the carton on the output conveyor (C3).   Once a carton is filled, the robot controller activates the drive system of C3 until an empty carton is at the correct location.   It then stops this conveyor. The parts on C1 and C2 arrive at random times (i.e., asynchronously).   Thus the robot must wait until it receives a signal indicating that a part is present before it can take the part and put it into a compartment in the carton on C3.   In addition to the signal indicating that a part is present, an additional signal informs the robot controller whether it is a good or a bad part.   Bad ones are placed in the reject bin, while good ones are placed in the next available location in the carton on the output conveyor.   Besides moving the robot to either of the two pickup points, the reject bin, or one of the compartments of a carton, the controller must be sophisticated enough to do all of the decision and control functions previously described (such as activating the conveyor and deciding what to do with a good or a bad part).

2.3.2 The Robot as a Peripheral Device

Figure 2.3.2 shows the same application as that described in the preceding section, with the exception that the robot is now considered to be a peripheral device. The cell controller instructs the robot what to do and when to do it. The robot system used in Figure 2.3.2 does not have to be as sophisticated as the one shown in Figure 2.3.1 because the cell controller is making the decisions and controlling the conveyors. The cell controller may be a minicomputer, a programmable controller, or even a dedicated microprocessor. In this case, the only requirements of the robot are that it be able to go to either pickup point, to the reject bin, or to one of the nine carton compartments as commanded by the cell controller. In fact, this controller may also be charged with keeping statistics on how many rejects occur from each line and how many boxes are filled per hour (on the average). This last set of tasks would usually not be associated with the control capability generally available with most robots.

2.3.3 Defining Robot Positions

In the systems shown in Figures 2.3.1 and 2.3.2, the robot was required to move to various positions in the workcell. These positions may be defined in a number of ways. For the most simple robot controller, the points are demonstrated; that is, the robot is moved to the positions it will be required to go to, and data corresponding to the positions of the joints are saved in memory. This process is referred to as *teaching*. Another method which requires a more sophisticated controller would make use of a physical layout of the workcell drawn to scale. The coordinates of the pickup points and box are entered into the controller and are used to determine the positions of each joint for the appropriate location in space. If this approach is used, the robot system must be accurate; otherwise, the positions attained by the manipulator may not be close enough (i.e., within a usable tolerance) to pick up a part. As may be inferred, the accuracy will depend on such factors as the resolution of the position-measuring device, stiffness of the mechanisms, computational errors, and thermal coefficients, to name a few.

2.4 FUNCTIONS OF A ROBOT SYSTEM

Section 2.3 identified some of the requirements of a robot system when used with other equipment in an application. Based on this discussion and some concepts associated with automated process control and machinery, we will now present a list of the attributes that we would like to have in a robot system to facilitate both

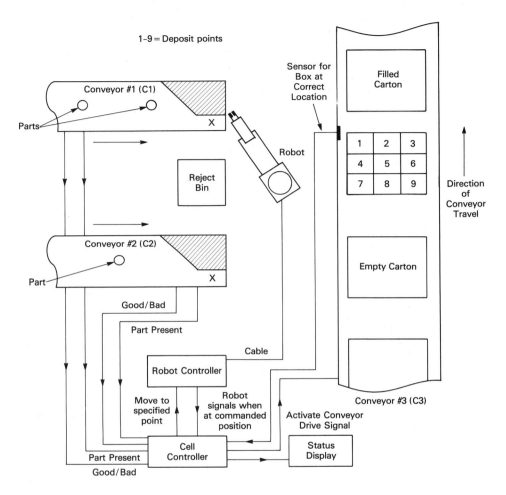

Figure 2.3.2.   The robot as a peripheral device.

the training (programming) of the robot and its use with ancillary equipment.   In addition, some features of the manipulator will also be noted.   As will be seen, these features and capabilities have a visible impact on both the architecture and complexity of a robot system.   In particular, the following capabilities are desirable:

Ability to Define Points (or Locations) in Space to Which the Robot Is to Go

- Demonstration (trainer moves the robot to a particular point and then "teaches" this location)
- Computation (controller computes offset distances from the current location to the new location, e.g., "move forward 3.6 inches")

- Interfacing to external sensors that define locations to which the robot should position itself; for example, in order to acquire a part (vision interface)
- Accepting off-line data (e.g., from a CAD/CAM system) which define points the robot is to move to as part of its program sequence

Ability to Move between Points in Various Ways

- Joint interpolated (all joints start and stop at the same time). This is sometimes called *coordinated motion.*
- Straight line (tool tip moves in a straight line while maintaining the same tool orientation).
- Continuous path (tool tip passes as close as possible to a series of taught or computed points while maintaining a constant velocity).
- Contouring (ability to draw circles or arcs, or move along a specified curve).
- Path profile specified whereby the acceleration, deceleration, and speed can be selected.

Program Control

- Delay before the next instruction is executed
- Ability to set, increment, or decrement counters or registers
- Ability to test the numeric condition of registers and branch to an instruction based on the result of the test
- Ability to display data (register values or positions)
- Ability to input data (for loop control, e.g., perform the same subroutine 20 times since 20 parts are present)
- Subroutine capability so that previously taught programs can be utilized in other larger programs

Control for the End Effector

- Command to open or close (simple pneumatic "bang-bang"-type gripper)
- Close or open a certain distance (position servo-controlled gripper)
- Close exerting a certain force (force servo-controlled gripper)

Provide Interfaces to Ancillary Equipment Such As Parts Feeders, Other Robots, and/or Vision Systems

- Interface via binary inputs and outputs so as to permit use of simple sensors, or turn-on or turn-off ancillary equipment. Typically, the interface should provide for various types of voltage levels (i.e., ac, dc, 24, 120, 240V, etc.). This interface may also be used to provide *handshaking** between the robot controller and the cell controller or even another robot.

---

*Handshaking is a technique whereby two or more systems exchange information to coordinate operations.

- Interface via serial link. This reduces the number of interconnects and permits communications to other devices using standard serial communications ports. Typically, this interface would be used to talk to a "host" computer. Various types of software protocols are necessary for maximum flexibility.

External Robot Control and Communications

- Remote specification of which program the controller is to run and an indication of when execution is completed. This provides the hook for the "cell controller" to tell the robot what to do.
- Remote control for safety features such as the provisions to halt the robot (regardless of what it is doing) or completely remove power from the unit.
- The ability to interface the robot with a factory network using a protocol such as GM's MAP (Manufacturing Automation Protocol) or SEMI's (Semiconductor Equipment and Materials Institute) SECS (Semiconductor Equipment Communication Standard). Protocols such as these are the basis of a standard for interconnecting various robots and other equipment to each other or to computers so that they may work in unison either as part of a workcell or within an entire factory.

Housekeeping Features

- Being able to store, retrieve, and delete program and point data
- Having editing capabilities (such as inserting or deleting program steps, reteaching a point, etc.)
- Specifying which program is to run
- Providing statistics for use by the trainer, such as how much memory is required for a program, how much mass storage space is available, or the date on which a particular program was created

Program Debug and Simulation

- Debugging facilities (such as single step and back step)
- Ability to run a program but to ignore input signals and/or prevent outputs from occurring
- Ability to get a trace of which instructions or steps were executed
- Ability to set "breakpoints"
- Ability to check the state of the inputs without running a program
- Ability to set the state of outputs outside program control

System Parameters

- Reliable calibration method, so that when power is removed, the same joint positions can be obtained when the system is restarted. This implies a way of absolutely defining each joint angle.
- Ability to move back to a point (i.e., one previously shown or demonstrated

to the manipulator) within a certain error. A measure of this is called *repeatability*.

- Ability to move to a computed point (never before attained) within a certain absolute error. A measure of this is called accuracy.
- Ability to follow a curve or move in a straight line within some known error envelope.
- Definition of payload versus performance (i.e., maximum motion speed or acceleration with a certain payload).
- Definition of "settling time" versus speed and payload (e.g., how long it takes the loaded tool to reach the desired location with a specified maximum acceptable deviation about this point).

Serviceability

- Ease of maintenance of the manipulator (change of actuators, links, sensors, etc.)
- Run-time diagnostics (to monitor that the system is operating properly, pinpoint general errors, and immediately stop the robot if a detectable error occurs, such as the jamming of an axis or loss of some functionality of the controller)
- Invoked self-diagnostics (to pinpoint faulty components, subassemblies, or subsystems easily)

From this list we observe that a robot system can be quite complicated. Besides acting as a sequencer to guide the manipulator through some predefined activity, it may be sophisticated enough to be able to alter its program sequence or performance in real time as dictated by its environment. That is, the robot must have the capability to interface to the world around it, either through the use of simple binary sensors or by a more complicated scheme such as vision.

Unlike fixed automation, its sequence of operations may be altered by its trainer or programmer. Since the sequence can be modified, provisions must exist to make the changes relatively simply. One possible approach is the use of a robot-specific programming language (Chapter 7 discusses some robot programming languages). Another consideration, which stems from the "flexible automation" concept, is that since the robot system may be easily reprogrammed, the designers cannot anticipate all the possible permutations of how the device will be used or even what it will be asked to carry. Thus specifications must be given so that the applications engineer or user can successfully predict what a particular robot system can and cannot do without a costly trial-and-error procedure.

Not all the features identified above are implemented in every robot controller. Certain subsets may be chosen to make a simpler or more "cost-effective" implementation. Regardless of its features, the bottom line for the successful use of a robot depends on:

- Meeting performance specifications
- Reliability
- Maintainability

Reliability is sometimes measured as "mean time between failures" (MTBF), and maintainability is indicated by "mean time to repair" (MTTR). Manufacturers are faced with a trade-off among cost, sophistication, ease of use, salability, flexibility, and market demand. A robot must be a relatively cost-effective solution to a manufacturing problem; otherwise, its use may not be justifiable compared to fixed automation or human labor. A robot having the most innovative controller or programming language which is mechanically unreliable becomes nothing more than an expensive laboratory toy.

2.5 SPECIFICATIONS OF ROBOT SYSTEMS

Having identified how a robot system may be used in a workcell and some of the features and functions that it requires to be useful, we will now examine the specifications that are typically used to describe commercial robots. Appendix A shows some of the commercially available robots and a list of specifications as supplied by their manufacturers. It should be noted that since there are currently no standards in the robot industry, the definitions used by different manufacturers may vary considerably. One manufacturer's method of measuring a parameter may be completely different from another's. Take, for example, the measurement of repeatability or settling time. If a noncontact sensor measurement is compared to that made with a contact sensor in determining the final location of a robot's tool tip, different results will occur. The contact sensor can actually damp out oscillations and may impose a force that the robot must overcome. On the other hand, the noncontact sensor does not disturb the arm dynamics. In addition, since manufacturers are constantly improving their systems, the same systems may perform differently depending on when they were produced.

The following discussion will attempt to define the questions that a potential user should ask when trying to decipher and compare the specifications supplied by robot vendors. The specifications follow the order listed in Appendix A.

The *load-carrying capacity* or payload specification does not define the additional weight that the manipulator can carry above the weight of its end effector or tool. Thus, when designing an application, both the weight of the tool and any parts it may carry must be considered since together they constitute the payload as seen by the robot. In addition, this specification does not define the shape of the gripper or payload. Two grippers weighing the same may have different inertias. The robot may be able to perform satisfactorily using one configuration but not the other. It is important for the user to know what the maximum inertia about each of the end effector axes is for a particular robot. The reason for this is that since inertia affects the torque required during acceleration, it may be possible to carry a larger inertial load if the acceleration is reduced.

The *repeatability* of a robot should not be confused with accuracy (see Chapter 3). It is a measure of the ability of a manipulator to return to a position in space where it had been previously. It is measured by going to that position in exactly the same way (i.e., over the same path, with the same payload, speed, acceleration, temperature, etc.) a number of times. Since most manipulators are designed to be slightly underdamped, so that they oscillate (in a damped manner) somewhat about the final position, it is necessary to wait for a short period before repeatability is measured. One must know if the published specifications include a delay, and if so, how much.

In addition, repeatability may be defined in three-dimensional space, or on a joint-by-joint basis. The actual measurement point is also important. If the measurements were taken at the tool-mounting flange of the manipulator, one cannot assume that the tip of a tool or a gripper attached to the mounting flange will have the same repeatability. Depending on the configuration of the manipulator, there may also be regions inside the workspace that exhibit different values of repeatability. Based on the preceding discussion, it is important to get a clear definition of how repeatability is measured by the manufacturer to ensure that the specification is properly interpreted. Chapter 3 defines repeatability and its measurement in more detail.

The *maximum tip speed, no load* is an attempt to define how fast the manipulator can move. As implied by this specification, it will be different if there is a payload. In addition, some other questions that arise are: Which joint or joints were moving when this measurement was made? Was the motion along a straight line? Is it an average value based on distance and time, or is it actually the maximum joint speed?

While we could continue to ask more questions about "maximum tip speed," the point to be made is that this number may be nebulous and requires clarification by the manufacturer. It should not be used in an attempt to obtain a cycle time for a robotic application without the proper inputs from a knowledgeable source.

The *coordinate system* specification defines the configuration of the robot. In Chapter 1 we provided a detailed discussion of the definitions. Certain configurations are more useful than others in specific applications, while in other instances this specification is of no consequence.

The *maximum movement* specification places bounds on the robot's workspace. It is important to note that this is usually measured at the tool-mounting flange. In addition, the maximum specifications may imply that some of the joints are fixed. Therefore, at the maximum values, it may not be possible to orient the tool in certain positions. The angular movement cited for certain joints may be defined for physical hard stops or positions defined by the controller's software.

The *type of drives* defines the joint actuator's source of power (e.g., electric, hydraulic, or pneumatic). Some applications may require one drive type over another.   For example, hydraulic systems, which provide the greatest force/volume characteristic, are prone to leak and may not be acceptable in the food or electronics industry.   Hydraulic fluid on a printed circuit board will inhibit the ability of a wave solder machine to solder the board properly and may result in a costly rework process.

The *control* defines the method used to control the axes (i.e., servoed versus nonservoed) and whether continuous-path or point-to-point motion is possible. Various applications may require one or more of these features.   For example, consider the application discussed in Section 2.3.   To transfer parts from the pickup point to the conveyor requires only point-to-point control.   Other applications, such as dispensing a bead of glue, may require the continuous-path feature (so as to draw a complex curve by defining points along it).   Other features, such as straight-line motion, may be included in this category.

Memory devices defines both internal memory types and mass storage capabilities.   Not explicitly indicated is what happens to the robot program if power is removed from the controller.   Is there a battery backup, or must the user rely on mass storage for program integrity?

Programming method defines the type of robot training via a keyboard, teach pendant, walk-through (in the case of painting robots), mechanical setup (for nonservo units), or CAD/CAM (off-line programming).

Memory capacity gives an indication of the number of program steps or points that the robot controller (without mass storage) can handle.   Typically, each position to which the manipulator moves requires a step. If I/O is used, it may be included in the same step or require another step. Also, some computer-controlled robots do not define their memory requirements in the same way.   Thus it is important to have a clear understanding of what this specification truly means and the limitations it may impose.

The remaining categories indicate potential or proven applications that the robot supplier cites as application examples.   One should be aware that certain types of robots are more suitable for certain applications than for others.   Typically, spray-painting robots are not useful in small-parts assembly or machine loading. In addition, some manufacturers have designed their robots to target specific markets, such as the electronics or semiconductor industries, which require certain special characteristics.   For example, robots used in these industries require a high degree of accuracy and a low particle emission.

What should be apparent from this discussion is that the user must know exactly what he or she wants to do with a robot and ask the correct questions of the manufacturer to ensure that the robot in question will perform properly.

2.6 SUMMARY

This chapter presented a systems approach to the architecture and use of a robot. Additionally, the common specifications used to describe commercially available units were discussed.   Some basic definitions and concepts were presented along with the foundation material for the remainder of the book.

2.7 PROBLEMS

2.1 Investigate the architecture of some commercial robots and their controllers.   Compare the functionality of the commercial units to the general subsystems shown in Figure 2.2.2.

2.2 Investigate the actual implementations of some commercial robots and their controllers. Discuss the advantages and disadvantages in these implementations (and architecture) if one wanted to implement features (such as those listed in Section 2.4) that are currently missing.

2.3 Investigate the various methods used by commercial robot manufacturers to program their controllers.   Consider software structure, languages or menu systems, dedicated controls, and so on.   Based on the list of the general features given in Section 2.4, define what commands are used to implement these features for one or two commercial units.   For example, how does one command a PUMA robot to move in a straight line, or how does one command a particular output line to be energized?

2.4 (Optional) As indicated, specifications are sometimes misleading.   To illustrate this, consider a 3-axis cylindrical coordinate robot with z fixed and θ and r capable of movement.

   a. If the θ axis is commanded to move a distance of 90° with an acceleration time of 250 ms, a constant velocity time of 500 ms, and a deceleration time of 100 ms, what is the average linear speed of the tip?   (That is, consider the distance moved along the circumference of the circle defined by a radius r and traced by the tip.)   What is the average speed if r is retracted to one-half of its distance ($r/2$)?

   b. What is the average linear speed for the conditions described in part a if the distance is defined by the length of a straight line connecting the initial and terminal points instead of using the distance of the circumference?

   c. What is the average joint speed (in rad/s)?

   d. Discuss which of these specifications is most useful; most useless; most impressive. Consider the answers from the perspective of a robot user, a robot designer, and a robot sales engineer.

2.8 FURTHER READINGS

1. Dorf, R. C., *Robotics and Automated Manufacturing*. Reston, Va.: Reston Publishing Co., Inc., 1983.

2. Engelberger, J. F., *Robots in Practice: Management and Applications of Industrial Robots*. New York: AMACOM division of the American Management Association, Inc., 1980.

3. *Manufacturing Engineering:*
   Farnum, Gregory, "Industrial Robots—The Next 10 Years," December 1985.
   Farnum, Gregory, "Robotic Assembly Today," November 1985.
   Schreiber, Rita R., "Robots and Electronic Manufacturing," December 1985.

4. SDP, Robot List (in Appendix A).

3

Mechanical Systems: Components, Dynamics, and Modeling

3.0 OBJECTIVES

In this chapter we present fundamental mechanical concepts as related to the field of robotics. Besides reviewing important basic theory it shows how elementary mechanical components may be connected together to form a system.

In addition, some topics unfamiliar to nonmechanical engineers, such as torsional resonance and the nonlinearities of physical components, are introduced. Finally, real-world components and engineering problems associated with applying them to robotic manipulators are discussed. Specifically, the topics that are covered include:

- Translational motion
- Rotational motion
- Motion conversion
- The problems with real-world components
- Modeling of mechanical systems
- Vibrations
- Kinematic chains
- End effectors
- Resolution, repeatability, and accuracy of a manipulator
- Forces encountered in moving coordinate systems
- Lagrangian dynamics as a method of modeling a manipulator

3.1 MOTIVATION

As indicated in Chapter 2, the field of robotics draws on a multitude of engineering disciplines. It is evident that mechanical engineering is prevalent in the design and control of the manipulator and also in applying the manipulator to situations in manufacturing such as material transfer.

When considering the mechanical design aspect of a robot, dynamics, kinematics, statics and even styling all play a vital role. Table 3.1.1 shows the relationship between forces and motion and these mechanical design considerations.

The purpose of this chapter is to provide a review of mechanical engineering concepts from a systems point of view, directed toward the robotic engineering problem, so that the reader can appreciate the considerations of Table 3.1.1. Additionally, it will introduce topics germane to the mechanics of robot manipulators. Thus although we do not cover the intricate details of mechanical design from a practical implementation standpoint nor do we derive every equation, we cover those concepts required for understanding, designing, and applying robots.

TABLE 3.1.1 MECHANICAL DESIGN
CONSIDERATIONS

	No forces	Forces
No motion	Styling	Statics
Motion	Kinematics	Dynamics

3.2 REVIEW OF ELEMENTARY MECHANICAL CONCEPTS

In this section we provide a terse overview of mechanical engineering concepts as related to the motion of robots. These concepts are used in the remainder of the chapter for derivations and also in Chapter 4.

3.2.1 Translation or Linear Motion

For our analysis and modeling purposes we will be concerned with rigid bodies. During motion, we may examine points on the body. If all the points move in lines parallel to one another, we term the motion *translational*.

In the case of robotic manipulators, certain joints may be constrained to move in straight lines. As mentioned in Chapter 1, they are called linear or prismatic joints. If one of these joints is considered individually, it may be modeled as a rigid body that moves in translation. Thus a looser definition of translational motion applied to a prismatic joint of a robot is *motion along a straight line*. In this section we discuss a number of elements associated with linear motion.

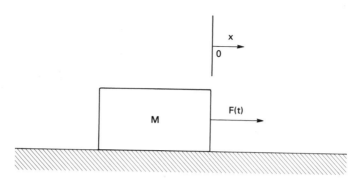

Figure 3.2.1   Force acting on a body of mass M.

3.2.1.1 Mass

Figure 3.2.1 shows a force, $F(t)$, acting on a body of mass M and the associated coordinate system.   The governing equation of motion developed by Newton is given by

$$F(t) = M\ddot{x}(t) \tag{3.2.1}$$

Note that the acceleration $(\ddot{x})$ is referenced to the coordinate system as shown. It is possible to choose the origin at any convenient location as long as it remains consistent.   By using calculus we may write expressions for the position (x) and the velocity $(\dot{x})$ of the body as functions of time.

Equation (3.2.1) is a form of Newton's law, which states that the algebraic sum of the forces acting on a body of mass M is equal to the mass times the acceleration experienced by the body.

$$\sum_{i=1}^{n} F_i(t) = M\ddot{x}(t) \tag{3.2.2}$$

It is important to distinguish between a mass (M) and weight (W).   The relationship is

$$M = \frac{W}{g} \tag{3.2.3}$$

where g, the gravitational constant $= 32.2$ ft/s^2 $= 9.80$ m/s^2.   Equation (3.2.3) indicates that weight is a measure of the gravitational force acting on a mass.

3.2.1.2 Springs

Linear springs are mechanical elements that "store" potential energy.   A spring has the property that its displacement from its equilibrium position is pro-

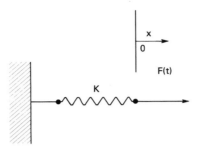

Figure 3.2.2   Force-spring system.

portional to the force applied to it.   Figure 3.2.2 shows the symbol for such an element, with an applied force and a coordinate system.   Note that one end of the spring is not free to move; this is considered to be at "ground" (Section 3.5 covers the case of a spring both ends of which are free to move).   The displacement of the spring is given by

$$F(t) - F_p = Kx(t) \tag{3.2.4}$$

where K is the spring constant in units of force/displacement, while F_p represents a preload tension, that is, a force required to keep the spring stretched a prescribed distance.   Under this condition, the spring is displaced from its original equilibrium point by a force equal to F_p, a new zero for the coordinate system is established, and any force in excess of F_p causes a displacement of distance x so that Eq. (3.2.4) is satisfied.   In practice, springs are not as linear as defined by Eq. (3.2.4).   However, in the case of small deformations the linear approximation is valid.

3.2.1.3 Friction

Translational systems also exhibit friction.   Recall that friction is a force that opposes motion.   Friction is usually nonlinear and dependent on such factors as velocity and pressure between the two moving surfaces and also on the surface composition.   For analysis friction can be broken down into three distinct components [8]:

- Viscous friction
- Static friction
- Coulomb friction

Viscous Friction.   Viscous friction is a retarding force that exhibits a linear relationship between the applied force and velocity.   A physical component that produces this relationship is the dashpot.   Figure 3.2.3 shows the symbol for a dashpot, an applied force, and a coordinate system.   The relationship between the

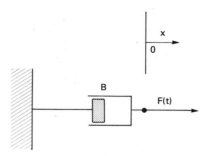

Figure 3.2.3   Dashpot and force system.

force and velocity for this element is given by

$$F(t) = B\dot{x}(t) \tag{3.2.5}$$

where B has the units of force/velocity.

Static Friction.   Static friction (sometimes called *starting friction*) prevents initial motion.   Once motion begins, this force "vanishes."   We will represent this friction by the equation

$$F(t) = \pm F_s\big|_{\dot{x}=0} \tag{3.2.6}$$

The sign will depend on the direction of motion and be such as to oppose the motion. F_s is the magnitude of the static friction force.

Coulomb Friction.   Coulomb friction (also called *running friction*) has a constant amplitude.   Its sign is dependent on the direction of the velocity.   The following equation will be used to represent running friction:

$$F(t) = F_c \, \text{sgn}\,(\dot{x}) \tag{3.2.7}$$

where F_c is the magnitude of the Coulomb friction and sgn is the signum function.

Figure 3.2.4 shows velocity versus force characteristics for systems exhibiting the frictions mentioned above.   Physical systems usually exhibit a combination of all three types of friction.   Therefore, when estimating the force required to accelerate a mechanical system, it is necessary to provide force sufficient to overcome all the frictional components as well as that required to produce the required acceleration.   During deceleration, it should be apparent that the frictional components will help to stop the system and a force of less magnitude than given by Eq. (3.2.1) will be needed.   If a constant-velocity portion of the motion exists, force will be needed to overcome both Coulomb and viscous components.   Appendix B relates this concept to the selection of motors.

A component used to reduce friction between surfaces is referred to as a *bearing*.   Bearings come in both linear (for translational motion) and rotary (for rotary motion) configurations.   Many different implementations are possible, each

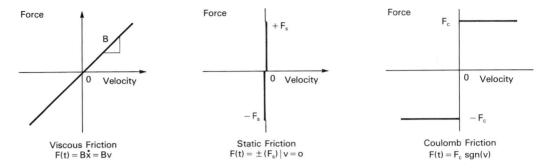

Figure 3.2.4 Force versus velocity characteristic for frictional forces. (Benjamin C. Kuo, Automatic Control Systems, 5e, © 1987, pp. 133, 143. Reprinted by permission of Prentice Hall, Inc. Englewood Cliffs, NJ)

of which exhibits certain characteristics.   Refer to [13] or any other mechanical engineering handbook for a description and classification of bearings.   For our discussions, it is important to realize that to implement mechanical systems, one must ensure that shafts which rotate are supported, or masses that move across a surface are both constrained to a single degree of freedom and have the minimum friction acceptable.   The use of bearings is one method of providing this constraint while minimizing the friction between the moving member and the surface to which its movement is referenced.

3.2.2 Rotational Motion

Rotational motion takes place about a fixed axis.   For the case of a rigid body, if all the points along some line remain fixed during a motion, the body is rotating about that line.   Since some robotic joints may be rotary (or revolute) in nature, they may be individually modeled as a rigid body rotating about some line or axis. Dynamical relationships and physical quantities associated with this type of motion are directly analogous to those of translational motion.

Thus linear displacements become angular rotations and are typically measured in radians.   The moment of inertia, the rotational analog of mass, quantifies the property of an element that stores the kinetic energy of rotational motion. The moment of inertia is highly dependent on the geometric properties of an object. Torque is analogous to force in a translational system and may be defined as the motive action for rotation.   Figure 3.2.5 shows a torque–inertia system with a coordinate system.   The equation describing the dynamics of this system is given by

$$T(t) \;=\; J\ddot{\theta}(t) \;=\; J\alpha(t) \tag{3.2.8}$$

where $T(t)$ is the instantaneous torque, $\ddot{\theta}(t)$ the angular acceleration, and J the body's moment of inertia through the axis of rotation.   Equation (3.2.8) states

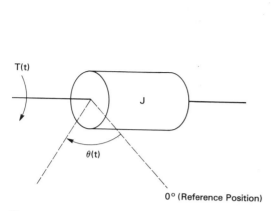

$\theta(t)$

0° (Reference Position)

Figure 3.2.5   Torque-inertia system.

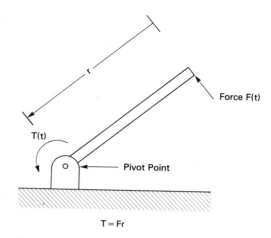

T = Fr

Figure 3.2.6   Torque generated by a force acting at a distance.

that if we apply a torque of magnitude T to a body having a moment of inertia, J, the body will experience an angular acceleration of $\ddot\theta$. Compare Eqs. (3.2.1) and (3.2.8).

Figure 3.2.6 illustrates another way of defining torque—as the product of force and the perpendicular distance from the pivot to the force vector. The applied force $F(t)$ can cause the arm to rotate in either a clockwise or a counterclockwise direction. Consider, for example, the case of a robotic manipulator carrying a load against the force of gravity. Figure 3.2.6 is an ideal model for calculating the torque which must be produced by a rotary joint to hold the object at rest.   In the situation where multiple forces are applied to a lever each at a known distance from the pivot point, a resultant torque may be calculated by summing all the torques with the appropriate signs.

Torsional springs provide a displacement proportional to the applied torque. Figure 3.2.7 shows a torque–spring system. The dynamics of this system are described by

$$T(t) = K\theta(t) \tag{3.2.9}$$

where K is the spring constant in units of torque/rotary displacement. This equation should be compared with Eq. (3.2.4).

Rotational systems also experience friction similar to translational systems. Equations (3.2.10), (3.2.11), and (3.2.12), respectively, describe viscous, static, and Coulomb friction in rotary systems.

$$T(t) = B\dot\theta(t) \tag{3.2.10}$$

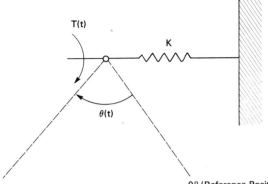

0° (Reference Position)   **Figure 3.2.7**   Torque-spring system.

$$T(t) = \pm T_s|_{\dot\theta = 0} \qquad (3.2.11)$$

$$T(t) = T_c \, \text{sgn} \, (\dot\theta) \qquad (3.2.12)$$

3.2.2.1 Moment of inertia: calculation

The computation of the moment of inertia is extremely important in the sizing of motors and in modeling the dynamics of a physical system. As stated in the preceding section, the moment of inertia of a body is highly dependent on its shape, and unlike mass, is not a unique property of the body but depends on the axis about which it is computed.

Equation (3.2.13) defines the moment of inertia about an axis for a system consisting of n discrete "point masses", $\{M_1, M_2, \ldots M_n\}$.

$$J = \sum_{i=1}^{n} M_i r_i^2 \qquad (3.2.13)$$

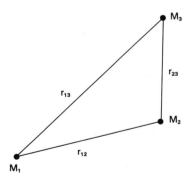

Inertia about axis through point masses M_1, M_2, and M_3 which are connected by rigid rods.

$J_{M_1} = M_2 \, r_{12}{}^2 + M_3 \, r_{13}{}^2$

$J_{M_2} = M_1 \, r_{12}{}^2 + M_3 \, r_{23}{}^2$

$J_{M_3} = M_1 \, r_{13}{}^2 + M_2 \, r_{23}{}^2$

Figure 3.2.8   Inertia about a point-mass system.

Here M_i is the mass of point i and r_i is the distance of the mass from the axis of rotation.   Figure 3.2.8 shows three point masses and the moments of inertia about axes passing through each one perpendicular to the plane of the paper.

For the case of bodies of complex geometric shape, Eq. (3.2.13) becomes the integral

$$J = \int_{\text{Volume}} \rho r^2 dV \qquad (3.2.14)$$

where ρ is the density of the body's material, dV is a differential volume element, and the integration is to be carried over the entire volume of the body.   In the case of homogeneous bodies (i.e., the density is the same throughout), ρ may be brought outside the integral sign.

Figure 3.2.9 shows some common geometric shapes and the moments of inertia about the three centroidal axes shown.   *Centroidal axes* are those passing through the centroid or "center of gravity" of the object.   Recall from basic physics that the center of gravity of an object is easily determined by balancing the object on a "knife edge" to find its balance point.   If this is done for a three-dimensional object along its length, width, and height (or other coordinate frame), the three balance points may be projected into the center of the body to define the actual center of gravity.

The moment of inertia of complex areas and shapes may be evaluated by the addition and/or subtraction of elemental areas with respect to the axis of rotation. For instance, to compute the inertia of a hollow sphere about any one of its centroidal axes, all that is necessary is to subtract the inertia defined by the inner radius from that defined by the outer radius.

Parallel Axis Theorem.   The parallel axis theorem provides the means of finding the moment of inertia of a body about an axis which is parallel to an axis that passes through the center of gravity of the body.   Mathematically, we have

$$J = J_c + Mr^2 \qquad (3.2.15)$$

where J_c is the moment of inertia of the body about the axis passing through its center of gravity (i.e., a centroidal axis), M the mass of the body, and r the distance between the two axes.   At this point it should be noted that formulas are tabulated in various books (see [1]) that allow calculation of the moment of inertia about an arbitrary axis not parallel to a centroidal axis; however, for purposes of illustration in this text, the parallel axis theorem of Eq. (3.2.15) will suffice.

EXAMPLE 3.2.1: USE OF PARALLEL AXIS THEOREM

Find the moment of inertia of a slender bar about an axis at one of its ends y', as shown in Figure 3.2.10.   From Figure 3.2.9 we know that the moment of inertia of a slender bar of length l and mass M about an axis through its center is given by

$$J = \tfrac{1}{12}Ml^2 \tag{3.2.16}$$

The center of mass is located in the middle of the bar at $l/2$ and has been designated as the y axis in Figure 3.2.9. Therefore, by Eq. (3.2.15) we obtain

$$J_{y'} = \tfrac{1}{12}Ml^2 + M(l/2)^2$$
$$= \tfrac{1}{3}Ml^2 \tag{3.2.17}$$

Figure		J_x	J_y	J_z
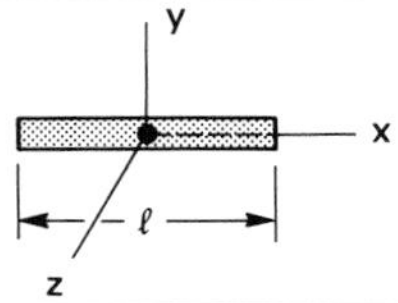	Slender Bar	0	Ml^2	Ml^2
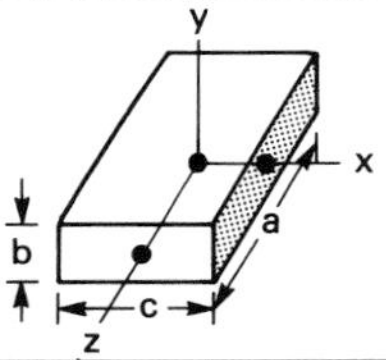	Rectangular Parallelepiped	$M(a^2 + b^2)$	$M(a^2 + c^2)$	$M(b^2 + c^2)$
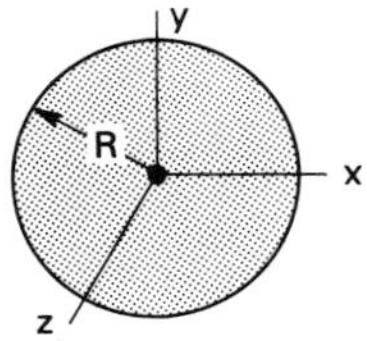	Thin Circular Disk	$\tfrac{1}{4}MR^2$	$\tfrac{1}{4}MR^2$	$\tfrac{1}{2}MR^2$
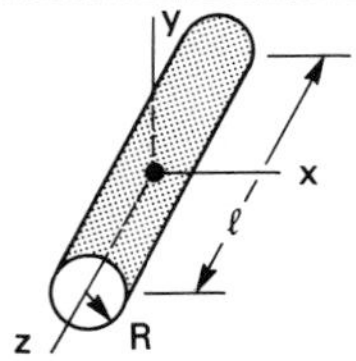	Right Circular Cylinder	$\tfrac{1}{12}M(3R^2 + l^2)$	$\tfrac{1}{12}M(3R^2 + l^2)$	$\tfrac{1}{2}MR^2$
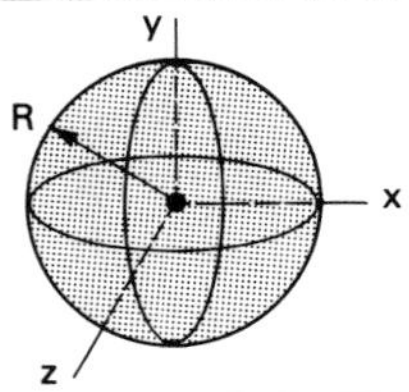	Sphere	$\tfrac{2}{5}MR^2$	$\tfrac{2}{5}MR^2$	$\tfrac{2}{5}MR^2$

Figure 3.2.9 Centroidal moments of inertia for some common shapes.

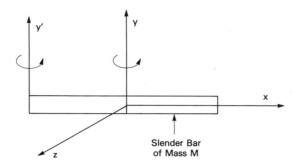

Slender Bar
of Mass M

|←——————— ℓ ———————→|

Figure 3.2.10   Figure for Example
3.2.1: use of parallel axis theorem.

Radius of Gyration.   The radius of gyration is the radial distance from any
given axis at which the mass of the body could be concentrated without altering
the moment of inertia of the body about the axis.   We define the radius of gyration,
k, by the following equation:

$$k = \sqrt{\frac{J}{M}} \qquad (3.2.18)$$

This equation is a solution of Eq. (3.2.13) for the radius, knowing the mass
and moment of inertia of the body.   It is important to understand that the radius
of gyration is not the center of gravity of the body.   The following example illustrates this concept.

EXAMPLE 3.2.2: RADIUS OF GYRATION AND CENTER OF GRAVITY

Compare the center of gravity and the radius of gyration for the moment of
inertia calculated about both axes of the rod in Figure 3.2.10.   From Example
3.2.1 we know that the center of gravity of Figure 3.2.10 is at $l/2$ and the
moments of inertia about the center axes and end axes are given by Eqs.
(3.2.16) and (3.2.17), respectively.   Using Eq. (3.2.18), we compute the
radius of gyration for each axis as follows:

$$k_y = l \sqrt{\tfrac{1}{12}} \qquad (3.2.19)$$

$$k_y' = l \sqrt{\tfrac{1}{3}} \qquad (3.2.20)$$

Equations (3.2.19) and (3.2.20) define the distance that a point mass
could be located from each of the indicated axes so that the same moment
of inertia would be generated if the bar were rotated about either of these
axes.   In both cases the radius of gyration is different than the center of
gravity.

Example 3.2.2 can also be used to illustrate a common error used in computing the moment of inertia about an axis. Although Eq. (3.2.13) correctly defines the moment of inertia of a point mass about an axis it cannot always be applied to obtain the moment of inertia of a body of arbitrary shape. Consider, for example, that we approximate the body as a point mass physically located at its center of gravity and then use this point to define the distance of the body from its axis of rotation. For the case of the rod shown in Figure 3.2.10, if we had used the center of gravity to compute the moment of inertia about the y' axis, the result would be

$$J_{y'} = \tfrac{1}{4}Ml^2 \tag{3.2.21}$$

Comparing this with the correct result of Eq. (3.2.17) shows that an error of 25% on the low side has been made. This error could cause serious problems in that the payload of a robot may be incorrectly calculated, thereby causing the system to be unable to perform adequately.

Based on the previous discussions, it should be obvious that the point-mass approximation of Eq. (3.2.13) should not be used arbitrarily to compute the inertia of an object. In some cases this approximation is sufficient. However, one must ensure that the error introduced does not produce misleading values. A more conservative approach is to decompose the body into elementary shapes as shown in a table of centroidal moments (e.g., those in Figure 3.2.9) and then use the parallel axis theorem [Eq. (3.2.15)] to compute the inertia of the object in question. Example 3.2.3 illustrates this procedure.

EXAMPLE 3.2.3: CALCULATION OF INERTIA FROM ELEMENTARY SHAPES

Figures 3.2.11 and 3.2.12 show a simplified parallel-jaw type gripper which has been modeled by three rectangular parallelepipeds, each consisting of a length, width, and height dimension. The density of the material, aluminum, is 1.56 oz/in.[3]. For the particular application being analyzed, the gripper is free to rotate about two perpendicular axes (z and y) as shown (i.e., the roll and pitch axes). Note that the z axis goes through the center of the gripper, while the y axis is some distance from the back surface. For the dimensions shown, compute the moment of inertia about both the z and y axes.

From Figure 3.2.9 we identify the axes associated with each rectangular member as shown in the exploded view of the gripper given in Figure 3.2.12. The dimensions a, b, and c correspond to the formulas given in Figure 3.2.9, and we identify the components by the subscript top, side, and bottom to delineate the members.

The contribution to the moment of inertia about the z axis is computed by first determining the moment of inertia of each member about the centroidal axis parallel to the z axis of the complete gripper and then using the parallel axis theorem. By summing the moment of inertia of the three members referenced to the z axis, we find the total moment of inertia about the z axis. Equations (3.2.22) through (3.2.24) show the value of each of the three members referenced to the z axis of the gripper.

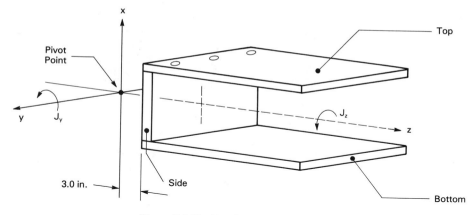

Figure 3.2.11 Parallel-jaw gripper model.

$$J_{z_{top}} = \tfrac{1}{12}M_{top}(b^2 + c^2) + M_{top}\,r_{zt}^2 \qquad (3.2.22)$$

$$J_{z_{bot}} = \tfrac{1}{12}M_{bot}(b^2 + c^2) + M_{bot}\,r_{zb}^2 \qquad (3.2.23)$$

$$J_{z_{side}} = \tfrac{1}{12}M_{side}(a^2 + c^2) \qquad (3.2.24)$$

Note that the parallel axis theorem was not needed to compute the contribution from the side member since its centroidal y axis was coincident with the z axis of the gripper. Therefore, the total moment of inertia about the z axis is given by

$$J_{z_{total}} = J_{z_{top}} + J_{z_{bot}} + J_{z_{side}} \qquad (3.2.25)$$

Utilizing the actual dimensions given in Figure 3.2.12 yields

$$J_{z_{total}} = 0.0252 \quad \text{oz-in.-s}^2 \qquad (3.2.26)$$

The moment of inertia about the y axis of the gripper is computed in a similar manner. In this case, however, the parallel axis theorem must be used for all three members since none of the centroidal axes under consideration are coincident with the y axis. Equations (3.2.27) through (3.2.29) define the moments of inertia due to each plate about the y axis.

$$J_{y_{top}} = \tfrac{1}{12}M_{top}(a^2 + b^2) + M_{top}\,r_{yt}^2 \qquad (3.2.27)$$

$$J_{y_{bot}} = \tfrac{1}{12}M_{bot}(a^2 + b^2) + M_{bot}\,r_{yb}^2 \qquad (3.2.28)$$

$$J_{y_{side}} = \tfrac{1}{12}M_{side}(a^2 + b^2) + M_{side}\,r_{ys}^2 \qquad (3.2.29)$$

$$J_{y_{total}} = J_{y_{top}} + J_{y_{bot}} + J_{y_{side}} \qquad (3.2.30)$$

Again substituting the dimension values of Figure 3.2.12 yields

$$J_{y_{total}} = 0.6389 \quad \text{oz-in.-s}^2 \qquad (3.2.31)$$

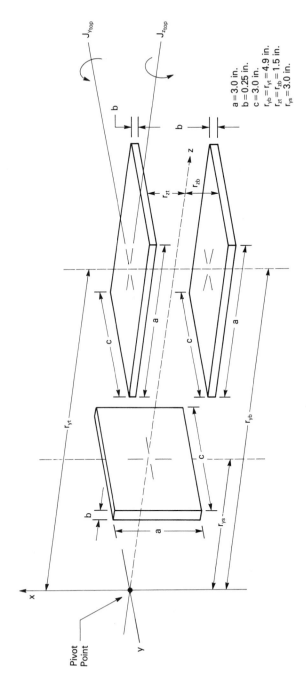

Figure 3.2.12 Exploded view of the parallel-jaw gripper model.

a = 3.0 in.
b = 0.25 in.
c = 3.0 in.
$r_{yb} = r_{yt} = 4.9$ in.
$r_{zt} = r_{zb} = 1.5$ in.
$r_{ys} = 3.0$ in.

An interesting trade-off in the design of a manipulator can be made between the moment of inertia seen by the joint's actuator and the ability of the joint's actuator to balance a load against gravity. Consider, for example, a joint similar to that shown in Figure 3.2.6 which is subject to gravitational forces. To prevent the joint from rotating downward, the joint's actuator must produce a torque (at the pivot point) which is opposite and equal to the torque generated by gravity. The concept of a counterbalance can be used to reduce this required torque, as this could be quite large when the manipulator is fully extended. Two commercial robots using this technique are the Maker series from United States Robots and the Merlin series from American Robots.

Counterbalancing consists of placing a counterweight (i.e., a known weight) on the opposite end of the arm at a known distance from the pivot point. It should be obvious that the torque load produced by the counterweight is opposite to that produced by the payload and therefore reduces the amount of torque necessary to keep the joint stationary when subject to gravitational loads. Although counterbalancing reduces the torque necessary for a static balance, the torque required for movement increases since the addition of the counterweight increases the moment of inertia about the pivot point.

Typically, there are two torque parameters for a joint's actuator: the peak torque rating and the root-mean-square (rms) torque rating. The former is an indication of the maximum value of torque that needs to be produced by the actuator, while the latter indicates how much torque can be produced continuously without destroying the actuator (Chapter 4 and Appendix B deal with this in more detail). Thus, in using a counterbalance, one must make a trade-off between the peak torque for acceleration (defined by inertia and friction) and the torque required for static balance and then weight these with a desired duty cycle and finally compare the results with the rms and peak values of the actuator.

It can be shown that in certain configurations (such as if the length from the pivot to the counterbalance is much shorter than the length from the pivot to the payload) the increase in inertia caused by counterbalancing can actually be relatively small. Therefore, this approach can be quite beneficial since it reduces the rms value of the torque to be provided by the actuator and does not greatly increase the peak value necessary for acceleration.

3.2.2.2 Moment of inertia: measurement

The moment of inertia of a physical body may be easily measured using either a torsional pendulum or a physical pendulum. Figure 3.2.13 shows a torsional pendulum. This arrangement consists of a wire or slender shaft of known spring constant from which is suspended an object of unknown inertia. The relationship between the object and shaft is such that the degree of freedom provided by the shaft is through the axis of the object about which we wish to determine the moment of inertia.

If the shaft or wire is twisted through a small angle and then released, the

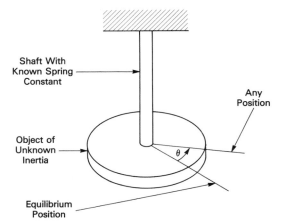

Shaft With
Known Spring
Constant

Any
Position

Object of
Unknown
Inertia

θ

Equilibrium
Position

Figure 3.2.13   Torsional pendulum.

object will oscillate about its original equilibrium position.   The relationship between the object's moment of inertia J, its frequency of oscillation f, and the spring constant of the shaft K is given by

$$f = \frac{1}{2\pi}\sqrt{\frac{K}{J}} \qquad \text{hertz} \tag{3.2.32}$$

In the real world the actual spring constant of the shaft may not be known with sufficient accuracy to perform a useful measurement.   To overcome this problem, the spring constant may be measured by using an object whose moment of inertia is known (such as a disk) and measuring its frequency of oscillation in the torsional pendulum arrangement.   Once K is determined, the moment of inertia of any object may easily be found.

The physical pendulum shown in Figure 3.2.14 is another practical method to determine the moment of inertia.   In this case, the object is pivoted about a fixed axis.   An imaginary line (of length h) from the pivot point to the center of gravity of the object is drawn.   If the object is displaced from its equilibrium position, this imaginary line will oscillate about its equilibrium position with a frequency defined by

$$f \approx \frac{1}{2\pi}\sqrt{\frac{Mgh}{J}} \tag{3.2.33}$$

3.2.3 Mechanical Work and Power

The concepts of work, energy, and power are essential in describing the dynamics of a system and sizing components.   The following discussion briefly touches on these concepts for both linear and rotary motion.

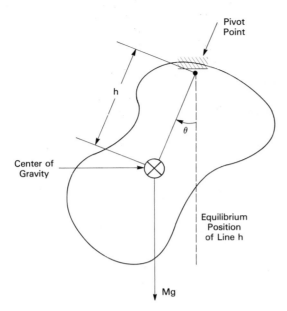

Figure 3.2.14 Physical pendulum.

Work may be defined in terms of a force F acting on a body in such a way that the force has a component along the line of motion of its point of application. The work done by the force is positive if the force acts in the direction of motion, while a negative value of work is obtained if the force is opposite to the direction of motion. Based on this definition, it should be apparent that an applied force that is perpendicular to the direction of motion does no work. Equation (3.2.34a) defines work in terms of a line integral.

$$W = \int_{s_1}^{s_2} F_s(s) \, ds \qquad (3.2.34a)$$

F_s is the component of the force along the path and is dependent on its location in space; thus it is a function of s. The limits of integration s_1 and s_2 define the initial and terminal points, and ds is an increment on the path.

For the simple case of a constant force in the same direction as the displacement, Eq. (3.2.34a) may be written as the product of the force and distance:

$$W = Fs \qquad (3.2.34b)$$

It can be shown that if a force is imparted to a body of mass M (causing it to accelerate), the work of this force equals the change in the kinetic energy of the body. This is expressed by Eq. (3.2.35), which can be derived from Eq. (3.2.34b).

$$W = \tfrac{1}{2}Mv_2^2 - \tfrac{1}{2}Mv_1^2 \qquad (3.2.35)$$

where v_1 is the initial velocity and v_2 is the final velocity of the body.

If the effect of gravity is included when we examine the work done by a resultant force, we may write

$$W = (\tfrac{1}{2}Mv_2^2 + Mgh_2) - (\tfrac{1}{2}Mv_1^2 + Mgh_1) \tag{3.2.36}$$

where W is the work due to all forces acting on the body except for that of gravity, which has been included in the potential energy terms, and h_2 and h_1 are the final and initial heights of the body with respect to some zero potential energy reference. The first expression in parentheses is the final value of the total mechanical energy, and the second is the initial value.

If the body is attached to a linear spring and experiences no change in potential energy due to gravity, the work done by the resultant forces on the body (excluding that done by the spring, which is expressed as a potential energy) may be written as

$$W = (\tfrac{1}{2}Mv_2^2 + \tfrac{1}{2}Kx_2^2) - (\tfrac{1}{2}Mv_1^2 + \tfrac{1}{2}Kx_1^2) \tag{3.2.37}$$

From Equations (3.2.36) and (3.2.37) we note that the work done by an external force (excluding gravity or elastic forces of a spring) acting on a body is equal to the sum of the change in the kinetic energy of the body and the change in the potential energy of the body. That is, the force must overcome the potential energy due to gravity and provide force for acceleration.

Power is defined as the time rate of doing work. Therefore, we may write

$$P = \frac{dW}{dt} \tag{3.2.38}$$

For the case of a force acting in the same direction as the velocity of a body, the average power may be defined as the product of the force and the average velocity of the body.

$$\overline{P} = F\overline{v} \tag{3.2.39}$$

The bar over the variables in Eq. (3.2.39) defines an average value.

The instantaneous power is the product of the force and the instantaneous velocity:

$$P(t) = F(t)v(t) \tag{3.2.40}$$

For the case of a frictional element the power is *dissipated as heat*. Consider the dashpot of Figure 3.2.3. Equation (3.2.5) defines the force that it exerts as a function of velocity. Assuming a constant velocity acting on the dashpot, the average power dissipated as heat is given by

$$P_{\text{dissipated}} = Bv^2 \tag{3.2.41}$$

In the case of rotational motion, there are energy relationships that are similar to those of the linear case. Work can be defined in terms of torque T, acting through a given rotational displacement. Thus, if a torque T, acts through an angle θ, the work performed is given as

$$W = T\theta \qquad (3.2.42)$$

Both mass and inertia have the ability to store kinetic energy. For the linear system we have

$$E_k = \tfrac{1}{2}Mv^2 \qquad (3.2.43)$$

The kinetic energy of a rotational system is given by

$$E_k = \tfrac{1}{2}J\omega^2 \qquad (3.2.44)$$

Similarly, the equations defining elastic potential energy for a linear spring and torsional spring, respectively, are

$$E_p = \tfrac{1}{2}Kx^2 \qquad (3.2.45)$$

$$E_p = \tfrac{1}{2}K\theta^2 \qquad (3.2.46)$$

Power for a rotational system is defined as the product of torque and angular velocity. Thus an equation analogous to Eq. (3.2.40) is

$$P(t) = T(t)\omega(t) \qquad (3.2.47)$$

3.3 MOTION CONVERSION

As described in Chapter 1, the joints of a robotic manipulator may be rotary or linear in nature. The components used to implement a practical joint usually consist of an actuator (e.g., motor) coupled to the physical joint by a *mechanical transmission*. This transmission is used to take the actuator motion and direct it to the joint in order to provide such characteristics as a change of rotational direction, a change of axis, torque multiplication (or reduction), and speed reduction (or multiplication). The transmission may be used to convert rotary motion to linear motion or to provide a "match" between the actuator and load in order that the maximum energy is transferred to the load. Some devices produce linear relationships between input and output motions and thus are well suited for use in closed-loop control. However, some of the more traditional mechanical schemes have highly nonlinear input/output relationships, which can create severe control problems.

We have used the word "transmission" to define all the components from the actuator to the physical joint. These components may consist of shaft couplers, multiple sets of gears, and lead screws, to name but a few. Although it is possible, and very desirable for reasons that will become clear throughout this chapter, to implement a joint by direct drive (i.e., actuator and joint are one component with no transmission), it may be impractical for cost or size reasons. One commercially available robot using a direct-drive approach is manufactured by ADEPT (see Chapter 4).

3.3.1 Rotary-to-Rotary Motion Conversion

Rotary-to-rotary conversion is associated with such components as gear trains, harmonic drives, belts and pulleys, or sprockets and chains. Each of these components will take the torque and speed provided by a rotary actuator connected to the input shaft and provide rotary motion on the output shaft. Depending on the type of component and its characteristics, the motion on the output shaft may be reversed, have an increased torque (with a reduced speed), or have an increased speed (with a reduced torque).

3.3.1.1 Ideal gears

To understand the process of rotary-to-rotary motion conversion, we will focus on the properties of an "ideal" gear train (the concepts developed are easily applied to the belt and pulley, or sprocket and chain). By the "ideal gear train" we mean a transmission composed of gears that are perfectly round, rotate on their true centers, and are also inertialess. The surface between the gears is also frictionless, thereby creating no losses.

Figure 3.3.1 shows the front and side views of a gear train. Either shaft may be designated as the input or output. The top gear has a radius of r_1 and has N_1 teeth, while the bottom gear has a radius of r_2 and N_2 teeth. Both gears have a rigid shaft attached to them with the torque and displacement of each shaft designated as T and θ, respectively. Note that for the case of two gears shown in Figure 3.3.1, the rotation of the shafts is in opposite directions.

By examining the same physical relationships between the two gears, equations relating the speeds, displacements, and torques on the two shafts may be obtained. Since the spacing between the teeth on each gear must be the same so that they mesh properly, the number of teeth on each gear is proportional to the radius of the gear. Thus

$$\frac{N_1}{r_1} = \frac{N_2}{r_2} \tag{3.3.1}$$

It should also be apparent that the number of teeth on a gear must be a whole number. However, the gear ratio N_2/N_1 may be fractional.

For the ideal gear train, there are no losses, and therefore the work done by the torque acting through an angular displacement on one shaft is equal to the torque acting through the corresponding angular displacement of the other shaft (see Section 3.2.3). Therefore,

$$T_1\theta_1 = T_2\theta_2 \tag{3.3.2}$$

The arc length of the distance traveled by one gear must equal that of the other. That is, the distance traveled along the surface of each gear is the same. Converting the arc length to angular displacement yields

$$r_1\theta_1 = r_2\theta_2 \tag{3.3.3}$$

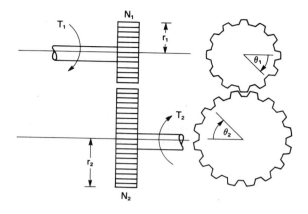

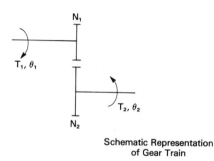

Schematic Representation
of Gear Train

Figure 3.3.1   Ideal gear train with parameters.

Finally, noting that since the two gear radii do not vary with time, if Eqs. (3.3.2) and (3.3.3) are differentiated with respect to time, their relationship still holds but with respect to $\dot{\theta}$ (i.e., the angular velocity ω^*) or $\ddot{\theta}$ (i.e., the angular acceleration, α). Using this concept, we may write

$$\frac{N_1}{N_2} = \frac{r_1}{r_2} = \frac{T_1}{T_2} = \frac{\theta_2}{\theta_1} = \frac{\omega_2}{\omega_1} = \frac{\alpha_2}{\alpha_1} \qquad (3.3.4)$$

Equation (3.3.4) can be used to investigate various properties of the "ideal gear train." For example, assume that the speed of both shafts is known and that the speed of shaft 1 is greater than the speed of shaft 2; then we know that the number of teeth on gear 2 is greater than that on gear 1. In addition, we also

---

*Although not shown explicitly, the reader should keep in mind that ω, α, and T are functions of time.

know the ratio N_1/N_2. Finally, if the torque on shaft 1 is known, we can compute the torque on shaft 2 by

$$T_2 = T_1 \frac{N_2}{N_1}$$

This particular relationship shows the speed reduction and torque multiplication property of a gear train.

A commonly used definition in motion conversion is that of the *coupling ratio*. Loosely defined, this ratio is the angular movement of the input compared to the load. For a rotational system, a coupling ratio of 2:1 defines a gear train in which two turns of the input shaft produce a single rotation of the output. Note that in this case the coupling ratio is the inverse of the tooth ratio, TR, which we define as (N_1/N_2).

It is interesting to note that the ideal gear train is similar to the ideal electrical transformer. In fact, one may transform a mechanical system containing a gear train into an analogous electrical network containing a transformer. In Section 3.5.4 we discuss this in more detail.

Employing the same concepts that were used to develop Eq. (3.3.4), the transfer relationship between the input and output shafts of a compound gear train (i.e., one consisting of more than two gears) may be derived.

Gear trains can be used to change "mechanical loads" in a manner that is similar to using a transformer to reduce or increase electrical impedances. For example, if a pure inertial load is placed on the output of a gear train as shown in Figure 3.3.2a, the input torque required to accelerate that load is given by

$$T_1 = \frac{N_1}{N_2} \alpha_2 J_2 \tag{3.3.5}$$

We may ask the question: What inertia is "seen" by the input shaft? Or in other words: What inertial load applied to the input shaft produces the same torque requirement as that of the original load? Figure 3.3.2b shows this equivalent system.

Assuming that T_1 accelerates an inertial load J_{eq} at an angular acceleration of α_1, we may write

$$\alpha_1 J_{eq} = \frac{N_1}{N_2} \alpha_2 J_2 \tag{3.3.6}$$

Using the relationships of Eq. (3.3.4), we may solve for the equivalent inertia J_{eq}.

$$J_{eq} = \left(\frac{N_1}{N_2}\right)^2 J_2 \tag{3.3.7}$$

For speed reduction and torque multiplication at the output of the gear train, the ratio N_1/N_2 is less than 1. The reflected inertia at the input shaft is seen to be less than that on the load.

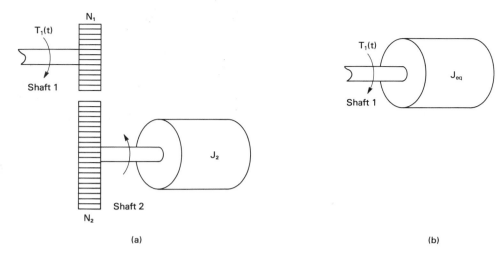

Figure 3.3.2 (a) Gear train with inertial load; (b) equivalent system.

Besides inertia, both the reflected viscous and Coulomb frictions are reduced by a gear train. Figure 3.3.3 shows two gears each having an inertia and friction on their shafts. The total torque as seen by the input shaft is given by Eqs. (3.3.8a) through (3.3.8c).

$$T_{\text{total}} = (J_1 + \text{TR}^2 J_2)\ddot{\theta}_1 + (B_1 + \text{TR}^2 B_2)\dot{\theta}_1 + T_f \qquad (3.3.8a)$$

$$T_f = F_{c_1} \text{sgn}\,(\theta_1) + \text{TR}\,F_{c_2}\,\text{sgn}\,(\theta_2) \qquad (3.3.8b)$$

$$\text{TR} = \frac{N_1}{N_2} \qquad (3.3.8c)$$

Note that both inertia and viscous friction are reduced (or increased) by the factor (N_1/N_2) squared, whereas Coulomb friction is reduced by the factor (N_1/N_2). Note also that Eq. (3.3.8a) is nonlinear.

By making TR less than 1, it is seen from Eqs. (3.3.8a) through (3.3.8c) that the gear train is effective in reducing the reflected inertial and viscous loads that must be accelerated by a motor (or other actuator). This is an attractive feature since the actuator does not actually have to produce the high torque needed at the output to drive the load but rather, a reduced value. Thus the actuator's size and torque capability can be significantly smaller than that required to drive the load directly. In robotic applications where large inertial loads must be accelerated, this property of reducing the inertia is often utilized in order to reduce significantly the size, weight, volume, and cost of the various joint actuators.

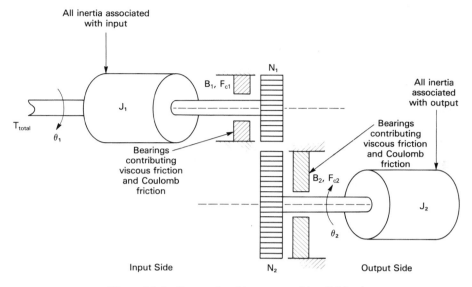

Figure 3.3.3   Gear train with torque and inertial loads.

Practical considerations (i.e., physical constraints) can place certain limits on gear-type transmissions. For example, the amount of torque that an actual gear train can transmit before the forces on its teeth become large enough to cause breakage may constrain the actual implementation. Also, the physical size of gears may present a problem. Consider, for example, the following thought experiment. Assuming that we wished a very high coupling ratio on the order of 100:1, then by Eq. (3.3.4) we see that the radius, r_2 of the output gear must be 100 times greater than that of the input gear. If the input gear had a diameter of $\frac{1}{2}$ in., the output gear would have a diameter of 50 in.! This may be quite impractical for a reasonably sized joint of a manipulator. To overcome this problem, multiple-pass gear trains can be used. This is a system of multiple gears mounted on different levels so that the distance from the input to the output shaft is minimized. While multiple-pass gearing may reduce the spacing between the input and output shafts, it increases the depth and is more complex, due to the increased number of parts.

The opposite situation is the case where we wish the actuator and output to be separated by a very large distance. If this is implemented by two gears, the physical size of the gears may become quite large and due to necessary rigidity, quite heavy. The resultant inertia of these gears may make this approach unacceptable. Even for a multiple-pass situation (with the gears lined up to take up the distance), the added complexity becomes a concern.

Two practical alternatives to these situations are the harmonic drive (for the case where input and output need to be as close as possible) and the pulley/belt or chain/sprocket (for the case where the input and output need to be separated by a large distance).

3.3.1.2 Harmonic drives

A cup-type harmonic drive and its components are shown in Figure 3.3.4. The three components of the harmonic drive are:

1. *Circular spline*: a rigid, thick wall ring with internal spline teeth. It is either a fixed or a rotating drive element.
2. *Wave generator:* an elliptical ball bearing assembly which includes a shaft coupling. It is the rotating input drive element.
3. *Flexspline:* a nonrigid cylindrical thin-walled cup which has two fewer spline teeth and a smaller pitch diameter than that of the circular spline. It is a fixed or rotating output drive element.

Harmonic drive gearing is a patented principle based on nonrigid body mechanics. It employs three concentric components to produce high mechanical advantage and speed reduction. The use of nonrigid body mechanics allows a continuous elliptical deflection wave to be induced in a nonrigid external gear, thereby providing a continuous rolling mesh with a rigid internal gear.

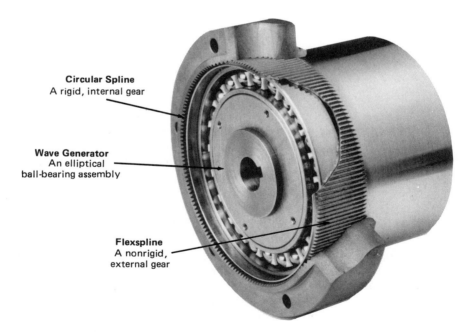

Circular Spline
A rigid, internal gear

Wave Generator
An elliptical
ball-bearing assembly

Flexspline
A nonrigid,
external gear

Figure 3.3.4 Cut-type harmonic drive and its components. (Reproduced with the permission of Harmonic Drive, a division of Quincy Technologies, Inc.)

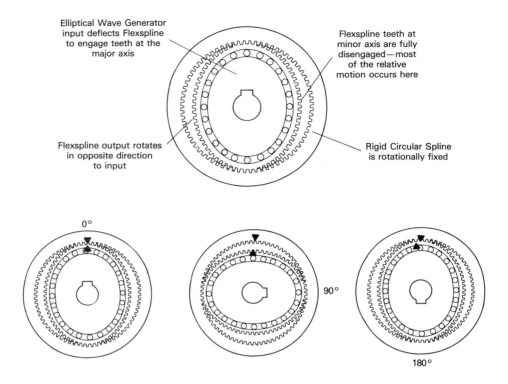

Figure 3.3.5   Analysis of harmonic drive motion. (Reproduced with the permission of Harmonic Drive, a division of Quincy Technologies, Inc.)

Since the teeth on the nonrigid flexspline and the rigid circular spline are in continuous engagement and since the flexspline has two fewer teeth than the circular spline, one revolution of the input causes relative motion between the flexspline and the circular spline equal to two teeth.   Thus, with the circular spline rotationally fixed, the flexspline will rotate in the opposite direction to the input at a reduction ratio equal to the number of teeth on the flexspline divided by 2.

The relative rotation may be understood by examining the motion of a single flexspline tooth over one-half an input revolution.   Figure 3.3.5 illustrates this. The tooth is fully engaged when the major axis of the wave generator input is at 0°.   When the wave generator's major axis is rotated to 90° (in the clockwise direction), the tooth is disengaged.   Full engagement occurs in the adjacent circular spline tooth space when the major axis is rotated to 180°.   This motion repeats as the major axis rotates another 180° back to zero, thereby producing the two-tooth advancement per input revolution.

Figure 3.3.6 shows how the harmonic drive component may be set up to perform six basic transmission functions by varying the input, output, and fixed element.

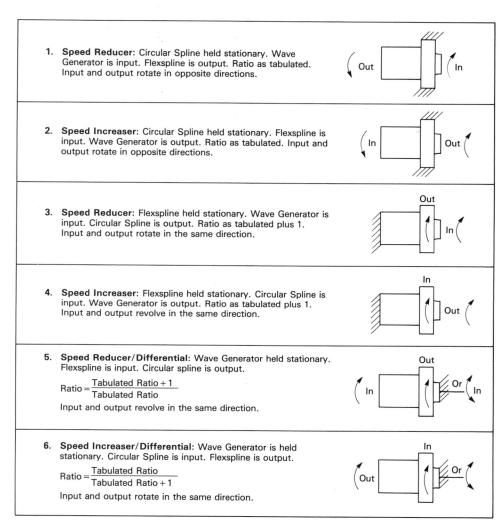

1. **Speed Reducer:** Circular Spline held stationary. Wave Generator is input. Flexspline is output. Ratio as tabulated. Input and output rotate in opposite directions.

2. **Speed Increaser:** Circular Spline held stationary. Flexspline is input. Wave Generator is output. Ratio as tabulated. Input and output rotate in opposite directions.

3. **Speed Reducer:** Flexspline held stationary. Wave Generator is input. Circular Spline is output. Ratio as tabulated plus 1. Input and output rotate in the same direction.

4. **Speed Increaser:** Flexspline held stationary. Circular Spline is input. Wave Generator is output. Ratio as tabulated plus 1. Input and output revolve in the same direction.

5. **Speed Reducer/Differential:** Wave Generator held stationary. Flexspline is input. Circular spline is output.

   $$\text{Ratio} = \frac{\text{Tabulated Ratio} + 1}{\text{Tabulated Ratio}}$$

   Input and output revolve in the same direction.

6. **Speed Increaser/Differential:** Wave Generator is held stationary. Circular Spline is input. Flexspline is output.

   $$\text{Ratio} = \frac{\text{Tabulated Ratio}}{\text{Tabulated Ratio} + 1}$$

   Input and output rotate in the same direction.

Figure 3.3.6 Transmission functions implemented by a harmonic drive. (Reproduced with the permission of Harmonic Drive, a division of Quincy Technologies, Inc.)

3.3.1.3 Belt-and-pulley systems

Figure 3.3.7 can be used to model an ideal belt-and-pulley or chain-and-sprocket system. The analysis of this system is similar to that of the ideal gear train. Note that in this case, the relative motions of both shafts are in the same direction. Timing belts and chains perform essentially the same function as gear trains. However, they allow the transmission of the rotational motion and torque over longer distances without the need for multiple gears. Figure 3.3.8 shows a

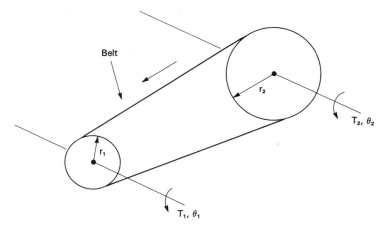

Figure 3.3.7   Ideal belt and pulley system with parameters.

timing belt and toothed pulleys.   As opposed to a belt and pulley, this method will not allow slippage between the belt and sprocket due to the teeth; however, it will exhibit a spring constant if the output is locked and a torque applied to the input.

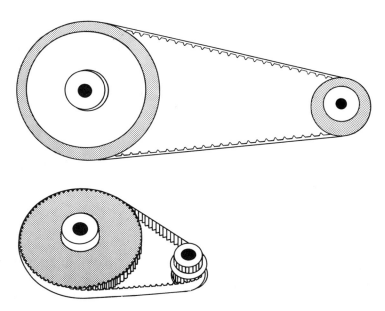

Figure 3.3.8   Timing belt and toothed pulleys.

3.3.2 Rotary-to-Linear Motion Conversion

Rotary-to-linear motion conversion is concerned with taking the rotational motion and torque from an actuator and producing a linear motion and force on the output. Components whose ideal model produces a linear relationship between the input shaft's rotation and the output's linear displacement are the lead screw, rack and pinion, and Roh'lix. Nonlinear relationships are available by slider-crank assemblies and cams.

3.3.2.1 Lead screws

Figure 3.3.9 shows a lead screw driving a payload along a single axis. In this configuration, the screw is fixed with its ends free to rotate. As the screw is turned, the nut moves along the shaft with the payload attached. Typically, the payload is supported and moves along a surface with a very low frictional component. A rotary displacement of the input shaft, θ_1, causes a linear motion of the payload, x. The *coupling ratio* is defined as the input shaft rotation per unit of linear motion. A typical rating of a lead screw is 10 turns/in. This is also referred to as the pitch (P). An alternative coupling ratio is given by the reciprocal of the pitch, which is called the lead. This factor (L) is defined as the axial distance the screw or nut travels in one revolution and may be given in such units as in./rev or cm/turn. A lead screw having a pitch of 10 turns/in. has a lead of 0.1 in./turn. Therefore, we may write the following relationships for the system of Figure 3.3.9:

$$\theta = Px \tag{3.3.9a}$$

$$x = L\theta \tag{3.3.9b}$$

These equations may be differentiated any number of times in order to obtain the relationships among linear velocity, acceleration, and jerk, and rotational velocity, acceleration, and jerk, respectively.

Just as in the case of rotary motion conversion, we are interested in how a load on the output is seen by the input. That is, is there an equivalent torque–inertia system for Figure 3.3.9? The derivation of the inertia reflected back to the input shaft is based on a balance of kinetic energy. For the linear motion of the payload's mass, M_L, we may write

$$E_k = \tfrac{1}{2} M_L v^2 \tag{3.3.10}$$

The corresponding kinetic energy of a torque–inertia system is defined as

$$E_k = \tfrac{1}{2} J_{eq} \omega^2 \tag{3.3.11}$$

Equating the kinetic energies of Eqs. (3.3.10) and (3.3.11) and solving for the inertia, J_{eq}, after relating the rotary and linear velocities by the pitch, yields

$$J_{eq} = \frac{M}{(2\pi P)^2} = \frac{W/g}{(2\pi P)^2} \tag{3.3.12}$$

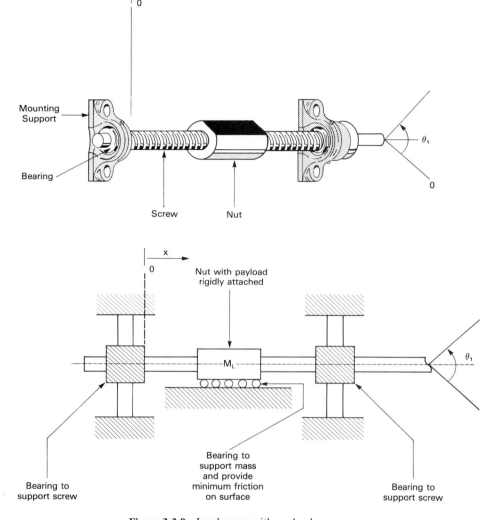

Figure 3.3.9   Lead screw with payload.

Equation (3.3.12) shows that the reflected inertia of a load driven by a lead screw can be reduced by choosing a lead screw with a greater pitch.

3.3.2.2 Rack-and-pinion systems

Figure 3.3.10 shows a representation of a rack-and-pinion gear train. The pinion is the small gear attached to the actuator, and the rack is a linear member with gear teeth on one side. The transfer relationship of such a mechanical system

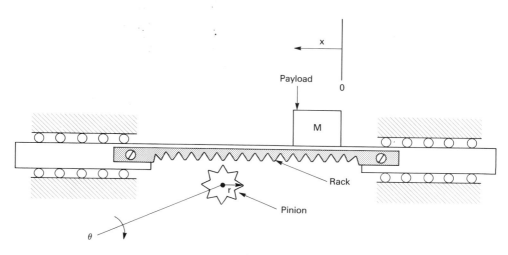

Figure 3.3.10 Rack-and-pinion driving load.

is defined as

$$x = 2 \pi r \theta \tag{3.3.13}$$

As defined by Eq. (3.3.13), the linear distance traveled is proportional to the input shaft rotation (in revs), with the constant of proportionality equal to the circumference of the pinion. That is, the linear distance traveled is equal to the distance traveled by the pinion. The reflected inertia, as seen by the input shaft, can be shown to be

$$J = Mr^2 \tag{3.3.14}$$

3.3.2.3 Belt and pulley driving a linear load

Another method of generating linear motion from rotary motion is shown in Figure 3.3.11. Note that both pulleys have the same radius. The pulley connected to the input is called the *drive pulley* and the second pulley is called an *idler pulley*. The distance traveled by the load is equal to the distance traveled by the drive pulley. Equation (3.3.13) also describes the transfer relationship of the belt and pulley system while Eq. (3.3.14) is the reflected inertia as seen by the input shaft.

In practical implementations of the lead screw, rack and pinion, and linear drive via belt and pulley, there are physical limitations on the linear motion. For example, if the linear motion exceeds its designated range, it typically hits an "end stop." This mechanical contact forces the rotary shaft to lock. In the event that a torque continues to be applied to the shaft in this locked position, it is possible for extreme forces to be generated, causing the devices to fail (i.e., the threads of the screw or nut strip, or gear teeth break). The practical solution to this problem is to place electrical limit switches some distance before the physical limit is reached. The signal from these switches is used to stop the actuator before a damaging force can be applied at the extreme limits of motion.

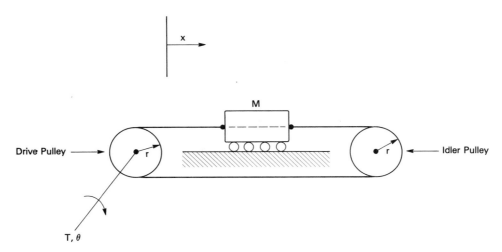

Figure 3.3.11   Rotary to linear conversion via belt and pulley.

3.3.2.4 The Roh'lix

An interesting mechanical device that overcomes this end travel problem is the Roh'lix (trade name).   Figure 3.3.12 shows the Roh'lix.   This device is based on the principle of rolling contact.   The drive shaft is the long shaft (as compared to the screw in a lead screw), while the "nut" is composed of six freewheeling bearings connected to idler shafts mounted in a block.   There are three shafts at

Figure 3.3.12   Roh'lix mechanism. (Reproduced with the permission of Zero-Max, a unit of Barry Wright, Minneapolis, MN.)

each end of the nut which are mounted at an angle relative to the axis of the drive shaft.   The block containing the idler shafts will move in either direction proportional to the rotation of the drive shaft.

An interesting property of this device is that when it reaches its limit of travel, instead of the input shaft jamming, it continues to rotate while the nut remains against the stop.   The shafts slip and no damage occurs.   The slip occurs when a preset thrust or overload point is equaled or exceeded.   The slip can be a disadvantage in a servo-controlled application, since if an obstruction of sufficient torque occurs, the input shaft will slip and the position calibration will be lost.   Therefore, the use of this device requires a feedback transducer on the linear portion as opposed to the driving member.

3.3.2.5 Slider cranks

The slider-crank mechanism is an extremely cost-effective means of converting rotary to linear motion.   Figure 3.3.13 shows a representation of a crank driving a linear stage (or load).   In this implementation, the crank portion is the wheel that rotates about its center and has a rod of fixed length mounted to a point on its circumference.   The other end of the connecting rod or link is attached to a linear stage which is constrained to move in only one dimension on a relatively frictionless surface.   At both its locations on the disk and linear stage, the connecting rod is free to rotate.   Thus the angle formed with the horizontal will change as a function of the disk's position.

As the disk travels from 0 to 180° in the counterclockwise direction, the linear stage moves a distance equal to $2r$.   If the disk continues to travel from 180° back

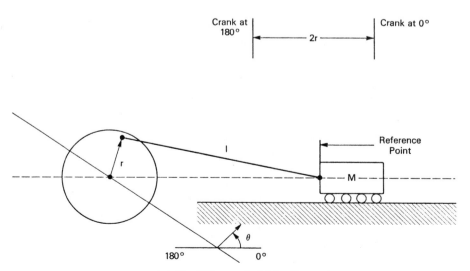

Figure 3.3.13   Slider crank driving linear stage.

to 0° (still in the counterclockwise direction), the load will move in the opposite direction over exactly the same linear distance. Note that there are two distinct angular positions of the input shaft that correspond to a single linear position of the load.

The angular positions that correspond to full extension and retraction of the linear stage are called dead-center positions or dead points and occur when the crank and connecting rod are collinear. If this system is used to convert linear to rotary motion (as is typically done in a gasoline reciprocating engine), a flywheel with sufficient inertia must be connected to the crank's axis of rotation. The reason for this is that when the connecting rod and crank are in line at a dead point, the direction of rotation may go either way. Thus the inertia contribution of the flywheel is necessary to bias the mechanism for proper operation.

If the input shaft is rotated continuously at some velocity, the motion of the linear stage is reciprocating. As can be observed by a careful examination of Figure 3.3.13, the torque seen by the input and the relationship between the input shaft's position and linear stage's position are nonlinear in nature.

3.3.2.6 Cams

Cams can also provide a nonlinear transfer of rotary to linear motion. The cam is an irregularly shaped machine which is driven by some type of motor. Touching the surface of the cam is a *follower* bearing to which the cam surface imparts motion. Figure 3.3.14 shows a cam with a follower while Figure 3.3.15 shows the accompanying displacement diagram. For the system shown in Figures 3.3.14 and 3.3.15, as the cam rotates, the follower touches its surface. Thus the rotary profile of the cam is imparted to the follower, which in turn causes a linear motion of the arm. The actual velocity, acceleration, and jerk of the linear motion can be defined by the shape of the cam.

It should be noted that cams are very common in automatic machines because they provide a simple method of providing almost any desired follower motion. By using a simple limit switch and motor, a cam may be driven one revolution and thus produce some desired motion of its follower. As can be seen from the displacement diagram, the follower may extend rapidly, dwell, retract partially, dwell, and then return to its original position as the limit stop is again reached.

The analysis and design of cams is abundant in engineering literature and the interested student should reference one of the many machinery design texts or mechanical handbooks. Cams are not generally used for implementing the linear joints of a programmable robotic manipulator, due to the inherent nonlinear transfer characteristics for position and torque. However, these and other classical mechanical devices are widely used in implementing the special motions required by end effectors or other peripheral equipment in workcells. In fact, many simple pick-and-place mechanisms use cams in their implementation.

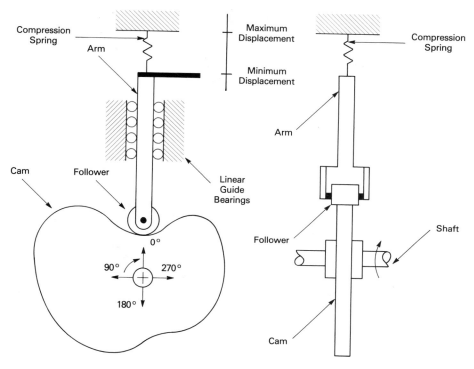

Figure 3.3.14 Cam with follower.

3.3.3 Linkages

Rather than include linkages in the rotary-to-rotary or rotary-to-linear conversion discussion, the subject is discussed as a separate topic. It should be noted that before the advent of computer-controlled machinery, the complete cycle of an automated machine was controlled by cam profiles and/or linkages.

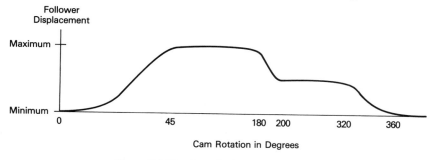

Figure 3.3.15 Cam displacement diagram.

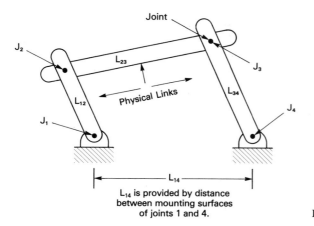

Figure 3.3.16   Four-bar linkage.

The four-bar linkage of Figure 3.3.16 is one of the most common in use. Most complex mechanisms can be studied or reduced in terms of the four-bar linkage [10].

Various permutations of moving or fixing certain links with respect to ground, along with the relationships between the lengths of the links, allow the basic four-bar linkage to perform different functions. For example, Figure 3.3.17 shows a situation in which links L_{12} and L_{34} are replaced by disks. The length of L_{14} is

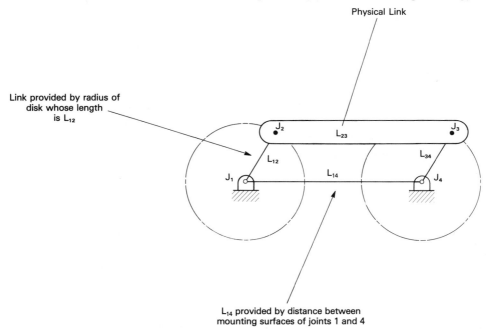

Figure 3.3.17   Four-bar linkage using disks for two links.

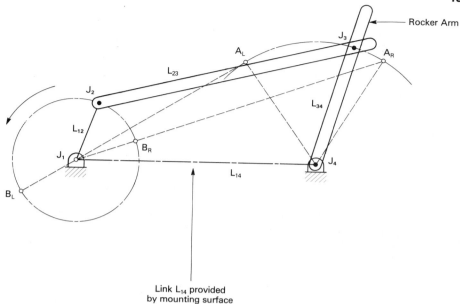

Figure 3.3.18   Crank and rocker.

equal to the distance between the centers of the two disks.   Link L_{23} is coupled
to the two disks in such a way that its motion is unconstrained in the plane of the
disks.   If the disk on the left is rotated at a constant velocity, the disk on the right
will also.   When L_{23} is coincident with the centers of the disks, a dead point occurs.
Here it is possible for the rotation of the follower to be in the direction opposite
to that of the driving disk.   As indicated in the discussion on slider cranks, an
inertial disk may be used to bias the motion in the desired direction.

The slider-crank mechanism discussed in Section 3.3.2 (see Figure 3.3.13) is
Another configuration of the four-bar linkage is called the crank and rocker.
This device transforms rotary motion into oscillating motion (see Figure 3.3.18).
Either link L_{12} or L_{34} can be the driving member.   If L_{12} is rotated about its pivot
point, then L_{34} oscillates back and forth between its two extreme points, A_R and
A_L.   If link L_{34} is the driver, it is necessary to bias the direction of rotation of link
L_{12} by an inertial disk since two dead points occur at the extreme points of link
L_{34}.

The slider-crank mechanism discussed in Section 3.3.2 (see Figure 3.3.13) is
a special case of the four-bar linkage.   By making link L_{34} infinitely long, joint J_3
is seen to move in a straight line.   By replacing link L_{34} by a slider, we obtain the
kinematic diagram of Figure 3.3.19, which is equivalent to the implementation
discussed previously.

The final application of the four-bar linkage that will be considered is the
pantograph of Figure 3.3.20.   This device is commonly used to reproduce drawings
of different scales.   The mechanism provides a parallel motion between the pen
and stylus, which, of course, are interchangeable.   An obvious use of this mech-

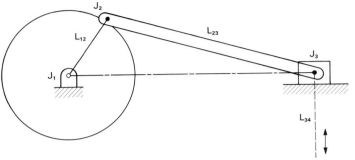

Figure 3.3.19 Slider crank from four-bar linkage.

anism (provided that there is space for the linkages to operate freely) is for producing remote-controlled motion of the stylus point by moving the pen point. Essentially, when used in this way, the mechanism is a teleoperator in two dimensions (see Chapter 1).

The mechanism of Figure 3.3.21 shows an interesting implementation of a single-jointed manipulator. As opposed to utilizing a rotary actuator to move the "arm" as was discussed in Figure 3.2.6, here the length of link 1 changes, causing link 2 to move. Link 2 is fixed in length and free to pivot on the vertical support, while link 1 could be implemented by a lead screw. One end of the screw is fixed to the vertical support but free to pivot, while the nut is coupled to link 2 with another pivoting device.

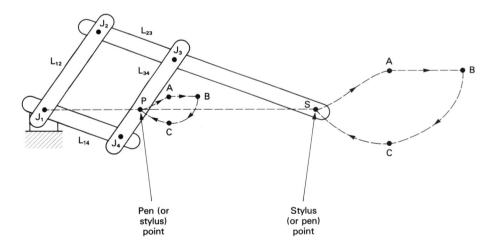

Pen (or stylus) point

Stylus (or pen) point

J_1 is fixed: J_2, J_3, J_4 are free to move in the plane of the drawing.

Figure 3.3.20 Pantograph.

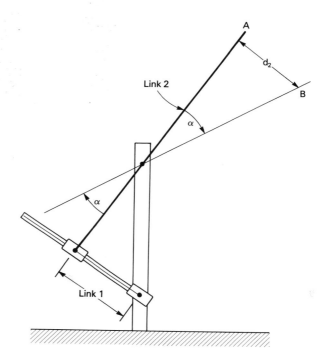

Figure 3.3.21   Linear drive for rotary joint.

3.3.4 Couplers

A coupler is a device used to connect shafts so that one can drive the other. Typically, it is used to connect an actuator or motor to the rest of the joint mechanism. In the ideal case, all that is required of the coupling is to transmit the torque, velocity, and position from the actuator to the rest of the system. However, since the world is imperfect, the job of couplers is much more difficult. As nothing ever lines up perfectly [e.g., shafts may be parallel but not collinear, or they may be skewed (so as to intersect at a point)], rigidly coupling the shafts would not be possible. Thus there is a need for a device that can compensate for these misalignments.

Although shaft couplers come in a variety of configurations and specifications, they are commonly used to solve the misalignment problems shown in Figure 3.3.22.

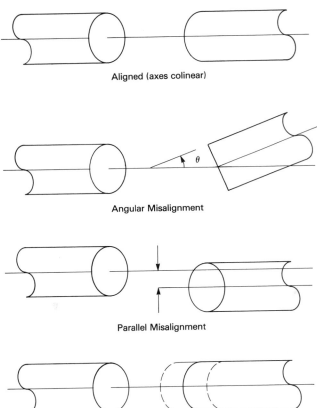

Aligned (axes colinear)

Angular Misalignment

Parallel Misalignment

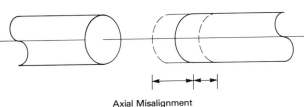

Axial Misalignment

Figure 3.3.22   Shaft misalignment.

Figure 3.3.23 shows a front and side view of an Oldham coupling, a device used to connect two shafts that are parallel but not collinear. The two outer plates that connect to the shafts each have a groove in them. The inner plate has a raised surface on each side which fits into the grooves on the outer plates. Since the raised surfaces are 90° apart, the coupler allows for misalignment of the two shafts in the plane of the plates. Since there is no relative motion between any of the three plates, this coupling permits a constant-velocity ratio to be transmitted. Of course, since the shafts are constrained to remain fixed along their axial direction, as one rotates, the coupler moves in x and y so as to impart the desired motion to the second shaft. The Oldham coupling is typically used to couple a motor's shaft to the wave generator of a harmonic drive.

The universal joint is used for connecting two shafts whose axes are not in line but intersect at a point. There are many possible implementations for a universal joint, as depicted in Figure 3.3.24. However, it does not work well if

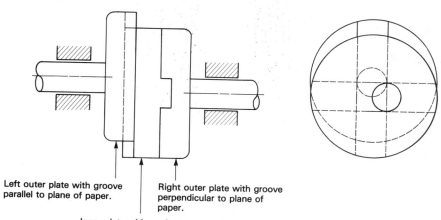

Left outer plate with groove parallel to plane of paper.

Right outer plate with groove perpendicular to plane of paper.

Inner plate with a raised surface on each side. The raised surfaces are at a right angle to each other.

Figure 3.3.23   Oldham coupling.

the angle δ is more than 45°. Preferably, the angle should be limited to less than 25° except when the speed of rotation is very slow and little torque needs to be transmitted.

The flexible shaft coupler shown in Figure 3.3.25 is a feasible alternative to both the Oldham coupler and the universal joint. These devices come in a variety of configurations, including the "accordion," "split-ring," and rubber, to name a few. Typically, shaft couplers are rated by the maximum amount of angular and parallel offset they can accommodate and also by the maximum torque they can transmit. In addition, when designing a system, we are interested in their inertia, their torsional spring constant, and their viscous damping component. We explore a model for the flexible coupler in Section 3.5.3.

3.3.5 The Concept of Power Transfer

In Section 3.3 we defined the equivalent inertia as seen by the input shaft, for both a gear train and lead screw. For the case of the gear train it was observed that the load inertia could be reduced by a factor of $(N_1/N_2)^2$ thus making the torque requirements of the actuator lower than if it had been forced to drive the load directly. Similarly, the inertia seen by the lead screw is also controllable by its coupling ratio. An obvious question is: What coupling ratio should be chosen? If the desired characteristic of the transmission is a speed or torque change, the

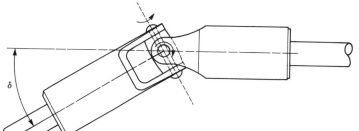

δ

Figure 3.3.24   Universal coupling.

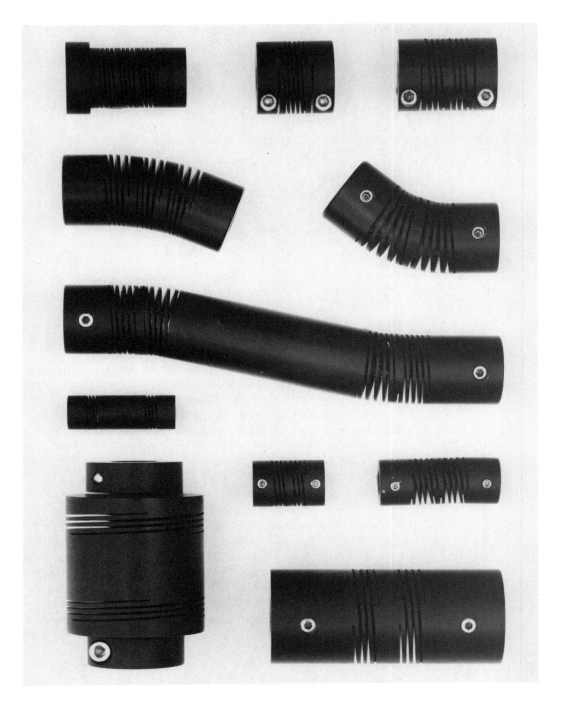

Figure 3.3.25 Flexible shaft couplers. (Reproduced with permission of Rocom Corp., Huntington Beach, CA.)

coupling ratio is chosen per the design specification.   However, if one wishes to minimize the energy dissipated in an actuator such as a dc motor (see Chapter 4), one may choose the coupling ratio based on the principle of an *inertial match*. Simply stated, this concept uses the coupling ratio to make the reflected load inertia equal to the inertia of the dc motor.   Thus, for a speed-reducing gear train, one should choose the ratio (N_1/N_2) by the following formula:

$$\frac{N_1}{N_2} = \sqrt{\frac{J_2}{J_1}} \qquad (3.3.15)$$

Here J_2 is the total inertial load on the output, while J_1 is the inertial load on the input due to the actuator.   Compare Eq. (3.3.15) with Eq. (3.3.7).   Similarly, one can define an optimal pitch for a lead screw which minimizes the energy dissipated in the actuator as

$$P = \frac{1}{2\pi} \sqrt{\frac{M_L}{J_1}} \qquad (3.3.16)$$

where J_1 is the inertia of the actuator driving the lead screw and M_L is the mass of the load.

Equations (3.3.15) and (3.3.16) may be derived by writing an equation for the total torque that must be provided by the dc motor, defining the current of the motor in terms of the required torque, and finally minimizing the power that must be dissipated in motor with respect to the coupling ratio.   It should be apparent that the expressions become more complex if constant and viscous friction terms are included in the derivation.

3.4 SOME PROBLEMS WITH REAL-WORLD COMPONENTS

In our previous discussions we considered only "ideal" mechanical components. However, real-world components do not necessarily behave as their ideal models, although they may approach this behavior in certain circumstances.   In addition, even though a component may be specified very tightly, its performance may be far from specification if it is used improperly.

Typically, it is reasonable to say that the closer an actual component behaves to its ideal model, the more costly it becomes.   This is usually because its manufacture requires more precise and careful steps.   Ideal behavior, while desired, may be cost-ineffective or actually unnecessary in certain cases.

If one understands the components and their interworkings in a particular application and the desired composite specification of the final product or system, reasonable trade-offs can be made against cost, complexity, and the errors that the components may introduce.   The following sections define some common errors that may be quantified for the mechanical components discussed previously or for general mechanical systems.

3.4.1 Efficiency

Efficiency η is defined as the ratio of the output power to the input power, or the ratio of the work output to the work input over the same period of time. For an ideal mechanism, the efficiency is 1 or 100%. In the case of real components, the work output is less than the work input, with the difference being dissipated in friction. Equation (3.4.1) defines efficiency.

$$\eta = \frac{\text{power out}}{\text{power in}} = \frac{\text{work output}}{\text{work input}} \qquad (3.4.1)$$

For the case of a gear train, we may restate Eq. (3.4.1) as the ratio of the actual output torque divided by the ideal output torque. Thus for a gear train having a tooth ratio TR of N_1/N_2, with $N_2 > N_1$ so that torque multiplication results, we obtain

$$\eta = \frac{\text{actual output torque}}{\text{input torque/TR}} \qquad (3.4.2)$$

Figure 3.4.1 shows a transmission consisting of a right-angle gear train having a tooth ratio of $\frac{1}{15}$ and a set of antibacklash gears having a ratio of $\frac{1}{5}$. A plot of the input versus the output torques for the assembly is shown in Figure 3.4.2. Note that the actual overall transmission has a measured efficiency of only 22% as compared with an ideal performance of 100%. Although it is possible for efficiency to be a function of speed, a first approximation would consider it to be dependent only on the forces encountered by the gear teeth, which are primarily frictional in nature. Since these forces are directly proportional to torques, the efficiency of

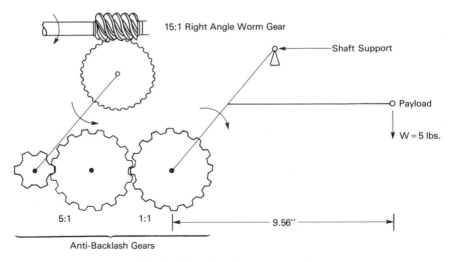

Figure 3.4.1 Complex gear assembly.

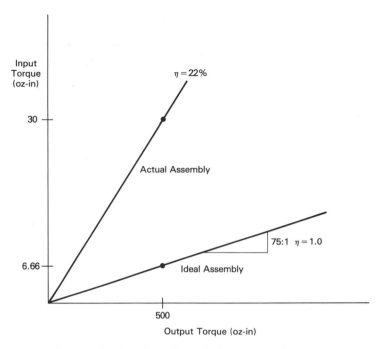

Figure 3.4.2   Actual and theoretical torque transfer curves.

the transmission can be approximated by measuring the resulting torque on the output for static torques applied to the input shaft.

The efficiency of any mechanical device becomes important in sizing actuators. It is no longer safe to assume that the output loads are reflected to the input shaft by a function of the gear ratio as defined in Eqs. (3.3.8a) through (3.3.8c), but one must now include the efficiency.   These equations then become

$$T_{\text{total}} = \left(J_1 + \frac{\text{TR}^2}{\eta} J_2 \right) \ddot{\theta}_1 + \left(B_1 + \frac{\text{TR}^2}{\eta} B_2 \right) \dot{\theta}_1 + T_f \qquad (3.4.3a)$$

$$T_f = F_{c_1} \operatorname{sgn}(\theta_1) + \frac{\text{TR}}{\eta} F_{c_2} \operatorname{sgn}(\theta_2) \qquad (3.4.3b)$$

$$\text{TR} = \frac{N_1}{N_2} \qquad (3.4.3c)$$

These equations reveal that any efficiency less than 1 (i.e., 100%) will increase the torque required to accelerate a given inertial load or overcome an external torque load.   It is important to note that efficiency does not affect the actual transfer ratio of the gears (or other transmission device) in terms of displacement, velocity, or acceleration, but greatly affects any torque-related property.

TABLE 3.4.1 EFFICIENCIES FOR
SOME TRANSMISSION
COMPONENTS

Device	Efficiency (%)
Lead screw	
Acme type	25–85
Ball type	70–90
Roh'lix	90
Worm gears	46–98
Harmonic drive	62–83

Typical ranges of efficiency for some of the mechanical devices described previously are given in Table 3.4.1. It should be noted that the efficiency tabulated is dependent on such factors as the coupling ratio, the material's coefficients of friction, and the angle used to define the gear teeth or the depth of cut and type of threads for screws.

An interesting side effect of efficiency is the ability to backdrive a transmission. For example, if one pushes on the linear stage connected to a lead screw, does the screw turn or does the stage remain locked? In general, the higher the efficiency of a transmission, the more likely it will be backdrivable.

Some transmissions are designed so that they are not backdrivable. In fact, this may be a desirable feature in certain designs. However, in the case of servo-controlled mechanisms, the ability to backdrive a transmission is extremely important. A nonbackdrivable transmission introduces a nonlinear transfer element into a mechanical system. Consider, for example, a position servo loop in which the motor is connected to a purely inertial load by means of a transmission. In theory, by adjusting various gains in the system, one should be able to control the amount of overshoot and settling time of the response of the system to a step input. If the transmission is not backdrivable, a nonlinear device is in the loop and the energy of the inertial disk cannot be transmitted back to the motor. Unless energy can be both "sourced" and "sinked" by the motor, the motor working in the loop will not have the ability to dampen out the motion. Thus the system begins to exhibit nonlinear behavior and may possibly become unstable, e.g., have a limit cycle.

3.4.2 Eccentricity

One of the phenomena that affect the displacement transfer characteristic of gears (or any rotating parts for that matter) is eccentricity. Figure 3.4.3 shows an ideal and an eccentric disk. The ideal disk spins on its true center. Therefore, if point *A* is identified on the circumference of the disk and the disk is then rotated, point *A* stays on the path identified by the circumference of the disk. A disk that exhibits eccentricity does not rotate on its true center but about some other point. As shown on the nonideal disk in Figure 3.4.3, *A* is the point on the circumference

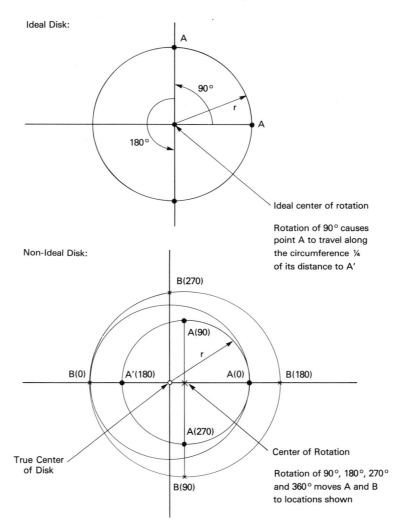

Figure 3.4.3   Disk on true center and disk on eccentric.

of the disk that is the closest to the center of rotation.   Note that as the disk is rotated, this point does not stay on the path that would be defined by a tracing of the circumference but rather follows a circular path defined by its radial distance from the off-center rotation point.   *B* is a point on the circumference of the disk which is farthest from the center of rotation.   It also traces out a circular path defined by its distance to the center of rotation.   Note that as the disk is rotated over one full revolution, the referenced points return to their original position.

Eccentricity is defined as the distance from the true center to the center of

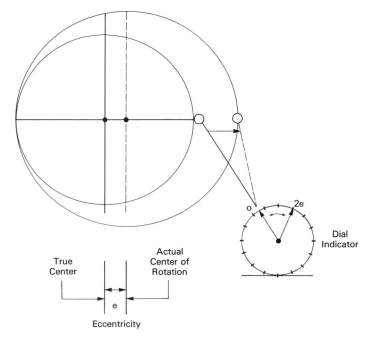

Figure 3.4.4 Measuring TIR of shaft or disk.

rotation.   It is easily measured by dividing the *total indicated runout* (TIR) by 2. Figure 3.4.4 shows a method of measuring the TIR or eccentricity of a shaft.

3.4.3 Backlash

Since rotating components can always have an error due to eccentricity, there are problems in interfacing components such as gears.   Consider for example, the case of two gears as shown in Figure 3.4.5.   Since both gears rotate about a point slightly off center, it is possible (depending on how the system is assembled) for point B on each gear to make contact on each revolution.   If one were to measure the torque on the input shaft for a full revolution, it would appear greater when the two high spots (occurring at point B) meet.   In addition, if the gears were adjusted so that the teeth meshed as perfectly as possible at the high spot, then at point A (the low spot), there would be space between the teeth.   Since the gears are not exactly meshed at A, if one of the shafts were held stationary, a small motion could be felt on the free shaft as it was rotated in both the clockwise and the counter-clockwise direction.   This phenomenon, called *backlash*, is depicted in Figure 3.4.6.

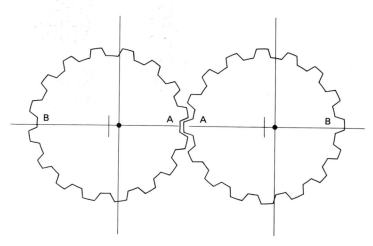

Figure 3.4.5   Two gears with eccentricity.

As can be seen, motion of the input does not cause the output to experience motion until the gap between the two members is removed.   This so-called *deadband* behavior contributes a nonlinearity to mechanical systems and can cause discontinuities in velocity and acceleration, leading to high mechanical stress of components.   Backlash can occur in both translational and rotational systems.

The manner in which backlash behaves in a dynamic system is dependent on the friction between the two contacting surfaces and also on the relative inertias of the input and output.   Consider a system in which the inertia associated with the output is much less than that of the input.   There will also be a large frictional load connected to the output.   Referring to Figure 3.4.6, if there is no contact between the two members when the input stops moving, the output will tend to stop since it has low inertia and a friction force to stop it.   When motion of the input begins, it will travel along the deadband until it makes contact with the output.   If the input member reverses direction, the output will stop until the input travels the distance necessary to take up the backlash.   Once the backlash is taken up, the output member instantaneously takes on the velocity of the input member and resumes moving.   Figure 3.4.7a shows the transfer relationship for this condition.

In the case where the output friction is negligible and the inertia of the output is considerable, the nonlinearity due to the deadband behaves differently.   In this case, the inertia of the output tends to keep it in contact with the input as long as the acceleration is in the direction to keep the members in contact.   When the acceleration of the input becomes zero, the output member does not stop immediately but coasts at a constant velocity, that is, at the maximum velocity attained

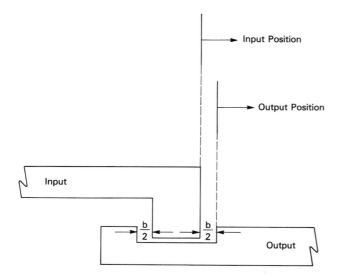

Figure 3.4.6 Backlash model. (Benjamin C. Kuo, *Automatic Control Systems*, 5e, © 1987, pp. 133, 143. Reprinted by permission of Prentice Hall, Inc., Englewood Cliffs, NJ.)

by the input. When the output has traversed a distance equal to the full backlash, it will be restrained by the opposite side of the input member. When contact is made, the output again assumes the velocity of the input. Figure 3.4.7b shows the input/output relationship for this second condition.

To minimize the problems of backlash, provisions must be made in the design of mechanical systems for adjusting one gear relative to the other, thereby affording complete control over backlash at initial assembly and throughout the life of the

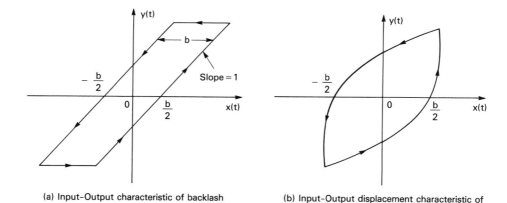

(a) Input-Output characteristic of backlash with negligible output inertia.

(b) Input-Output displacement characteristic of a backlash element without friction.

Figure 3.4.7 Input-output relationships for backlash model. (Benjamin C. Kuo, *Automatic Control Systems*, 5e, © 1987, pp. 133, 143. Reprinted by permission of Prentice Hall, Inc., Englewood Cliffs, NJ.)

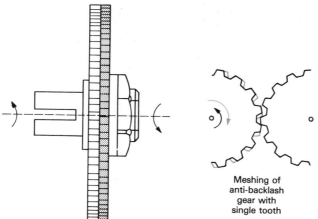

Meshing of
anti-backlash
gear with
single tooth

Figure 3.4.8   Anti-backlash gear assembly.

gears.   A number of techniques for doing this exist.   For example, if one gear is purposely mounted on an eccentric shaft, the meshing of the teeth can be adjusted periodically as the gear surfaces wear down.   In addition, at initial assembly, one can match the high spot of one gear to the low spot of the second.   This attempts to cancel out the eccentricities.   Of course, this approach is good only for integer ratios, since other ratios will not guarantee consistent line up of the low and high spots of the two gears.

Antibacklash gears are another method used to reduce the problem.   In this approach, the gears are composed of two sets of teeth that slide over one another. Spring loading is used to ensure that the gear teeth of the antibacklash gear spread apart and tend to fill the space between the teeth of the gear with which they are in contact.   The double-teeth arrangement of an antibacklash gear is shown in Figure 3.4.8.

3.4.4 Tooth-to-Tooth Errors

Another problem experienced by real gear trains is tooth-to-tooth errors (affecting their accuracy).   These arise from errors in the manufacturing process (i.e., no teeth are identical).   The three major contributions are errors in tooth thickness, tooth profile (shape), and the roundness of the gear.   If an "ideal gear" were used to drive a gear under test, a plot similar to that of Figure 3.4.9 would be obtained. It is seen that the general characteristic of the error is sinusoidal.   The small sine waves superimposed over the longer-period wave are due to the errors of the individual teeth, whereas the longer period is due to the eccentricity of the gear blank.   Taken together, they define the total composite error, which is also the total indicated runout (TIR) (see Section 3.4.2) defined for rotating members.

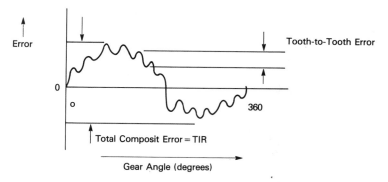

Figure 3.4.9   Gear error over one revolution.

3.4.5 Other Errors

From the discussion in previous sections it should be evident that robotic manipulators can experience various errors due to inaccuracies in their mechanical components.   Besides the errors due to nonlinearities, perpendicularity can affect the positioning accuracy of a manipulator.   Consider, for example, a two-axis cylindrical coordinate manipulator as shown in Figure 3.4.10.   Ignoring the errors in the drive components, if the link that positions θ is eccentric, the distance from its center of rotation to the tip of the extension arm may not be known exactly since the extension arm may be placed on either the high or low spot or anywhere in

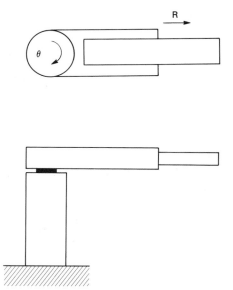

Figure 3.4.10   Schematic of two-axis cylindrical manipulator.

between.   In addition, if the beam is not perfectly perpendicular to the column, the path it traces when the column rotates is not parallel to the base but can be skewed.   As more axes and links are added, the errors tend to increase and the actual position of the tip becomes more difficult to determine in an exact fashion. Add to these problems drive component errors and sensors that have their own inherent errors and may be mounted on the shaft of an actuator instead of the actual link output, and the problem of determining the tip location is compounded even more.   Other problems that affect this situation are abuse of the manipulator (what changes if the robot accidentally hits something?) and the problem of changing a component when standard or necessary maintenance is performed.

As implied in the preceding discussion, the art of making a manipulator is a trade-off in cost, complexity, and performance with real-world problems that may be impossible to model correctly.   Despite this seemingly complex maze of tolerances and errors, the reader should understand that robots do work.   The errors are still present, but the overall device is nevertheless useful and quite reliable. By understanding the limitations of the mechanism and using it to take advantage of its capabilities, we have in a sense the triumph of human beings over real-world machines.

3.4.6 Vibrations

Mechanical vibration of a system is motion that repeats itself over a finite period of time.   Every system has associated with it a natural frequency.   The system will vibrate at its natural frequency if excited by an impulse.   These free vibrations are due to internal forces and are *transient* in nature.   Forced vibrations result when a system is acted on by an external periodic force.   These vibrations are *steady state* in nature and go away only when the forcing function is removed.

Resonance is a phenomenon that occurs when the frequency of a forcing function is the same as or near the natural frequency of the system.   This vibration can be destructive in nature and should be avoided.

Since all physical systems consist of mass (or inertia) and elasticity, these elements can be used to model vibrations that occur in all physical systems.   As will be seen in Section 3.5, mechanical systems may be modeled by lumped springs, masses (or inertias), and dampers.   By setting up an appropriate model and deriving its dynamic behavior in terms of a differential equation, the natural frequency and response to any forcing function may be obtained.

It is important to distinguish between vibrations of physical structures and the vibrations caused by the interaction of mechanical elements.   In robot design, and physical structure of the mechanism, that is, the links and mountings for the actuators and castings, must be sufficiently stiff and should not resonate at frequencies that can be excited by motions of the manipulator.   The mechanical elements (actuators, couplers, and transmissions, to name a few) themselves must be chosen to transmit the torques or forces required, and when connected together as a system should not possess resonances in the desired range of operation.

To study a structure, *finite-element* or *modal* analysis may be used. The objective of the finite-element method is to subdivide a structure into an assemblage of many smaller elements, such as beams, plates, shafts, and so on. A finite number of degrees of freedom are chosen to model the structure based on the number of these elements. The overall equations of motion of the structure are constructed from equations describing the motions of each of the individual elements, plus all the boundary conditions at the connection points between elements. These equations do not necessarily contain mass values, spring constants, and damping coefficients, which are associated with the lumped elements on the structure. Once the mathematical model is built, the equations of motion can be solved using computer methods, and various studies showing the response of the structure to stimuli can be done. Finite-element modeling has two main disadvantages: (1) a large model is extremely complex and is expensive (in both time and labor) to develop and simulate on a computer, and (2) the model can be inaccurate if the finite elements do not approximate the real-world conditions well enough. Thus even though a model is developed, it is necessary to test the structure dynamically to compare the physical system with the model.

Modal analysis (based on transfer function techniques) has become increasingly popular since the late 1970s. This technique is based on measuring the transfer functions between a single impulse point and multiple points on the structure. Modal parameters (i.e., the frequencies of vibration and their amplitude) may be obtained from these multiple measurements. Finally, a line drawing connecting the measurement points (and defining an outline of the structure) can be animated to show an accurate picture of the deflections that may occur for the various structural frequencies.

To study the dynamics due to the mechanical components associated with the motion of a joint, it is possible to use a lumped-parameter approach. In this technique (detailed in Section 3.5) one models the joint and link of a manipulator as a spring−inertial−damper system. The actual structure is lumped into one mass or inertia. This simple model is sufficient to point out potential system problems [such as resonance frequency as a function of payload (see Section 3.5.3.1)] and gives the designer feedback on the choice of component selection.

3.4.6.1 Critical shaft speed

Another concept in the study of vibrations is the critical shaft speed, defined as the speed that causes the shaft to vibrate excessively in a direction perpendicular to its center of rotation. In fact, if allowed to operate at this speed, permanent deformation or structural damage may occur. An example of this is the case of a disk where the center of gravity of the disk is different from the center of gravity of the shaft on which it is mounted, due to balancing problems (as in the case of balancing automobile tires) (see Figure 3.4.11). If the shaft and disk are rotated, the centrifugal force generated by the heavier side will be greater than that generated by the side directly opposite it. The shaft will tend to deflect toward the heavier side, causing the "center" of the disk to rotate in a small circle. This condition holds true up to a certain speed at which excessive vibrations occur. At

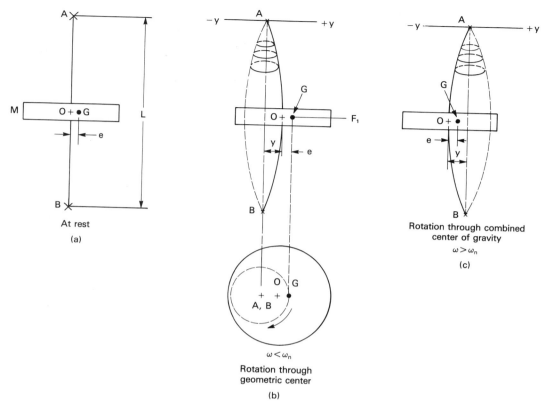

Figure 3.4.11   Critical speed rotation of disk on shaft. (From *Kinematics and Dynamics of Machines*, by George H. Martin, Copyright 1969, McGraw-Hill Book Company, NY, NY.)

this speed, the axis of rotation changes from the geometric center of the system to an axis through their combined center of gravity. The shaft itself is then deflected so that for every revolution, its geometrical center traces a circle about the center of gravity of the rotating mass. As the speed increases again, the rotation reverts back to the geometric center and the vibrations cease. This phenomenon can be thought of as resonance.

Critical speed becomes an important factor in robot design because if shafts are used to transmit torque or to allow an actuator to be placed some distance from a transmission, we must ensure that the angular velocities demanded of the actuator fall below the critical shaft speeds. The lead screw also exhibits this phenomenon. Besides material and construction parameters of the lead screw, the types of support have a direct effect on the value of the first critical frequency. Figure 3.4.12 shows four possible ways of supporting a lead screw. It can be shown that the critical frequency is a function of C_s, the "end fixability factor" (see Figure 3.4.12); D, the mean diameter of the screw; L, the distance between bearing

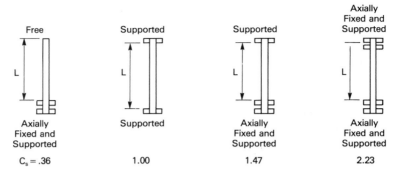

Figure 3.4.12   Bearing support of lead screw.

supports; and a constant k, which defines the material and construction properties of the screw under consideration.

$$f_c = kC_c \frac{D}{L^2} \tag{3.4.4}$$

Of course, a safety factor $(0.8f_c)$ should be included to ensure that operation is below the critical speed since this is a first order model.   Thus in applying a lead screw for prismatic joint design, it may be necessary to compromise the maximum linear speed and reflected inertia in view of the lead screw's critical shaft speed.

3.5 MODELING OF MECHANICAL SYSTEMS

The objective of modeling mechanical systems is to provide a way to express mathematically the relationships between such quantities as the input torque to a system and the physical positions of various components.   The combination of elementary mechanical components such as lumped inertias or springs results in a "mechanical network" which may be analyzed by differential equations, Laplace transforms, or simulation techniques.   In addition, electrical analogs of the variables and parameters of the mechanical network can be formed.   These analogous circuits can provide a very simple and cost-effective method of physically modeling a mechanical system without the need to fabricate mechanical components or directly solve the differential equations.   If a test signal (such as a sine wave or step function) is applied to the analogous electric circuit, the response of the mechanical network may be predicted; also, if the values of the electrical components are varied, the effects of parametric changes in the mechanical system may be observed.

3.5.1 Elements, Rules, and Nomenclature

The four basic elements that are used to model linear mechanical systems are shown in Figure 3.5.1.   The figure shows the corresponding translational and rotational components along with their force or torque equations.   From top to bottom the network elements represent the moment of inertia or mass, rotary or

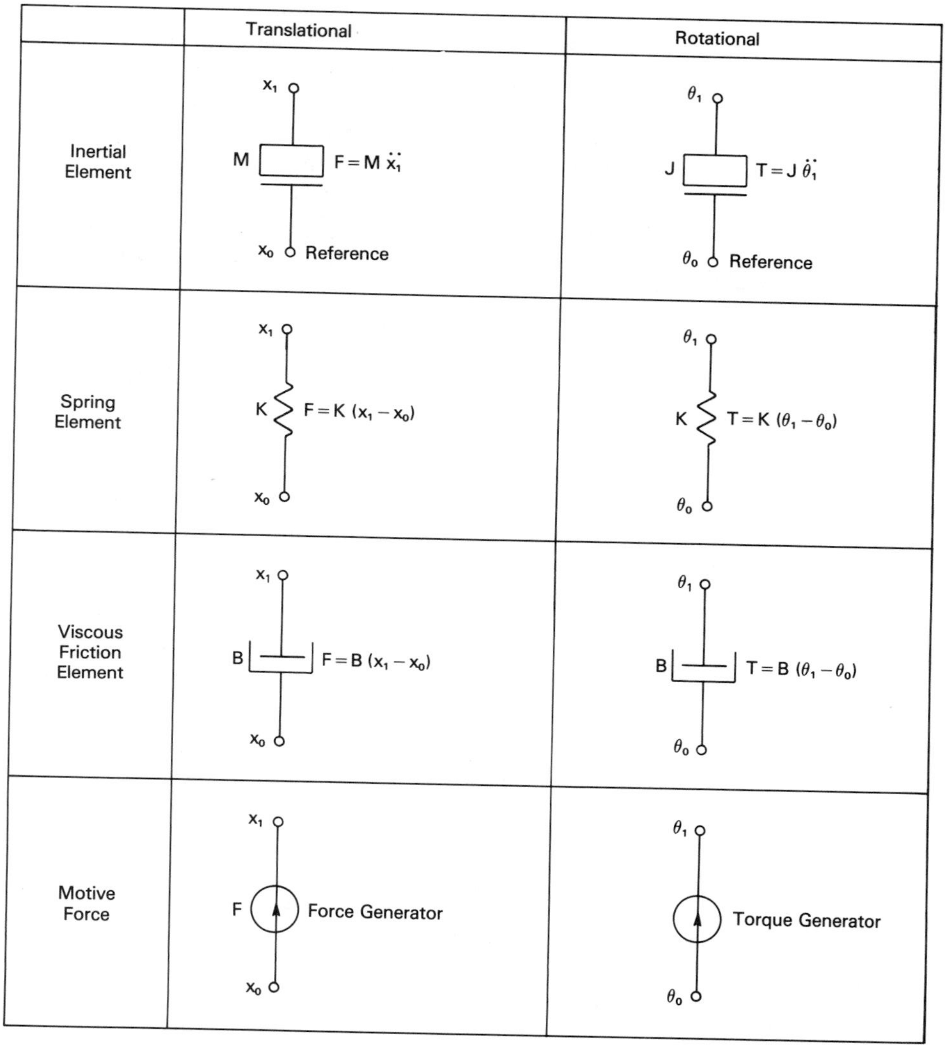

	Translational	Rotational
Inertial Element	x_1 M $\quad$ $F = M\,\ddot{x_1}$ x_0 $\circ$ Reference	θ_1 J $\quad$ $T = J\,\ddot{\theta_1}$ θ_0 $\circ$ Reference
Spring Element	x_1 K $\quad$ $F = K\,(x_1 - x_0)$ x_0	θ_1 K $\quad$ $T = K\,(\theta_1 - \theta_0)$ θ_0
Viscous Friction Element	x_1 B $\quad$ $F = B\,(x_1 - x_0)$ x_0	θ_1 B $\quad$ $T = B\,(\theta_1 - \theta_0)$ θ_0
Motive Force	x_1 F $\quad$ Force Generator x_0	θ_1 Torque Generator θ_0

Figure 3.5.1 Basic elements of a mechanical network.

linear springs, and rotary or linear dashpots.   Recall that the dashpot represents a viscous friction component.   Each of these elements is "linear," and therefore combinations of them produce a *linear system*.* The final element shown represents an independent torque or force generator which is the motive force for the network.

---

*Linear systems exhibit the properties of additivity and homogeneity which are mathematically defined as follows for the operator H, the constants α and β, and the time dependent variables $x(t)$ and $y(t)$. If $H[\alpha x(t) + \beta y(t)] = \alpha H[x(t)] + \beta H[y(t)]$ the system is linear.   Systems having Coulomb and running friction violate this relationship and are therefore nonlinear.

The two nodes of the network elements may be interpreted in the following manner. The mass of Figure 3.5.1 has the node x_1 associated with the motion of the mass, while the node x_0 is associated with the motion of the reference (note that the reference may be fixed or moving). The inertial element has node θ_1 associated with its rotational position and node θ_0 as its reference. The linear spring's nodes each represent a displacement from its equilibrium position. Thus their difference times the spring constant defines the reaction force of the spring. The torsional spring is defined similarly to the linear spring. The nodes of the dashpot (viscous friction element) represent the relative velocities of the two ends of the element.

Drawing a mechanical network simplifies the writing of a system's differential equations. Initially, nodes are defined for each position or angular displacement, with reference positions taken from the static equilibrium positions. Then the appropriate element is connected between these nodes so that the ends of the element are located at the nodes that define motion of that element. For example, inertia elements are connected from the reference node to the node representing the inertia's position, while springs and viscous elements are connected between nodes representing the position of the ends of the elements. Torque or force equations are written for each node by equating the sum of the torques (or forces) at each node to zero.

Although not considered in detail, nonlinear elements may be included in a mechanical network. Since closed-form solutions of networks containing nonlinearities may not be attainable, the most practical method is by computer simulation techniques.

3.5.2 Translational Examples

The following examples show how a mechanical network may be generated for a translational mechanical system. The examples are intended to familiarize the reader with the methodology of converting a mechanical system into a mechanical network and writing nodal equations.

EXAMPLE 3.5.1: MECHANICAL NETWORK FOR SPRING–MASS–DAMPER SYSTEM

Figure 3.5.2 shows a mechanical system consisting of a linear spring (that may be expanded or compressed) connected to a mass that has a viscous frictional component between itself and the reference. This may be considered as a simple model for a prismatic joint driven by a linear motor. The joint consists of the mass and viscous friction (due to bearings), the spring models a coupler, and the force the linear motor. Example 3.5.2 shows a more realistic model for this same system. The objective is to draw a mechanical network to facilitate the writing of the nodal equations.

The first step is to identify the nodes of the system. The reference,

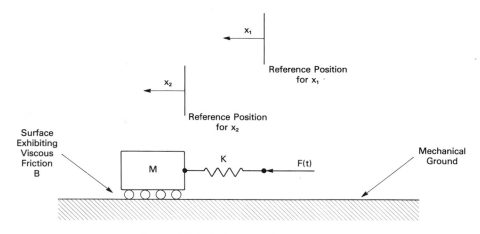

Figure 3.5.2 Spring-mass-damper system.

node x_1, and node x_2 are identified.   Recall that the nodes are associated with the positions of the elements.   We note that the spring is connected between nodes x_1 and x_2, the mass between node x_2 and the reference, since we are interested in the position of the mass with respect to the reference, and finally, the viscous friction element between node x_2 and the reference. The friction element is connected here because its friction is dependent on the relative velocity of the mass and its reference.   Figure 3.5.3 shows the corresponding mechanical network.

The node equations for the network are obtained by noting that the sum of the forces at each node must equal zero.   Thus we obtain the following two equations in the Laplace domain:

$$\text{node } x_1\text{:}\quad F(s) = K[X_1(s) - X_2(s)] \tag{3.5.1}$$

$$\text{node } x_2\text{:}\quad (s^2M + sB + K)X_2(s) - KX_1(s) = 0 \tag{3.5.2}$$

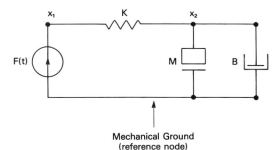

Mechanical Ground
(reference node)

Figure 3.5.3 Network for spring-mass-damper system.

Once the nodal equations are written, it is possible to combine them algebraically and write any desired transfer function or obtain a differential equation by inverting the Laplace transforms.

EXAMPLE 3.5.2: MOTION OF TWO MASSES ELASTICALLY COUPLED

Figure 3.5.4 shows two masses each having a different viscous frictional component with respect to the surface on which they are moving. The two masses are coupled together by an elastic member. This system is a more realistic model of a linear motor driving a prismatic joint. Note that the motor has a mass and viscous friction associated with it. As will be seen in Chapter 4, the force will be proportional to current, and the acceleration that can be produced will be dependent on all the components.

The coupling between the masses is modeled by a spring in parallel with a dashpot. This is a reasonable model for a physical coupler that is not purely elastic in nature (Section 3.5.3.1 expands on this topic).

The mechanical network is shown in Figure 3.5.5. Note that there are only three position nodes, including the reference. The nodal equations are defined as the following differential equations.

node x_1: $F(t) = (M_1\ddot{x}_1 + (B_1 + B_2)\dot{x}_1 + Kx_1) - Kx_2 - B_2\dot{x}_2$ (3.5.3)

node x_2: $(M_2\ddot{x}_2 + (B_2 + B_3)\dot{x}_2 + Kx_2) - Kx_1 - B_2\dot{x}_1 = 0$ (3.5.4)

Note that $F(t)$, x_1, x_2, and their derivatives are functions of time. By combining Eqs. (3.5.3) and (3.5.4) one may solve for $x_2(t) = f[F(t)]$, the position of the prismatic joint. With this information the settling time of the prismatic joint for various applied force profiles may be investigated.

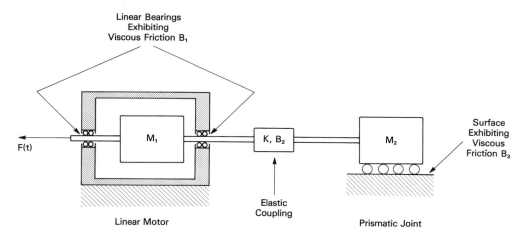

Figure 3.5.4 Linear motor driving a prismatic joint.

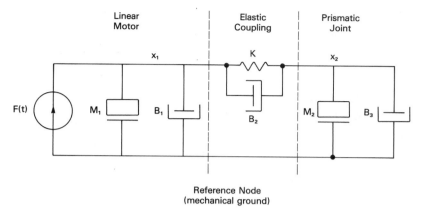

Figure 3.5.5 Network for Fig. 3.5.4.

3.5.3 Rotational Examples

The principles of network analysis are similar for rotational systems, with the exception that the nodes are now angular displacements and the nodal equations are with respect to torques. Two examples will be considered in this section. The first is a single node spring–mass–damper system, while the second is a practical model of a harmonic drive transmission.

EXAMPLE 3.5.3: ROTATIONAL SPRING–INERTIA–DAMPER SYSTEM

Figure 3.5.6 shows a simple rotational spring–inertia–damper system. This system consists of a disk connected to mechanical ground by a thin shaft. This thin shaft is a torsional spring. A viscous friction, B, acts on the disk while the applied torque, $T(t)$, also acts on the inertial disk. The reader should compare this to Figure 3.2.12, the torsional pendulum.

This topology can be used to model a stepper motor (see Chapter 4), which is settling at an equilibrium position, or a rotary joint of a manipulator. In the latter case, one can assume that the motor driving the joint is locked at some position (this is the mechanical ground in the figure), the torsional spring, K, models the stiffness of the servo loop and any couplings from the motor to the load, while the inertia disk models the load, and the viscous friction models the bearings of the system. After obtaining the system equations, one could investigate the results of a torque disturbance applied to the end of the link driven by the rotary joint in terms of the change in position, the settling time, or other parameters of interest.

Figure 3.5.7 shows the mechanical network for the system. Note that there is only one node besides the reference in the system. This is because we are interested in the position of the inertia with respect to the reference,

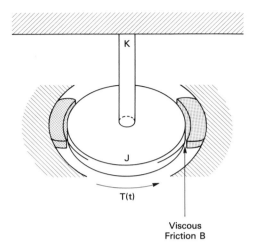

Figure 3.5.6 Rotational spring-inertia-damper system.

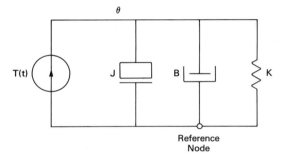

Figure 3.5.7 Network for spring-inertia-damper system.

the spring is connected to the inertia and the reference, and the viscous friction is defined by the velocity of the inertia with respect to the reference.   Summing the torques at node θ_1 we obtain the following equation in the Laplace domain:

$$\text{node } \theta_1: \quad T(s) = (s^2 J + sB + K)\theta_1(s) \tag{3.5.5}$$

EXAMPLE 3.5.4: MULTIPLE-ELEMENT ROTATIONAL SYSTEM

The following is a model of the harmonic drive component of Section 3.3.1.2 utilized to drive an inertial and viscous friction load.   This models a typical robot actuator–transmission–load.   Figure 3.5.8 shows a schematic representation of the harmonic drive and its load.   In this case the harmonic drive is represented as an ideal gear train. On its input is an inertia that is due to the wave generator, which is a component of the harmonic drive.   On the output, we note that there is a spring connecting the actual inertial load to

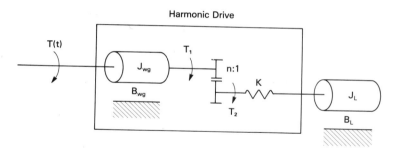

Figure 3.5.8   Multiple-element rotational system: harmonic drive with load.

the transmission.   This is the spring constant that can be observed if the input of the harmonic drive is locked and the output displacement versus applied torque is measured.   The actual harmonic drive exhibits different spring rates depending on the load torque.   However, to simplify the model we will assume a constant spring rate, thus making the spring a linear component. This is physically possible if the load torques do not exceed the values that change the spring rates.   In addition, this particular model does not include any of the nonlinear effects found in most gearing.

The mechanical network is shown in Figure 3.5.9.   Note that we have included a block labeled n between nodes θ_1 and θ_2.   This block is a torque multiplier and defines the relationship between torques T_1 and T_2.   In addition, we will also have to take into account the positional relationship between θ_1 and θ_2 as defined by the coupling ratio.   The nodal equations are given by:

$$\text{node } \theta_1: \quad T(s) = (s^2 J_{wg} + s B_{wg})\theta_1(s) + T_1(s) \tag{3.5.6}$$

$$\text{node } \theta_2: \quad T_2(s) = K[\theta_2(s) - \theta_3(s)] \tag{3.5.7}$$

$$\text{node } \theta_3: \quad K[\theta_3(s) - \theta_2(s)] + (s^2 J_L + s B_L)\theta_3(s) = 0 \tag{3.5.8}$$

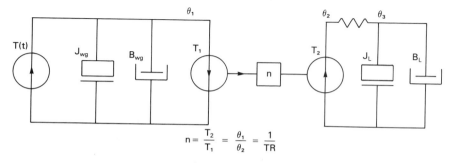

Figure 3.5.9   Mechanical network for harmonic drive with load.

The equation that relates nodes θ_1 and θ_2 is

$$n = \frac{T_2}{T_1} = \frac{\theta_1}{\theta_2} = \frac{1}{TR} \tag{3.5.9}$$

Once a set of equations describing the network [such as Eqs. (3.5.6) through (3.5.9)] are defined, it is possible to define transfer functions or solve for one variable in terms of another.

3.5.3.1 Torsional resonance

In most mechanical systems, the velocities of the moving parts are not all the same since real-world mechanical parts are elastic in nature and physical devices such as couplers may be intentionally introduced in a system. Since the magnitude and direction of the velocities of the various components may be different as a function of frequency, the system may store a large amount of mechanical energy, which results in very noticeable vibrations. This phenomenon is called torsional resonance.

The model of the harmonic drive transmission presented in Example 3.5.4 takes this phenomenon into account. The node associated with the actuator is coupled to the node associated with the load by a spring. This type of model is called a "lumped-parameter" model since we assume infinite stiffness of all shafts other than that represented by the torsional spring and all inertias are concentrated at specific places. Many complex systems may be analyzed by these techniques; however, when distributed parameters are assumed to be lumped, the validity of the model must be assessed.

In general, the torsional resonance phenomenon appears anywhere two inertial loads are coupled by a resilient member, as shown in Figure 3.5.10. Thus the use of a coupler in a mechanical system can lead to resonance, and, as indicated, most gear trains or transmissions will have a finite stiffness associated with their outputs. This stiffness may be due to an antibacklash arrangement or something similar to the flexible cup characteristic of the harmonic drive.

Torsional resonance becomes quite important in the design of high performance (i.e., high bandwidth, fast, accurate) servo loops associated with the parameters required by robot manipulator systems. It is a well-established fact that if the mechanical resonance frequency occurs inside or near the servo bandwidth, the loop's stability is degraded even to the point of sustained oscillation. To eliminate this problem, designers may specify extremely stiff mechanical components so as to ensure that the resonance is outside the required servo bandwidth, or if the performance parameters may still be met, the servo bandwidth may be set well below the resonance frequency of the mechanical structure.

To further explain the torsional resonance phenomenon, we will develop a simple model for a two-mass–spring system. A network used to model the system of Figure 3.5.10 is shown in Figure 3.5.11. The model consists of a torque source,

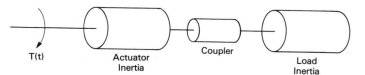

Figure 3.5.10   System that will exhibit torsional resonance phenomenon.

an inertia J_a, which is part of the actuator, the inertia J_L, which is the load, and a spring in parallel with a dashpot to model a resilient coupler. The dashpot element of the coupler represents the inherent molecular damping in the spring material, while the spring component represents the finite stiffness of the coupler. Typically, the molecular damping of the coupler is much larger than that associated with the external viscous friction of most drives. This damping term controls the shape of the resonance response.

By the analysis techniques presented in the preceding section, the transfer function of the actuator's shaft velocity, $\Omega_a(s)$, to input torque is found to be

$$\frac{\Omega_a(s)}{T(s)} = \frac{s^2 J_L + sB + K}{s[s^2 J_a J_L + sB(J_a + J_L) + K(J_a + J_L)]} \qquad (3.5.10)$$

Equation (3.5.10) includes the effects of loading at node θ_a due to the coupling and the load inertia. Note that if the coupling had been assumed to be perfectly rigid, the network would consist of the torque source across the two inertial loads and the corresponding transfer function would be given by

$$\frac{\Omega_a(s)}{T(s)} = \frac{1}{s(J_a + J_L)} \qquad (3.5.11)$$

A careful examination of Eq. (3.5.10) reveals that the numerator consists of two zeros which are typically complex conjugates, and the denominator consists of three poles: a single pole at the origin and a set of complex conjugate poles. Figure 3.5.12 shows a Bode plot for Eq. (3.5.10). Note the characteristic -20

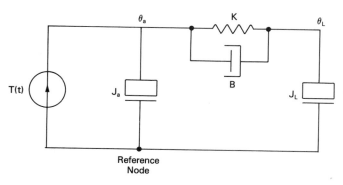

Figure 3.5.11   Mechanical network for torsional resonance phenomenon.

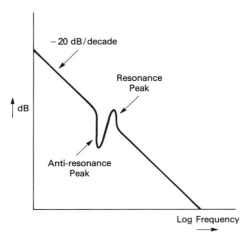

Figure 3.5.12 Bode plot for Eq. (3.5.10).

dB/decade slope until the sharp notch occurs. This notch is called an *"antireso-nance"* and is due to the complex conjugate zeros. The peak following the anti-resonance is the *"resonance peak"* and is due to the complex poles. The resonance is more damped than the antiresonance and always occurs at a higher frequency.

Algebraic manipulation of Eq. 3.5.10 yields

$$\frac{\Omega_a(s)}{T(s)} = \frac{s^2 + (B/J_L)s + (K/J_L)}{sJ_a\{s^2 + sB[(J_a + J_L)/J_aJ_L] + K[(J_a + J_L)/J_aJ_L]\}} \qquad (3.5.12)$$

The constant term in the numerator quadratic of Eq. (3.5.12) corresponds to the square of the antiresonance frequency. Note that this is an undamped frequency. This is the same frequency of oscillation that will occur if the load is disturbed with the actuator locked. Referring to Figure 3.5.11, this situation is modeled as a short to the reference node from both the torque generator and actuator inertia. This is sometimes referred to as the *locked-rotor resonant frequency* and can easily be measured in a physical system. Of course, the frequency measured in a physical system will include the effects of damping. Similarly, the denominator quadratic's constant term corresponds to the square of the resonance frequency which is also the "free rotor and load resonant frequency." Once again a physical measurement will yield a damped frequency, but knowledge of the envelope allows one to extract the undamped natural frequency. In addition, it can be shown (see [2]) that the damping ratio for the antiresonance is significantly lower than the damping ratio of the quadratic factor corresponding to the resonance. Additional information on this topic can be found in [2], [7], [16] and [18].

A common problem in robotic design is that the payload may be variable; that is, a robot may be capable of carrying a payload of up to say 25 lb located some distance from its tool plate. If one considers the variations in inertia from a no-load to a fully loaded condition, the effect on the bandwidth may be staggering. In addition, antiresonance and resonance frequencies may move considerably.

TABLE 3.5.1   MECHANICAL AND ELECTRICAL ANALOGS

Translational		Rotational		Electrical	
F	Force	T	Torque	i	Current
$\dot{x}$	Velocity	ω	Angular velocity	v	Voltage
M	Mass	J	Moment of inertia	C	Capacitance
K	Linear spring constant	K	Torsion spring constant	$\dfrac{1}{L}$	Reciprocal inductance
B	Linear viscous friction constant	B	Rotational viscous friction constant	$\dfrac{1}{R} = G$	Conductance

3.5.4 Electrical Analogs

As stated previously, an analogy can be made between the components of a mechanical network and an electrical network. Force and torque are analogs of current, while velocity is the analog of voltage. The nodes of the mechanical networks that were analyzed corresponded to linear or angular displacements. Therefore, to make an electrical analog, one must write the mechanical equations in terms of node velocities. This is easily done by including a derivative operator with each node variable and factoring it out of the node's coefficient. Table 3.5.1 shows the analogs between translational or rotational mechanical elements and electrical components.

A transformer may be considered the electrical analog of a gear train with angular velocity and torque analogous to voltage and current, respectively.

To use the electrical analogy for analyzing a mechanical network all one has to do is make the substitutions indicated in Table 3.5.1 and then analyze the circuit. As should be evident from the table, the electrical components are considered as admittances (which in fact simplifies nodal analysis).

It is possible that the values of the electrical components that may result from the transformations may be impractical or impossible to obtain. This presents no problem if the network is to be analyzed by hand or a network analysis program such as PCAP or SPICE. In the case that one wishes to build the corresponding network physically, scaling techniques can be used to bring values into a reasonable range.

The following example illustrates the concept of a mechanical and electrical analog.

EXAMPLE 3.5.5: MECHANICAL–ELECTRICAL ANALOG

Draw the electrical analog for the mechanical network shown in Figure 3.5.5. The network elements of mass, viscous friction, and spring rate are replaced by capacitors, resistors, and inductors. The values of the electrical components are given by their admittance and equated to the cor-

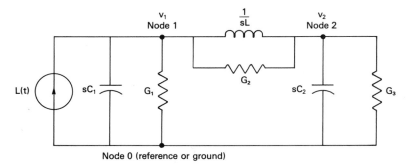

Figure 3.5.13   Electrical analog for mechanical network of Fig. 3.5.5.

responding mechanical values.   The resultant electrical network is shown in Figure 3.5.13.

The node equations for the electrical network are given as:

$$\text{node 1:}\quad I(s) = \left[sC_1 + (G_1 + G_2) + \frac{1}{sL} \right] V_1(s) - \left(\frac{1}{sL} + G_2 \right) V_2(s) \qquad (3.5.13)$$

$$\text{node 2:}\quad \left[sC_2 + (G_2 + G_3) + \frac{1}{sL} \right] V_2(s) = \left(\frac{1}{sL} + G_2 \right) V_1(s) \qquad (3.5.14)$$

The corresponding nodal equations for the mechanical network written in terms of node velocities ($V_1(s)$ and $V_2(s)$ corresponding to nodes X_1 and X_2) are given by:

$$\text{node } X_1:\quad F(s) = \left[sM_1 + (B_1 + B_2) + \frac{K}{s} \right] V_1(s) - \left(\frac{K}{s} + B_2 \right) V_2(s) \qquad (3.5.15)$$

$$\text{node } X_2:\quad \left[sM_2 + (B_2 + B_3) + \frac{K}{s} \right] V_2(s) = \left(\frac{K}{s} + B_2 \right) V_1(s) \qquad (3.5.16)$$

A comparison of Eqs. (3.5.13) and (3.5.14) with Eqs. (3.5.15) and (3.5.16) shows the relationship between the electrical and mechanical variables.

3.6 KINEMATIC CHAINS: THE MANIPULATOR

Kinematics is the study of motion without regard to forces or other factors that influence the motion.   So far in this chapter we have been focusing on individual components and the mechanical concepts and considerations that are required to design or analyze a mechanical system in terms of dynamics.   This section focuses briefly on the configuration that a manipulator may take and attempts to unite some of the concepts presented thus far so that the reader can see how everything fits together.

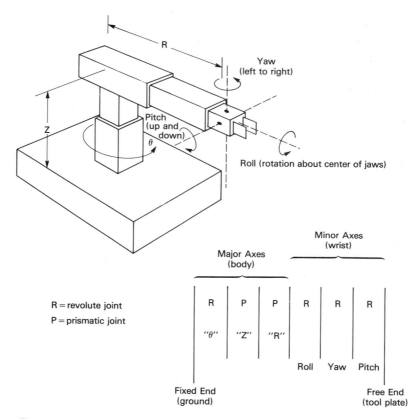

Figure 3.6.1   Joint classification of a cylindrical coordinate robot having a wrist with three rotary joints.

An industrial robot may be thought to consist of a group of "rigid bodies" called links connected together by joints. The links are interconnected such that they are forced to move relative to one another in order to position the end effector. The two types of joints used in commercial manipulators are the revolute and prismatic joints. Revolute joints allow pure rotation of one link about the joint axis of the preceding link, while prismatic joints allow a pure translation between links. Actuators are included in the mechanism to enable the motion of the joints; see Figure 8.6.1.

The links of a manipulator are connected to no more than two others (via joints) so that closed loops are not formed. The links and joints of a manipulator form a kinematic chain which is open at one end and connected to ground at the other. The end effector or hand is connected to the free end and the control objective of the robot system is to position the end effector at a desired location in space.

One way of classifying a robot is by defining the type and order of joints. Using this classification, a cylindrical robot having a wrist capable of rotation in three degrees would be designated RPPRRR.   Figure 3.6.1 shows this classification and identifies the fixed and free ends of the manipulator.

The classification of Figure 3.6.1 says nothing about how the joints are connected together: Are they parallel or perpendicular; do their axes intersect? An additional classification based on the link parameters discussed fully in Chapter 8 clarifies this point.   However, for our current discussion it is sufficient to say that the axes of the joints are either perpendicular or parallel to one another. In addition, typically, the first three joints position the end effector in space while the last three joints are used to orient the tool.   An obvious physical consideration for the configuration of a manipulator is that if the first three joints are to position the end effector in space, we would like them to be able to move in three-dimensional space.   Therefore, configurations consisting of joints that do not permit three degrees of freedom are not used except in the case of special manipulators such as a *SCARA*, which intentionally has only two degrees of freedom by its first two joints (major linkage) and depends on rigidity in the z or vertical direction as one of its unique characteristics.   Recall that the degrees of freedom of a system depend on the number of variables or coordinates that are needed to describe its position.   If one considers the possible combinations of R and P joints for the first three links and eliminates any that produce redundant motions, there are only 12 useful distinct configurations

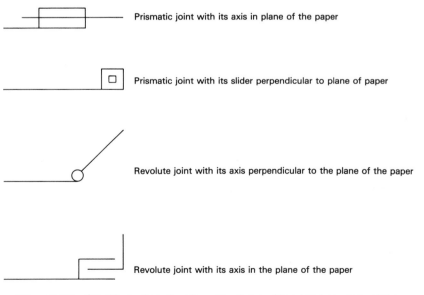

Figure 3.6.2   Graphic symbols for kinematic chains. (Reprinted courtesy of the Society of Manufacturing Engineers. Copyright © 1983, from the ROBOTS 7/13th ISIR Conference Proceedings.)

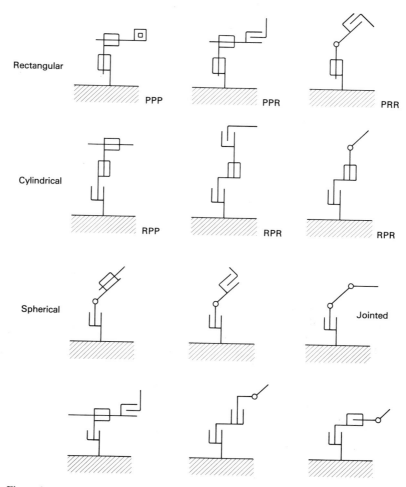

Figure 3.6.3 Graphic representation of 12 useful major linkages. (Reprinted courtesy of the Society of Manufacturing Engineers. Copyright © 1983, from the ROBOTS 7/13th ISIR Conference Proceedings.)

out of a total of 36 [11].   Figure 3.6.2 shows graphical symbols used to depict joints, and Figure 3.6.3 shows the 12 useful major linkages.

Some possible configurations for the wrists are shown in Figure 3.6.4.   Note that it is possible to have less than three degrees of freedom in a wrist.   In fact, many robot applications may be performed satisfactorily with considerably less than six degrees of freedom (or joints).

The kinematic structure defines a "workspace" or "work volume" in which the robot can position itself.   Some kinematic configurations are more suited than others for particular tasks.   Figure 3.6.5 shows the workspaces which may be

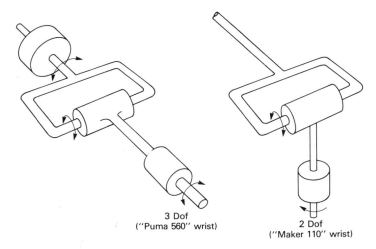

3 Dof
("Puma 560" wrist)

2 Dof
("Maker 110" wrist)

Figure 3.6.4 Some possible wrist configurations.

accessed by the major joints for some kinematic configurations. The possibility exists that some points in the work volume may be addressable by more than one set of joint configurations; this is typical in a jointed arm such as the PUMA. With the addition of the wrist axes, we add more degrees of freedom and now besides being able to position the end effector at a point in space we have the ability to control its orientation. Besides multiple addressing of points, some points and orientations in a robot's work volume may be unachievable. Ideally, if the manipulator is able to position itself to an arbitrary point in space, the tool should be able to be oriented along any radius pointing inward to an imaginary sphere encircling the point. Unfortunately, due to physical limitations of the ranges of joints and problems such as joints not being able to pass through themselves, the ideal case is not attainable.

In addition, it is important to note that the control strategy of a robot may require complex motions in three-dimensional space, and that some kinematic configurations may have easier Cartesian-to-joint solutions than others. Finally, a major consideration for servo response is the stiffness of the entire structure. Prismatic joints are inherently stiffer than rotary but may not be able to give the manipulator the dexterity it needs.

By now the reader should be able to visualize how the devices discussed in Section 3.3 can be used to implement rotary and prismatic joints. Of course, good mechanical design techniques must be used to ensure that the components can handle the loads due to the payload and forces exerted by the mechanism. The kinematic formulation does not consider the weights or inertias of the links and the actuators for the joints that may be housed inside the links. In addition, no link is perfectly rigid, nor do all joints behave ideally. Thus the robot manipulator must be analyzed in terms of its dynamic properties and component nonlinearities as well as its kinematic properties.

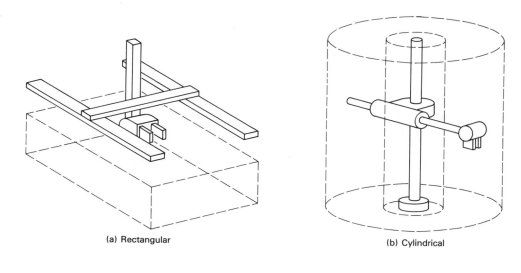

(a) Rectangular   (b) Cylindrical

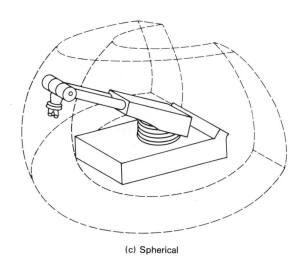

(c) Spherical

Figure 3.6.5   Workspaces for some kinematic configurations.

3.7 END EFFECTORS

An end effector is the term used to describe the tool or device attached to the end of the wrist of a manipulator.   The end effector itself may be the complete payload of the robot, or it may be a mechanism used to hold one or more parts.   As stated previously, one of the major objectives of robot control is to position the end effector at some desired location in space by controlling the position of each of the manipulator's joints.

End effectors range from commercial devices such as pneumatic grippers to special tools for applications ranging from arc welding and spray painting to the handling of silicon wafers in a corrosive acid environment. Due to the diverse applications of robots, end effectors are usually customized for a particular application and may include additional components such as force measuring devices or mechanical linkages that can be locked in various positions. They may be multifunctional; that is, the same tool may be used to handle a part as it changes shape during a machining or assembly process. It is also not uncommon for a gripper to be designed for multiple-part pickup in lieu of a single part. This may actually be necessary to cost justify using a robot if the cycle time of manipulating a single part does not provide the needed throughput.

When considering the design of the end-of-arm tooling for a manipulator, certain parameters of the manipulator and the surrounding environment of the workcell must also be addressed. The end effector itself has weight and moments of inertia; thus these properties, along with the weight and shape of the payload, must be considered to ensure that the specifications of the robot are not violated.

The configuration of the parts being handled is quite important. It is imperative that the parts be consistent (within their tolerance range). It is really not fair to expect a robot end effector to be capable of handling anything that comes in front of it. If it was designed to pick up cubes with sides of 1 in., it may not be capable of working with cubes ½ in. on a side. Since the piece parts on which the gripper must operate may vary (based on acceptable manufacturing tolerances), the design of the end effector should be capable of functioning continuously within the normal level of variations encountered in a manufacturing environment.

The end-of-arm tool not only has to interface to the robot, but must not interfere with any of the peripheral devices in the workspace. It may be necessary to orient the end effector a certain way when using it to insert a part into a machine. In fact, it may not be possible to have the end effector mounted directly on the tool plate of the robot since the opening to the machine may not be large enough to accommodate the actual end of the robot. In this case, very long thin fingers may be used to hold the part so that the end of the robot and the actual mechanical gripping mechanism do not enter the machine. Of course, one must ensure that the use of long fingers to hold a part does not introduce moments beyond the specification of the robot.

When robots are used for such operations as insertion, that is, joining parts with close mating tolerances, accuracy becomes a very important parameter. This is sometimes referred to as the peg-in-hole problem. If the parts are off-center they will jam, and one can easily visualize that it is impossible to insert one into the other. Even putting a chamfer on one or both parts may not work unless there is some compliance between the two parts. Compliance may actually be provided by the manipulator itself since it is not infinitely stiff; however, if this is still not sufficient, a *remote center compliance* (RCC) device may be used between the tool plate and end effector. The RCC provides the necessary give or wrist action to allow the close-fitting parts to be mated. This device is described in detail in Section 3.7.2.

In today's automated factories, reliability is of paramount importance. Since reliability decreases with increasing complexity, it is better to try to keep the end effector simple in both its design and function. One may be able to design a gripper to perform multiple tasks; however, this may be expensive to design or purchase and may prove costly to maintain. If possible, modification of the parts or process may actually simplify the requirements of the end effector, making a simpler and more reliable system.

In addition to gripping a part, the end effector can include a sensor to determine if a part is present. The addition of a simple sensor can make a gripper a relatively intelligent device. Consider, for example, a simple gripper that has a sensor in it which tells if there is something between its jaws. This could be as simple as a light and phototransistor. If the robot is commanded to go and get a part, the manipulator will position the tool to the correct spatial location and then check the gripper's sensor before closing the gripper. If the part is present, the gripper is commanded to close and the cycle proceeds; however, if the part is not present, the robot can then take some predefined action based on this exception and perhaps prevent other machinery from cycling without a part.

More sophisticated end effectors are also possible that include such features as servo-controlled axes with force control, and sensors to measure the dimensions or weight of an object. Chapter 5 discusses these sensors in detail. In the following sections we explore some gripper configurations and the remote-center compliance device.

3.7.1 The Gripping Problem

There are essentially three classifications of grippers [9]:

1. Those that come in contact with only one face of the object to be lifted and use a method such as vacuum, magnetism, or adhesive action to capture the object.

2. Those that use two rigid fingers to grip an object. This type makes contact with the object at two specific points and may or may not deform the object.

3. Those that deform and attempt to increase the contact area between the gripper and object. This type includes multijointed fingers or a device operating on a principle similar to a balloon being inflated inside a glass.

These three classifications are referred to as systems using *unilateral action*, *bilateral action*, or *multilateral action* [9]. Figures 3.7.1 through 3.7.3 show examples of each of these systems.

Of course, certain considerations must be made in using each of these devices.

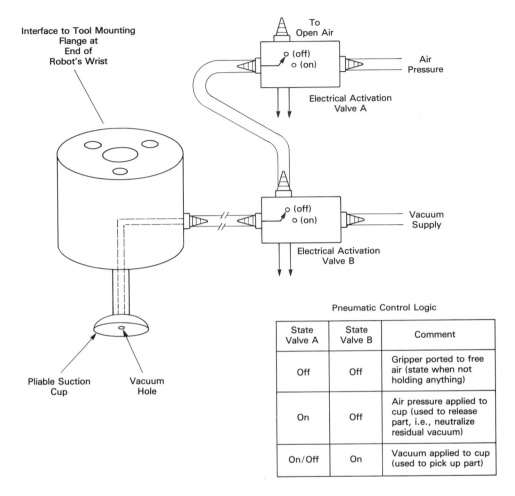

Figure 3.7.1 End effectors with unilateral gripping action.

In the case of unilateral action, the end effector must also be capable of releasing the part as well as picking it up. If a vacuum tool is used, the vacuum may have to be purged with air to release the part; otherwise, if the seal between the part and vacuum cup is sufficient, the part may stay on the tool even when the vacuum is shut off until some external force overcomes the residual vacuum. Similar considerations exist when using an adhesive or magnetic approach. In fact, it may be necessary to have another device "kick" the part from the pickup tool.

For bilateral action, it may be necessary to include a material such as a piece of rubber on both points of contact to increase the coefficient of friction; otherwise,

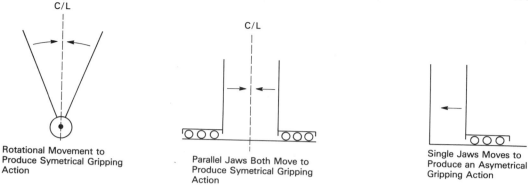

Figure 3.7.2 End effectors with bilateral gripping action.

the part may fall out of the gripper. The implementation of bilateral systems can invoke such devices as rack and pinions or various mechanical linkages to achieve the required motions. Typically, bilateral devices are binary in nature, that is, either opened or closed. For this simple case, a pneumatic actuator is commonly used. The more sophisticated case of a parallel-jaw device capable of closing with variable force or at defined distances would probably make use of such technology as a servo-controlled rack-and-pinion device (see Figure 5.10.1).

Implementation of multilateral devices uses such technology as pneumatics to inflate and deflate long thin bags which can articulate a joint, wires that stretch or compress as a function of electric current running through them, or very complicated mechanical systems. This particular type of gripper is the subject of current research, but to date no commercially feasible "three-fingered" gripper is available.

An interesting implementation of a multilateral gripper is shown in Figure 3.7.4. This particular system uses a fluid-filled chamber to force the outer layer to conform to the part in the jaws. Although the multilateral system looks very inviting, one must also consider some of the problems associated with the control of these types of devices. For example, if we consider an implementation similar to that of Figure 3.7.4, one apparent observation is that if the device does not "inflate" uniformly, it can cause the part it is attempting to capture to rotate. This may be an undesirable side effect. This simple thought experiment makes one realize the sophistication of the human being who is capable of picking up an irregular object and not causing it to rotate.

3.7.2 Remote Center Compliance Devices

Remote center compliance devices are commonly used for applications requiring the assembly of close-fitting parts, for example, the insertion of a peg into a hole. A little thought can show how difficult this so-called simple task can be. If the

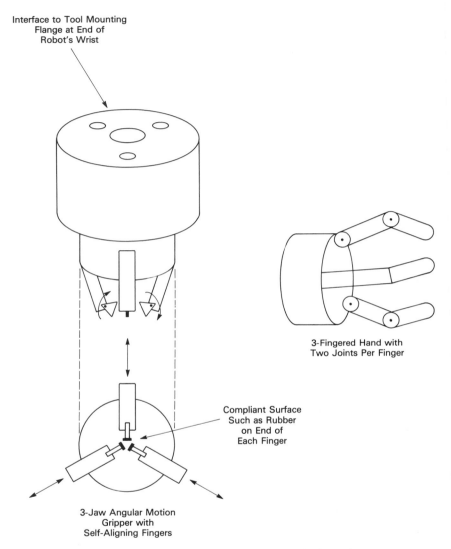

Interface to Tool Mounting
Flange at End of
Robot's Wrist

3-Fingered Hand with
Two Joints Per Finger

Compliant Surface
Such as Rubber
on End of
Each Finger

3-Jaw Angular Motion
Gripper with
Self-Aligning Fingers

Figure 3.7.3   End effectors with multilateral gripping action.

centerlines of the peg and hole are not coaxial, the parts will not mate.   Two
distinct possibilities arise.   The first is the case of the centerlines being parallel
but not coaxial.   This is a translational misalignment.   The second is the case
where the centerlines intersect but are not parallel.   This is a rotational misalignment.   Of course, both translational and rotational misalignments can occur simultaneously.

Consider that we have a rotational misalignment (i.e., a peg goes partially

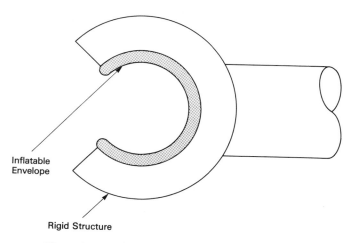

Inflatable
Envelope

Rigid Structure

Figure 3.7.4 Implementation of a multilateral gripper.

into the hole).   If a force is used to attempt to push the peg into the hole, jamming may occur.   If excessive force is used, something will have to give and it is possible for damage to the parts or the robot to occur.   Figure 3.7.5a shows the case of a peg being pushed into a hole.   Figure 3.7.5b shows one way of solving the peg-

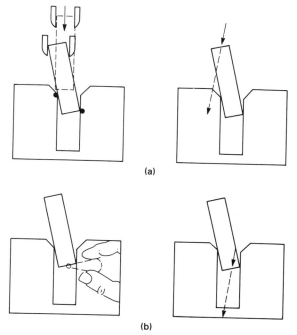

Figure 3.7.5: (a) peg being pushed into hole; (b) peg being pulled into hole. (Courtesy of J. Rebman and Lord Corporation, Cary, NC.)

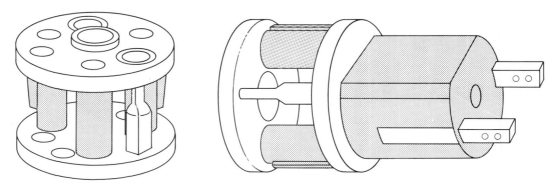

Figure 3.7.6   A remote center compliance device.

in-hole problem; this involves "pulling" the peg into the hole rather than pushing. The pulling allows the peg to rotate and align itself rather than jam against the hole, as in the case of pushing.

Remote center compliance devices (RCCs) solve the peg-in-hole problem by effectively causing the peg to be pulled into the hole, as depicted in Figure 3.7.5b. The RCC device provides the ability to compensate for both translational and rotational errors.   The rotation is about a point called the *center of compliance* or *elastic center*.   This center is a point that is remote from the unit.   Figure 3.7.6 shows a typical RCC device.   This device consists of two parallel plates which are separated by two rigid rods firmly attached to one plate but with a ball joint on the other.   In addition, three elastic members are also placed between the plates which keep them separated and parallel.   The rods and plates are arranged so that one plate is fixed and the other one has limited rotation and deflection.

Figure 3.7.7 shows the result of using an RCC device to insert a peg into a chamfered hole.   The compliant center is located at a point on the peg since by definition it is remote from the RCC device itself.   In the case of lateral error (as shown in Figure 3.7.7a), the axial force exerted by the peg on the chamfer causes a lateral force to be applied to the peg.   This force applied to the elastic center will cause only a translational motion.   This horizontal force causes the peg to translate into the hole by causing the RCC device's bottom plate to move parallel to the top plate.   If the axis of the hole is not parallel to that of the peg as shown in Figure 3.7.7b, the peg will partially go into the hole and jam.   Two points of contact are made, one by the leading edge of the peg and the other by the edge of the chamfered hole.   These two points of contact define forces which in turn generate moments about the compliant center.   A rotation about this compliant center will cause the peg to rotate and line up with the hole.   In this case the plates of the RCC device are no longer parallel.

The remote center compliance devices that we considered are passive in nature.   However, it should be noted that it is possible, although perhaps not cost-effective, to implement this concept using servos, sensors, and some type of mechanism.

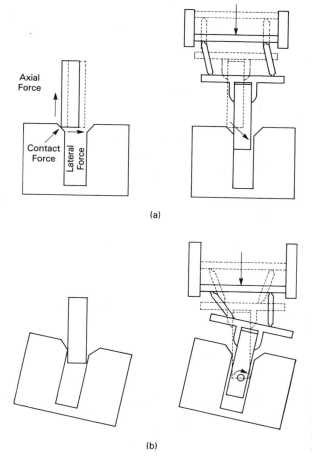

(a)

(b)

Figure 3.7.7 RCC for inserting a peg into chamfered hole: (a) insertion with laterial error; (b) insertion with rotational error. (Courtesy of J. Rebman and Lord Corporation, Cary, NC.)

3.8 RESOLUTION, REPEATABILITY, AND ACCURACY OF A MANIPULATOR

The terms *resolution, repeatability,* and *accuracy* are used by robot manufacturers to define the positioning capability of their manipulators. Unfortunately, there is no standard test that all robot manufacturers perform to come up with these numbers. This problem arises from the fact that repeatability and accuracy depend on a great many variables, including payload, velocity, temperature, direction of travel, stiffness of the arm, and so on. Additionally, does the specification hold for a single joint's motion or for a multiple joint move? Was the measurement in three dimensions or one? What type of sensor was used to obtain the data: one that contacted the robot and perhaps affected its performance or a noncontacting sensor? Is resolution defined for joint motions or Cartesian motions? Is the resolution the same over the entire workspace?

By now the reader should have an appreciation of some of the mechanical effects that can cause differences in robot positioning such as mechanical deflections and the nonlinearities of gearing, to name a few. These combined with sensor error and computations can greatly affect the positioning ability of the tip of the robot. The definitions we will use for accuracy, repeatability, and resolution are based on [17] and provide a good model to account for contributions from mechanical, sensor, and computational effects.

To begin our discussion, consider a mechanism that moves in one dimension parallel to the paper. Control resolution will be defined as the smallest incremental change that the control system (usually a servo) can distinguish. In addition, we assume completely ideal conditions, such as no deadband, computational problems, or sensor errors. Figure 3.8.1 shows a series of equally spaced points representing where our mechanism may be commanded to go; the distance between these points is the control resolution. The control resolution can be computed by dividing the total distance that can be traveled by the total number of discrete positions that the mechanism can resolve. If we include the effects of mechanical inaccuracies or sensor errors, we find that instead of the mechanism stopping on each point associated with the control resolution, there is a zone about the ideal point where it may stop. This has been shown as a symmetrical zone in Figure 3.8.1; however, depending on the implementation of the mechanism, it may be biased more in one direction than in the other. Spatial resolution is defined as the worst-case distance between two adjacent positions, as illustrated in the figure. Resolution is important for both record-and-playback applications and off-line programming. Essentially, it defines the ability of the manipulator to be able to reach positions close enough during initial training or defines the smallest move that the robot can make, which affects the ability of the manipulator to move in nonjoint spaces such as Cartesian coordinates.

Accuracy is a measure of the ability of a manipulator to approach an arbitrary point in space, previously never approached by the manipulator. This concept is extremely important for off-line programming, where the coordinates that define the position that the robot is to move to are obtained from a data base. Figure

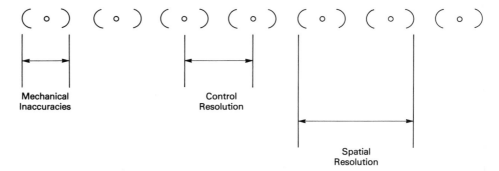

Figure 3.8.1   One-dimensional representation of control and spatial resolution.

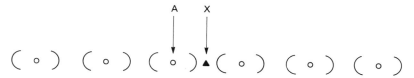

Figure 3.8.2 One-dimensional representation of accuracy and resolution.

3.8.2 shows an arbitrary position, X in space, along with the control resolution positions and the mechanical inaccuracy about each position. If the mechanism is commanded to go to the position designated by X, it will go to the position closest to X as defined by the control resolution, point A in Figure 3.8.2. However, since there are mechanical inaccuracies in the system, the actual location of the mechanism is somewhere in the "zone" about A. Accuracy is then taken as the worst-case distance from the arbitrary point to where the manipulator may be positioned. Examination of Figure 3.8.2 indicates that if one were to define the worst-case value for accuracy, it is equal to one-half the spatial resolution. Obviously, some points would have a smaller value, but this figure is the worst that the manipulator could be off from a commanded position and should be the accuracy figure quoted in the specifications.

Accuracy is usually divided into local and global. *Global accuracy* defines the accuracy over the entire workspace of the robot, while *local accuracy* is concerned with accuracy in the neighborhood of specified points. One of the contributing factors to global accuracy is axis misalignment. The kinematic model assumes that axes are positioned ideally; however, the manufacturing process includes tolerances for mechanical assemblies, and therefore axes thought to be perpendicular may in fact be off and cause considerable errors in certain regions of the workspace.

Repeatability is defined as the ability of a manipulator to reposition itself at a position to which it was previously commanded. Repeatability is dependent on many factors; by the strictest definition, measurements must be made with the same payload, velocity, acceleration, direction of approach, and ambient temperature. The repeatability specification determines whether the manipulator will be able to reach previously demonstrated positions close enough to do the job the second and succeeding times during a repetitive operation. The initial position may be demonstrated, or obtained from a data base, but in either case, the manipulator's control will associate a control resolution point with the position. Remember that due to mechanical inaccuracies, the mechanism will position itself somewhere in the zone associated with the point. If the arm is moved away and then commanded to go back to this location, it will go to the same position or another position in the zone associated with the control point. The distance from the initial position of the manipulator to the position it attained on the second try is a measure of its repeatability. Figure 3.8.3 illustrates this definition.

If one were to command the manipulator to return to a particular point a

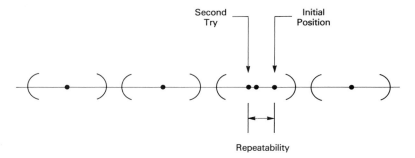

Figure 3.8.3 One-dimensional representation of repeatability.

great many times and plot the position that the manipulator attained on each try, you would obtain a locus of points that fell within the mechanical inaccuracy zone. This is a measure of the repeatability of the manipulator. One way of defining repeatability is to bracket the hits on either side of the zone and measure this distance. Note that the locus of hits may be shorter than the distance defining the mechanical inaccuracy zone. Half of the bracketed distance is a measure of the repeatability.

Long-term repeatability is concerned with the repeatability of the manipulator over a large time frame (on the order of months). In general, long-term repeatability is influenced by factors such as mechanical wear.

Short-term repeatability is associated with changes that occur over a short time, typically hours or shifts. It is concerned with the changes in performance as the robot "warms up" or as the ambient temperature changes during the day.

The National Machine Tool Builders' Association (NMTBA), the association for manufacturing technology, has released definitions for both linear and rotary accuracy and repeatability. To provide standardization, these are based on a standard temperature of 68°F and are specified for only one axis at a time. The NMTBA definitions are based on statistics and are equivalent to the definitions mentioned above. The definitions of unidirectional and bidirectional repeatability are given as follows:

1. *Unidirectional repeatability* shall be defined as the expected dispersion on each side of the mean resulting from a series of trials when approaching any given point under the same conditions.
2. *Bidirectional repeatability* shall be defined as the expected dispersion on each side of the mean resulting when the approach to any given point is programmed from both directions in a series of trials.

Figure 3.8.4 shows the unidirectional repeatability, while Figure 3.8.5 shows a possible definition of bidirectional repeatability. Typically, the number of trials at any given point is taken as seven in the NMTBA formulas. In addition, the

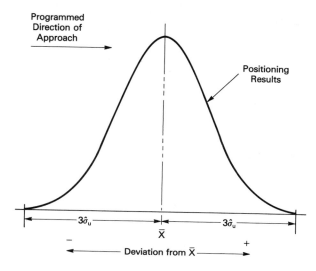

Figure 3.8.4 NMTBA definition of unidirectional repeatability. (Reproduced with the permission of NMTBA, the association for manufacturing technology.)

dispersion of 3σ takes into account 99.74% of the data assuming a Gaussian distribution.

The NMTBA definition of accuracy at a point is defined as being the sum of the signed value of the difference between the mean and the target at any point plus the value of the dispersion at that point which gives the largest absolute sum. Figure 3.8.6 illustrates this definition.   If enough data points for both accuracy and repeatability along a given axis of motion are obtained, a plot can be made showing the distance along an axis versus the accuracy at each point.   Figure 3.8.7

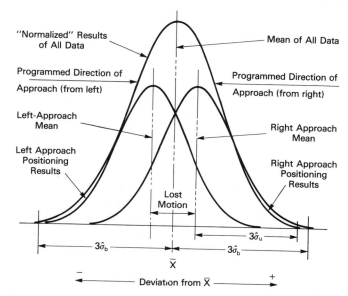

Figure 3.8.5 NMTBA definition of bidirectional repeatability. (Reproduced with the permission of NMTBA, the association for manufacturing technology.)

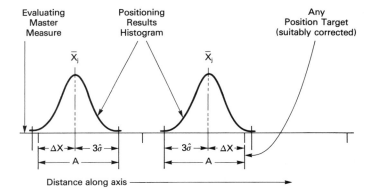

$$A_b = \Delta X_b \pm 3\hat{\sigma}_b$$
$$A_u = \Delta X_u \pm 3\hat{\sigma}_u$$

Definitions are as follows:

A_b = accuracy value at any point when using a programmed bidirectional approach

A_u = accuracy value at any point when using a programmed unidirectional approach

ΔX_b = difference between mean and target (a perfect master measure) when using a programmed bidirectional approach; a signed number

ΔX_u = difference between mean and target when using a programmed unidirectional approach; a signed number

$3\sigma_b$ = expected dispersion on each side of mean when using a programmed bidirectional approach

$3\sigma_u$ = expected dispersion on each side of mean when using a programmed unidirectional approach

NOTE: Length units must be consistent, i.e., inches, thousandths, etc.

In the event the choice of unidirectional or bidirectional programmed approach is not stated, the unidirectional approach is assumed to be specified.

Figure 3.8.6 NMTBA definition of accuracy. (Reproduced with the permission of NMTBA, the association for manufacturing technology.)

shows a plot of the position of an axis versus the accuracy at any given point. The envelope defines the worst-case system accuracy, and a safety zone is included to define the specified accuracy of the mechanism.

So far our discussion on accuracy and repeatability has been based on a single dimension. The remainder of the discussion will focus on some of the considerations for defining multidimensional repeatability and accuracy and should serve to give the reader some insight into the many-faceted problem of coming up with a set of standard specifications.

Most commercial manipulators have discrete control resolutions for each of their axes. That is, the smallest move that the axis can make is defined by its control resolution. This gives rise to workspaces in which the tip of the manipulator can only be positioned at the intersection of the discrete control positions. Figure 3.8.8 shows the incremental workspace for both a Cartesian and cylindrical robot system consisting of only the three major axes. In either of these two workspaces the only positions that the mechanism may attempt to go to are the points which are the intersections of the solid lines. Thus there are only certain positions in space that the robot may actually attain.

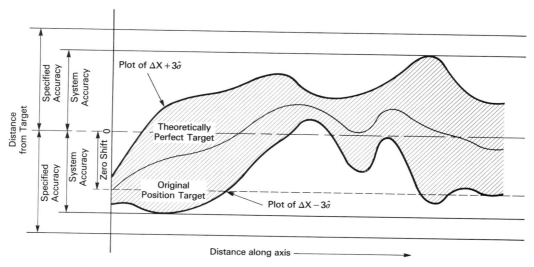

Figure 3.8.7   NMTBA definition of accuracy versus position for an axis. (Reproduced with the permission of NMTBA, the association for manufacturing technology.)

For the case of the Cartesian workspace in Figure 3.8.8, if we assume that each intersection point is attainable without any error, then the worst-case accuracy of this system may be defined to be:

$$A = \sqrt{\left(\frac{dx}{2}\right)^2 + \left(\frac{dy}{2}\right)^2 + \left(\frac{dz}{2}\right)^2} \tag{3.8.1}$$

Equation (3.8.1) says that the worst-case distance from where we want the manipulator positioned to its actual position (i.e., the intersection of the solid lines) is the distance from the center of the cube to one of its corners. Note that in this case, it is possible for the positioning algorithm to choose one of eight possible positions.

The cylindrical workspace has severe limitations on being able to reach certain points defined by Cartesian coordinates. As the r dimension increases, the actual spatial displacement in the θ direction increases even though the control resolution of the θ axis remains fixed. Thus one may generalize that the accuracy of a cylindrical manipulator is better with the r axis retracted than when it is extended.

Add to this concept of *positioning granularity* the mechanical inaccuracies that make the control resolution into spatial resolution and we may define a *solid region* about each composite control point of where the actual position of the manipulator goes. The actual position of the manipulator in this solid region would be dependent on various factors. The distance from the center of this region to the point farthest away from it defines the worst-case repeatability for this three-dimensional figure.

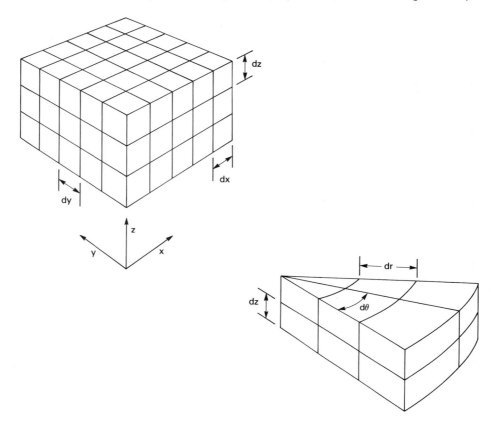

Figure 3.8.8 Incremental Cartesian and cylindrical workspaces.

Another problem that comes into play in defining both the accuracy and repeatability of the control system depends partially on how computer control is implemented. When the servos cause each axis to move so as to reach a final destination, there is sometimes a tolerance associated with the difference from the desired final position and the actual final position of each joint. This tolerance is necessary to account for settling-time considerations. Thus the computer may consider that the manipulator is positioned at the designated set point if all the axes are within, say, ± 8 encoder pulses of the desired final position (see Chapter 5). Some axes may be exactly on target, others off by one or two pulses, and others at the tolerance limit. Since the control to each axis is still active after the computer acknowledges that the axes are within an acceptable tolerance, the manipulator's control system slowly forces each axis to move to its final position with zero error. The time between the first report that the manipulator is within the tolerance range, and when it stops moving may be on the order of a few seconds.

Thus if measurements are made on a manipulator or if a specification is quoted, it is important to know "when" the measurement was made with respect to the motion dynamics in addition to the actual value of the specification.

Consider, for example, the following scenario, which may cause some problems in a robot application. Initially, the robot is "taught" by demonstration a location in space. When the robot is commanded to return to that position, it is found to be off by an amount less than or equal to the tolerance. However, if a delay is included before the measurement is made, the distance from the initial taught point to the location the manipulator reaches is reduced. This illustrates the "tolerancing" effect on the repeatability of a manipulator used in a playback mode.

Finally, one must consider the effects of the measurement methods used to obtain data for repeatability and accuracy. Although an electronic "dial indicator" can be used for some measurements, the fact that the manipulator touches the measuring device causes a reaction force and damping thereby generating errors in the data. A three-dimensional noncontact sensing device such as a laser or capacitance-type sensor provides the best snapshot of the robot's positioning performance and can give a "feel" for settling time.

3.9 FORCES ENCOUNTERED IN MOVING COORDINATE SYSTEMS

So far, the forces and other variables describing motion that we have discussed have all been assumed to be in an inertial reference frame. One of the problems in sizing the components of a robot or ensuring that a robot can properly do a task involves estimating the forces or torques that will be applied to the various joints' actuators. Recall that to move a jointed manipulator's end effector in an arbitrary straight line may require very complex motions of the joints that make up the kinematic chain. The forces or torques required to accelerate each joint include contributions from reflected inertias, gravity, friction, and the effects of Coriolis and centripetal forces. The following discussion is modeled after [15].

Figure 3.9.1 shows two coordinate systems, XYZ and xyz. We will assume that system XYZ is fixed and is our reference; that is, all forces and positions will be defined with respect to this system. System xyz's origin is displaced from the fixed system and vector $\mathbf{R}$ describes this translation. Vector $\mathbf{R}$ is a function of time, since we will assume that the relationship between the two coordinate systems is not fixed. Thus $\dot{\mathbf{R}}$ and $\ddot{\mathbf{R}}$ denote the velocity and acceleration of the origin of the moving coordinate system relative to the fixed origin.

Furthermore, assume that system xyz is free to rotate about its origin. Point P is defined as a particle or a fixed point on a mechanism referenced to xyz; it may represent the center of gravity of the manipulator's payload. Vector $\mathbf{r}$ is the position vector from the origin of xyz to P.

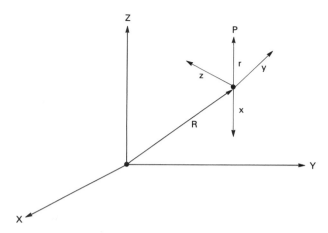

Figure 3.9.1 Moving coordinate frames.

It can be shown [15] that the total forces (as seen in the fixed system) which act on the particle P can be represented by the following equation:

$$F = M\ddot{R} + M\ddot{r} + M\dot{\omega} \times r) + 2M(\omega \times \dot{r}) + M(\omega \times (\omega \times r)) \qquad (3.9.1)$$

All the quantities in Eq. (3.9.1) are vectors, with the exception of M, which represents the mass of the particle at point P. The quantity ω is the angular velocity of the xyz system with respect to the fixed system XYZ. The left side of Eq. (3.9.1) is the resultant of all forces acting on the particle as seen by an observer in the fixed system. The first two terms on the right of the equation define the force on the particle due to its mass and acceleration relative to the fixed frame; the third term is sometimes called linear acceleration and typically goes to zero unless there is acceleration between the fixed and moving frames. The fourth term on the right of Eq. (3.9.1) represents the Coriolis force, while the last term represents centripetal force.

As shown by Eq. (3.9.1) there are many possible contributions to the total force required to accelerate a particle. This concept of a moving coordinate system (i.e., frame) may be used to model the forces or torques required to be produced by the actuator of a joint. In fact, one may group several joints and links together, such as those composing the wrist and define the forces acting on the link on which they are ultimately connected. Since each joint of a manipulator typically possesses one degree of freedom, the equations may simplify if frames are chosen correctly.

It is important to realize that it may be quite impossible to model accurately the actual performance of a manipulator this way; however, it may be sufficient to recognize that these types of forces exist and approximate their magnitude so as to get an idea of the "headroom" necessary in choosing an actuator so that the effects of these components do not degrade the performance of the robot.

In Chapter 8 we provide the basis for the mathematics used to describe the position of the manipulator's end effector with reference to a fixed coordinate system.

3.10 LAGRANGIAN ANALYSIS OF A MANIPULATOR

Up to this point in our analysis, we have considered the dynamics of a single axis. That is, we have assumed that the link driven by a joint actuator had an inertia or mass associated with it, had frictional terms (both linear and nonlinear in nature), and may have been affected by the force of gravity. Additionally, while the inertia may have been affected by subsequent links, we could approximate it in terms of a minimum or maximum value. There was no mention of other reactive forces that could occur if other joints and links moved simultaneously. Thus the sizing of our actuator was based solely on worst-case numbers for friction, inertia (acceleration force), and the effect of gravity. Unfortunately, there are other reactive forces which can occur if joints move simultaneously and which may influence the sizing of the actuator and the headroom needed in the control strategy.

One method of analyzing the complex nature of the serial link chain is to utilize Lagrange's equation. While this treatment will reduce to Newton's law, it provides a method for an easier formulation of the problem. Lagrange's equation is based on the concept of generalized coordinates and generalized forces. Generalized coordinates in terms of robotics are typically angles and distances for revolute and prismatic joints respectively. Generalized forces become the torques or forces associated with the joint actuators. As in any physical system, the number of generalized coordinates needed to describe the position of a mass is the number of degrees of freedom of that mass. Since this method is based on the analysis of particles or systems of particles, the modeling technique is based on a lumped parameter model of the manipulator with point masses representing the mass of links and their internal components with the point masses located at the center of gravity of the link.

The Lagrangian, L, is defined as the difference between the kinetic and potential energies of all of the particles of the system expressed in generalized coordinates.

$$L = E_k - E_p \qquad (3.10.1)$$

Lagrange's equation for a system with both conservative (derivable from a potential such as gravity) and nonconservative forces is given as

$$F_i = \frac{\delta}{\delta t} \left(\frac{\delta L}{\delta \dot{q}_i} \right) - \frac{\delta L}{\delta q_i} \qquad (3.10.2)$$

where q_i is the generalized coordinate associated with the force F_i. As mentioned before in the case of rotary joints, the generalized forces become torques, T, and the generalized coordinates become angular displacements, θ. For prismatic joints the generalized forces are F, and the generalized coordinates correspond to linear displacements. It is important to note that the linear displacements are not necessarily in Cartesian coordinates; they are typically along the axis of extension.

Our objective in this section is to give the reader some insight into this analysis technique but by no means provide the ultimate discussion. We will present two

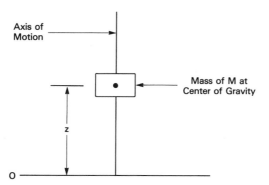

Axis of
Motion

Mass of M at
Center of Gravity

z

O

Figure 3.10.1   Single prismatic joint
working against gravity.

extremely simple systems in order to define terminology.   References are provided
at the end of this section for those who wish additional information.

The easiest way to gain an understanding of Eqs. (3.10.1) and (3.10.2) is to
use them to analyze a single joint manipulator as shown in Figure 3.10.1.   This
"manipulator" consists of a single prismatic axis that can move up and down against
gravity.   As can be seen from the figure, there is only one coordinate, z, needed
to define the position of the manipulator.   The kinetic energy can be expressed
as:

$$E_k = \tfrac{1}{2} M \dot{z}^2 \tag{3.10.3}$$

while the potential energy is given as:

$$E_p = M g z \tag{3.10.4}$$

The Lagrangian becomes:

$$L = \tfrac{1}{2} M \dot{z}^2 - M g z \tag{3.10.5}$$

and the force directed along the positive z axis needed to move the mass for some
given acceleration is given by:

$$F_z = M \ddot{z} + M g \tag{3.10.6}$$

These results are the same ones that would be derived using Newton's equations.
They state that the force needed to accelerate the mass upwards is equal to the
sum of its weight and its mass times the desired acceleration.

So far, Lagrangian analysis has not produced any more information than we
could have obtained with our previous analysis techniques.   Now let us consider
a more complicated system.   Figure 3.10.2 shows a schematic of a cylindrical
coordinate robot based on Figure 3.4.10.   Specifically, the plane in which the θ
and r axes move is shown.   Note that in this case, neither axis is under the influence
of gravity, and therefore the potential energy term of the Lagrangian equals zero.
For this system, the generalized coordinates chosen will be the angle θ, corresponding to the position of the rotary joint and the length, r, corresponding to the
location of M_R.   The point masses, M_θ and M_R are the masses for the links as-

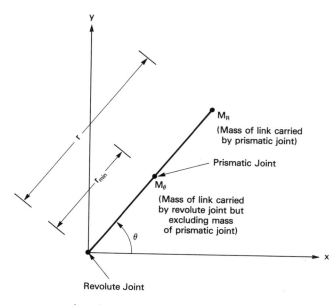

Figure 3.10.2 Plane of motion for the r and θ axes of a cylindrical coordinate manipulator.

sociated with the rotary and prismatic joints respectively.   The kinetic energy associated with mass M_θ is given by:

$$E_K = \tfrac{1}{2} M_\theta \, r_{\min}^2 \, \dot\theta^2 \tag{3.10.7}$$

This may be obtained by writing the position of the mass in terms of the Cartesian coordinates x and y, taking the derivative to obtain the respective velocities, and then noting that the velocity directed along the r axis is the square root of the sum of the squares of the x and y components.   As expected, the only variable in Eq. (3.10.7) is in terms of the generalized coordinate θ, since this single variable (along with the constant radial distance $r_{\min}$) is sufficient to describe the position of the point mass.

The kinetic energy associated with mass M_R is:

$$E_K = \tfrac{1}{2} M_R(\dot r^2 + r^2\dot\theta^2) \tag{3.10.8}$$

As one would expect, Eq. (3.10.8) is a function of the two generalized coordinates, r and θ.

The Lagrangian is given by:

$$L = \tfrac{1}{2} M_\theta r_{\min}^2\dot\theta^2 + \tfrac{1}{2} M_R(\dot r^2 + r^2\dot\theta^2) \tag{3.10.9}$$

The generalized force associated with the θ axis is the torque given by:

$$T = M_\theta r_{\min}^2\ddot\theta + M_R \, r^2\ddot\theta + 2M_R \, r \, \dot\theta \, \dot r \tag{3.10.10}$$

while the generalized force associated with the r axis is given by:

$$F = M_\theta\ddot r - M_R \, r \, \dot\theta^2 \tag{3.10.11}$$

Examination of Eq. (3.10.10) shows that the first two terms are what would be expected from the point mass approximation of inertia. Each term contributing to the inertia has a different radial distance from its corresponding mass to the axis of rotation. The last term is the Coriolis force seen by the actuator of the θ axis due to combined velocities of both the θ and r axes. Note that the Coriolis force will be zero if both axes do not move simultaneously. Based on the desired performance of the system, one can get a numerical value for the contribution of the Coriolis force which could be conceivably large or small with respect to the two other terms.

The first term of Eq. (3.10.11) shows the contribution to the force due to acceleration of the point mass while the second term is a centripetal force which occurs at the joint actuator attached to the r axis due to a velocity at the joint controlled by the θ actuator. This term indicates a force applied in a radial direction directed inward and proportional to the distance r that the axis is extended.

This simple yet illustrative example shows that by using Lagrange's equation, we were able to determine the required torque and force on the joint actuators and account for forces due to the interaction of the motions of the point masses. At this point, it may be interesting to investigate a generalized expression that results from the use of Eq. (3.10.2). For discussion purposes we will assume there are two generalized coordinates (q_1 and q_2) and therefore two generalized forces (F_1 and F_2). The generalized forces for each of the joint actuators that arise from dynamic interactions between the links may be expressed as:

$$F_1 = D_{11}q_1 + D_{12}q_2 + D_{111}q_1^2 + D_{122}q_2^2 + D_{112}q_1q_2 + D_{121}q_2q_1 + D_1 \qquad (3.10.12)$$

$$F_2 = D_{21}q_1 + D_{22}q_2 + D_{211}q_1^2 + D_{222}q_2^2 + D_{212}q_1q_2 + D_{221}q_2q_1 + D_2 \qquad (3.10.13)$$

Note that there are essentially three types of terms in Eqs. (3.10.12) and (3.10.13):

- Those due to acceleration
- Those due to the product of velocities
- Those due to gravity

The coefficients may be identified as follows. The first subscript indicates the generalized force while the second (and possibly third) indicate where the cause of the force is originating. Table 3.10.1 defines the coefficients.

The effective mass or inertia is the value of mass or inertia one would compute if all the inertial loads were reflected to the axis of interest. This should be apparent from Eq. (3.10.10). The coupling inertia or mass states that acceleration at the jth axis will cause a force to be applied to axis i. Thus we see that the concept of inertia is not as simple as presented in Section 3.3.2. It may be interesting to note [19] that if coupling inertias are small with respect to the effective joint inertias, the manipulator may be treated as a series of independent mechanical systems. In this case, the analysis techniques of Section 3.2 may be used to define the dynamics of each joint-link pair. Coupling inertia terms appear in kinematic configurations where two point masses are coupled by rotary joints (such as for a SCARA or a jointed arm).

TABLE 3.10.1 COEFFICIENTS OF
GENERALIZED FORCE EQUATION

Coefficient	Description
D_{ii}	Effective inertia or mass
D_{ij}	Coupling inertia or mass
D_{iii}, D_{ijj}	Centripetal term
D_{iij}, D_{iji}	Coriolis term
D_i	Gravity term

The centripetal terms show the contribution to the force at axis i due to a velocity occurring on either axis i or j. The Coriolis terms show the contribution of force due to a combined velocity of the i and j axes.

As one can see, the use of Eq. (3.10.2) can result in a nonlinear differential equation [see Equations (3.10.10) and (3.10.11)]. The dynamic equations are such that the torque or force for a given actuator is defined explicitly for the movements of all the joints. These equations can be used to model the performance of a manipulator so that the dynamics are included; however, for the actual implementation of a commercial robot, it is usually adequate to provide actuators with sufficient headroom, a control system capable of driving the actuators, and a profile generator that supplies the control system with the appropriate signals (such as position versus time). The concept of the profile generator and multi-axis control is discussed in Chapter 8.

Thus one objective of using Lagrangian analysis is to get a handle on the magnitude of the forces or torques that can be required by each of the joint actuators. By knowing the maximum velocities of each joint (either from the requirements of an application or a desired performance specification) and the maximum payload that the manipulator can carry (along with the weights of the links and their internal components) one can perform a worst-case analysis to obtain the magnitude of a torque or force that a joint's actuator must supply. This permits the proper sizing of an actuator and can also give insight into the dynamic range of the control signal.

For more information on Lagrange's equation, the reader should consult [15], [21], or any other theoretical mechanics textbook. For a more detailed application of the Lagrangian to robotic manipulators, [20] provides an excellent overview while [19] presents a method coupling the Lagrangian analysis technique to the concepts of homogeneous transforms (which are presented in Chapter 8).

3.11 SUMMARY

This chapter provided an overview of some of the considerations required for the successful design or applications of robots from a mechanical systems point of view. Topics in dynamics and modeling of linear systems were considered with respect to obtaining the motions required by revolute and prismatic axes of a manipulator. In addition, some of the limitations and properties of real-world components were discussed to make the reader more aware of the practical problems that occur when robots are applied. Finally, a brief introduction to the modeling of the actual dynamics of a manipulator was presented. Based on this discussion, the reader should be aware of the complexity of the dynamics of a manipulator but should also understand that detailed analysis of this type can be used to predict worst-case situations, and excruciating analysis may not be necessary to design a working manipulator.

3.12 PROBLEMS

3.1 Compute the torque required to balance a 5-lb load mounted to the end of a bar similar to Figure 3.2.6 as a function of the angle that the bar makes with the horizontal. Assume that the length of the bar ranges from a minimum of 36 in. to a maximum of 72 in. For the minimum and maximum lengths, plot the torque as a function of angle. What is the average value? What is the rms value?

3.2 A rotational system consists of an inertial load J, a viscous friction coefficient B, a static friction component F_s, and a Coulomb friction component F_c. It is desired to accelerate the system at a constant angular acceleration of 1 rad/s per second for t_a seconds and then run at a constant velocity of 1 rad/s. Sketch the angular acceleration and velocity as a function of time; also determine the torque that must be supplied as a function of time.

3.3 The inertia of a disk-shaped component of radius r, height h, and weight w is given as J_1. Suppose that it is desired to reduce the original inertia by a factor of 0.5. What is the new radius and weight with h fixed? What is the new height and weight with r fixed?

3.4 For the case of a sphere (see Figure 3.2.9) find the error between the exact computation of inertia and the point-mass approximation as a function of r and R, where r is the distance from the sphere's centroidal z axis to another axis parallel to it and R is the radius of the sphere. In other words, how far from the axis of rotation does the sphere have to be before the point-mass approximation yields an acceptable error of 1%? Of 10%?

3.5 a. For the gripper of Example 3.2.3, compute the inertias about the z and y axes if a payload of 5 lb (weight) is inserted inside the gripper. Assume that this mass completely fills the gripper when inserted.

   b. Repeat part (a) if the same payload exists but extends 1 in. outside the gripper in all dimensions.

3.6 a. Derive the equation for the total inertia about a pivot point due to a payload of weight W located a distance r from the pivot and a counterbalance of weight W' located a distance r' from the pivot (refer to Figure 3.2.6).

   b. What relationships between r and r' must exist for the inertia due to the counterbalance to be negligible?

   c. What happens to this relationship if the inertia of the payload and counterbalance cannot be approximated by Mr^2?

3.7 Comment on how a spring could be used to counterbalance a robotic manipulator. Write the equations for a configuration similar to Figure 3.2.6.

3.8 Assume that a simple manipulator consists of a prismatic joint mounted on a pivot such that when fully retracted the tip of the robot is 30 in. from the pivot, and when fully extended, the tip is 60 in. from the pivot. The prismatic joint is pivoted perpendicular to the floor. It will be assumed that the joints and links of the robot are weightless and inertialess by themselves; however, the payload will vary from a minimum of 5 lb to 25 lb.

    a. Compute the torques necessary to statically balance the joint for the four loading cases (5 lb retracted, 5 lb extended, 25 lb retracted, 25 lb extended).

    b. Compute the inertias for all four cases.

    c. Design a counterbalance so that the "optimal" torque rating is obtained for all four cases in terms of *static balance torque* and *minimum inertia*. Comment on the relationship between the distance of the counterweight from the pivot and its effect on the inertia.

    d. Compare rms and peak torques for the four cases with and without the counterbalance if the joint is moved from a horizontal position to perpendicular with respect to the ground. The motion is described by a trapezoidal velocity profile having equal acceleration, deceleration, and constant-velocity periods.

3.9 a. Using a cable and a disk of known inertia, find the spring rate of the cable by means of a torsional pendulum.

    b. Check the validity of the cable's spring rate by using Eq. (3.2.32) and your spring rate to determine the inertia of another object, such as a slender rod whose inertia you can calculate.

3.10 Make a wooden model of a more complex metal structure. Use a physical pendulum to determine the inertia of the wooden model. Determine the inertia of the actual part by modifying the value of inertia you measured by a ratio of the densities of both materials.

3.11 Derive Eq. (3.2.35).

3.12 Compute the work done by a torque that is used to accelerate a weight of 5 lb connected to a rod of 30 in. through a distance of 90° with a terminal velocity of 90° per second. What is the power?

3.13 For a gear train similar to Figure 3.3.1, assume that $N_2/N_1 = 100$. If an input torque of 5 oz-in. is applied to shaft 2 with a speed of 10 rpm, what is the output torque and speed? Repeat the problem if $N_1/N_2 = 100$.

3.14 Employing the same concepts that were used to develop Eq. (3.3.4), the transfer relationships between the input and output shafts of a compound gear train (i.e., one consisting of more than two gears) may be derived.

    a. Derive the transfer relationship between the input and output shafts of a gear train consisting of three gears. The input gear has N_1 teeth, the middle or idler gear has N_2 teeth, while the output gear has N_3 teeth.

    b. Assume that a system consists of two gear trains, each similar to the one shown in Figure 3.3.1. The output of the first gear train is coupled by a rigid shaft to the input of the second gear train so that both gears are affixed to the same shaft. The number of teeth of the first gear train from input to output is N_1/N_2. For the second gear train the input gear has N_2 teeth while the output has N_3 teeth. Compare the overall transfer relationship to the three-gear system of part a.

3.15 For the system shown in Figure 3.3.3, assume that the inertia of the gears must also be included. The inertia of the gear on the input shaft is J_{g_1}, while that of the gear on the output shaft is J_{g_2}. Derive a series of equations similar to Eqs. (3.3.8a) through (3.3.8c) which include these terms.

3.16 For the belt-and-pulley system of Figure 3.3.7, derive an expression similar to Eq. (3.3.4).

3.17 Derive an equation for the total inertia as seen by the input for the system shown in Figure 3.3.11. Include the inertia of the pulleys. What happens if the radii of the two pulleys are not equal?

3.18 Derive the relationship for the input shaft position and the linear stage's position of the slider crank of Figure 3.3.13. Find an expression that defines the torque on the input needed to accelerate the mass on the linear stage from any initial position to any final position over a single-valued motion.

3.19 Comment on the design of a simple pick-and-place mechanism using cams. Define a simple motion such as rising above a point 1 in., traveling forward 2 in., and then descending 1/3 in. Draw the cam profiles and suggest a simple mechanism for accomplishing this task.

3.20 a. For the mechanism shown in Figure 3.3.21, derive the relationship between the length of link 1 and the angle of link 2 with the vertical.
  b. If a 5-lb point mass is hung at the end of link 2, what force must be exerted by link 1 so that the system is statically balanced?
  c. Assume that link 1 is implemented by a lead screw and nut having a pitch of 5 turns/in. What torque must be exerted to statically balance the mechanism?
  d. Compare this mechanism to that of an implementation as designed in Figure 3.2.6. What are the advantages and disadvantages of each in terms of such quantities as range of motion, required torque for static balance, and so on?

3.21 a. Derive the optimum coupling ratio with respect to minimum power dissipation in a dc motor for the case of a dc motor with an inertia J_1 driving an inertial load J_2, a viscous load B_2, and a constant torque load T_2 through a gear train.
  b. What happens if the inertia of the gears is not neglected?
  c. Do a sensitivity study based on changes in the coupling ratio and the loads J_2, B_2, and T_2.

3.22 a. Consider a 2-axis cylindrical coordinate robot with an r-axis whose extension range falls within 25 to 45 in. Assume that the r-axis rotates parallel to the floor so that the effect of gravity may be neglected. Let J_L be the inertia seen by the rotary joint which is a function of r.
  b. Using the same model as shown in Figure 3.5.11, compute the locked rotor and free rotor resonance frequencies. Examine the results if J_a is equal to the minimum value of J_L, J_a is much smaller than the minimum value of J_L, and J_a is much greater than the minimum value of J_L. Comment on how a gear ratio should be selected based on the foregoing results.

3.23 Draw analog circuits for the mechanical networks in Sections 3.5.2, 3.5.3, and 3.5.3.1.

3.24 Model the torsional resonance phenomenon with an electronic circuit and explore the relationship between J_a, J_L, K, and B.

3.25 a. Derive an equation similar to Eq. (3.8.1) for the r-θ-z workspace for Figure 3.8.8.
  b. Draw a set of curves that define accuracy over the entire workspace as a function of r, θ, and z.

3.26 Include the effects of mechanical inaccuracies in Eq. (3.8.1).

3.27 Compute the inertia of the gripper in Example 3.3.3 by assuming that the gripper is a solid, and subtracting the inertia of elementary shapes so that the remaining figure looks like the gripper.

3.28 For the Lagrangian analysis of the θ-r manipulator given in Section 3.10, specifically Equations (3.10.10) and (3.10.11), assume a trapezoidal velocity profile for the motion of each axis so that the acceleration, constant velocity, and deceleration times are the same. Investigate the relationship between the maximum velocities of the θ actuator and r actuator so that:

   a. Coriolis force is minimized.
   b. Coriolis force is maximized.
   c. Centripetal force is minimized.
   d. Centripetal force is maximized.

3.29 Extend the analysis of Figure 3.10.2 given by Eqs. (3.10.10) and (3.10.11) to include a third degree of freedom, i.e., in the z direction. Comment on each of the terms in the force equation for z. Do Eqs. (3.10.10) and (3.10.11) change?

3.30 Assume that Figure 3.10.2 is in the y-z plane. Perform the Lagrangian analysis taking into account the potential energy of both point masses.

3.31 a. For each of the kinematic configurations shown in Figure 3.6.3 sketch the incremental workspace (similar to Figure 3.8.8).

   b. Derive an expression for the accuracy of a point defined in Cartesian coordinates as a function of the joint variables.

3.32 For the slider crank mechanism shown in Figure 3.3.13

   a. Derive an equation for the position of the reference point of the payload as a function of the crank angle, θ.

   b. Sketch the position of the payload as a function of time if the crank is rotated at a constant velocity (i.e., do not include effects of acceleration and deceleration).

   c. For a trapezoidal velocity profile applied to the crank for a total crank displacement of 180° (so that the reference point moves from one extreme to the other) and equal times for acceleration, constant velocity, and deceleration:
   Sketch
   1. position versus time,
   2. velocity versus time,
   3. acceleration versus time, and
   4. jerk versus time
   for both the crank angle θ and the position of the reference point.

   d. Compare the frequency content of the corresponding signals for the crank angle and reference point position. That is, examine the frequency content of the trapezoidal profile defining the velocity of the crank and the profile defining the velocity of the reference point.

3.33 Assume that the $r - \theta$ manipulator shown in Fig. 3.10.2 and described by Eqs. (3.10.10) and (3.10.11) has the following parameters:

$$r_{min} = 2.0 \text{ in.}$$
$$M_\theta = 1.0 \text{ lb weight}$$
$$M_R = 1.0 \text{ lb weight.}$$

   a. If the tip of the manipulator is moved from the x,y coordinates (0, 10) to (10, 0) in a straight line (as described in Section 8.8.2 and Figs. 8.8.1, 8.8.2, and 8.8.3) plot the contributions to the total torque and force as well as the total torque and force curves. Use 33.3 ms for the acceleration, deceleration, and constant velocity time of the velocity profile.

b. Repeat part a if joint interpolated motion is used instead of straight-line motion.

c. Compare the force and torque requirements for straight-line versus joint interpolated motion.

d. Repeat parts a, b, and c for a motion from (0, 10) to (5, 0).

3.34 For the manipulator shown in Fig. P.3.34

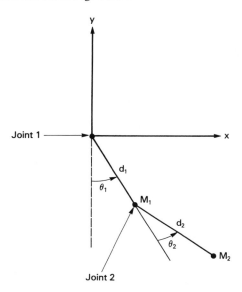

a. Use Lagrangian analysis to find the torques required by the actuators for joint 1 and joint 2. Note that the x-y plane is parallel to the floor and no gravitational effects have to be considered.

b. Repeat if the effects of gravity must be considered.

3.13 REFERENCES AND FURTHER READING

1. Beer, F. P., and Johnston, E. R., *Vector Mechanics for Engineers: Statics*, 2nd ed. New York: McGraw-Hill Book Company, 1972.

2. Bigley, W. J., and Rizzo, V. *Resonance Equalization in Feedback Control Systems*, ASME publication, 78-WA/DSC-24. New York: American Society of Mechanical Engineers, 1978.

3. Colson, J. C., and Perreira, N. D., "Kinematic Arrangements Used in Industrial Robots," *Conference Proceedings of the 13th International Symposium on Industrial Robots and Robots 7*, Vol. 2, 1983.

4. Dorf, R. C., *Modern Control Systems*, 3rd ed. Reading, Mass: Addison-Wesley Publishing Company, Inc., 1981.

5. Dorf, R. C., *Robotics and Automated Manufacturing*, Reston, Va: Reston Publishing Co., Inc., 1983.

6. D'Azzo, J. J., and Houpis, C., *Linear Control System Analysis and Design*, 2nd ed. New York: McGraw-Hill Book Company, 1981.

7. Humphrey, William M., *Introduction to Servo Mechanism System Design*. Englewood Cliffs N.J.: Prentice-Hall, Inc., 1973.

8. Kuo, B. C., *Automatic Control Systems*, 4th ed. Englewood Cliffs, N.J.: Prentice-Hall, Inc., 1982.

9. L'Hote, F., Kauffmann, J., Andre', P., and Taillard, J., *Robot Technology*, Vol. 4, *Robot Components and Systems*, Englewood Cliffs, N.J.: Prentice-Hall, Inc., 1983.

10. Martin, George H., *Kinematics and Dynamics of Machines*, rev. printing. New York: McGraw-Hill Book Company, 1969.

11. Milenkovic, V., and Huang, B., "Kinematics of Major Robot Linkages," *Conference Proceedings of the 13th International Symposium on Industrial Robots and Robots 7*, Vol. 2, 1983.

12. McLean, W. G., and Nelson, E. W., *Engineering Mechanics*, 3rd ed., Schaum's Outline Series. New York: McGraw-Hill Book Company, 1978.

13. Oberg, E., Jones, F. D., and Horton, H. L., *Machinery's Handbook*, 22nd ed. New York: Industrial Press, Inc., 1984.

14. Sears, F. W., and Zemansky, M. W., *University Physics*, 4th ed. Reading, Mass: Addison-Wesley Publishing Company, Inc., 1970.

15. Spiegel, Murray R., *Theory and Problems of Theoretical Mechanics*, Schaum's Outline Series. New York: McGraw-Hill Book Company, 1967.

16. Wilson, D. R., *Modern Practice in Servo Design*. Elmsford, N.Y.: Pergamon Press, Inc., 1970.

17. *ICAM Robotics Application Guide, AFWAL-TR-80-4042, AFWAL/MLTC*, Volume 2, April 1980. Wright Patterson Air Force Base, Ohio, 45433.

18. *DC Motors, Speed Control, Servo Systems*, An engineering handbook by Electro-Craft Corp., 5th edition. Hopkins, Minn.: Electro-Craft Corp., 1980.

19. Paul, Richard C., *Robot Manipulators: Mathematics, Programming and Control*. Cambridge, Mass.: The MIT Press, 1981.

20. Brady, M., Hollerbach, J., Johnson, T., Lozano-Perez, T., and Mason, M., eds., *Robot Motion Planning and Control*. Cambridge, Mass.: The MIT Press, 1982.

21. Goldstein, Herbert, *Classical Mechanics*, 2nd ed. Reading, Mass.: Addison-Wesley Publishing Company, 1981.

4

Control of Actuators in Robotic Mechanisms

4.0 OBJECTIVES

In this chapter we present the practical aspects of controls as they relate to robots, with the emphasis being placed on how robotic actuators are driven to achieve a desired performance. It will be assumed that the reader has a basic knowledge of "classical" control theory, and hence topics such as Laplace transforms and stability theory will not be developed. However, these and other concepts will be used in discussing typical and reasonable control models that are applicable to robot systems. The material will be presented from the standpoint of a servomechanism rather than from the more traditional theory of controls approach.

Specifically, the topics that will be covered are as follows:

- Closed-loop position servo
- Frequency response of a typical joint position servo
- Frequency-domain compensation techniques applied to this servo to meet joint specifications
- Effects of gravity and friction on robot performance
- Role of position and velocity feedback
- Elimination of position error
- Robotic actuators including dc servomotors (brush and brushless types), stepper motors, direct-drive motors, linear motors, and hydraulic and pneumatic devices

- Power amplifier configurations, including linear and pulse-width-modulation devices
- Differences between voltage and current amplifiers and their effects on the operation of a servo

Also included is a brief discussion of both optimal and adaptive control and how these more modern disciplines *might* be utilized to control a robotic actuator.

4.1 MOTIVATION

One of the major objectives of a robot is to move its manipulator (which may or may not be carrying a special-purpose tool) from one point to another one in an accurate and repeatable manner. Usually, *position* is important, so that it is not unreasonable to assume that the individual joints are controlled by a position servo. Since this is true in most of the robots currently available, we will present the details of the position servo. Included will be a discussion of the advantages and disadvantages of the pure digital, pure analog, and the hybrid (i.e., one that uses both analog and digital components) servo.

A typical robot has a master computer that is responsible for sending the appropriate position commands (often referred to as "set points") to each of the joints (axes). This information is used by a separate computer (or microprocessor) to command the individual joint to move in the desired manner (see Figure 4.1.1).

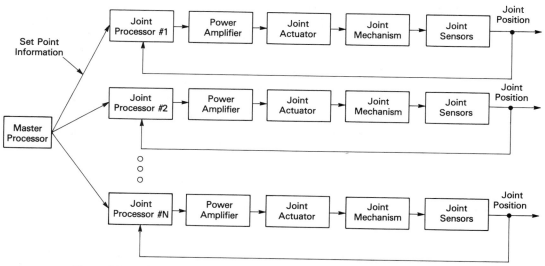

Figure 4.1.1.   Common robot control architecture. It will be seen later that usually, the sensors are actually mounted at the actuator output.

Often, the joint processor is used to provide some or all of the following functions: (1) digital summation of the set points and actual position so as to obtain a *position error*, (2) interpolation of the master's set points, and (3) digital filtering (or compensation) of the joint. In any event, this processor produces the required commands to the axis actuator [e.g., either servo or other types of motors or hydraulic (or pneumatic) devices], which will then cause the particular joint to move. Information about whether and/or how it is actually moving may or may not be used. In the latter instance, the axis is said to be operating in an *open-loop* manner, whereas in the former case, the term *closed-loop* applies. The major emphasis in this chapter is on closed-loop control (of a robotic joint) because the overwhelming majority of the robots currently available utilize such a control structure. Only those simple robotic devices driven by stepper motors or simple pneumatic or hydraulic actuators may actually utilize open-loop control.

It is important to understand that the term "closed-loop" as applied to robots does not mean that the loop is closed back to the master computer (see Figure 4.1.1). In reality, current control practice requires that information about the axis motion be fed back only to the corresponding *joint processor*. The master is informed only when the move is completed or if an emergency situation arises (e.g., the manipulator encounters an unexpected obstacle). It is to be expected, however, that as more powerful microprocessors become available at reduced cost, this situation will certainly change.

In general and with relatively few exceptions, the loads that must be moved by a typical robotic joint actuator vary considerably as the manipulator moves throughout its workspace. Most certainly, such variations should be taken into account when trying to determine the proper control action. The fields of adaptive and/or optimal control would be appropriate in this context, and it is for this reason that these topics are discussed (although briefly) later in the chapter. However, current robot design practice usually ignores such variations and utilizes a *worst-case* (load) approach. Consequently, the discussions contained in the succeeding sections of this chapter will assume that the *load is constant*. As the reader will come to appreciate shortly, the results of this admittedly gross approximation are surprisingly good (but obviously not optimum from the point of view of motion speed and power required). It is reasonable to expect, however, that more modern control techniques will become feasible with the advent of lower-cost and more powerful microprocessors.

4.2 CLOSED-LOOP CONTROL IN A POSITION SERVO

The block diagram of a typical closed-loop control system is shown in Figure 4.2.1. Here some desired function or (position) command is the input and the response (or actual position) of the system or joint is the output. A controller and an amplifier are used to drive a motor which then drives a load (e.g., the joint). Knowledge of how the joint is moving is provided by one or more sensing devices

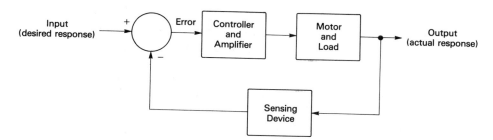

Figure 4.2.1.   General closed-loop control system.

(e.g., an optical encoder or tachometer; see Section 4.5) and this information is used to produce an error signal, which, in turn, drives the controller/amplifier, and so on.   For reasons that will become apparent shortly, a typical position servo will actually use two sensing signals: position and velocity.

Let us now consider the specifics of the pure analog position servo shown in Figure 4.2.2.   In this diagram, θ_d and θ are the desired and actual joint positions with $\omega(t)$ $(= \dot{\theta})$ being the angular velocity of the joint.   Also, K_p and K_g are position and velocity gains, with A and K_m the amplifier and motor gains, respectively.   Here a single-pole model for both the power amplifier and the dc servomotor is assumed (a more complete model for the latter device is developed in Section 4.3.1).   Since angular position θ is related to the integral of the angular velocity $\omega(t)$, an integrator is shown in the diagram.   The reader should understand that this is for modeling purposes only since in practice, θ is actually obtained through the use of a sensor (e.g., an optical encoder), as will be discussed in Section 4.5.

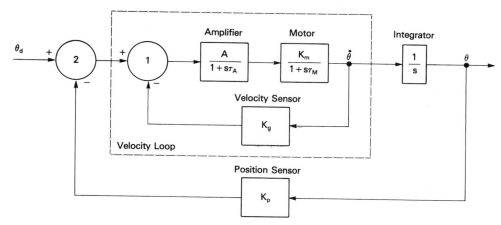

Figure 4.2.2.   Typical analog position servo showing the velocity and position loops.   The latter consists of the velocity loop, position sensor, integrator, and summing junction 2.

With respect to Figure 4.2.2, it is interesting to note that since angular veocity is fed back to summing junction 2, the position error can be viewed as a *velocity command* signal to the block marked "velocity loop." In fact, it is not uncommon to *specify* the shape of the velocity versus time curve (i.e., the velocity profile). The command position signal is then adjusted so as to produce this profile and drive the joint to the desired final position.

4.2.1 No Velocity Feedback

First consider the case where there is no "tach" or velocity feedback; that is, the velocity sensor (e.g., a tachometer) is removed so that $K_g = 0$ and hence the velocity loop in Figure 4.2.2 is open. Then the overall open-loop transfer function for the system in this figure becomes

$$GH\,(s) \,=\, \frac{AK_pK_m}{s\,(1\,+\,s\,\tau_A)\,(1\,+\,s\,\tau_M)} \tag{4.2.1}$$

Typical values for the reciprocals of the motor and amplifier time constants τ_M and τ_A are 10 to 20 rad/s and 6000 to 60,000 rad/s, respectively. Defining the open-loop gain constant to be

$$K \,=\, AK_pK_m \tag{4.2.2}$$

the root locus of this system is shown in Figure 4.2.3.

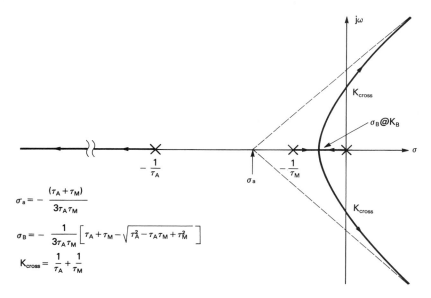

$$\sigma_a = -\,\frac{(\tau_A+\tau_M)}{3\tau_A\tau_M}$$

$$\sigma_B = -\,\frac{1}{3\tau_A\tau_M}\left[\tau_A+\tau_M-\sqrt{\tau_A^2-\tau_A\tau_M+\tau_M^2}\,\right]$$

$$K_{cross} = \frac{1}{\tau_A}+\frac{1}{\tau_M}$$

Figure 4.2.3. Root locus for the system described in Eq. (4.2.1).

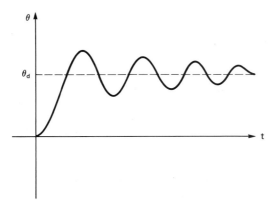

Figure 4.2.4. Typical damped sinusoidal response to a position step command.

Using standard root-locus techniques, it is found that the system will become unstable if $K > K_{\text{cross}}$ ($= 1/\tau_A + 1/\tau_M$), that is, the poles of the closed-loop system will move into the right-half portion of the s-plane and the response of the system will increase without bound ("blow up"). Also, for K less than K_{cross} and greater than K_B (the gain where branches of the root locus leave the real axis), the poles of the closed-loop transfer function are complex conjugates (with negative real parts), so that the system exhibits underdamped performance. That is, a step command in θ_d will cause θ to respond as shown in Figure 4.2.4. Usually, this type of behavior is undesirable, as it will not produce the fastest moves for a robot joint. That is, the *final* steady-state position will not be reached in the shortest time. Also, significant stresses on the mechanical components may be produced due to the rapid acceleration and deceleration required, as this final position is overshot (or undershot) and the servo is forced to make several corrections to bring the joint back to the desired point. To reduce or eliminate such response, it is necessary to provide the axis (joint) with some type of damping in order to reduce or eliminate the oscillations entirely. A certain amount of damping is inherent in the components themselves (e.g., motor and gear frictions) and in some instances may be sufficient to produce an acceptable response characteristic (i.e., either critically damped or just slightly underdamped). When this is not the case, however, another source of damping must be employed. This usually takes the form of viscous friction [i.e., a friction torque that is proportional to angular velocity $\omega(t)$] and is obtained from "tach" or velocity feedback.

4.2.2 Position Servo with Tach Feedback

Now let us restore the tach feedback in Figure 4.2.2, that is, consider the case where K_g is not zero. It will be demonstrated shortly that doing this will produce the desired damping in the position loop. Using standard block diagram simplification techniques, the open-loop transfer function for the joint with tach and position feedback is found to be

$$GH(s) = \frac{AK_m(K_p + sK_g)}{s(1 + s\tau_M)(1 + s\tau_A)} \qquad (4.2.3)$$

Note that velocity feedback causes a zero to be added to the overall open-loop transfer function (at $s = -K_p/K_g$). Using Eq. (4.2.2), the root locus for this system (plotted as a function of K) is shown in Figure 4.2.5. It is observed that in both cases, the system is now stable for all (positive) values of K. In the second case shown in Figure 4.2.5b, however, there is a dominant set of poles. Physically, this means that the pole due to the amplifier has little effect on the closed-loop response of the system. Under these conditions, the joint-position servo can be approximated by a *second-order* system, and the closed-loop transfer function becomes

$$T(s) = \frac{\theta(s)}{\theta_d(s)} = \frac{AK_m}{s^2\tau_M + s(1 + AK_gK_m) + AK_mK_p}$$

or

$$T(s) = \frac{AK_m/\tau_M}{s^2 + s(1 + AK_gK_m)/\tau_M + AK_mK_p/\tau_M} \qquad (4.2.4)$$

It can be shown that for a second-order system, a general form of the closed-loop transfer function is given by

$$T(s) = \frac{\omega_n^2}{s^2 + 2\zeta\omega_n s + \omega_n^2} \qquad (4.2.5)$$

where ω_n is the undamped natural radian frequency and ζ is the damping coefficient.* Comparing Eqs. (4.2.4) and (4.2.5), the damping coefficient for the joint with position and tach feedback is then found to be

$$\zeta = \frac{0.5(1 + AK_gK_m)}{\sqrt{AK_mK_p\tau_M}} \qquad (4.2.6)$$

Several important conclusions can be drawn from this equation:

1. The more tach feedback (i.e., as K_g is increased), the more damping there will be in the position servo. Thus the joint response will tend to become *less underdamped as K_g is increased, and vice versa.*

2. The more position feedback (i.e., as K_p is increased), the less damping there will be. Thus the joint response will tend to *become more underdamped as K_p is increased, and vice versa.*

---

*The system transient response as a function of ζ is underdamped for $0 < \zeta < 1$, critically damped for $\zeta = 1$, and overdamped for $\zeta > 1$.

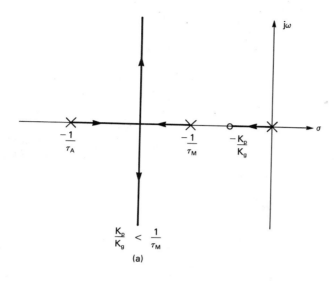

$$\frac{K_p}{K_g} < \frac{1}{\tau_M}$$

(a)

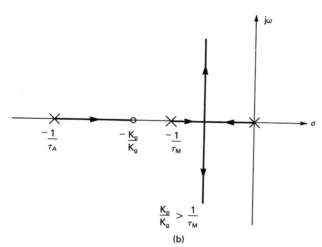

$$\frac{K_p}{K_g} > \frac{1}{\tau_M}$$

(b)

Figure 4.2.5. Root locus for a position servo with velocity feedback: (a) $K_p/K_g < 1/\tau_M$; (b) $K_p/K_g > 1/\tau_M$.

Thus it is seen that tach feedback has the effect of making the individual joint servo more stable and less oscillatory. It would also appear from the above that using the smallest value of K_p would keep the system from being underdamped. Although this is true, it will now be shown that other considerations limit the minimum value of position feedback gain.

4.3 THE EFFECT OF FRICTION AND GRAVITY

As mentioned previously, one of the major goals of a robot is to move a tool or a part from one point to another in an accurate and repeatable manner. Anything that prevents this goal from being achieved is clearly undesirable and must, therefore, either be compensated for or eliminated. In practically all electromechanical, pneumatic–mechanical, or hydraulic–mechanical systems, friction in various components will create a position error. Also, gravity will produce a position error of varying magnitude in one or more joints for most robots. To see more clearly the effects of these "disturbances," it will be necessary first to consider a more complete model of a dc servomotor.

4.3.1 Modeling the DC Servomotor

One of the common methods of driving a robotic joint is through the use of a dc servomotor, a fractional-horsepower motor with a stationary magnetic field that is generated by a permanent magnet. [For this reason it is sometimes called a permanent-magnet (PM) motor.] No power is used in the stator structure of this device and the field is constant over a wide range of armature currents. A PM motor usually requires less cooling than do other types of dc motors (e.g., a shunt motor). Other advantages of the PM motor over the wound types are:

1. **High stall torque.** This is an important characteristic during joint acceleration and also when the manipulator is required to keep a load stationary.
2. **Smaller frame size and lighter weight for a given output power.** This is especially important in a robot, where the design may require the motor to be moved within the joint itself.
3. **The speed–torque curve of a PM motor is linear,** as shown in Figure 4.3.1. If these curves were nonlinear, a potentially significant amount of extra computational effort might be required of the joint processor to ensure that a given amount of torque (or current) was produced by the motor.

A linear circuit model of a dc servomotor armature is shown in Figure 4.3.2. Here R_a is the total armature resistance (including that due to the brushes), R_L is a resistance that represents the magnetic losses in the armature (and is normally $\gg R_a$ since these losses are usually small at the frequencies of interest), L_a is the armature inductance, and E_g is the back EMF produced when the armature rotates in a dc magnetic field. This last term is proportional to the angular velocity $\omega(t)$ of the armature (i.e., $\dot{\theta}$). That is,

$$E_g = \omega(t)K_E \tag{4.3.1}$$

where K_E is referred to as the back EMF constant of the motor. Applying ele-

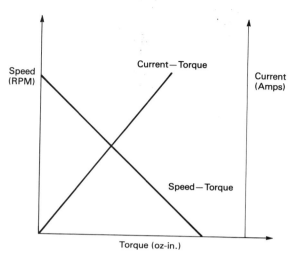

Figure 4.3.1. Speed and current versus torque for a DC servomotor.

mentary circuit theory to Figure 4.3.2 and using Eq. (4.3.1) gives

$$V_{arm} = R_a I_a + L_a \dot{I}_a + \omega(t) K_E \tag{4.3.2}$$

I_a and $\dot{I}_a$ represent the armature current and its time derivative.

As shown in Eq. (4.3.1) and Fig. (4.3.1) the torque generated by the armature moving in a PM field is linearly related to the current. Thus

$$T_g = K_T I_a \tag{4.3.3}$$

where K_T is the torque constant. This generated torque is required to accelerate an inertia (usually consisting of the motor armature itself and an external load), overcome any viscous damping torque due to the motion of the armature, and to

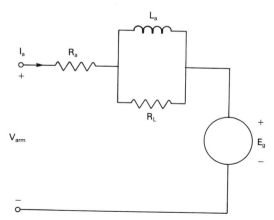

Figure 4.3.2. Circuit model of a DC servomotor armature.

overcome any external load torque (e.g., due to either gravity or static friction). Thus we may write

$$T_g = (J_M + J_L)\,\dot{\omega}(t) + B\omega(t) + T_f + T_{gr} \qquad (4.3.4)$$

In this equation, J_M and J_L are the armature and reflected load inertias (which for convenience will be combined and written as $J_T = J_M + J_L$), B is the armature viscous damping coefficient, T_f is the friction torque of the motor and the load (including the gears, etc.), and T_{gr} is the gravitational torque load. This model assumes that all components of the rotary system turn in phase (i.e., there is no torsional resonance in the system; this topic is discussed in Section 3.5.3.1). Note that if the load has any viscous friction, this can be added to B.

Combining Eqs. (4.3.1) through (4.3.4), the model of a servomotor (including the friction and gravity terms) is found to be the one shown in Figure 4.3.3. Using block diagram reduction techniques, the transfer function of the motor becomes

$$G_m(s) = \frac{\Omega(s)}{V_{\text{arm}}(s)}$$

$$= \frac{K_T/L_a J_T}{s^2 + [(R_a J_T + L_a B)/L_a J_T]s + (K_T K_E + R_a B)/L_a J_T} \qquad (4.3.5)$$

It is important to note that the disturbance torques T_f and T_{gr} do not appear in this expression because they are treated as additional *inputs*. They will be utilized, however, later in this section.

In developing the results above, it has been assumed that the load inertia (reflected back to the motor shaft) does not vary with time, so that J_L is a constant. This nontrivial assumption is valid for a large number of applications. Unfortunately, however, the actuation of a robotic joint is usually not one of them. In fact, the reflected inertia of most of the axes of a robot will normally fluctuate significantly while the manipulator is moving. An example of such behavior can be seen in Figure 4.3.4, where the inertia variation for each of the six joints of a JPL-Stanford arm is shown. It is observed that the inertia of joint 1 (i.e., the

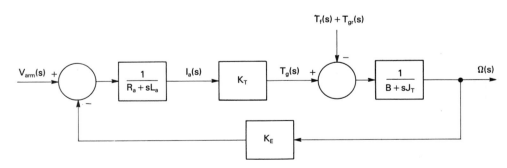

Figure 4.3.3. Block diagram of DC servomotor including gravitational and friction disturbance torques.

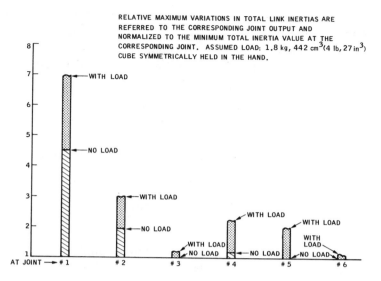

RELATIVE MAXIMUM VARIATIONS IN TOTAL LINK INERTIAS ARE
REFERRED TO THE CORRESPONDING JOINT OUTPUT AND
NORMALIZED TO THE MINIMUM TOTAL INERTIA VALUE AT THE
CORRESPONDING JOINT. ASSUMED LOAD: 1.8 kg, 442 cm^3(4 lb, 27 in^3)
CUBE SYMMETRICALLY HELD IN THE HAND.

Figure 4.3.4. Inertia variations for each of the six joints of a JPL-Stanford arm. Both variations with and without a load (carried by the gripper) are shown. (Courtesy of A.K. Bejczy and Jet Propulsion Laboratories, Pasadena, CA.)

"trunk") varies approximately 4.5 times under no load conditions and by as much as 7 times when the manipulator is carrying a 4-1b load.   Joints 2, 4, and 5 are seen to have significant variations also.   Only axes 3 and 6 have relatively constant inertias over their entire range of travel.   This behavior is typical of other robots, although the specific inertia fluctuations may be different.   Despite the fact that the inertia of a robotic joint usually does undergo dramatic changes as a function of the manipulator position, it is common design and control practice to ignore this and to assume that J_L is a constant.   The resultant computational simplification permits a "worst-case" approach to be used whereby the motor and associated mechanical linkages and gears are selected for the maximum load (i.e., inertia) conditions.   It is apparent that this may not produce the lowest-cost design.   Alternatively, the system is "derated" under full load so that maximum acceleration is not permitted under this condition.   Hence, smaller, less costly components may be used.

Regardless of which of these design processes is selected, the control of the individual joints of a robot is usually accomplished by assuming that there is no time variation of the load inertia.   Compromises in performance are the inevitable consequence (e.g., reduced speed and/or payload capability of the manipulator). It is reasonable to expect, however, that as microprocessors become more powerful and cost-effective, adaptive schemes that compensate for the large changes in reflected inertia will be used.   A corresponding improvement in performance will undoubtedly result.   This is discussed briefly in Section 4.5.5.

EXAMPLE 4.3.1

Let us find the poles [i.e., the roots of the denominator polynomial of $G_m(s)$] of a commercially available dc servomotor. For example, the parameters of an Electrocraft Corporation E530 motor are:

$$K_T = 10.02 \text{ oz-in./A}$$

$$K_E = 7.41 \text{ V/1000 rpm}$$

$$R_a = 1.64 \ \Omega \text{ (including brush resistance)}$$

$$L_a = 3.39 \text{ mH}$$

$$B = 0.1 \text{ oz-in./1000 rpm}$$

$$J_M = 0.0038 \text{ oz-in.-s}^2$$

Assuming that there is no inertial load coupled to the motor shaft, $J_T = J_M$ in Eq. (4.3.5). Since the gravity and friction terms do not affect the motor poles, we may also assume that T_f and T_{gr} are both zero, so that Eq. (4.3.5) is applicable. To utilize this equation, K_E and B first must be converted to V/rad/s and oz-in./rad/s, respectively, in order to have a consistent set of units. This is accomplished by dividing each of the given parameters by 104.72. Thus

$$K_E = 0.0708 \text{ V/rad/s}$$

$$B = 9.55 \times 10^{-4} \text{ oz-in./rad/s}$$

These parameters can now be substituted into Eq. (4.3.5). The actual transfer function for the Electrocraft motor is found to be

$$G_m(s) = \frac{7.778 \times 10^5}{s^2 + 484.027s + 5.516 \times 10^4}$$

As mentioned above, the roots of the denominator polynomial will give the motor poles. Thus

$$s_1 = -183.621 \text{ rad/s}$$

$$s_2 = -300.406 \text{ rad/s}$$

both of which are observed to be negative and real. This is quite often the case for commercial servomotors, implying that the open-loop response of such motors (to a step voltage on the armature) will be, in general, *over-damped*.

The model of the motor used in Figure 4.2.2 can now be replaced by the one just developed in Figure 4.3.3. This is shown in Figure 4.3.5. Using standard

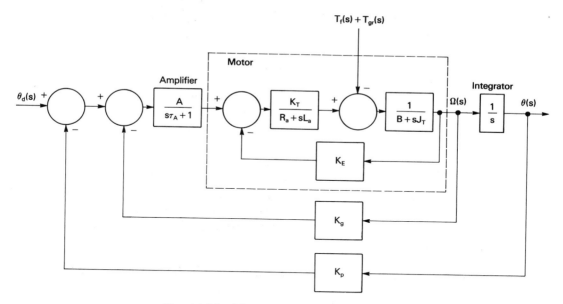

Figure 4.3.5.   More complete model for a position servo.

block diagram reduction techniques, the actual joint position transform $\theta(s)$ in terms of the command input transform $\theta_d(s)$ and the friction and gravity terms is found to be

$$\theta(s) = G(s)_{\text{equiv}}\left\{\theta_d(s) - \frac{[T_f(s) + T_{gr}(s)](s\tau_A + 1)(R_a + sL_a)}{AK_T}\right\} \qquad (4.3.6)$$

where

$$G(s)_{\text{equiv}} = \frac{AK_T}{s(s\tau_A + 1)[K_E K_T + (sJ_T + B)(R_a + sL_a)] + AK_T(sK_g + K_p)}$$

$$(4.3.7)$$

Here, $T_f(s)$ and $T_{gr}(s)$ are the Laplace transforms of the friction and gravity torque disturbances on the joint, and $J_T = J_L + J_M$, as before.

Using these results, we are now able to investigate the effects of friction and gravity on position error that can be expected to occur for a typical robot joint. Two cases can be considered.

4.3.2 Final Position with no Friction or Gravity Disturbance

Let the command position be a step of amplitude θ_d and suppose that there is no friction or gravity torque. Then $\theta_d(s) = \theta_d/s$. Substituting this into Eqs. (4.3.6) and (4.3.7), the final value theorem can be used to find the steady-state value of the position, θ_{final}. Thus

$$\theta_{final} = \lim_{s \to 0} s\theta_d(s)G(s)_{equiv}$$

or

$$\theta_{final} = \frac{AK_T}{AK_TK_p} \theta_d = \frac{\theta_d}{K_p} \tag{4.3.8}$$

This result indicates that the actual joint position will be the desired position divided by the position gain. Scaling (i.e., multiplying) the command position signal by K_p will permit the actual joint position to reach θ_d with no error. Alternatively, a different control structure can be used which permits the position loop to be closed with unity feedback. [For example, K_p can be placed in the forward loop, i.e., in cascade with the power amplifier (see Problem 4.7).] This scheme will result in zero final position error without the need for the scaling.

4.3.3 Final Position with Nonzero Friction and/or Gravity Disturbance

Assume that the joint is commanded to remain in the same position so that θ_d is constant (for simplicity, assume it to be zero) and that friction and/or gravity produces a disturbance to the system. Also define

$$T_L(s) = T_f(s) + T_{gr}(s)$$

Furthermore, let this disturbance torque be a step of amplitude T_L so that $T_L(s) = T_L/s$. Using Eqs. (4.3.6) and (4.3.7) [with $\theta_d(s) = 0$] and again applying the final value theorem, we find that

$$\theta_{final} = -\frac{T_LR_a}{AK_TK_p} \tag{4.3.9}$$

This result indicates that the friction and/or gravity disturbance will cause the steady-state joint position to be in error (recall that the desired position was zero). This error is sometimes referred to as *hysteresis*. If it is assumed that the power amplifier has a fixed gain A, it is observed that the right-hand side of Eq. (4.3.9) can be made as small as desired by increasing the position gain K_p. Often, the maximum hysteresis for a joint is specified (e.g., as θ_{max}). If this is so, then

$$\left|\frac{R_aT_L}{AK_TK_p}\right| \le \theta_{max}$$

and

$$K_p \ge \frac{R_aT_L}{AK_T\theta_{max}} \tag{4.3.10}$$

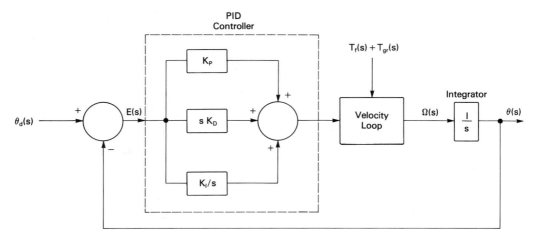

Figure 4.3.6. Position servo with a PID controller.

The minimum value of position gain that will keep the position error due to a torque disturbance below a certain value is seen to be

$$K_{P_{\min}} = \frac{R_a T_L}{A K_T \theta_{\max}} \tag{4.3.11}$$

If the specified value of the hysteresis is small, the value of K_p will be large. In fact, this value may be so large that the joint response will either be unstable or highly underdamped (recall the discussion in Sections 4.2.1 and 4.2.2).   Under these circumstances, it will be necessary to use a good deal of tach feedback in order to improve the performance of the joint.   In a later section of this chapter it will be seen that this is not always possible to do.   Thus some compromise in response may be necessary.

There is another way to reduce or eliminate the position error, however. This is accomplished by adding an integrator to the control structure.   Such a scheme is referred to as PID (standing for *proportional, integral, derivative*) control. A position servo utilizing such a controller is shown in Figure 4.3.6.   In this figure the block marked "velocity loop" will usually consist of the amplifier and servomotor.   In order to see how the addition of the integral term affects the final value of the joint position, we consider the following example.

EXAMPLE 4.3.2

To simplify the transient response calculations, let us assume that the inductance and viscous damping of the servomotor are zero and that the amplifier pole can be neglected (because it occurs at such a high frequency). Then the position servo of Figure 4.3.6 becomes that shown in Figure 4.3.7.

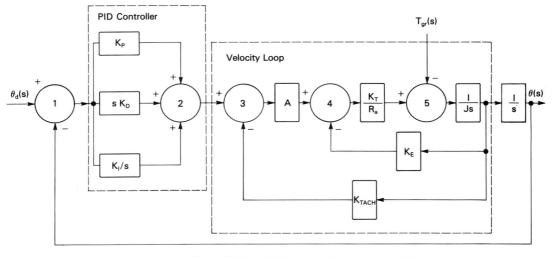

Figure 4.3.7.   Position servo for Example 4.3.2.

Here summing junctions 2, 3, and 4 are assumed to be unity-voltage-gain devices (e.g., noninverting op amps). Consequently, the tachometer constant K_{tach} has the units of volts/rad/s. The specific parameters of the velocity loop are

$$A = 10.0 \text{ V/V}$$

$$R_a = 1.62 \ \Omega$$

$$J_T = 0.0067 \text{ oz-in.-s}^2$$

$$K_T = 10.7 \text{ oz-in./A}$$

$$K_E = 0.0754 \text{ V/rad/s} = 7.8 \text{ V/1000 rpm}$$

$$K_{\text{tach}} = 0.056 \text{ V/rad/s} = 6 \text{ V/1000 rpm}$$

Note that J_T represents the sum of the motor and reflected load inertias. (See Section 3.3.1 for a discussion of reflected inertia calculation.)

Let the desired final position of the joint be $\pi/2$ radians, so that $\theta_d(s)$ $= 1.57/s$. In Figure 4.3.8 it is observed that the joint is initially at $\theta = 0$ rad, so that the gravitational force produces no additional load on the axis motor. However, as θ increases with time due to the command signal, the gravitational disturbance also increases and is, in fact, proportional to the sine of θ. For a particular joint geometry, we will assume that the magnitude of this disturbance is 21 oz-in. so that the time variation of the gravitational

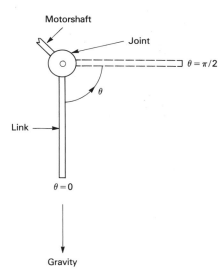

Figure 4.3.8. Joint used in Example 4.3.2. showing the effect of gravity as a function of angle.

torque will be

$$T_{gr}(t) = 21 \sin \theta(t) \qquad (4.3.12)$$

The results of a computer simulation of the step response of the joint modeled in Figures 4.3.7 and 4.3.8 are given in Figures 4.3.9 through 4.3.13. The first of these figures shows the system with proportional control only. It is observed that as K_p increases, the overshoot increases, as expected. Recall the discussion in Section 4.2.2. In addition, there is also a steady-state error due to the gravitational torque disturbance. As mentioned previously, increasing K_p reduces this error but at the expense of overshoot. This is clearly demonstrated in the figure.

The effect of adding derivative control is shown in Figure 4.3.10 for a fixed proportional term (i.e., $K_p = 20$). As expected, the larger the damping (i.e., K_D), the smaller the overshoot. It is seen that it is possible to obtain a response with practically no overshoot (i.e., critical damping), but the steady-state error is still present and does not vary with K_D.

Figure 4.3.11 demonstrates the effect of adding an integral term in the controller. Here proportional plus integral (*PI*) control eliminates the steady-state error with slightly increased overshoot. A PID controller is used in Figure 4.3.12. In this case the step response for different values of damping is shown for $K_p = 20$ and $K_I = 75$. It is observed that it is now possible to obtain a zero steady-state error (i.e., the desired final position is actually

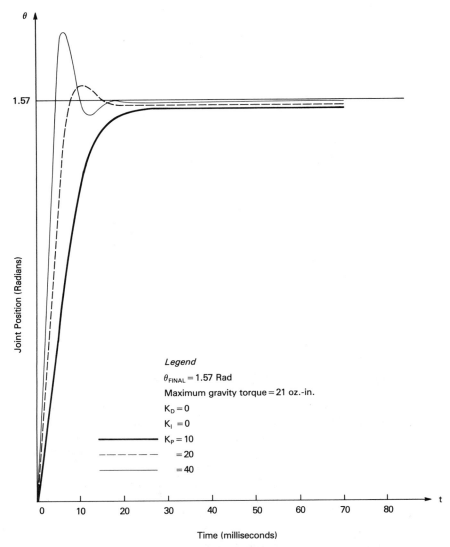

Figure 4.3.9. Step response of the system in Fig. 4.3.7 for proportional control only.

achieved) with no overshoot for $K_D = 0.02$. For completeness, a PID controller using different values of the proportional term is shown in Figure 4.3.13.

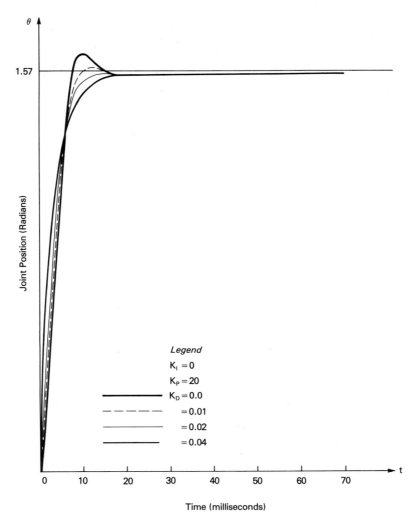

Figure 4.3.10. Step response of the system in Fig. 4.3.7 for proportional plus derivative (PD) control.

A rule of thumb for selecting the parameters of the PID controller is as follows:

1. With $K_I = K_D = 0$, adjust K_p until the system step response is either critically (or *slightly* under-) damped.

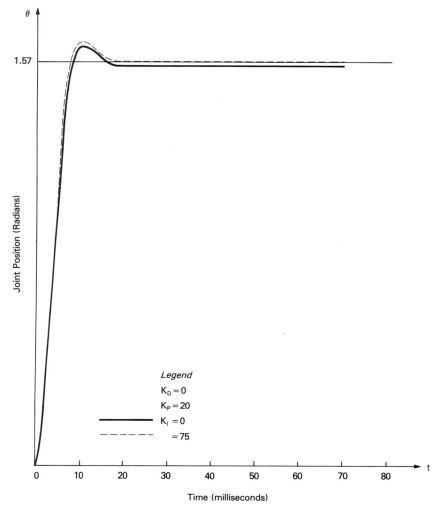

Figure 4.3.11. Step response of the system in Fig. 4.3.7 for proportional plus integral (PI) control.

2. For the value of K_p just found, increase K_I until the steady-state error is either zero or has reached an "acceptable" value.   (Normally, $K_I > K_p$.)

3. Increase K_D until the system step response is again either critically or slightly underdamped.

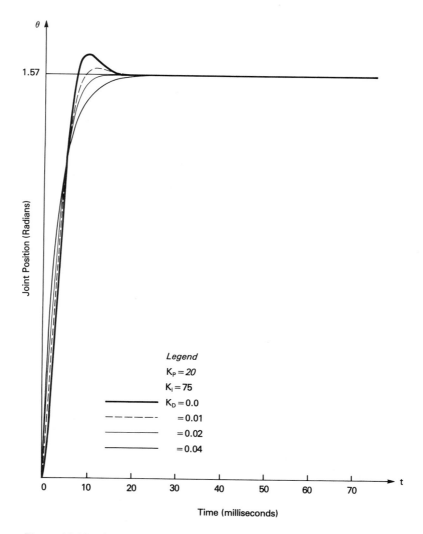

Figure 4.3.12. Step response of the system in Fig. 4.3.7 for proportional plus integral plus derivative (PID) control.

It is important to note that the PID controller can be synthesized using analog components (e.g., op amps). Alternatively, the individual joint processors can produce the required proportional, derivative, and integral terms with appropriate gain factors by operating on the error signal (see Appendix C).

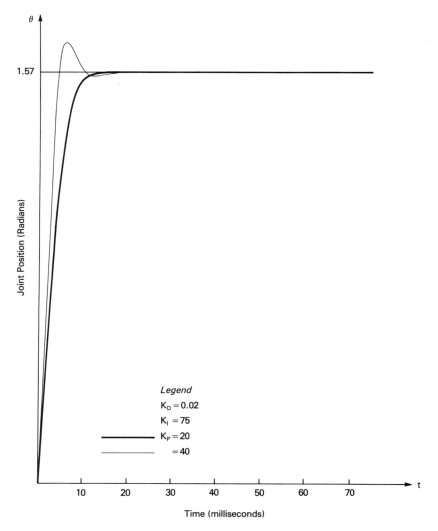

Figure 4.3.13. Step response of the system in Fig. 4.3.7 for PID control with two different values of K_p.

4.4 FREQUENCY-DOMAIN CONSIDERATIONS

It is often convenient and useful to look at the response of a robot axis servo from the frequency-domain point of view. If the time or transient response of the joint is inadequate (e.g., the joint will not faithfully track the command position signal), plots of the magnitude and phase of the frequency transfer function versus frequency will usually reveal where the problem lies and compensation can then be added to correct it (these graphs are referred to as *Bode plots*).

4.4.1 Bode Plots

Let us illustrate the frequency-domain approach through the use of an example. We begin by considering the tach or velocity portion of the position servo shown in Figure 4.3.5, consisting of an amplifier and servomotor. Assuming that the amplifier bandwidth is 1000 Hz (i.e., $\tau_A = 1/6280$) and that the motor in Example 4.3.1 is used with an inertial load of 0.007 oz-in.-s^2 [i.e., J_T in Eq. (4.3.5) is 0.0108 oz-in.-s^2], the tach open-loop transfer function is given by

$$GH(s) = \frac{14.1AK_g}{(1 + s/6280)(1 + s/44.14)(1 + s/439.73)} \tag{4.4.1}$$

The frequency transfer function (FTF) is obtained from this equation by substituting $j\omega$ for s, where ω, the radian frequency (having the units rad/s) is equal to $2\pi f$ (f in hertz). The magnitude of the FTF, expressed in decibels [i.e., $20 \log_{10}(|FTF|)$] and its corresponding phase angle both drawn versus $\log_{10}\omega$ are called Bode plots. This is shown in Figure 4.4.1 for the tach open-loop FTF.

In Figure 4.4.1a both the straight-line approximation and the continuous plots are given for $AK_g = 1$. The former is obtained by following a set of simple rules:

1. The FTF is placed in Bode form as shown in Eq. (4.4.2):

$$GH(j\omega) = K_{\text{Bode}} \frac{\displaystyle\prod_{i=1}^{M} (1 + j\omega/\omega_{z_i})}{\displaystyle\prod_{k=1}^{N} (1 + j\omega/\omega_{p_k})} \tag{4.4.2}$$

where ω_{z_i} and ω_{p_k} are called the "break" frequencies corresponding to each of the M zeros and N poles of $GH(s)$, respectively. Note that the transfer function in Eq. (4.4.1) is already in Bode form.

2. The magnitude of $GH(j\omega)$ expressed in dB is then

$$dB = 20 \log (K_{\text{Bode}}) + \sum_{i=1}^{M} 10 \log \left[1 + \left(\frac{\omega}{\omega_{z_i}} \right)^2 \right] - \sum_{k=1}^{N} 10 \log$$

$$\times \left[1 + \left(\frac{\omega}{\omega_{p_k}} \right)^2 \right] \tag{4.4.3}$$

Where "log" implies "log to the base 10." Note that multiple poles and zeros are permitted so that all of the ω_{p_k}'s and/or ω_{z_i}'s need not be distinct.

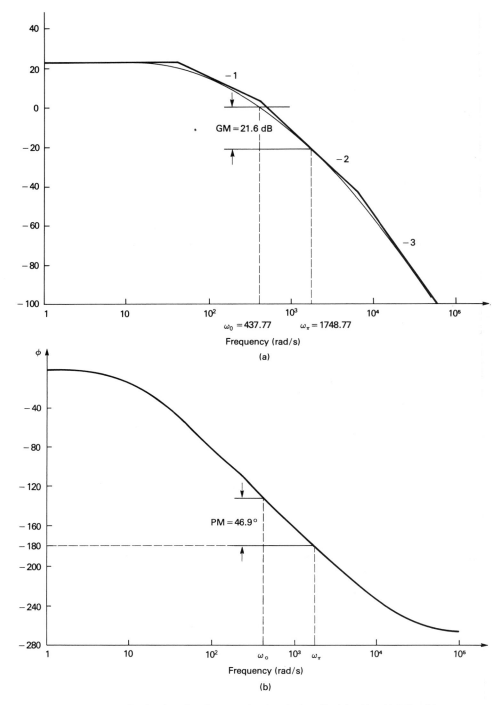

Figure 4.4.1. Bode plots for the open-loop tach described by Eq. (4.4.1) with $AK_g = 1$ and $K_g = 1$. The gain and phase margins are also shown: (a) magnitude (in dB) vs. $\log_{10} \omega$; (b) phase angle ϕ (in degrees) vs. $\log_{10}\omega$.

3. For each of the terms $10 \log (1 + [\omega/\omega_{break}]^2)$ in Eq. (4.4.3), the following is assumed:

$$10 \log \left[1 + \left(\frac{\omega}{\omega_{break}} \right)^2 \right] = \begin{cases} 0 \text{ for } \omega < \omega_{break} & (4.4.4a) \\[2ex] \pm 20 \log \dfrac{\omega}{\omega_{break}} & \text{for } \omega > \omega_{break} \quad (4.4.4b) \end{cases}$$

In Eq. (4.4.4b) the plus (minus) sign is used for the break frequencies corresponding to the zeros (poles) of $GH(s)$. As an example, in Eq. (4.4.1), the pole at 44.14 produces a term $- 20 \log (\omega/44.14)$ for $\omega > 44.14$.

4. The straight-line approximation to the magnitude plot is obtained from Eqs. (4.4.4a) and (4.4.4b) by drawing two lines, one having zero slope (in the range $\omega < \omega_{break}$) and the other having a slope of ± 20 dB/decade of frequency (in the range $\omega > \omega_{break}$) for each of the terms in Eq. (4.4.3). These two lines will intersect at $\omega = \omega_{break}$. The resultant set of straight-line segments is then summed algebraically and the constant $20 \log (K_{Bode})$ is added to the above *piecewise linear* curve, causing it to be *translated up or down*. Note that for a pth-order (i.e., multiple) zero or pole, the slopes will be $\pm 20P$ dB/decade.

The exact magnitude plot can be obtained from Eq. (4.4.3) by evaluating this expression at representative frequencies or from the straight-line approximation itself. The latter is accomplished by noting that for $\omega < 7\omega_{break}$ or $\omega > 7\omega_{break}$, the actual and straight-line curves approach one another. (Thus the latter is, in reality, an asymptotic approximation to the former). The maximum deviation between the actual and approximate curves occurs at $\omega = \omega_{break}$ and is equal to 3 dB multiplied by the difference in slopes (in multiples of 20 dB/decade) on either side of the break.

It is important to note that the results above are valid provided that:

1. Successive break frequencies are separated by a factor of about 7, so that they do not "interact" with each other. If this is not so, the error between the actual and straight-line plots at the break frequency will not be 3 dB times the difference in slopes.

2. All of the zeros and poles of the open-loop FTF in Eq. (4.4.2) are real numbers. If instead, some are complex quantities (appearing in complex conjugate pairs), the actual plot may be significantly different from the straight-line approximation in the vicinity of ω_{break}. In fact, the true plot may actually exhibit a resonance. This situation will not be considered here. However, the reader who is interested in further information is referred to any one of the several references listed at the end of this chapter.

It is also possible to develop a set of simple rules for obtaining the approximate phase versus frequency curve. This will not be discussed in this chapter, however. The reader can, if desired, find an excellent discussion of the subject in Dorf [1].

4.4.2 Gain and Phase Margins

Two measures of relative stability for a system, the gain and phase margins (GM and PM), can be obtained directly from the Bode plots and are defined as

$$\text{GM} = -20 \log (|\text{FTF}|) \text{ at } \omega_\pi$$
$$\text{PM} = \pi - \text{angle of FTF at } \omega_o \tag{4.4.5}$$

where ω_π is the radian frequency for which the phase angle of the FTF (for the *open-loop system*) is π. Also, the FTF is unity (i.e., 0 dB) at ω_o. Physically, the GM tells the designer how much the gain (i.e., K_{Bode}) can be increased before the closed-loop system becomes unstable. Similarly, the PM indicates how much additional phase lead can be tolerated before instability results. Usually, if both the GM and PM are positive, the closed-loop system will be stable. If either or both are negative, the closed-loop system will probably be unstable. A commonly used design objective is to adjust (i.e., compensate) the open-loop system so that its GM = 20 dB and the PM = 45°.

Applying the definitions in Eq. (4.4.5), the GM and PM for the tach loop in Figure 4.4.1 are 21.6 dB and 47°, respectively (see Problems 4.10 and 4.11). These results imply that the tach closed loop is stable provided that $AK_p < 12$. (This can most easily be seen by understanding that increasing the gain AK_g by NdB is equivalent to shifting the entire magnitude plot vertically NdB. Although the phase plot is not affected by the gain change, the GM and PM *will* change.)

4.4.3 Approximate Closed-Loop Frequency Plot

An approximation to the frequency response for the closed tach loop can quickly be found from the open-loop plot by noting that for the general feedback system in Figure 4.4.2, the closed-loop transfer function is given by

$$T(j\omega) = \frac{G(j\omega)}{1 + G(j\omega)H(j\omega)} \tag{4.4.6}$$

When the gain of the open-loop transfer function is large [i.e., $G(j\omega)H(j\omega) \gg$

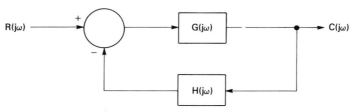

Figure 4.4.2. General form of a closed-loop system.

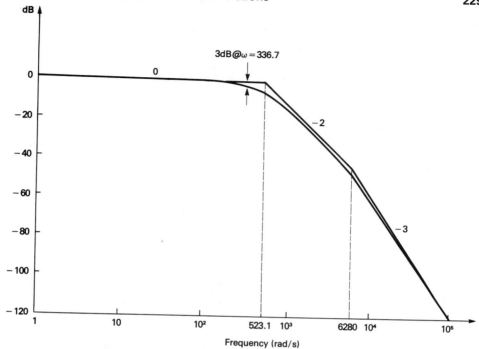

Figure 4.4.3. Approximate magnitude plot for the closed tach loop in the example in Section 4.4.1. The result is obtained from Fig. 4.4.1a using the approximations in Section 4.4.3.

1], $T(j\omega)$ is approximately equal to $1/H(j\omega)$, whereas it is approximately equal to $G(j\omega)$ when $G(j\omega)H(j\omega) \ll 1$.* The easiest way to determine the approximate closed-loop response using these ideas is to obtain the piecewise linear plot and then, from this, draw in the continuous frequency curve. The approach to follow is that for all frequencies such that the magnitude of $G(j\omega)H(j\omega) > 1$, the straight-line plot is found from $1/H(j\omega)$ using the rules presented in the preceding section. It is worthwhile noting that if $H(j\omega)$ is a constant, as is often the case, this portion of the response will be a *horizontal* line at $20 \log [1/|H(j\omega)|]$. For all other frequencies, $T(j\omega)$ is $G(j\omega)$ and the piecewise linear plot is obtained from this function.

Making use of this technique for the open-loop transfer function of Figure 4.4.1 results in the piecewise linear approximation shown in Figure 4.4.3. Note that the horizontal line to the left of 523.1 rad/s is at the 0-dB level because it was assumed that both K_g and AK_g were unity (see Figure 4.4.1). If this is not so (e.g., $AK_g = 1$ but $K_g = 5$), the horizontal line in Figure 4.4.3 is at -14 dB for $\omega \leq 1166.4$ rad/s. The rest of the curve stays the same (see Problem 4.12).

---

*Strictly speaking, this approximation is valid only for closed-loop systems that are *overdamped*. However, even in the case of closed-loop systems that are underdamped, it is possible to obtain a rapid "ballpark" estimate of the frequency response.

Recalling that the break frequencies occur at the poles or zeros of the transfer function, the approximate tach closed-loop transfer function can be obtained from this figure. Thus

$$G_{tach}(s) = \frac{1}{(1 + s/523.1)^2(1 + s/6280)}$$ (4.4.7)

Note that the double pole at 523.1 occurs because the slope of the straight-line plot changes from 0 to -2 at this frequency. It is important for the reader to understand that the continuous curve shown in Figure 4.4.3 is what one would get for the FTF of Eq. (4.4.7). However, it is *still an approximation of the actual closed tach loop response*.

4.4.4 Bandwidth and Tracking Error Considerations

Often, the *bandwidth* of a system is defined as the frequency where the response is down from its peak value by 3 dB. Using this definition, it is seen from Figure 4.4.3 that the closed-loop tach bandwidth is about 336.7 rad/s or 53.6 Hz. Note that the motor bandwidth was 44.14 rad/s or 7.03 Hz. Thus it is observed that closing the loop has significantly increased the bandwidth of the system. This is, of course, a well-known result.

A more important point to remember is that the bandwidth limits the maximum speed with which a system can respond to an input signal. That is, if the input requires a rapid change (with respect to time), the system must have sufficient bandwidth to follow (or "track") this command. Otherwise, the response will significantly lag behind the command input, thereby producing a large "tracking error."

An example of such behavior is shown in Figure 4.4.4. Here the effect of moving the single system pole is observed. In Figure 4.4.4a, the pole is located at $s = -2$ (bandwidth = 2 rad/s), whereas in Figure 4.4.4b it is at $s = -0.5$ (bandwidth = 0.5 rad/s). Note that the tracking error is much smaller for the first case (where the bandwidth is larger).

Amplifier saturation and sluggish performance are possible consequences of large tracking error. Moreover, in robotic applications, this quantity is often monitored in order to determine whether the manipulator is actually moving (i.e., has it hit an unforeseen obstacle or is the load too large to handle?). If the joint normally exhibits a large tracking error because of insufficient bandwidth in the servo loop, the criterion for automatically halting the manipulator motion must be relaxed. This can have serious safety consequences and can also permit damage to electrical and/or mechanical components. In addition, if the tracking error of all of the joints of the robot is not the same, it will be virtually impossible for the robot to move its end effector in a straight line.

In more mathematical terms, if it is required that at least a certain percentage of the total energy contained in a signal be "passed" by a system in order to reproduce it faithfully at the output, and the frequency at which this occurs is f_o,

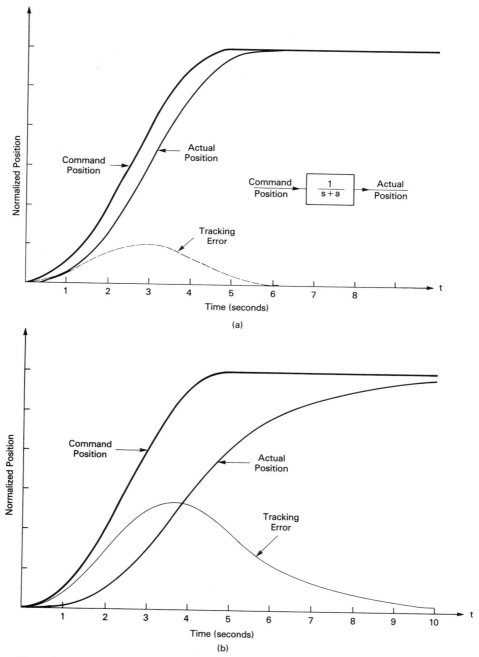

Figure 4.4.4.   Illustration of the effect of system bandwidth on tracking error: (a) $a = 2$; (b) $a = 0.5$.

then the system bandwidth must be at least equal to f_o to ensure that the signal is transmitted without significant distortion. The common definition of bandwidth (i.e., 3 dB) is used because the "half-power" frequency occurs when the response of the system is down by 3 dB. However, it should be noted that the bandwidth based on this definition may not produce an adequate reproduction of the signal and something far more stringent may be required. For example, if it is found that 98% of the energy content of the input must be transmitted (instead of 50%) in order to produce adequate tracking of the signal, the required system bandwidth will be much higher than that predicted by the 3-dB criterion.

As the system in Figure 4.3.5 is a position servo, we are really interested in the frequency response of the position loop. Therefore, once again referring to this figure and also using Eq. (4.4.7), the open-loop position transfer function is found to be

$$GH(s)_{\text{position}} = \frac{K_p}{s} G(s)_{\text{tach}}$$

$$= \frac{K_p}{s\left(1 + \dfrac{s}{523.1}\right)^2 \left(1 + \dfrac{s}{6280}\right)} \qquad (4.4.8)$$

The magnitude and phase plots (for $K_p = 1$) are shown in Figure 4.4.5.

The gain and phase margins for this loop are found to be 59.1 and 89.8°, respectively. From a stability point of view, it appears that the position loop will behave more than adequately. However, closing the loop will produce a 3-dB position bandwidth of less than 0.16 Hz (1 rad/s). Thus, the joint will only be able to track slowly changing position commands (e.g., those that require about 1 s to reach about 63.2% of their final value). This response may be inadequate for the joints of a high-speed robot, so it is necessary to increase the position loop bandwidth.

4.4.5 Compensation of a Position Servo

A simple way to raise the closed-loop bandwidth is to increase the position gain constant K_p. Such a procedure is referred to as *gain compensation*. For example, if $K_p = 70$ (37 dB), the GM = 22.2 dB and the PM = 74.4°. Using the approximation described in the preceding section for closing the loop, the half-power bandwidth is about 10.8 Hz (67.5 rad/s), so that the joint will respond to position commands that reach 63.2% of their final value in about 15 ms (1/67.5). For most robotic joints, this is more than an adequate response time. For example, the U.S. Robots' Maker 100 (e.g., see Figure 1.9.6) has a bandwidth of about 4 to 28 Hz (for each of its joints). The reason for the range is that joint bandwidth is a function of the load and also the geometric configuration [i.e., whether or not the manipulator is fully extended (which produces a lower bandwidth)].

If for some reason a bandwidth of 10.8 Hz is not large enough, the position

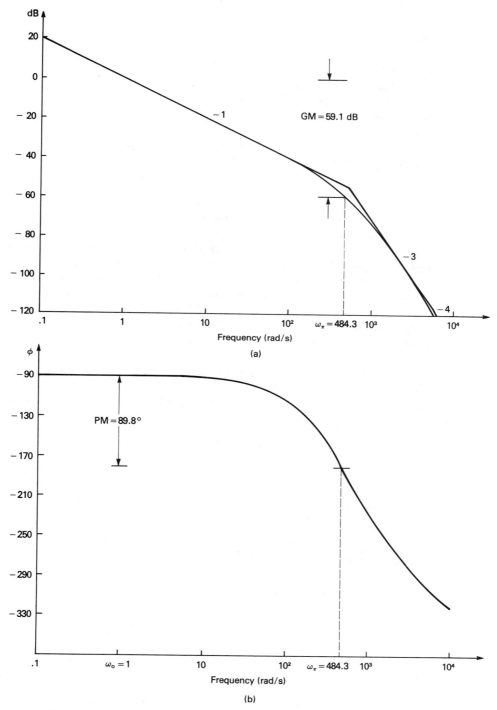

Figure 4.4.5. Bode plots for the open position loop described by Eq. (4.4.8) with $K_p = 1$: (a) magnitude plot; (b) phase plot.

gain constant can be further increased. For example, if $K_p = 40$ dB, the GM = 19.1 dB and the PM = 68.2° with the closed-loop bandwidth about 14.9 Hz. From a stability point of view, this value of K_p would be about as large as one would be able to use. A still larger bandwidth could only be obtained by other means (e.g., lead compensation).

As an example of this technique, consider the transfer function of a lead compensator described in

$$T_{\text{lead}}(s) = \frac{1 + sa}{1 + sb} \tag{4.4.9}$$

where $a > b$. Although the compensator can be placed in either the tach or position loop, suppose that we use it in the former. The position servo of Figure 4.3.5 becomes that shown in Figure 4.4.6. The break frequencies $1/a$ and $1/b$ must be selected so that the compensator increases the closed tach loop bandwidth. From Figure 4.4.1 it is seen that the zero should be placed somewhat before the second break frequency of the open-loop tach FTF (i.e., 439.73 rad/s) and the pole after the 0-dB crossover frequency (i.e., 523.1 rad/s, e.g., $1/a = 250$ and $1/b = 2500$). The piecewise linear approximation to the magnitude plot for the open tach loop with and without this compensation is shown in Figure 4.4.7. Also shown in the figure is the approximation to the magnitude plot for the closed tach loop.

The closed-loop tach bandwidth is found (using the approximation in Section 4.4.3) to be 147.6 Hz (up from 53.5 Hz) and the GM = 19.2 dB and PM = 73.4°. From Figures 4.4.5 and 4.4.6, the position open-loop FTF is found to be

$$GH(j\omega) = \frac{K_p}{j\omega(1 + j\omega/1094.8)(1 + j\omega/2500)(1 + j\omega/6280)} \tag{4.4.10}$$

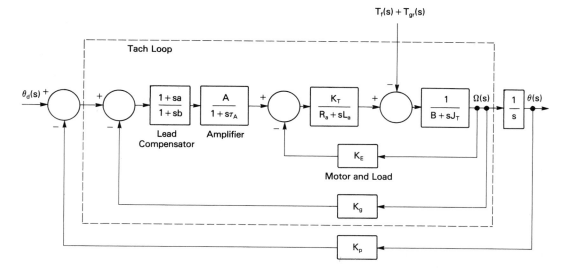

Figure 4.4.6.   Position servo with lead compensator in the tach loop.

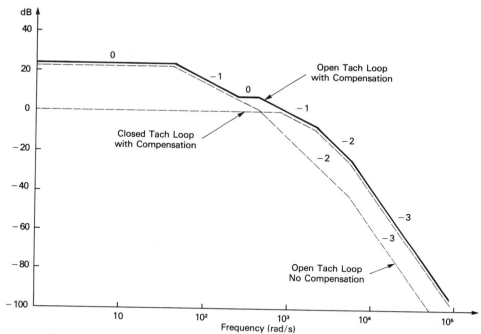

Figure 4.4.7. Piecewise-linear approximation for tach loop of system in Fig. 4.4.6 with and without compensation. Also shown is the closed-loop tach response.

Using this equation with $K_p = 40$ dB, the (position) gain and phase margins are found to be 27.6 dB and 81.6°, respectively, with the closed-position loop bandwidth now 15.7 Hz. Thus the position servo bandwidth has been slightly increased by using lead compensation. However, since the phase margin is also higher, it is also possible to increase the position gain constant (to 47.6 dB). Doing this results in a GM, PM, and closed-loop bandwidth of 20 dB, 70.6°, and 36.0 Hz, respectively. Clearly, the use of simple lead compensation has more than doubled the bandwidth while not affecting the relative stability of the closed-loop system. Thus the joint could, if required, respond to command signals having significantly higher frequency content than before (i.e., it could be made to move considerably faster. See Problems 4.15 and 4.16.)

Before leaving this topic, it is important to understand that a large bandwidth *may not always be desirable.* For example, it is well known that high-frequency interference (noise) will adversely affect the performance of a system if its bandwidth is too large. Also, where a digital-to-analog converter (DAC) is employed, it is critical that the servo not respond to each individual piecewise constant update from the DAC. Thus it is often necessary to place an *upper limit* on the system bandwidth in order to ensure proper performance.

In looking at the performance of a robotic joint from a frequency-domain point of view, we tacitly assumed a particular *structure* for the servo.   Actually, it is possible to configure the servo in many different ways, and this is the subject of the next section.

4.5 CONTROL OF A ROBOTIC JOINT

In the previous sections, it was assumed that the position and velocity information was "available" with no thought being given as to how one actually obtains this information.   Also, it was assumed that all components in the position servo were analog devices.   Although this is useful in the analysis of servo loops, it is not usually the case in practice.   What is normally done in a robotic joint servo is to utilize either a digital approach where all sensory information is obtained and processed in a digital fashion, or else *both* analog and digital techniques are used to obtain and process information.

Regardless of the scheme employed, however, the command signal to the joint servo of a robot is invariably obtained from a microprocessor (i.e., the "master") and is, therefore, digital in nature.   It is important to understand that this implies that the input to the joint is not a continuous-time function but is, instead, a "sampled" signal which is updated (i.e., changed) only periodically (e.g., every 25 ms) by the master or a special math coprocessor.   Such an approach is taken because the master must send information to *all* the joint servos (e.g., six in a six-axis robot).   Consequently, it must have enough time to complete the various computations required in the path planning algorithm and then to communicate this information to the individual joints.   We will call the update (also referred to as either a sample or *set point*) time T_s.

Although it is quite feasible to convert the digital position command into an analog signal by using a digital-to-analog converter, this is not often done.   Instead, the individual joint processors themselves perform an interpolation between two consecutive set points output by the master.   For example, if $T_s = 25$ ms, the *interpolated* set points might occur every 3.33 ms ($T_s/8$), implying that the master update interval is divided into eight subintervals.   As a consequence, considerably smoother manipulator motion is produced.

In practice, it is possible to obtain both position and velocity data in either an analog or a digital fashion using the same or separate devices (i.e., sensors) for monitoring these signals.   We now consider several different possibilities.

4.5.1 Digital Position and Analog Velocity: Separate Sensors

A configuration that is often encountered in equipment used today by the machine tool industry (e.g., *xyz* incremental motion devices) utilizes separate sensors for monitoring velocity and position. An optical encoder (or some other digital position sensor; see Section 5.3.2.2) mounted on either the motor shaft or output (i.e., workpiece or tool tip) is used to produce a digital representation of the current position. Velocity information is provided through the use of an analog tachometer similarly mounted. Since the information from the latter device must be converted into digital data in order to be utilized by the joint processor, an analog-to-digital (A/D) converter may have to be employed.

For the joint servo in a robot, the command signal from the master computer is digital in form. As mentioned previously, the joint processor uses this input together with the position and/or velocity information to produce an error signal which is used either directly or indirectly as the input to the servo (power) amplifier. Two approaches are possible.

In the first case, the joint processor uses *both* the position and velocity information to produce a velocity error signal, which is in turn, used as the drive

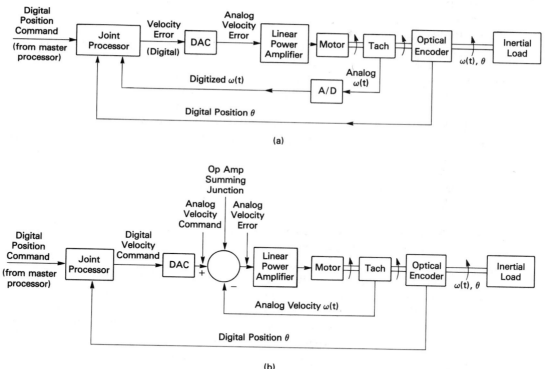

(a)

(b)

Figure 4.5.1. Two realizations of a joint position servo utilizing digital position and analog velocity information: (a) both ω and θ are digitized and fed back to the joint processor to obtain the velocity error signal; (b) only the digitized value of θ is used to produce the velocity command. Then, this is compared to the analog value of ω to obtain the velocity error signal.

signal for the power amplifier. In this instance, the tach signal must be digitized so that an A/D converter is required (note that this produces a *sampled* version of the continuous tach signal). After the joint processor compares the sampled versions of the current actual position and velocity with the command position (received from the master), the resultant error signal is either output to the amplifier directly (through a peripheral interfacing device such as a PIA which outputs TTL-level pulses) if a pulse-width-modulated (PWM) amplifier is employed (see Section 4.11) or through a D/A converter is a linear servo amplifier is being used.

In the second case, the joint processor uses only the sampled command and actual position information to determine the sampled *position* error. This difference in command and actual positions, or "delta position," at each sampling instant T_s may be thought of as a *velocity command* signal to the tach loop. (To see this, the reader should recall that velocity is approximately equal to "delta position"/ "delta time" and here "delta time" = T_s.) The velocity command is usually converted into analog information using a D/A converter (although it may not be necessary if a PWM amplifier is employed). An *analog* velocity error is created by comparing this signal with that fed back directly from the tachometer. The comparison is usually performed using analog hardware (e.g., an operational amplifier). Both of these cases are illustrated in Figure 4.5.1. Note that the second case shown in Figure 4.5.1b substitutes an op amp for an A/D converter. A cost saving can be realized by taking this approach since the tach provides a continuous flow of velocity information, whereas the A/D converter only gives the information at the sampling times. If it is necessary to have closely spaced samples of the tach signal in order to maintain proper tracking of the joint and/or smoother manipulator performance, a high-speed A/D converter may be required at a significant cost compared to that of an op amp (e.g., tens of dollars as compared to about a dollar for the op amp; recall that for an N-axis robot, this cost differential would be multiplied by N).

4.5.2 Measured Digital Position and Derived Digital Velocity: Single Sensor

The previous technique of obtaining position and velocity information required the use of two separate sensors: an encoder and a tachometer. An obvious disadvantage of this is that the extra device (e.g., the tach) will increase the cost of the robot joint considerably (typically by $50 to $100). In certain instances, the extra expense may be warranted and, in fact, this procedure is regularly used by manufacturers of *x-y* motion devices. However, two extremely important considerations in a robot often dictate against using a separate tach—the additional space and weight.

Quite often, it is necessary to design a robotic joint so that the motor is actually moved within that joint or some other one. If a tach is to be used, it is normally attached directly to the motor shaft so that the combined weight and volume of the package is increased (sometimes by as much as 100% over the motor alone). Since weight and volume in a robot joint directly influence the size and cost of many mechanical and electrical components, anything that increases these quantities must be justified from a performance point of view. In certain instances it is possible to obtain the same dynamic performance without resorting to the extra piece of hardware. Consequently, there may be a considerable savings in volume and weight, and it is therefore not surprising that many robots today do not employ analog tachometers.

How can we apparently "have our cake and eat it too" since, in an earlier section of this chapter, we learned about the need for velocity feedback? That is, how can the velocity information be obtained without using a tachometer? The answer to this question is found by recalling that velocity is the time derivative of position and also that a *first-order* approximation to a derivative is given by

$$\omega(t)_{approx} = \frac{\Delta\theta}{\Delta t} \qquad\qquad (4.5.1)$$

In practice, position is measured by reading the encoder count at specific (sampling) instants of time (e.g., either T_s or the time between *interpolated* set points). If the joint processor takes the current position count and substracts it from the one obtained at the previous time, this difference represents $\Delta\theta$. Usually, the time between samples is fixed, so that ΔT [in Eq. (4.5.1)] is always equal to one sampling instant. Thus $\Delta\theta$ obtained in the manner above is, in reality, directly proportional to an approximation of the angular velocity, $\omega(t)_{approx}$.

As might be expected, there is a "price" to be paid for doing the above. That is, the approximation to the tach signal may be inadequate, and difficulties may result. An example will illustrate the potential problem.

EXAMPLE 4.5.1

We wish to calculate the effective resolution of the digital position and velocity information when a 200-line optical encoder is placed on the shaft of a motor that is used to drive a particular joint of a robot. Suppose that, in addition, a 50:1 gear reduction is used to couple the axis output to this motor and that it takes 40 motor revolutions to cause the axis to move from one motion limit (or extreme) to the other (these are often referred to as *stops* or *limit stops* and be of either the hardware or software variety). Consequently, the axis can move a total of 288° ([40/50] × 360). Thus the encoder will generate a total of 8000 (200 × 40) lines or counts over the entire range of travel, which implies that the joint processor (and associated digital hardware) must have the ability to count this number. That is, they must be able to handle a 13-bit number representing the joint position. In this case the smallest change in angular position of the axis that can be obtained is 0.036° (288/8000) and the position resolution is said to be 0.036°.

Now let us suppose that the scheme for deriving the velocity signal described above is employed. If it is assumed that the *maximum* velocity of the joint is 20 motor revolutions per second ($=$ 4000 lines/s), then if the master processor is sending out position updates (set points) every 25 ms, the maximum change in position count between any two samples will be about 100 counts. Thus the maximum velocity will be represented by a 7-bit number and will have a resolution of 0.072°/s. More important, it is recognized that the difference between the current and previous position counts will be much smaller during the acceleration and deceleration phases of the motion when the joint is moving at velocities well below the maximum. As a consequence, the approximate velocity derived in this manner during these times will be significantly less than 7 bits (e.g., 1 or 2 bits to begin with).

The effects of reduced velocity resolution and quantization are illustrated in Figure 4.5.2a and b. Here the ideal quantized velocity is obtained by taking the difference between the current and previous samples of the *ideal continuous* position signal [$= \Delta$ (actual position)] and then quantizing the result (i.e., "integerizing" it). Often, however, the actual position is quantized *before* the differencing operation between current and previous position is performed. This produces the quantized velocity curve shown in the figure. It is observed that although the ideal velocity signal is a continuous function of time, the approximate velocity derived from the encoder (position) signal is piecewise constant or a "staircase" function.

Now it will be recalled that the tach gain must be increased in order to increase the damping and hence reduce oscillations in the joint. Therefore, the 7-bit number representing the joint velocity will have to be multiplied by some factor. For example, suppose it is found that a K_g of 16 is required in order to produce the proper performance. Because of the quantizing effect of the procedure described above, the initial value of the amplified tach signal will be zero and will then jump several sampling instants later to 16 and stay at that level for several more sampling times (see Figure 4.5.2). The effect of a staircase damping term in the velocity loop may well produce undesired oscillations in the joint response and could even cause it to become unstable. Thus even though increasing the tach loop gain should reduce the oscillations, it is seen that quantizing may actually cause the problem to become worse.

Several techniques can be employed to reduce or eliminate this potential problem. These include:

1. Increasing the resolution of the encoder by using "times 4 logic" (see Section 5.3.2.2) and/or increasing the number of lines on the encoder disk. (The second method can significantly increase the cost of the encoder.)

2. Reducing the maximum value of K_g (which may cause the joint to be stable but still more underdamped than is desirable—recall the discussion of Section 4.2.2).

3. Obtaining a better approximation to the velocity, that is, using a second- or

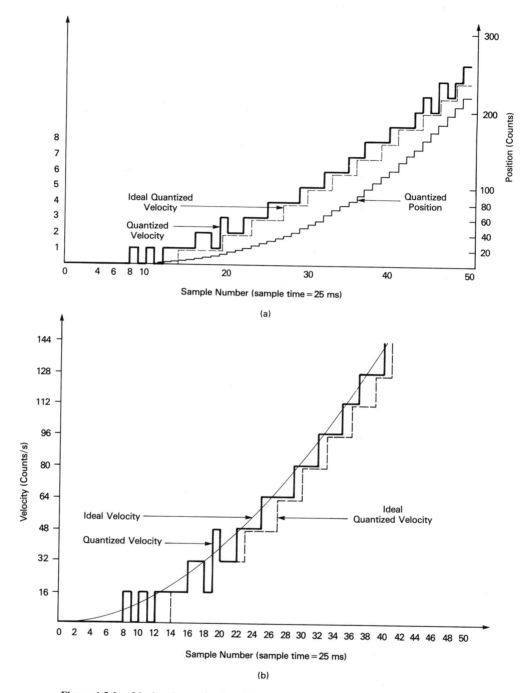

Figure 4.5.2. Ideal and quantized position velocity: (a) quantized position and δ position (velocity); (b) scaled ideal and quantized velocity signal (scaling factor = 16). [Note: quantized velocity = δ (quantized position); ideal quantized velocity = integer value (delta [actuator continuous position]).

higher-order approximation to the derivative (which may require significantly increased computation time).

4. Using two or more different values of K_g: for example, a small one during acceleration (and deceleration) and a larger one during constant velocity.

It is worth noting that the problem of reduced velocity signal resolution can also arise during the constant-velocity portion of the joint motion. The reason for this is that it is often necessary to operate a robot joint (or joints) at less than maximum velocity; for example, when the maximum payload is being carried by the gripper or in a spray-painting application where the thickness of the paint is, in part, determined by the linear speed of the sprayer. In this instance the number of bits of velocity information is reduced and a situation similar to that which is observed during acceleration and deceleration of the joint can occur.

4.5.3 Measured Analog Velocity and Derived Analog Position: Single Sensor

Although a tachometer is usually much larger and heavier than an encoder, the relatively high cost of the latter device might make eliminating it from the system quite an attractive idea. If this was done, the tach would have to be used to provide both the velocity and position information. An op amp integrator could be used for this purpose since position is the integral of velocity.

When position is obtained indirectly by sensing velocity as described above, inevitable difficulties arise. These might prevent one of the major requirements of a robot from being met (that it be able to move to points within its workspace in an *accurate* and *repeatable* manner). Consider two problems that are directly attributable to the op amp:

1. Op amp drift as a function of time. Repeatability during the course of a work shift might be affected significantly.
2. Nonideal integration. It is clear that accuracy would suffer in this case.

Another problem with using a tach to provide both velocity and position information is that it is not perfectly linear, that is, the output voltage is not linearly related to shaft angular velocity. This is particularly true at low velocities (i.e., below 100 rpm).

If these problems could be overcome, the analog position and velocity information could be used directly, thereby requiring D/A conversion of the digital position command signal. In this case it might be possible to eliminate the joint processor provided that adequate smoothing of the D/A output could be achieved with analog circuits (e.g., a filter). Alternately, A/D conversion of both analog signals would permit them to be compared with the command position in the joint processor in a manner similar to that described in Section 4.5.1. Clearly, the success of this configuration would depend on being able to get accurate and stable (with respect to time) position information.

4.5.4 Measured Analog Position and Derived Analog Velocity: Single Sensor

As already mentioned, the cost of the encoder and tachometer is not insignificant. Consequently, eliminating both of these sensing elements and replacing them with something less costly is an extremely attractive idea. In fact, there is an inexpensive method of measuring the position of a rotating shaft, that is, by using a rotary potentiometer (or "pot") which produces a voltage that is proportional to shaft angle. This device is discussed more fully in Section 5.2.1.

If position is measured in the foregoing manner, the velocity information can most easily be obtained by differentiating the pot output. For this purpose, an op amp can be used. A problem with this approach is that a differentiator is a high-pass filter, and hence it tends to preemphasize electrical noise in the system. Another potential problem is that pots are inherently noisy devices with relatively limited life. This is due to the fact that the wiper (i.e., the moving member) must physically touch the resistive element. The contact deteriorates with time, becoming dirty and nonuniform, and produces signals that can be exceedingly unreliable. Consequently, of the four configurations discussed, this is the one that is the most difficult to use.

To summarize this section thus far, four different techniques of obtaining position and velocity information that can be used to control the joint of a robot have been presented. Only the one described in Section 4.5.2 is purely digital in nature. The remaining three are hybrid schemes involving some analog and some digital aspects (although in the last two, only the command signal is digital). Of the four, the first two are commonly employed in a variety of incremental motion applications (e.g., x-y tables). Many of the robots on the market, as of this writing, use the all-digital approach, primarily because of the added cost, weight, and volume of a separate tachometer.

At this juncture, a point worthy of mention has to do with the actual placement of the position and/or velocity sensors. It was stated previously that these devices can be placed either on the motor shaft or at the drive output. The major advantage in monitoring the work output is that, in principle, it permits careful control of the end effector by providing information about what is actually happening at the tool tip. Thus disturbances and oscillations should be handled quite well. On the other hand, using sensors on the motor shaft means that we "control" only the motor output. The tool tip is "unobservable." We make the assumption that *the joint output and hence the tool tip follows the motor exactly*, implying that there is almost perfect rigidity among mechanical members. Clearly, this is not correct and yet, most robots, and for that matter, most incremental motion devices, utilize this approach.

The most important reasons why this should be so are as follows:

1. It may be physically difficult to mount the encoder and/or tach on the output end of the drive. This is particularly true in the wrist axes, where space is at a premium.

2. Mechanical nonlinearities such as gear backlash, Coulomb friction, and non-rigid linkages may produce a "limit cycle" or "ratcheting" type of oscillation. It is possible to compensate for such an instability by injecting a high-frequency or "dithering" signal into the servo. Despite this, it is not common practice to monitor position and velocity at the output of a robot joint probably because the mechanical components tend to deteriorate with use. As a consequence, the robot could become suddenly unstable, creating a significant safety hazard.

3. Increased position resolution resulting from the multiplication of the encoder resolution by the speed reduction ratio of the coupling device (e.g., a gear train). See Chapter 5 for a more in-depth discussion of this important point.

4. Reduced problems due to "digital jitter" (i.e., digital limit cycle). This is discussed more fully in Section 5.5.

Before leaving the topic of the control of robotic joints, a brief discussion of adaptive control and optimal control is in order. The former topic is discussed in the next section.

4.5.5 Adaptive Control

Up to this point it has been assumed that all gains and parameters associated with the joint servo of a robot are *fixed with time*. In fact, this is true for virtually every manipulator currently being manufactured. A number of robot controllers *do* permit the user to specify the load (weight) before performing any move operations. Parameters that give the best (compromise) performance are then downloaded from a table located in memory. Once these quantities are set, however, they are usually *not modified* unless the machine is informed (by the operator) that the load has been changed.

Previously it was stated that robot controllers and actuators are almost always designed assuming that inertial loads are constant, even though this is far from valid, as was seen, for example, in Figure 4.3.4. As a consequence, motors must be more conservatively selected (i.e., larger) and performance may suffer. That is, it might not be possible to move the end effector as fast as desired with a given load, and/or the motion might not be as smooth as required. If, however, one could dynamically compensate for inertial load variation by constantly adjusting servo parameters, improved operation characteristics would no doubt result.

Although not currently done, one approach for improving the operation of a robot would be to utilize *adaptive control* whereby the system response is monitored and parameters and/or gains modified continuously in time so as to produce the "best" or "optimum" results, e.g., the fastest possible motion from one point in space to the other with the least amount of vibration (for a specific actuator). Most often, this adaptive approach produces an overall system that is both nonlinear and time varying, so that the analysis and synthesis problems tend to be quite complicated.

In the case of a robot, where movement is restricted to its workspace and the motion profile is well defined and known *a priori*, it might be feasible to design *preprogrammed* time variations of controller parameters to achieve an instantaneous optimum control at all times. Such an approach, referred to as *preprogrammed adaptive control*, has been used for years in the field of missile guidance. It is noted that this scheme would be used *in addition to normal feedback procedures*.

In many instances, memory and/or computational time constraints would not permit the system parameters to be varied continuously with time. In such cases, a *zoned adaptive* approach could be adopted whereby the entire workspace would be divided into a *finite* number of distinct subspaces or zones. Then gains and parameters could be changed in a predetermined manner as the manipulator moved from one of these previously defined zones into another. Clearly, this technique would require far less memory and computation time than the continuous one. Since parameter values would change abruptly as the system entered a new zone, care would have to be taken to prevent such an action from introducing unwanted vibrations.

The adaptive procedures described above are relatively simple and depend on being able to predict the motion of the manipulator beforehand. It is possible, however, that unforeseen variations in the system and/or the work environment (e.g., noise), could occur which could not be handled by the preprogrammed adaptive scheme working in conjunction with a standard feedback controller. In this case, another adaptive approach that has been applied to other systems is potentially applicable to robots. Specifically, a measure of the system operation, called a *performance index* (PI), is constantly monitored and system parameters adjusted so as to optimize in some manner (e.g., either minimize or maximize) the PI. Once again, the adaptive control loop formed in this way would operate in addition to the normal feedback scheme used to control position. Note that the AGC (automatic gain control) in radio, which adjusts the receiver gain so that its output level is relatively constant over a wide range of input signal amplitudes, is an example of such an adaptive system.

A robot control system with a general adaptive controller added is shown in Figure 4.5.3. Here the adaptive control is used to compensate for time variations in the robot's transfer function which are produced by the environment and unpredictable component variation due to wear and manufacturing irregularities. By monitoring the performance index (this is sometimes referred to as "the identification process"), the effects of these factors on the robot's operation is measured continuously. The decision logic is then used to evaluate this information and determines how to modify the controller's parameters so as to produce an optimum response (e.g., in the desired position). Alternatively, the modification process

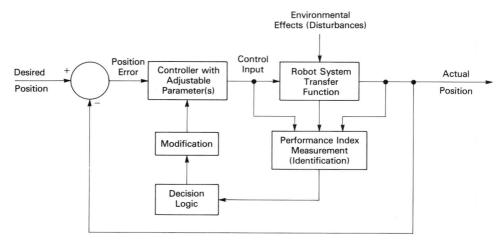

Figure 4.5.3. Adaptive controller added to a normal position feedback control loop (represented by the system transfer function) of a robot.

can be used to adjust the control input to the robot itself to produce "good" performance. In general, adaptive controller design involves the choice of a *physically meaningful and computationally tractable PI*, selection of a controller topology and the set of appropriate parameters to be adjusted, and finally, a method of dynamically adjusting those parameters.

Another variation of the adaptive approach, shown in Figure 4.5.4, is called a *model reference adaptive control system*. In this instance it is assumed that a known (reference) model of the actual system exists that has the desired response characteristics to the command signals. Then an observed response error signal is generated by comparing the output of the model to that of the actual system. Note that the latter is often "corrupted" by noise. An adaptive controller utilizes a function of the error (i.e., some previously defined PI) to modify parameters so as to obtain a more optimum response. Obviously, the success of this scheme is quite dependent on being able to model the robot system accurately. Model reference adaptive control has been used successfully in a variety of applications, including aircraft autopilots.

Despite the obvious advantages of being able to compensate for system component variations and environmental disturbances, adaptive control of commercial robots is not now utilized. The major reasons for this are as follows:

1. *Significant cost of developing the adaptive control hardware and software.* Manufacturers of manipulators do not have the personnel to handle the added work load that the implementation of adaptive control would require. Also, the robots currently being produced are perceived (by the companies) to be "good enough" for the applications of today. Thus industry finds it difficult to justify the additional cost associated with adding such technology. (Fortunately, university and government laboratories have not been as shortsighted).

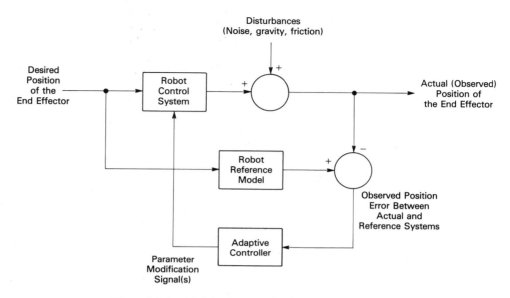

Figure 4.5.4.   Model reference adaptive control of a robot.

2. *The added computational burden placed on existing microprocessor systems.*   Even though most robots currently manufactured utilize distributed processor architectures, there is barely enough computing power to handle the work load now.   Since adaptive control is often quite computationally intensive, it is clear that robot controllers would be hard pressed to complete all the required computations during each update.   In the future, however, as more powerful microprocessors become cost-effective, it is safe to assume that inadequate computation time will no longer be a deterrent to implementing adaptive robotic control.

4.5.6 Optimal Control

Another important area of control seeks to make a system perform in the "best" manner possible.   This discipline is referred to as *optimal control* because the designer attempts to find the input (or inputs) that *optimize* the system in some sense.   In general, it is necessary to determine the set of inputs that produces the desired performance (e.g., drives the system from an initial point in space to a second point).   Both of these points are generally specified.   In addition, it is usually required that these inputs be such that a specified *performance index* be minimized or maximized (i.e., optimized).   Finally, it may be necessary to perform this optimization in the presence of one or more *constraints* on the inputs and/or the system states and/or outputs.   For example, the *magnitude* of the input could be limited by the dc supply voltage.

There are a number of different classes of optimal controls. An example of one is *time optimal control*, where the engineer seeks to find the control that drives the given system from point *A* to point *B* in the *shortest time* while satisfying all input constraints (e.g., magnitude, fuel, etc.). A second example, the *energy optimal control*, involves finding the input that moves the system in a desired manner (i.e., over a specified "trajectory") and minimizes the energy expended by the control.

The literature contains numerous applications of optimal control to physical systems. However, little work has been done in the area of robots, probably because of the extremely heavy, real-time computational load incurred when attempting to determine the control parameters. In this respect, the earlier discussion on adaptive control is pertinent. A few researchers *have* looked at trying to optimize *overall motion time* for a robot. Besides the computational burden placed on the already overworked low-level microprocessor systems currently utilized in robots, mechanical constraints place severe *physical* limitations on manipulator speed. For example, linkages, gears, harmonic drives, and so on all have torque, speed, and/or acceleration limitations that might be difficult to model completely so that a meaningful optimal control could be found. Thus it is not too surprising that this type of optimization has not been stressed in robot design.

Some interest has centered on the time optimal control of certain robots that do not need to perform coordinated motion (i.e., simultaneous movement of two or more manipulator axes in order to arrive at the desired endpoint). In this instance the individual joints are moved *sequentially* in time, and what is sought is the sequence of separate joint motions so that the overall motion time is minimized [2].

As in the case of adaptive control, it is likely that as the computational power of robot controllers increases, it will become more feasible for commercial devices to utilize optimal control. Before this becomes a reality, however, it will also be necessary for novel mechanical structures and materials to be developed that permit high-performance manipulators to withstand the extreme stresses encountered while an optimal trajectory is being executed. Although these developments will probably not occur in the short term, optimally controlled robots will, no doubt, one day be a reality.

Thus far, it has been assumed that the device causing a robotic axis to move is a dc servomotor. Although this is a valid assumption for a number of computer-controlled robots, there are other types of actuators currently employed. In the next sections we consider several of the more commonly used techniques.

4.6 STEPPER MOTORS

It is possible to construct a motor in which the rotor is able to assume only discrete stationary angular positions. Rotary motion occurs in a stepwise manner from one of these equilibrium positions to the next, and as a consequence such a device is called a *stepper motor*. Although to date, these actuators have been used in only a few robots (e.g., the Merlin manufactured by the American Robot Corporation), they have been employed in a variety of applications, the most notable of which is in the field of computer peripherals (e.g., printers, tape drives, capstan drives, and memory access systems). Steppers have also been used in equipment related to the areas of process control, machine tools, and medicine.

There are several general characteristics of a stepper motor that have made it the actuator of choice in such a large number of applications:

1. The device can be operated in an *open-loop* manner with a positioning accuracy of ± 1 step* (assuming that the rotor angular velocity is low enough so that no steps are lost during a move). Thus if a certain angular distance is specified, the motor can be commanded to rotate an appropriate number of steps, and the mechanical elements coupled to the shaft will move the required distance.

2. The motor exhibits high torque at small angular velocities. This is, of course, useful in accelerating a payload up to speed.

3. The motor exhibits a large holding torque with a dc excitation. Thus it has the property of being a "self-locking" device when the rotor is stationary. In fact, the rotor can move only when the terminal voltage *changes* with time.

In addition to these characteristics, there are other advantages that often make designers of various pieces of equipment select the stepper motor over the dc servomotor:

- The stepper is directly compatible with digital control techniques. Consequently, it can readily be interfaced with digital controllers and/or computers.
- It exhibits excellent positioning accuracy, and even more important, errors are *noncumulative*.
- Since open-loop control can be employed with the motor, it is often unnecessary to use a tachometer and/or an encoder. Thus cost is reduced considerably.
- Motor construction is simple and rugged. There are usually only two bearings, and the motor generally has a long, maintenance-free life. For this reason it is a *cost-effective* actuator.
- The stepper can be stalled without causing damage (due to overheating).

Several of these traits make the stepper motor potentially useful in certain types of robots. An obvious one would be the relatively low cost with respect to the dc servomotor. A second property of the stepper that makes it an extremely

---

*We will shortly see that this "full-step" value can be significantly reduced by employing a technique called "microstepping."

attractive choice in the actuation of a robot joint is its dc holding characteristic. The reason for this is that a common failure mode for the power amplifier that is used to drive a motor is for an output transistor to short from collector to emitter. The result of this is to apply the full dc voltage to the motor armature. If a servomotor is being used, the joint will "run away" until an obstacle or mechanical limit stop is encountered. In either case, an extremely dangerous situation will result. However, if a stepper motor is being used, the application of full supply voltage will cause the motor *to move one step and hold*. Thus no runaway condition will exist, and the stepper is seen to be safer under this type of failure.

Although safety and cost are two factors that make the stepper an attractive alternative, the open-loop-control feature, which contributes to its cost advantage over the servomotor, may actually turn out to be a disadvantage. The reason for this is that since there is no position or velocity feedback, the joint controller or master processor does not know whether the manipulator is actually moving. For example, if the manipulator encounters an unforeseen obstacle in its workspace during a move (e.g., another machine tool that is being used in the same workstation), the controller will nevertheless continue to output the calculated number of pulses required to cause a specific arm motion. When all these have been sent, it will "think" that the desired position has been reached and hence no more pulses will be produced. However, the arm may actually be stalled somewhere else in the workspace. Thus large position errors can result.

It is possible to prevent this type of error from occurring by utilizing some type of position sensor. A simple, inexpensive device can often be employed for this purpose since it is not actually used in the control process but is present only to inform the controller that the manipulator is actually moving. However, an intermittent and/or momentary type of stall (e.g., one caused by increased bearing friction due to wear) will not be helped by such a sensor. In this case a "good" position and/or velocity detector would be required. As a consequence, some of the stepper's cost advantage would disappear. (The Merlin robot, in fact, utilizes position sensors to "close the loop.")

Position errors can also be caused when one attempts to move the rotor too rapidly. In this instance the motor cannot respond fast enough to the pulses coming from the controller so that each pulse does not produce an actual step in the shaft position. This phenomenon of "dropped steps" will cause the robot arm to reach a final position that is in error. The "fix" in this case is to reduce the maximum rotor velocity and/or the load (i.e., the inertia reflected back to the motor shaft). Clearly, such a compromise causes a degradation in performance.

Another potential difficulty with a stepper motor in a robot application is that the stepwise motion can excite significant manipulator oscillation. Since no velocity feedback is used, the only way to improve the response is to employ a much more elaborate controller that is capable of *microstepping* (explained below). This increases the cost of the drive electronics significantly, further eroding the cost differential between the stepper and servomotor. At this writing it is unfortunately true that the potential problems in using the stepper motor as a robotic actuating device outweigh the advantages.   However, this may change in the future.

4.6.1 Principles of Stepper Motor Operation

There are two basic varieties of stepper motors that can be constructed: (1) the variable reluctance (VR) type, and (2) the permanent magnet (PM) type.   Although the PM style is most often used in a broad range of applications today, the operation of the VR stepper is easier to understand and therefore we consider its operation in this section.

The structure of a typical VR stepper motor is shown in Figure 4.6.1.   It is observed that unlike the servomotor, both the stator and the rotor are *toothed* structures.   Fundamental to the operation of this motor is that the rotor and stator do not have the same number of teeth.   For example, the stator shown in Figure 4.6.1 has eight (located every 45°) and the rotor has six (located every 60°).   In addition, each stator tooth has a coil wound on it with oppositely placed coils (e.g., A and A') being grouped together and referred to as a *phase*.   In this example it is seen that there are four phases (labeled A, B, C, and D).

The operation of the VR stepper is quite simple, being based on the principle of "minimum reluctance" whereby a magnetic structure always attempts to reorient itself so as to minimize the length of any air gap in the magnetic path.   One can think of the magnetic device moving so that the magnetic field can find the path of "least resistance."   Thus for the example shown in Figure 4.6.1, when phase A is energized, rotor teeth 1 and 4 (R4,1) will align with stator teeth 1 and 5 (S5,1) and will remain in this position as long as the coils in the same phase are energized. This is said to be a stable equilibrium point and represents "one step" of the motor.

It is important to understand that as long as the excitation remains on coils A–A' there is a *holding torque*, so that if any applied external torque is less than

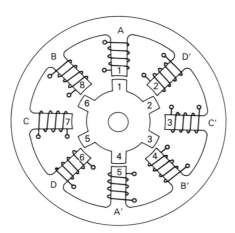

Figure 4.6.1.   Basic structure of a variable reluctance-type stepper motor (Redrawn with permission of the Superior Electric Co., Bristol, CT.)

this value, no motion will occur.   Also, increasing the current through this phase will *not* cause the rotor to move but will, in actuality, increase this holding torque. Thus the motor will tend to "lock" more under increased current excitation.   This should be compared to the servomotor, where increasing the excitation will tend to make the rotor turn faster.

Now suppose that the excitation is removed from phase *A* and placed on phase *B*.   R6,3 will align with S8,4 as shown in Figure 4.6.2a.   It can be seen that the rotor has moved clockwise through an angle of 15° (60 − 45 = 15).   This process can be repeated for phases *C*, *D*, and then back to *A* (see Figures 4.6.2b, c, and d).   In each case, a 15° step occurs with a complete sequence of phase excitations (e.g., *A*, *B*, *C*, *D*, *A*), producing a rotation of 60°.   Consequently in this example, it requires six such cycles to cause one complete rotor revolution, and we would therefore describe this as a "24-step/revolution" motor.

The reader may easily verify that making the phase excitation sequence *A*, *D*, *C*, *B*, *A* will produce counterclockwise rotation of 60°.   This sequence reversal is quite easy to implement in practice.   It can be demonstrated, in general, that change of direction requires at *least* three phases.

The step angle is related to both the number of stator teeth N_s and rotor teeth N_r.   Specifically, it can be shown that

$$\text{step angle} = 360° \frac{|N_s - N_r|}{N_s N_r} \tag{4.6.1}$$

and that the

$$\text{number steps/rev} = \frac{N_r N_s}{|N_s - N_r|} \tag{4.6.2}$$

Physically, rotors of all stepper motors exhibit an underdamped response in moving from one step to another (e.g., see Figure 4.2.4).   This can most easily be seen by recognizing that when the excitation is changed to an adjacent phase, the rotor travels toward the new equilibrium point.   Although the accelerating torque is zero when the rotor and stator teeth are in alignment, the *angular velocity* of the rotor is not zero.   As a result, an overshoot of the equilibrium position occurs.   There will now be a torque on the rotor that will accelerate it back toward the equilibrium point that was just passed (i.e., in the opposite direction).   In fact, this process may actually be repeated several times before the rotor comes to rest. In general, the magnitude and duration of the damped oscillation is dependent on the step angle (i.e., the larger the angle, the larger the overshoot).   In certain applications, such behavior may not be acceptable (e..g, a robot), whereas in others it will be perfectly all right (e.g., a printer).

Stepper motors can be made with a wide range of steps/rev.   From the standpoint of cost, a practical upper limit is 200 and produces a step angle of 1.8°. For many applications, this relatively small angle is quite adequate (e.g., the rotor oscillation will not interfere with the device operation).   However, where finer step-angle resolution is required (note that this is one way to reduce the angular overshoot problem described above), other techniques can be used.

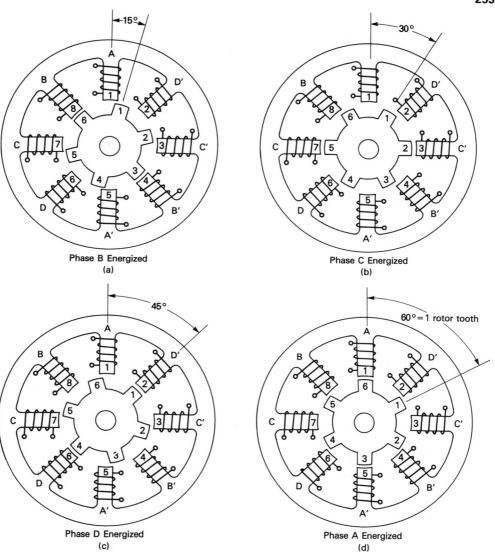

Figure 4.6.2. VR stepper motor excited in the sequence BCDA. The motor is assumed to start in the position shown in Fig. 4.6.1. (Redrawn with permission of the Superior Electric Co., Bristol, CT.)

4.6.2 Half-Step-Mode Operation

In the previous discussion, one phase was energized at a time and produced a step angle given by Eq. (4.6.1). This method of operation is called *full-step mode*. Now suppose that two adjacent phases (e.g., *A* and *B*) are energized *simultaneously*.

It is logical that an equilibrium point is created somewhere between the two full-step points (as determined by separately exciting phases *A* and *B*). In fact, if the electrical properties of the coils in the two phases are identical and if the same excitation amplitude is applied to both sets of coils, the new equilibrium point will be about halfway between the full-step points. This process can be repeated for phases *BC*, *CD*, and *DA* so that additional "halfway" equilibria can also be obtained. It should now be clear to the reader that if the phase excitation sequence is *A*, *AB*, *B*, *BC*, *C*, *CD*, *D*, *DA*, *A*, and so on, the rotor will make twice the number of clockwise moves as before (i.e., with respect to the full-step mode), and thus the stepper in Figure 4.6.1 will now have 48 discrete equilibrium points per revolution. The name given to such an operation is, not surprisingly, *half-step mode*. Since the rotation angle per step has been cut (approximately) in half, the angular overshoot of the rotor in moving from point to point is reduced. Reversing the phase excitation sequence will cause the rotor to turn in the counterclockwise direction, as before. The switching circuitry needed to produce half-step operation is somewhat more complicated and is therefore more costly than the relatively simple full-step electronics.

4.6.3 Microstep Mode

A little thought should convince the reader that there is nothing "sacred" about exciting two adjacent phases equally (e.g., both with *V* volts). In fact, it is possible to use an excitation voltage anywhere between 0 and *V* in order to energize the second phase. In this case the stable point will occur at some location (but not halfway) between the two full-step equilibria. This scheme produces a mode of operation generally referred to as *microstepping*. Most often, the microstep size is determined by dividing the angular distance of a full step by an integral power of 2 (e.g., 2, 4, 8, 16, or 32; this produces the smallest computational burden on the stepper motor controller). Although microstepping requires considerably more complex switching circuitry to implement so that the cost is quite a bit higher than that required for full-step operation, its use generally produces smoother low-speed operation of the motor. In a robot application, this is an important consideration since oscillation at the desired final point is usually unacceptable.

4.6.4 Additional Methods of Damping Rotor Oscillations

There are other ways to damp out the rotor oscillation described in Section 4.6.1: for example, by adding a viscous inertia (often called a Lanchester damper) or by using either a friction disk or eddy-current damper. Although these techniques achieve the desired goal, they also add inertia and may therefore adversely affect the transient response of the rotary system (this will not be true if the system inertia is already high).

An electronic technique exists that avoids the problem created by added inertia. Called *bang-bang damping*, the idea is to accelerate the motor in the normal way.   However, before the rotor reaches its desired position, the phase excitation sequence is reversed, causing the rotor to decelerate more rapidly.   If the phase reversal is timed correctly, the rotor can be made to come to rest at the equilibrium point with almost no overshoot.   Clearly, the timing of the reversal is critical and, it turns out, the switching instants are a function of system parameters (e.g., friction and load inertia).   In a robot, where the inertia of any joint usually varies significantly with position (and hence with time during any move), a very sophisticated scheme 's required to sense these changes and then modify the phase reversal times accordingly.

4.6.5 Permanent-Magnet Stepper Motors

As stated previously, the PM stepper motor is the most comonly used type.   It consists of a multiphased stator and a two-part, permanent-magnet rotor.   Just as with the VR stepper, both of these structures are also toothed (see Figure 4.6.3). The major difference in this case is that the opposite ends of the rotor are north and south poles of a permanent magnet with the teeth at these ends being offset by half a tooth pitch.   Although the operation of this type of motor will not be discussed any further, it is worthy of mention that the PM stepper can be operated in full, half, or microstep mode.   Table 4.6.1 indicates the major differences between the two classes of steppers.

4.6.6 Stepper Motor Drives

As indicated in Table 4.6.1, the PM stepper rotor position is dependent on the polarity of the phase excitation.   Consequently, a *bipolar* signal is required to achieve bidirectional control.   With respect to Figure 4.6.3, it is seen that this

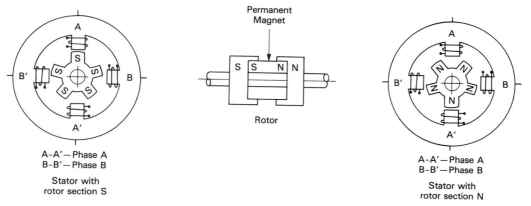

Figure 4.6.3.   A simplified diagram of a permanent-magnet stepper motor.   (Redrawn with permission of the Superior Electric Co., Bristol, CT.)

TABLE 4.6.1 DIFFERENCES BETWEEN PM AND VR STEPPER MOTORS

Characteristic	PM motor	VR motor
1. Motor	Magnetized	Not magnetized
2. Rotor position	Depends on stator excitation polarity	Independent of stator excitation polarity
3. Rotor inertia	High due to magnet	Low (no magnet)
4. Mechanical response	Not as good (due to high inertia)	Good (low-inertia device)
5. Inductance	Low due to rotor offset	Generally high for same torque rating
6. Electrical response	Faster current rise (due to low inductance)	Slower current rise (due to higher inductance)

operation is accomplished with only *two* phases, whereas the reader will recall that the equivalent VR stepper required four phases (see Figure 4.6.1).

Using a double-ended power supply, the motor in Figure 4.6.3 can be driven in the full-step mode with the switching arrangement shown in Figure 4.6.4a. It is observed that two *tristate* switches, SW1 and SW2, are required. This figure shows phase *A* positively energized and phase *B* off. A possible method of synthesizing such a device is shown in Figure 4.6.4b. The "fly-back" diodes are normally used to protect the power transistors from the "inductive kick" that occurs when an open circuit is suddenly placed in series with an energized inductor (e.g., Q_1 is switched off). Without this protection, it is possible to apply a voltage well in excess of the transistor's collector–emitter breakdown value during the switching interval.

For the motor in Figure 4.6.3, it can be shown that each step is 18° and that there are therefore 20 steps/rev (see Problem 4.22). The excitations and simple logic signals to the transistors that produce four rotor steps (i.e., 72°) are given in Figure 4.6.5. It is assumed that each step takes the same amount of time implying that any load attached to the rotor is moving at a constant velocity. During acceleration or deceleration of a load, the step spacing would, of course, vary with time.

In actual operation, a microprocessor (e.g., the master) would determine the number of steps needed to cause a load to be moved a certain distance. This would be done for each joint in a robot application. The processor would then transmit the information, together with direction and step timing data, to a discrete digital hardware package. The latter would keep track of the total number of steps moved and would implement the appropriate switching sequence. Clearly, this would represent open-loop joint control, with its inherent problems. Recall

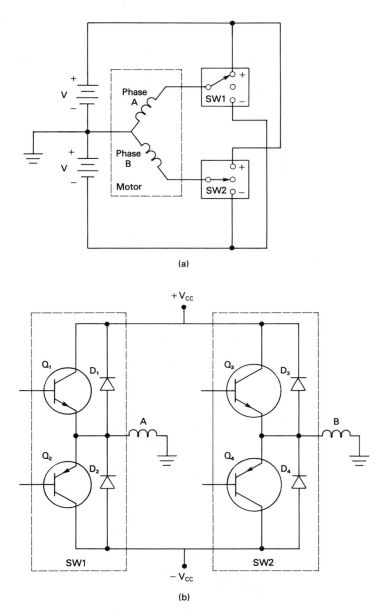

Figure 4.6.4. (a) Simplified bipolar two-phase switch stepper motor drive; (b) possible realization of the circuit in part (a).

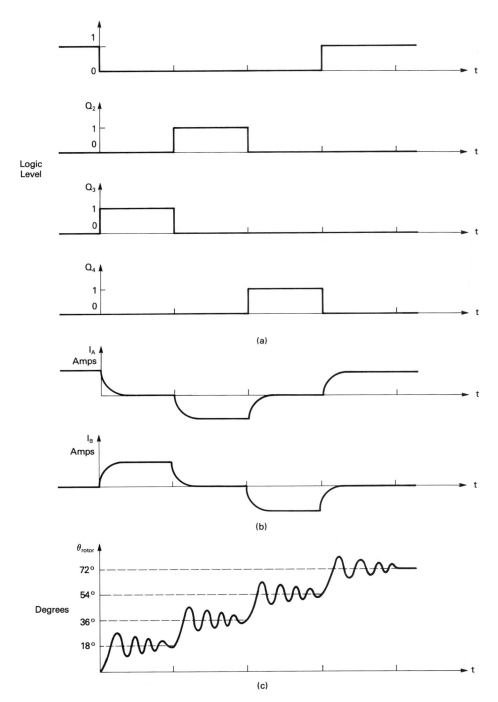

Figure 4.6.5. (a) Transistor logic signals; (b) motor phase currents; (c) rotor motion for the motor in Fig. 4.6.3 and the drive in Fig. 4.6.4a.

the discussion in Section 4.6. Closed-loop control could also be implemented by using one of the methods described in Section 4.5. As mentioned previously, however, this would negate some of the cost advantage of the stepper over the servomotor.

EXAMPLE 4.6.1

Suppose that we wish to move a robot joint a total distance of π radians in 400 ms using a stepper motor. The joint "sees" a reflected load inertia J_L = 0.004 oz-in.s². (The load is coupled to the motor shaft through a 10:1 gear train.) It is proposed to use a stepper with the following specifications:

$$J_M = 0.003 \text{ oz-in.-s}^2$$

$$\text{step angle} = 1.8° \text{ (200 steps/rev)}$$

$$\text{rms torque} = \text{rated continuous torque} = 7 \text{ oz-in.}$$

$$\text{maximum step rate} = \text{slew rate} = 4000 \text{ steps/s}$$

In addition, a triangular velocity profile is to be assumed (see Figure 4.6.6). Here t_a and $(t_g - t_a)$ are the acceleration and deceleration times, respectively, and are assumed to be equal in this case. The acceleration and distance curves resulting from this velocity are also shown in parts b and c of this figure and ω_{pk} and α_{pk} are the peak angular velocity and peak angular acceleration of the motor shaft, respectively. The problem is to determine whether the given motor will be able to meet all the motion requirements for a joint move of this type.

Because the load is coupled to the motor shaft through a 10:1 speed reducer, the *motor* must move 10π radians (i.e., 10 times the distance of the actual joint output). From Figure 4.6.6b and c, it is seen that

$$\omega_{pk} \frac{t_g}{2} = 10\pi$$

and

$$\alpha_{pk} = \frac{\omega_{pk}}{t_a}$$

Since $t_a = t_g/2 = 200$ ms, ω_{pk} and α_{pk} needed to make this move are found to be

$$\omega_{pk} = 50\pi = 157.1 \text{ rad/s}$$

$$\alpha_{pk} = \frac{157.1}{0.2} = 785.5 \text{ rad/s}^2$$

The corresponding peak acceleration and deceleration torques in this case

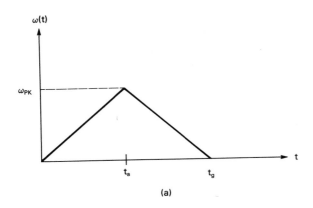

(a)

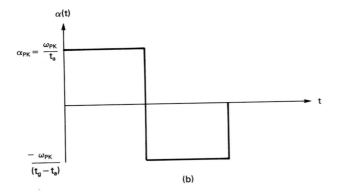

(b)

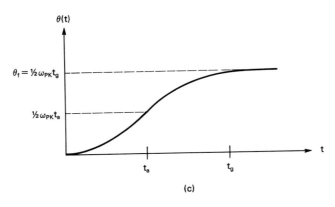

(c)

Figure 4.6.6. Assumed velocity, acceleration, and position profiles for Example 4.6.1. Shown are the motor shaft angular: (a) velocity; (b) acceleration; (c) position.

are equal.* Thus

$$T_{accel} = (J_L + J_M)\alpha_{pk} = 5.5 \text{ oz-in.}$$

Since the acceleration curve in Figure 4.6.6b is piecewise constant, the rms and peak torques are equal.* Consequently, it is seen that the proposed motor is adequate from a torque point of view. However, a single step is 1.8° or 0.031416 rad. The resultant peak angular velocity is therefore

$$\omega_{pk} = \frac{157.1}{0.031416} = 5000 \text{ steps/s}$$

Clearly, the speed requirement exceeds the maximum slew rate of the motor by 25%. In fact, it is probable that if we use it for the proposed application, steps will be dropped and accuracy will suffer.

There are two things that can be done to meet the requirements of the problem. One involves using a motor that has the same torque rating but a higher slew rate. The other necessitates relaxing one of the specs. For example, suppose that it is permissible to make the move in slightly more than 400 ms. Then the trapezoidal velocity profile shown in Figure 4.6.7 could be employed. Using this profile with ω_{pk} = 4000 steps/s and the acceleration and deceleration times still assumed to be equal, it is found that the acceleration torque is still 5.5 oz-in, t_a = 160 ms and the overall move time T = 410 ms (see Problem 4.23). Thus only a small time penalty results from using a constant-velocity segment of 90 ms during the move.

4.6.7 Linear Stepper Motors

An interesting variation of the conventional rotary stepper motor is the Sawyer-principle linear stepper motor. Invented in 1969, this patented device is manufactured by Xynetics Corporation, Santa Clara, California, and consists of two major mechanical components. The first, a movable armature which is referred to as a *forcer*, is suspended over the second or fixed stator (also called a *platen*) (see Figure 4.6.8). A bearing is used to ensure that there is a constant space between the armature and the stator. In contrast to a conventional rotary stepper which has a *closed* geometry, the platen's length is *variable* and depends on how far it is desired to move a load attached to the forcer. This configuration also differs from the rotary stepper in that the payload is *directly driven* by the motor and no mechanical advantage can be obtained through the use of a transmission.

As may be observed in Figure 4.6.8, the forcer consists of a permanent magnet (PM) and two electromagnets (EM) with four poles (two per electromagnet). The

---

*Note that this is true only because we have assumed that both the viscous damping and friction torques of the motor are zero. If either or both of these loss terms are not neglected, asymmetry in the acceleration and deceleration torque profiles will occur.

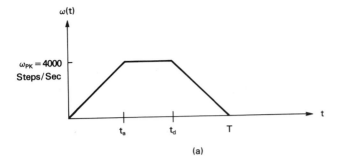

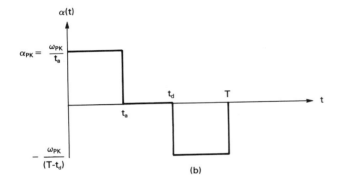

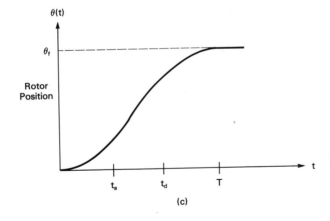

Figure 4.6.7. Alternate set of profiles (assuming a trapezoidal velocity) for Example 4.6.1. Shown are the motor shaft angular: (a) velocity; (b) acceleration; (c) position.

faces of the poles are grooved to form the pitch of the motor. Grooving of the platen produces a similar pattern. As will be seen shortly, the use of grooves allows finer resolution steps. In addition, when the spaces between both sets of grooves are filled with a nonmagnetic material, the resultant flat surfaces can be used to construct an *air bearing* between the bottom of the forcer and the top of

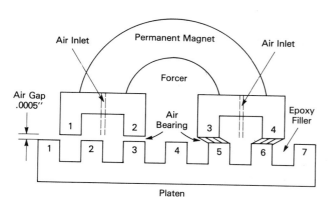

Figure 4.6.8. Components of a two-phase linear stepper motor. An air bearing is formed between the bottom of the forcer and the top of the platen. (Redrawn with permission of General Signal Corp., Santa Clara, CA.)

the platen. This is accomplished by supplying air under pressure from tiny holes located in the forcer. The air bearing produces a surface with negligible starting and running friction.

The permanent magnet causes the platen and the unenergized forcer to be drawn together (except for the space provided by the bearing). Therefore, it is possible to position the platen so that the forcer travels above or below it. With no current flowing, the PM flux closes its path through the air gap, platen, and the poles of the electromagnets. The flux splits equally at both EM poles since the *magnetic* paths have approximately the same reluctance (e.g., see Figure 4.6.9a, poles 3 and 4). If current is switched through the electromagnets, commutation occurs. In general, the flux produced by the permanent magnet is about equal to that produced in the magnetic circuit by the current flowing through the windings. Thus as the current changes, the flux swings from a maximum value to almost zero.

The commutation, together with the relative positions of the forcer and platen teeth, causes forces to be produced which are perpendicular to the teeth and parallel to the platen. Since the teeth of the EMs are arranged in *spatial quadrature* from one pole face to the next, the PM's flux can be commutated by the electromagnets and emerges at polefaces whose teeth are misaligned with respect to those of the platen. The result is a tangential force that causes the forcer and platen teeth to move in such a manner as to minimize the gap (i.e., reduce the reluctance). This force produces motion along the length of the platen. A normal force also exists which pulls the forcer and platen toward one another, thereby providing the preload for the air bearing.

Figure 4.6.9a–d is used to illustrate the principles of operation outlined in the following paragraphs. In each of the figures, the direction of current and flux flow is indicated by the arrows. If electromagnet A (EMA) is energized, maximum flux density occurs at pole 2 and alignment is as shown in Figure 4.6.9a. When EMA is deenergized and EMB is energized, the maximum flux density occurs at pole 3 and the minimum density at pole 4. The attractive force at pole 3 causes the alignment of this pole with the platen's tooth on the *right*. Therefore, the motion is one-quarter of a tooth to the right and the motor and the forcer have

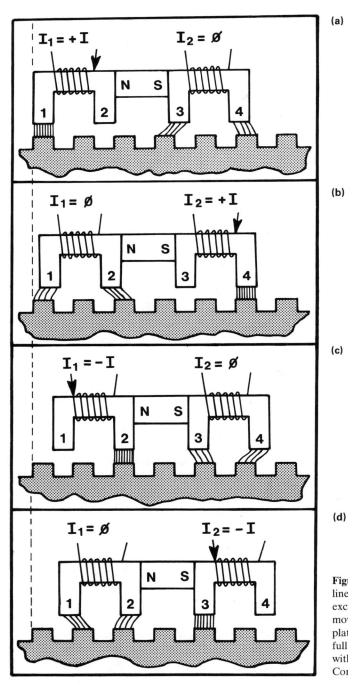

(a)

(b)

(c)

(d)

Figure 4.6.9. Motion in a two-phase linear stepper motor. Each new phase excitation produces a quarter pitch movement of the forcer relative to the platen. Thus a total motion of ¾ of a full pitch is indicated. (Reproduced with permission of General Signal Corp., Santa Clara, CA.)

the spatial relationship shown in Figure 4.6.9b. If EMB is deenergized and then EMA energized (with a current flow opposite to that shown in Figure 4.6.9a), motion again occurs to the right since pole 1 now has the maximum flux density, pole 2 the minimum, while poles 3 and 4 have the flux supplied by the PM. The forcer now resides at the location shown in Figure 4.6.9c. Finally, with EMA deenergized and EMB energized (also in the opposite direction from before), pole 4 has the maximum flux density, pole 3 the minimum, and poles 1 and 2 the flux supplied by the PM. To complete the cycle, EMA is again energized as in Figure 4.6.9a and the system has moved a distance of *one tooth* (i.e., the *pitch*) of the platen (the equivalent of full step mode in a rotary stepper). The frequency of the current cycling establishes the velocity with which the forcer moves.

Obviously, the positions of the forcer relative to the platen are discrete in nature if the current is cycled as described above. Used in this manner, the linear stepper has a full-step resolution defined by the spacing of the teeth on the poles. A typical pitch is 0.040 in. Thus for the sequence shown in Figure 4.6.9, the resolution is one-quarter of the pitch or 0.010 in. These positions are sometimes referred to as *cardinal steps*. To obtain a finer resolution between steps, it is possible to use current values that are *between* those used in the full-step mode. That is, the motor is operated in *microstep mode* (see Section 4.6.3).

It is also possible to construct this type of motor so that it consists of two orthogonally oriented forcers assembled on one motor frame. To complement the forcer, the platen is constructed of square teeth in a *waffle* pattern, as shown in the two-axis linear stepper of Figure 4.6.10. This configuration allows motion in both the *x* and *y* directions or along any vector in the *x-y* plane.

As indicated previously, the linear stepper is a *direct-drive motor*. This implies that the control resolution and force needed to position and move the load are defined solely by the motor's capabilities. Thus for any application requiring resolution better than that of the tooth pitch, a controller capable of microstepping to the desired resolution must be used. Additionally, the speed–force curve for the motor–driver combination must be examined carefully to ensure that the motor *can* produce the required forces over its operating speed range.

The linear stepper exhibits the same resonance phenomena and loss of synchronization (i.e., loss of steps) as the rotary stepper. However, two characteristics tend to complicate the task of controlling this motor. First, the device is inherently "springy." The armature resides in a force detent that is the width of a tooth interval. As the motor is subjected to inertial and drag forces, it experiences setbacks that displace it from the center of the detent, thereby creating the thrust necessary to overcome these external forces. If the setback reaches the limit of the detent, the motor will slip and lose synchronization. The second characteristic that makes control difficult is the highly resonant nature of the armature, which results from it being supported on a frictionless surface (the air bearing). Thus there is very little damping of the spring–mass system formed by the armature mass and force detent function. Obviously, this condition makes the motor sus-

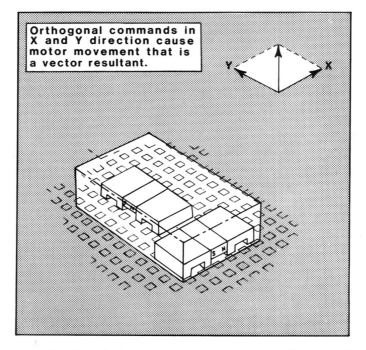

Figure 4.6.10. Two-axis linear stepper motor and platen. (Reproduced with permission of General Signal Corp., Santa Clara, CA.)

ceptible to excitations near its natural resonance frequency. Consequently, it is slow to settle from unwanted perturbations that might occur during its motion.

Figure 4.6.11 shows a plot of motor force versus displacement and is useful in explaining qualitatively the dynamic behavior of the linear stepper. In this figure it is assumed that the motor has been commanded to move to position P, a stable equilibrium point. Thus, in the steady state, no force will be exerted on the armature when it is at this location. If the armature is now moved to position A or B, it will be subjected to a restoring force that accelerates it back toward P. The farther away it is from P, the greater will be the restoring force until location C or D is reached. These points are $\frac{1}{4}$ of a tooth pitch (as defined by the mode of operation, i.e., full, half, or microstep) from the equilibrium. If the armature is moved still farther away from P toward U_1 or U_2, the restoring force will be in the proper direction (i.e., toward the equilibrium point), but its magnitude will *decrease*.

Points U_1 and U_2 are *unstable* equilibria and, as can be seen from Figure 4.6.11, no force is exerted when the armature resides in these locations. However, if the motor is slightly displaced from either of these locations, it will be accelerated away from them and continue to move (until a stable equilibrium point is reached). Thus, as long as the armature is not outside the (open) position interval between

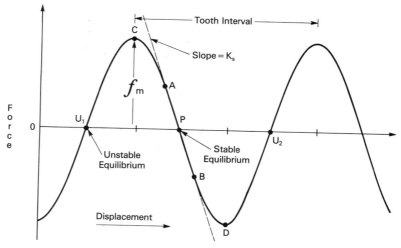

Figure 4.6.11. Motor force versus displacement for a linear stepper motor (3). (Redrawn with permission of J. Nordquist and P. Smit, Xynetics Products, a division of General Signal, Santa Clara, CA.)

U_1 and U_2, it will remain captive about P. Note that just as in the case of a rotary device, the linear stepper is inherently underdamped, which causes the armature to oscillate about P before it finally comes to rest. In this regard, the reader should recall the discussion of Section 4.6.1.

It should be clear from the above that the magnetic field configuration exerts force on the armature *only when it is displaced from the commanded equilibrium position*. Thus for the inertial and frictional forces to be overcome, the motor must actually lead or lag the commanded position. An examination of Figure 4.6.11 reveals that the driving function must be such that the error between the desired and the actual armature positions is no more than $\frac{1}{4}$ of a tooth pitch. If this difference is larger, synchronization may be lost.

As the linear stepper is normally operated in an open-loop control configuration, it is important that the driving (forcing) function not excite the motor's resonant frequency. For example, the microstepping rate must be carefully selected so that it is not at or near the stepper's natural frequency. In addition, oscillations in the motor can be induced even by seemingly smooth forcing functions. This happens because the signal contains enough energy at the motor's resonant frequency point to produce the "ringing." Figure 4.6.12 shows the command and actual armature positions as a function of time for a triangular velocity profile. Also indicated in the figure is the error (i.e., the difference) between these quantities. Of particular note here is the appearance of oscillations in the error signal. It is important to understand that the ordinate represents the lag between the actual and commanded positions. The reader should recall that this quantity is a measure of the force exerted by the magnetic structure on the armature. It can also be

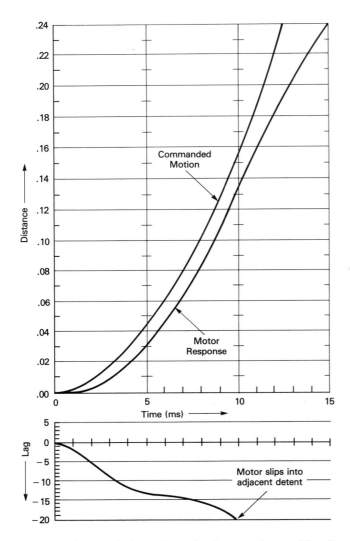

Figure 4.6.12. Motor response and error for a triangular velocity profile and 8.7 g's initial acceleration with rolloff (3). (Redrawn with permission of J. Nordquist and P. Smit, Xynetics Products, a division of General Signal, Santa Clara, CA.)

observed that when the lag reaches -20 mils, the armature is outside the open position interval (U_1, U_2) and hence synchronization is lost. Clearly, with proper selection of the motor, load, and acceleration, it is possible to utilize a triangular velocity profile without dropping steps.

A technique that has recently been developed for preventing ringing in linear steppers is called *burst*. A forcing function is defined which is based on making the difference between the command and actual armature positions (i.e., the error) a *constant* (lag) during the acceleration phase of the motion. Figure 4.6.13 indicates a linear stepper driven in this manner. It is observed that there is an initial step

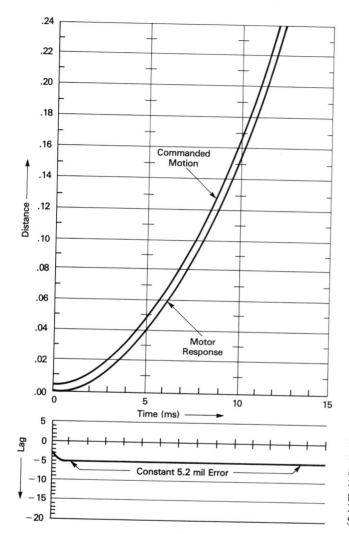

Figure 4.6.13. Motor response and error for a triangular velocity profile, 8.7 g's initial acceleration with rolloff, and 5.2 mils of burst (3). (Redrawn with permission of J. Nordquist and P. Smit, Xynetics Products, a division of General Signal, Santa Clara, CA.)

difference of 5.2 mils.   This is referred to as the "burst."   Note that with such a forcing function, the motor does not ring and there is no danger of dropping steps. Normal implementation of the technique requires that the burst be applied at the breakpoints (i.e., at the beginning and end of acceleration and deceleration phases of the motion).   Further detail about the procedure may be found in the literature [3].

The speed–force curve for a typical linear motor is shown in Figure 4.6.14. From the curve it is apparent that while the speed of this motor can be quite high,

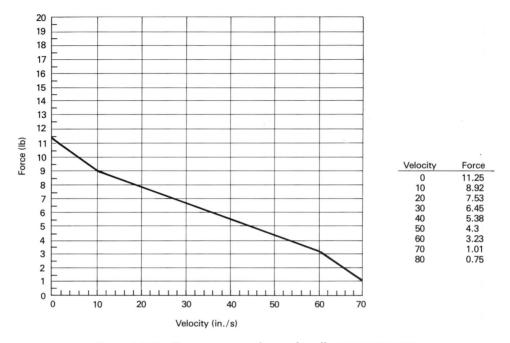

Velocity	Force
0	11.25
10	8.92
20	7.53
30	6.45
40	5.38
50	4.3
60	3.23
70	1.01
80	0.75

Figure 4.6.14. Force versus speed curve for a linear stepper motor.

there is a corresponding decrease in available force in much the same way that the torque of a rotary device decreases as the angular velocity increases. One approach to utilizing such motors in a given application is to size them based on the force required at the maximum speed. This force must never be exceeded over the entire operating range of the motor. For example, referring to Figure 4.6.14, since the maximum force available at 50 in./s is 4.3 lb, then to ensure proper operation, no more than 4.3 lb of force should be used during acceleration and deceleration. As can clearly be seen, this conservative approach does not utilize the full capability of the motor.

Another approach permits a given motor to be used more efficiently. Here, the velocity profile is *piecewise linear* (i.e., has multiple slopes) during the acceleration and deceleration phases. This technique can reduce overall motion time considerably. For example, if the acceleration phase is divided into three regions, say 0 to 20 in./s, 20 to 40 in./s, and finally, 40 to 50 in./s, one can use corresponding maximum forces of 7.53, 5.38, and 4.3 lb. For a load of 16 lb, this results in the velocity profile shown in Figure 4.6.15. It is observed that by more fully utilizing the capabilities of the motor, it is possible to reduce the acceleration time by over 25% (0.36 instead of 0.481 s).

The attempt to better utilize the force capabilities of these motors is called

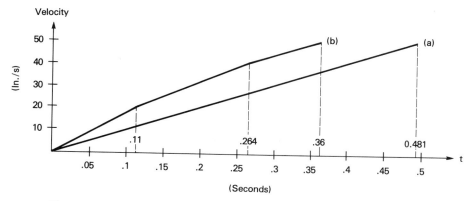

Figure 4.6.15.   Acceleration portion of velocity profile for accelerating a 16-lb. load to 50 in./s: (a) maximum force permitted is 4.3 lb. (at 50 in./s); (b) piecewise-linear profile using forces of 7.53 lb. in the range of 0 to 20 in./s, 5.38 lb. in the range of 20 to 40 in./s, and 4.3 lb. from 40 to 50 in./s.

acceleration roll-off.   Its implementation may take the form discussed above or an exponential acceleration waveform can be used.   In either case, the basic idea is to adjust the acceleration and deceleration during any move, utilizing the force–velocity curve of the motor, so that the sum of the inertial and drag forces is maintained at a constant value.   It is important to understand that this technique does not prevent characteristic ringing associated with all steppers.   As mentioned above, this must be handled by carefully selecting the forcing function.   An example of a motor that uses both acceleration roll-off and burst is shown in Figure 4.6.13.

Some of the major advantages of linear stepper motors are their ruggedness, high reliability, and simplicity, all due to the low component count needed to implement these devices.   Additionally, they have the ability to achieve extremely good accuracy and repeatability while moving loads at very high speeds and over large distances without the use of transmissions.   Also, the air bearing that is a major part of the linear stepper is virtually free from wear so that the motor generally requires little or no maintenance.

The use of this type of motor to implement a linear drive stage results in a cost that is typically in the top end of the price range for conventional implementation (i.e., a stage made up of a lead screw, sliders, dc servomotor, and feedback elements).   Other disadvantages besides cost are the complexity of the drive electronics needed to reduce ringing, achieve microstepping, and produce the high pulse rates, and the reduction in force that occurs with increased velocity.

At this writing, no commercial industrial robots utilize the linear stepper motor.   However, as cost is reduced, it is expected that direct-drive capabilities of these actuators may well make them attractive in applications requiring the high reliability that results from minimum component count.

4.7 BRUSHLESS DC MOTORS

In electrically actuated robots, brush failures in the dc servomotors used on the joints account for a major source of downtime. These devices wear, causing the effective terminal resistance of the armature to increase significantly, thereby reducing the efficiency of the servo. Increased heating and torque reduction are two of the major consequences. In addition, as the motor turns, arcing between the brushes and commutator segments occurs due to the sudden interruption of current in the particular armature coil being commutated.* Besides contributing to mechanical deterioration of the brushes themselves, which can limit their use in "clean room" applications (e.g., in the handling of semiconductors), this situation also prevents robots so actuated from being used in explosive environments. Finally, the electromagnetic interference (EMI) produced by the electrical spark can also create reliability problems for other electronic devices working in the vicinity of the robot.

In recent years, dc motors have been developed which avoid many of the difficulties attributable to the brushes of a standard servomotor. As shown in Figure 4.7.1, the brushless dc motor (BDCM) can be viewed as an "inside-out" version of a standard dc servomotor.† It is observed that the *rotor* of the brushless device contains the permanent magnets (two in this case, thereby producing a *four-pole* motor) whereas the stator consists of the coil segments and iron.

Since there is no *mechanical commutation* of the coils in a BDCM due to the elimination of the brushes and commutator bars, a method of properly energizing the stator coil segments must be provided. This is often accomplished by placing inside the motor itself solid-state devices (e.g., Hall effect bipolar sensors) that determine the actual position of the magnets as the rotor turns. A simple logic circuit then processes the information provided by these sensors, thus enabling the appropriate stator coil to be excited. As an example, consider the eight-pole (four north and four south), three-phase winding BDCM and the electronic commutation scheme shown in Figure 4.7.2. The output of a Hall effect sensor is *high* (logical 1) when the *south pole* of a permanent magnet is in close proximity to it. The output is low (logical 0) if the magnet's north pole is passing by. With the sensors placed approximately 120 *mechanical* degrees apart and the four magnets 90 mechanical degrees apart as indicated in Figure 4.7.2, it is easily demonstrated that the outputs of the three sensors are the waveforms shown in Figure 4.7.3. These signals can be processed by a simple logic circuit to determine the position of the magnets at any instant of time. This information is then utilized by the driver circuit to cause the appropriate motor windings to be energized.

Ideally, it is possible to produce a torque output that is constant with respect

---

*For this reason, most servomotors divide the armature coil into many segments (e.g., 7 to 32) so as to reduce the inductance and hence the "kick" produced during commutation.

†Although this is not the only design for such a device, it is the most common form.

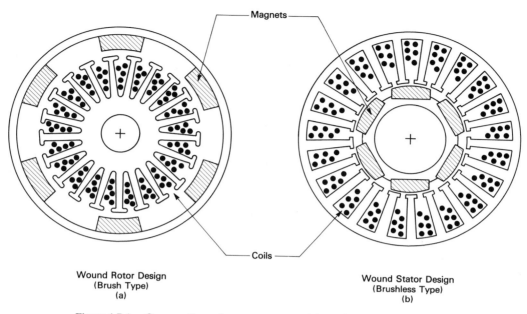

Figure 4.7.1. Cross sections of two servomotors: (a) standard DC (brush-type); (b) brushless DC-type. (Redrawn with permission of Electro-Craft Corp., Hopkins, MN.)

to angular position. To see this, consider the three-phase driver circuit shown in Figure 4.7.4 and assume that the BDCM windings are arranged in a *wye* configuration. If a *constant current* is applied to each winding and the rotor is moving at a constant angular velocity, the torque produced by each of the phases as a function of the angular position of the motor shaft θ is shown in Figure 4.7.5. It is important to understand that the total or net torque T_{gen} produced by the motor is the algebraic sum of the three torques T_{a-b}, T_{a-c}, and T_{b-c}. Clearly, if we permit a constant current to flow in each of the three phases, T_{gen} will *not be constant* as is desired.

However, suppose that the information from the Hall effect sensors' output (Figure 4.7.3) is used by the logic circuit to produce the transistor base drive signal sequence indicated in Figure 4.7.6. Also assume for the moment that despite the presence of an inductance, a constant current I *immediately* flows through each transistor in the three-phase bridge circuit when the corresponding base signal jumps to logical 1 and that the collector current instantaneously drops to zero when the drive signal is logical 0. Then it is easy to understand why T_{gen} will be independent of angular position. For example, from 0 to 60°, Figure 4.7.6 indicates that Q_1 and Q_5 are conducting so that only coils A and B are energized. Thus, during this spatial interval, $T_{\text{gen}} = T_{\text{max}}$ since the torque produced by coils $A-C$ and $B-C$ is zero (the current in these coil pairs is zero). During the next 60°

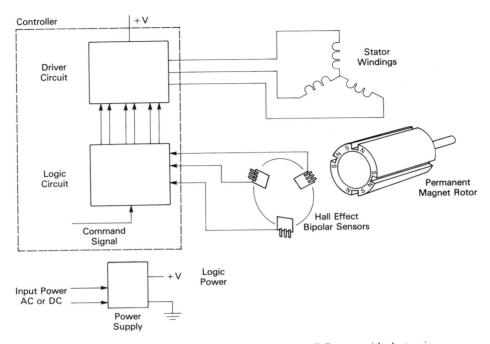

Figure 4.7.2. Eight-pole, three-phase winding brushless DC motor with electronic commutation scheme utilizing three Hall effect sensors. (Redrawn with permission of Micro Switch, a Honeywell Division, Freeport, IL.)

period, Q_1 and Q_6 conduct and so only coil pair $A-C$ is energized. Thus T_{gen} is still T_{max}. It is easy to extend this analysis to other 60° intervals.

The reader should understand that because the three-phase bridge circuit is driving inductive loads (i.e., the stator coils), the phase currents cannot instantaneously change when the transistors are rapidly turned on or off by the digital

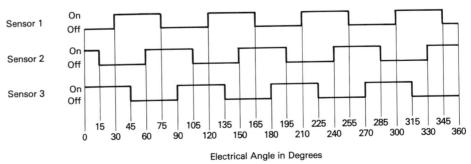

Figure 4.7.3. Outputs of three Hall effect sensors used in the 8-pole, 3-phase BDCM of Fig. 4.7.2. (Redrawn with permission of Micro Switch, a Honeywell Division, Freeport, IL.)

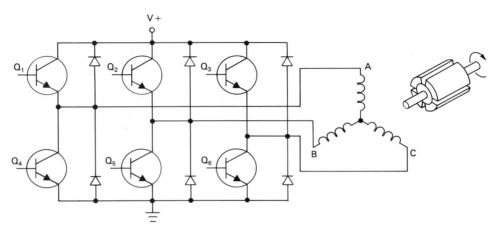

Figure 4.7.4.   Three-phase driver circuit for the BDCM with a wye configuration stator.   (Redrawn with permission of Electro-Craft Corp., Hopkins, MN.)

signals from the logic circuit.   Instead, these currents *exponentially* rise or fall in time (and hence angular position), which causes T_{gen} to vary somewhat with θ (i.e., there will be some *torque ripple*).

As mentioned in Section 4.6.6, it is important to include "flyback" diodes whenever an inductive load is rapidly switched in order to prevent the large collector-to-emitter voltage produced by the "inductive kick" occurring at switchoff from destroying the power transistor.   This is the purpose of the diodes shown in

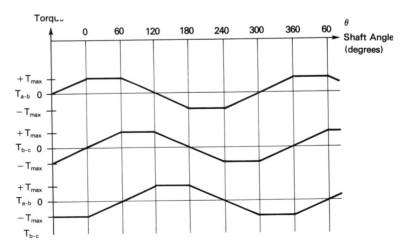

Figure 4.7.5.   Torque versus motor shaft angle θ for a BDCM assuming a *constant current* in each of the three phases.   (Redrawn with permission of Electro-Craft Corp., Hopkins, MN.)

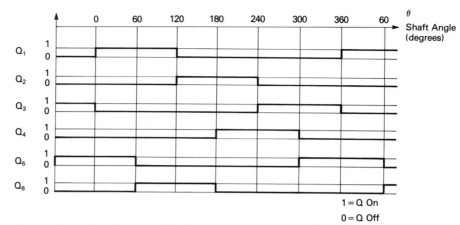

Figure 4.7.6. Transistor base drive (logic) signals for the three-phase driver circuit shown in Fig. 4.7.4. (Redrawn with permission of Electro-Craft Corp., Hopkins, MN.)

Figure 4.7.4. If fast recovery diodes are utilized for this purpose (e.g., Schotky devices), the fall time of the phase currents is very rapid and the torque ripple described above is reduced somewhat (see Problem 4.31).

Despite its obvious advantages over standard servomotors, currently few if any commercial robots utilize brushless dc motor technology. The primary reasons for this are the added cost of the driver circuitry and the lack of familiarity with the procedures needed to make the motor operate properly. However, it is expected that as the demand for higher-performance manipulators increases, the cost of the electronics will fall and more manufacturers will employ BDCMs in the subsequent generation of industrial robots.

4.8 DIRECT-DRIVE ACTUATOR

One of the major problems with commercial robots is that at certain speeds (usually but now always low), mechanical resonances are excited and exceedingly rough motion results (a so-called *"palsy"* is exhibited). Although some of the difficulty can be traced to the structure of the manipulator itself, it has been found that one of the primary causes of poor motion is the mechanical devices used to couple the motion of the actuator to the output of each joint mechanism. For example, the harmonic drive, which is currently used extensively for this purpose, contributes significantly to the low-speed performance degradation due, in part, to its compliance (i.e., "springiness") and also to machining errors which are inherent in the design and cannot be entirely eliminated. (See Section 3.3.1.2 for a more complete discussion of this component.)

A design that does not employ such mechanical units is obviously desirable (i.e., a direct connection between actuator and load is indicated; this is referred to as a *direct-drive* approach).   Nevertheless, despite the acknowledged difficulties with coupling devices such as the harmonic drive, they are still utilized extensively in manipulators.   The justification for this is that "torque multiplication" and increased position resolution that such components afford are absolutely critical in the successful design of robots.   Without these attributes, motors would have to be extremely large, bulky, and quite costly since they normally produce maximum torque at speeds too high to be of any use in a direct-coupled application (e.g., thousands instead of tens of rpm).   In addition, it would be necessary to employ very high resolution encoders that also would be costly (this is, in fact, one of the primary arguments against a direct-drive design).

In the early 1980s, however, a new motor was developed which does permit a practical direct-drive robot to be constructed.   This novel actuator, manufactured by Motornetics Corporation and called a Megatorque motor, produces extremely large torques (e.g., 35 to 1000 ft-lb) at low angular velocities (e.g., 30 rpm) without the need for a speed reducer.   In addition, a position-sensing element that is an integral part of the motor has been developed and permits the resolution of a robot based on such a motor to be at least as good as those manipulators that currently utilize more traditional sensors (e.g., optical encoders).

In effect, the Megatorque motor is a three-phase synchronous device that is operated as a brushless dc actuator, i.e., electronic commutation is employed. Unlike the BDCM discussed in the preceding section, however, this one is a *variable-reluctance* device and consequently does not contain a permanent magnet (see Figure 4.8.1).   The heart of the motor is a series of laminations that combine the rotor and stator.   One such lamination is shown in Figure 4.8.2.   It is observed from this figure that there is a thin annular rotor mounted between two concentric stators.   Both stators react with the rotor, thereby producing a significant torque multiplication (over a single stator design).   The large number of magnetic teeth located on the rotor and the two stators is also instrumental in large torque production.

The three-phase magnetic field is produced by 36 stator windings (18 on each of the two stators).   There are 150 teeth on each of the stators and the rotor which perform as motor poles in this design.   Torque is produced by sequentially energizing these poles.   For a single rotor revolution, there are 150 ac cycles, which, in effect, creates a 150:1 gear reduction with the corresponding torque multiplication.*   It is interesting to note that without the toothed rotor/stator combination, a more conventional motor would require 300 poles per phase or 900 windings in order to yield the same performance!

Another advantage of the sandwiching of the rotor between the two stators is that the magnetic flux travels over an extremely short path, as shown in Figure

---

*It is well known that increasing the number of poles of a motor tends to increase the generated torque while reducing the speed.   However, the output power is unaffected, so this scheme is equivalent to placing an *ideal* gear train (or other speed reducer), that is, one with a 100% efficiency, on the motor's output shaft.

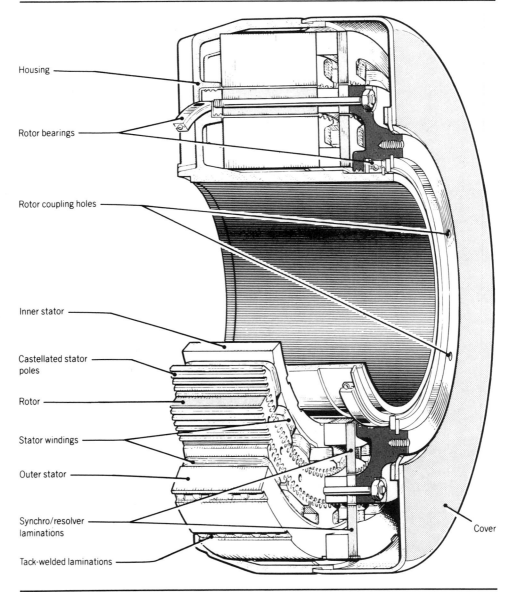

MEGATORQUE™
MOTOR SYSTEM

Housing

Rotor bearings

Rotor coupling holes

Inner stator

Castellated stator poles

Rotor

Stator windings

Outer stator

Synchro/resolver laminations

Tack-welded laminations

Cover

TM trademark of Motornetics Corporation

MOTORNETICS CORPORATION   480 Tesconi Circle   Santa Rosa, CA 95401

Figure 4.8.1. Cross section of a Motornetics, Inc. "Megatorque" direct drive motor. (Reproduced with permission of B. Powell and the Motornetics Corp., a subsidiary of NSK, Santa Rosa, CA.)

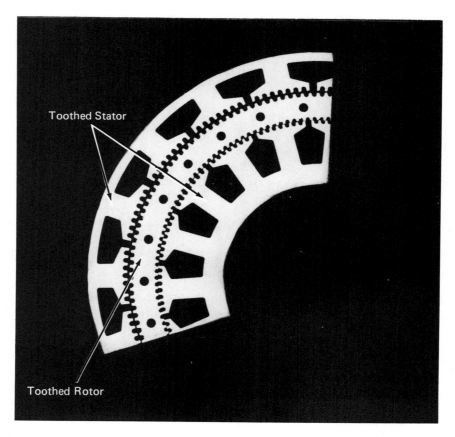

Figure 4.8.2.   A quarter section of the laminations of a toothed megatorque motor showing two stators sandwiched around a toothed rotor. (Reproduced with permission of B. Powell and the Motornetics Corp., a subsidiary of NSK, Santa Rosa, CA.)

4.8.3.   It is observed that the flux from one stator passes radially through the thin rotor into the other stator. (This is to be compared with the more conventional motor design, where a flux path of 180° through the rotor is typical.) Such a configuration lowers the magnetic resistance, and hence the motor has a high torque-to-input power ratio (high flux per ampere-turn).

Unlike the more conventional dc servomotor, the rotor of the Megatorque motor does not carry any current. As a consequence, there is little heating of the rotating member and therefore heat dissipation problems are minimized. This is a particularly important attribute in robot applications, where the actuator is often operated in a stall condition [e.g., when a load is being held (against gravity) in one place]. Any heat that *is* produced (in the stators) is easily conducted away by the case. In addition to essentially removing the temperature limitations which

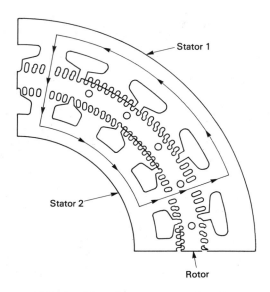

Figure 4.8.3. Flux path for megatorque motor laminations. The center portion (the rotor) moves with respect to the outer and inner laminations (stators 1 and 2). (Reproduced with permission of B. Powell and the Motornetics Corp. a subsidiary of NSK, Santa Rosa, CA.)

are associated with other motors, the direct-drive device does not have the demagnetization problems associated with many of these actuators. Thus there is no danger of causing a permanent degradation of the motor's electrical performance due to a large (and perhaps inadvertent) current spike.* All that happens in the Megatorque unit is that the iron laminations are driven into saturation, a situation that is completely reversible by reducing the flux (current excitation).

When combined with an integral position-sensing element also developed by Motornetics (see Section 5.2.4 for a discussion of this resolver-like device), a successful direct-drive robot is feasible. It is important to understand, however, that not every robotic configuration can utilize this new actuator technology. The primary reason is that the Megatorque motor is extremely *heavy*, and thus it must be incorporated into a manipulator that does not require the actuator to be carried by the particular axis. One such design is the SCARA-type robot (see Section 1.3.2.3) where the two major axis motors can be placed so that their weights are supported by the manipulator *structure* rather than by a motor-produced torque. The first commercial direct-drive robot was demonstrated at the Robots 8 conference held in Detroit in June 1984 by the Adept Corporation. Called the Adept 1, this SCARA class manipulator incorporates two Megatorque motors and shows impressive low- and high-speed performance.

---

*When the armature current in a dc servomotor exceeds a specified maximum value, the magnetic field intensity of the alnico magnet will be decreased. This *demagnetization*, although not a catastrophic failure, causes the effective torque constant K_T to decrease so that the motor is not as "powerful." Looked at from a different point of view, demagnetization requires more current for the same torque production. Note that rare earth and ceramic magnet motors are not as sensitive to overcurrent-caused performance degradation.

In describing a direct-drive approach to robotic actuation, we have limited the discussion to the Megatorque motor since this is the only *commercial* device currently applicable to robotics. It should be mentioned, however, that at this writing, research efforts at both MIT and Carnegie–Mellon University are under-way. It is certainly possible that in the future, additional commercial direct-drive motor designs will emerge and that more robots will utilize this potentially attractive technique.

4.9 HYDRAULIC ACTUATORS

When a robot is required to move sizable loads (e.g., greater than 10 lb) in a rapid and precisely controlled manner, the use of a servomotor as the actuator may become impractical because of its high cost, weight, and volume. In this case a hydraulic actuating device is most often employed, the decision being based on a rule of thumb which states that "for actuators of greater than 5 to 7 horsepower, hydraulics should be considered." (It should be pointed out, however, that at this writing, new design and manufacturing techniques are making the use of servo-motors more feasible in large robots.) These types of devices produce high po-sitional stiffness, deliver large forces, and can be designed so that they are over-damped. Moreover, such features can all be incorporated into a fairly compact physical package. We now consider the behavior of such actuators.

The heart of a hydraulic actuator is the *servovalve* (also called a pilot valve), a device having the characteristic of producing an oil flow that is proportional to the current supplied by a servo amplifier. This oil flow, in turn, causes a cylinder/piston arrangement (sometimes called "the actuator" or "hydraulic motor") to move producing either linear or rotary displacement. Thus any load coupled to the piston shaft will be moved in the same manner.

The schematic of a two-stage flow control servovalve (of the deflector jet type) is shown in Figure 4.9.1. Here P represents a (constant) supply pressure and R is the return or drain. The tubes C_1 and C_2 at the bottom of the figure are called control ports and are attached to the cylinder/piston as shown in Figure 4.9.2. It should be noted that other types of flow control configurations exist (e.g., double nozzle and jet pipe valves). However, their behavior is essentially the same as the deflector jet valve, so they will not be discussed here.

Let us consider the operation of the second stage of the deflector jet servo-valve. If the valve spool is moved to the right, the high-pressure supply port on the right is connected to control port C_2, causing it to become an outlet (see Figure 4.9.2a). It is also seen that the left-hand supply port is closed off and that C_1 is connected to the drain, making it act as an inlet port to the servovalve. Figure 4.9.2a reveals that with the situation above, the pressure on one side of the piston will be P and on the other side, P_1. Since $P > P_1$, the mass M that is coupled to the piston shaft will move to the left.

Moving the spool to the left will reverse the actions described above. In this

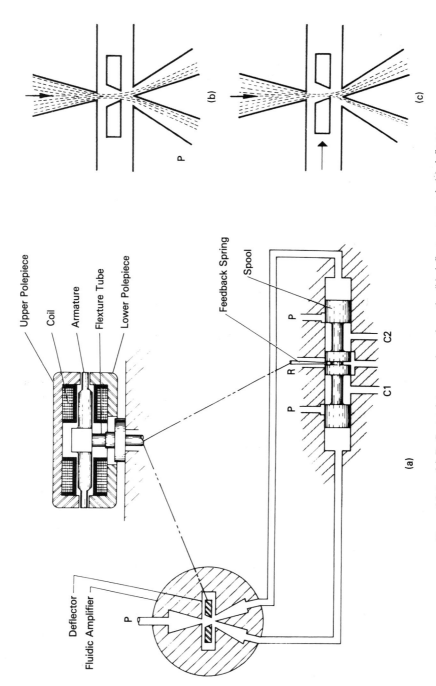

Figure 4.9.1. (a) Deflector jet servovalve schematic; (b) deflector centered; (c) deflector displaced. (Reproduced courtesy of Moog, Inc., East Aurora, NY.)

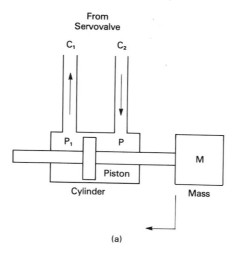

(a)

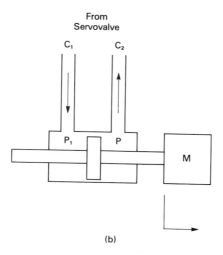

(b)

Figure 4.9.2. Hydraulic cylinder piston arrangement: (a) valve spool (in the servovalve) moves right. $P > P_1$ which causes mass M to move left; (b) valve spool moves left. $P_1 > P$ and M move to the right.

case, Figure 4.9.2b indicates that the larger pressure (P_1) will be on the left side of the piston, so that the mass will move to the right.

The first stage of the servovalve consists of a small dc permanent-magnet torque motor and a deflector jet hydraulic amplifier. These components are used to produce the valve spool motion. The armature, coil, and pole pieces of the motor are shown in Figure 4.9.1 together with the fluidic amplifier. When current flows in the coil, the armature rotates either clockwise or counterclockwise (in the plane of the paper), causing the flexure tube to move the deflector. This action connects the supply pressure to either the right- or left-hand far surface of the spool and results in a difference in pressure between these two surfaces, thus producing linear motion of the entire spool assembly. Reversing the direction of the current will result in the opposite motion of the spool. Extremely small currents, on the order of tens of milliamperes, are needed to produce this motion and cause the load and supply pressures to be equal. Smaller currents will produce lower load pressures. It is interesting to note that recently, stepper motors have been used in place of the torque motor.

The deflector jet servovalve shown in Figure 4.9.1 uses a cantilevered spring to monitor the position of the spool. This information is "fed back" internally and is used to produce a load torque on the flexure tube. Thus a positional equilibrium of the spool is achieved for this scheme. This can be most easily seen by noting that for a given input current, the motor armature, and hence the spool, continues to move until the countertorque produced by the spring balances the torque developed by the motor. It is important to note that the feedback described here is internal and not from the load (i.e., the cylinder/piston). Thus, for a *constant current* supplied to the servo amplifier, the valve motion, and hence the load motion, will be *constant*.

The static characteristics of a typical flow control servo valve are shown in Figure 4.9.3. Here Q is the flow in gallons/minute, P_s the supply pressure in psi, ΔP the differential pressure across the piston, i the input current (in milliamperes) to the torque motor, and i_r the rated servovalve current. This figure reveals that full supply pressure can be transmitted to the load with currents that are a fraction (on the order of 1 to 5%) of the rated value.

Servovalves having a wide range of operating parameters are available. For example, it is possible to obtain ones that have:

1. Flow rates from 2 to more than 500 gal/min. This corresponds to a power range of approximately 3 to more than 700 hp.

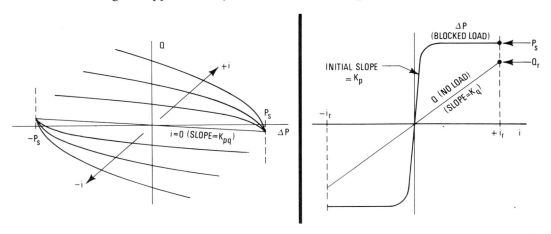

Figure 4.9.3. Static characteristics of a typical flow-control servovalve. (Reproduced courtesy of Moog, Inc., East Aurora, NY.)

2. Supply pressures in the range 1000 to 3000 psi. Higher pressures produce better weight and power efficiencies.

4.9.1 The Hydraulic Position Servo

In practice, the hydraulic actuator described above can be used to replace a dc servomotor in a control loop. Hydraulic velocity, pressure (or force) control, and position servos can be configured in much the same way as they are with all electrical components. For robot applications, a position servo is usually employed for reasons mentioned previously. The idealized schematic and block diagram (with dynamic effects neglected) for such a device are shown in Figure 4.9.4. Although this servovalve/actuator would be used as an element in the control loop of a prismatic joint, rotary joint control could also be achieved if the cylinder/piston was replaced by an actuator that converted the fluid flow from the valve into rotary motion (i.e., a hydraulic motor).

In Figure 4.9.4, Q_o and Q are the servovalve no-load and output flows, respectively, A is the piston working area, and K_{fx} the position transducer gain. In addition, K_{pq} and K_q are, respectively, the servovalve "droop" (due to internal fluid leakage) and flow gain with K_a and K_s actuator stiffness terms. Other terms are defined below.

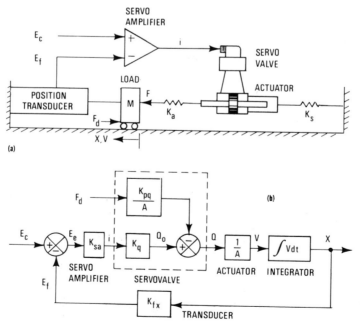

Figure 4.9.4. Typical hydraulic position servo: (a) schematic diagram; (b) idealized block diagram (dynamic effects neglected). (Reproduced courtesy of Moog, Inc., East Aurora, NY.)

The operation of this servo is similar to that already discussed for the servomotor. A position command voltage E_c from the master microprocessor is compared to the actual position as represented by E_f, the output voltage of some type of position transducer (e.g., a "pot" or in the rotary case, an optical encoder). (Use of D/A conversion is assumed where necessary but is not explicitly shown here.) The resultant error signal is amplified by the servo amplifier (having a gain K_{sa}) to produce a drive *current i*. As explained above, this current causes the servovalve and actuator assembly to produce a force F which acts on the load mass M. As a result, the load will move toward the desired position (with velocity V in the x direction) until the error E_e is zero.

As mentioned in a previous section of this chapter, there will usually be two disturbances present when a position servo is used to control a robot joint (i.e., friction and gravity). In Figure 4.9.4 these forces are represented by the term F_d. The reader will recall that such disturbance forces will cause a steady-state position error. Thus compensation techniques must be employed to reduce or completely eliminate this unacceptable error (e.g., integral control; see Section 4.3.3).

In view of our discussion concerning the need to limit final position overshoot during a joint move, it is reasonable to ask the question: What prevents the load from oscillating, that is, how is damping produced in the hydraulic position servo since Figure 4.9.4 does not reveal the presence of any velocity loop? This question does not have a single answer since a variety of techniques can be used. For example, due to the viscous nature of oil, it is possible to make certain structural modifications in the design of the servovalve which permit the required damping to be achieved. One such change causes the flow–pressure curves shown in Figure 4.9.3 to become nearly linear and to have increased slope. It can be shown (although this will not be done here) that doing this directly increases the damping coefficient ξ. Thus, for a given application having a specified load, it may be possible to select the servovalve so that the proper amount of damping is provided.

A technique that is often used to realize an increase in the slope of the flow–pressure curves involves deliberately introducing an appropriate leakage (or bypass) path in the valve/actuator combination. Either laminar or sharp-edged orifice bypass designs can be employed, with the latter giving greatly increased damping at very low pressure drops. The major disadvantage of increasing the slope in this manner is that the static stiffness of the hydraulic system is reduced. Thus disturbance forces or torques such as gravity will produce a significant position error. Consequently, servovalves of this type are not the best choice for robot joint control.

The problem of static stiffness reduction can be overcome by several additional modifications to the valve structure (e.g., utilizing a spring-centered capacitance piston). The resulting device is called a DPF servovalve and has the desired linear flow–pressure characteristic with the corresponding high damping coefficient *and* the required stiffness.

It is also possible to achieve a similar damping characteristic electrically. This is accomplished by using load force feedback. Actuator force output or pressure differential can be monitored using an appropriate transducer. The electrical signal from such a device is then fed back to the servo amplifier. It can be shown that when the load on the actuator increases, the valve current will decrease, thereby producing a corresponding reduction in flow. In fact, the feedback causes the *effective* flow–pressure curves to be very nearly linear (under load), which is the desired result. Alternatively, a velocity sensor can be used to provide the extra damping, as discussed previously.

Finally, a simple, nonelectrical method of obtaining the required damping is available. That is, by externally attaching a passive viscous damper, similar to an automobile shock absorber, to an appropriate hydraulically actuated member, it is possible to limit endpoint oscillation significantly. A number of commercially available robots today utilize this approach on several of their axes. It is important to understand, however, that such an approach is not always practical for every robot joint because of space and weight restrictions. This is particularly true in the case of wrist axes.

4.10 PNEUMATIC SYSTEMS

Unlike the components found in hydraulic systems, pneumatic devices make use of a fluid medium that is highly compressible. This fluid is usually air and has the advantage that it is both readily available and nonflammable. Most frequently, hydraulic devices use oil in a *closed system* so that any leak necessitates the replacement of the fluid and creates a fire hazard. Moreover, pinhole leaks in the high-pressure hydraulic hoses represent a significant safety hazard since the fluid escaping from these holes is almost invisible but can nevertheless punch a hole through many materials and also do quite a bit of damage to a human operator. In normal operation, however, pneumatic devices are quite often deliberately vented to the atmosphere, with the working medium (i.e., the air) actually being replaced at the completion of each work cycle, and the relatively low pressure makes them quite a bit safer to use.

There are other differences between pneumatic and hydraulic components. For example, oil is a highly viscous medium and air is not. Consequently, pneumatic systems tend to exhibit highly underdamped dynamic behavior, whereas hydraulic systems tend to be fairly well damped. Another major difference between the two has to do with stiffness. Air-actuated systems are highly compliant because of the compressibility of the working medium. Oil is not as easily compressed, so that hydraulic devices tend to be far less compliant. By increasing the operating pressure, however, it is possible to overcome some of the problems associated with pneumatic systems. Thus it is not surprising to find that some smaller, fairly simple pick-and-place-type robots are pneumatically actuated. For example, Seiko makes a variety of this type of device. Also, in the early 1980s, one manufacturer, International Robomation/Intelligence (IRI), did produce a five-axis, computer-controlled, air-actuated, *servo-controlled* robot which was capable of moving 25-lb payloads at speeds of up to 20 in./s (see Figure 1.3.20). Although IRI no longer produces this robot, possibly because its performance was not as "positive" as that of hydraulic or servomotor-driven units, its very existence indicates that some of the difficulties previously encountered with pneumatic systems have been overcome.

Let us briefly consider the operation of a simple pneumatic proportional controller shown in Figure 4.10.1. The heart of the device is a two-stage amplifier. In the first stage, often referred to as a *flapper valve amplifier*, the error linkage position x controls the pressure P_2, which, in turn, determines the position y of a metering valve for the second-stage amplifier. Called an *air relay*, this second amplifier is capable of handling large airflows.

An error signal e that causes the error linkage to move to the right produces a corresponding *decrease* in the pressure P_2. As a result, the metering valve moves *up*, thereby bleeding less air to the atmosphere and making the output pressure P_o (i.e., the pressure to the system being driven) increase and approach the supply pressure P_1. It is observed that this action extends the feedback bellows, which moves the error linkage to the left (z increases). The converse is true when the error linkage moves to the left so that the system pressure is made to decrease, thereby causing the feedback bellows to contract (z decreases). Positional equilibrium of both the metering valve and feedback bellows is obtained through the use of springs having spring constants K_2 and K_f. (A_2 and A_f are the areas of the metering valve and feedback bellows, respectively.)

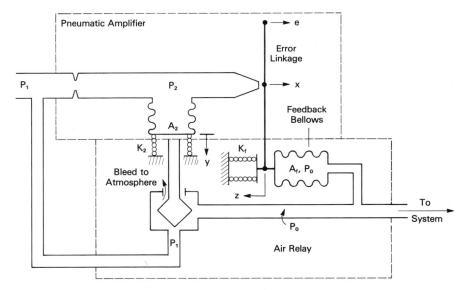

Figure 4.10.1. Pneumatic proportional controller. (Reprinted from F. Raven, *Automatic Control Engineering*, second edition, McGraw-Hill Book Company, NY, 1968, p. 537.)

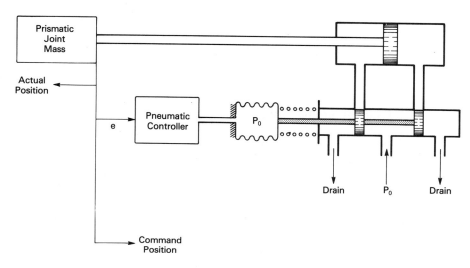

Figure 4.10.2.   Pneumatically controlled prismatic joint.

Other pneumatic controllers can be constructed by adding additional features to the basic proportional device in Figure 4.10.1.   For example, derivative action is produced by inserting a properly shaped restriction in the feedback line between the system output pressure P_o and the feedback bellows.   Integral control is obtained by adding another bellows on the left-hand side of point z.   Clearly, it is also possible to combine these actions to obtain a pneumatic PID controller.

Any one of the devices above can be used to drive a prismatic joint of a robot, as shown in Figure 4.10.2.   The four-way valve and cylinder/piston arrangement operate in a manner similar to the corresponding hydraulic device described in the preceding section.   It is also possible to drive a rotary joint using a vane-type pneumatic motor.   This device produces a torque that is proportional to the controller output pressure P_o and is independent of the shaft speed.

As mentioned above, pneumatically actuated robots are usually limited to small devices that are used to move light payloads from point to point without regard to the actual trajectory.   (The IRI robot was a notable exception.)   High-pressure compressed air permits fast, accurate, and generally simple moves (e.g., noncoordinated) with mechanical limit stops often being utilized to stop the individual joints.   However, where large loads, requiring a controlled trajectory, must be handled, either servomotors or hydraulic actuators are normally employed.

4.11 SERVO AMPLIFIERS

As shown in Figure 4.2.2, a servo amplifier must be used to convert the low-power command signals that come from the master computer and are then "processed" by the joint computer to levels that can be used to drive the joint motor.   An amplifier that can provide the necessary logic and drive for a stepper motor was described in Section 4.6.6. In this section we consider possible configurations that can be used to drive a dc servomotor. Specifically, pulse-width-modulated (PWM) and linear amplifiers incorporating voltage feedback or voltage and current feedback will be discussed.

4.11.1 Linear Servo Amplifiers

Two basic classes of linear servo amplifiers exist: (1) the H type and (2) the T type. These are shown in Figures 4.11.1 and 4.11.2, respectively. The first of these, the H, is sometimes called a bridge amplifier, and has the advantage of requiring a *single* or unipolar dc supply. However, it is not always easy to operate in a linear fashion, and because the motor must be "floated" with respect to the system ground, current and/or voltage feedback is not easy to achieve. In actual operation, one set of diagonally opposite transistors is turned on [e.g., Q_1 and Q_4 (or Q_2 and Q_3)]. It can be seen that if the first of these sets is made to conduct by applying a positive control voltage to channel 1 (and grounding channel 2), the armature voltage $V_{AB} \leq +V$ and the motor will turn (e.g., in the clockwise direction). When the control signals on channels 1 and 2 are reversed, the second set of transistors conducts, thereby making $V_{AB} \geq -V$. The motor will now turn in the opposite or counterclockwise direction. The actual size of the armature voltage, and hence the motor speed, will depend on the amount of base current supplied by the control circuitry that precedes the power amplifier stage (e.g., a preamplifier, not shown in Figure 4.11.1).

The second general type of servo amplifier, the T, requires a bipolar dc supply, as shown in Figure 4.11.2. However, it is easy to drive and since the motor *does not have to float* with respect to ground, current and/or voltage feedback is easy to implement. Since complementary power transistors are employed (see Figure 4.11.2), a single bipolar control signal can be used to turn on either Q_1 or Q_2, thereby making V_{AB} either $\leq +V$ or $\geq -V$ and producing the desired bidirectional rotation. In the T configuration, it is important to bias the transistors so that Q_1 and Q_2 are *not* both on at the same time since output transistor failures are likely to occur if this happens (i.e., they conduct simultaneously).

An undesirable characteristic of a T servo amplifier is the "deadband" or "crossover distortion" that exists around zero output voltage. This produces an armature drive voltage that is a nonlinear function of the servo amplifier input for small positive and negative inputs signal. The problem can be reduced by keeping *both* transistors on around the zero-voltage region. From what was said above, it is clear that care must be taken to prevent simultaneous operation of the transistors from occurring when large currents flow.

It is important to note that the amplifiers in Figures 4.11.1 and 4.11.2 do not have any type of flyback protection shown. However, this is absolutely essential since the inductance in the servomotor armature can produce an "inductive kick" when the power amplifier transistors are either suddenly all turned off or when

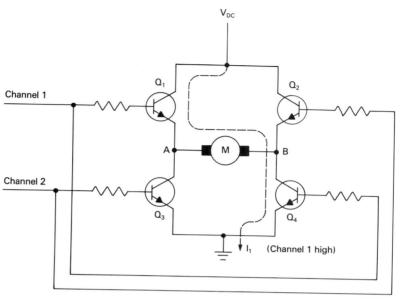

Figure 4.11.1. "H" type servo amplifier (one power supply required). Path of current I_1 is shown for channel 1 high and channel 2 low. (Redrawn with permission of Electro-Craft Corp., Hopkins, MN.)

the motor is "plugged" (i.e., the armature voltage is rapidly reversed to provide dynamic breaking). Recall the discussion of Section 4.6.6. Thus regardless of which configuration is used, flyback diodes (or some other method of protecting the power transistors from breakdown) must be placed across the collector–emitter terminals of the output transistors. Failure to do this risks a collector-to-emitter

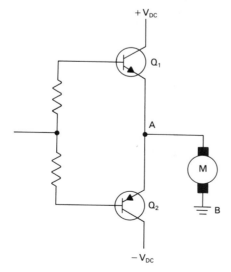

Figure 4.11.2. "T" type servo amplifier (two power supplies required). (Redrawn with permission of Electro-Craft Corp., Hopkins, MN.)

short circuit, which, as explained previously, can cause a runaway condition—an especially dangerous occurrence in robotic applications.

There are other factors that must be considered when working with linear servo amplifiers: for example, the power dissipation capability of both the power transistors and the associated heat sinks, the need to provide some type of *active* cooling (e.g., by using a fan), and the need to protect both the power transistors and the motor from current overloads by using current limiting. The last of these factors is particularly important in robotic applications since it is not at all uncommon for a stall to occur in the middle of a move due to the manipulator coming into contact with a foreign object that has accidentally found its way into the workspace. Clearly, one or more motors will stall in this case and some type of protection is absolutely essential in order to prevent amplifier and/or motor damage or destruction. Current limiting is one such technique, although fusing of motors and software methods (i.e., implementation of "timeout" criteria; see Section 4.4.4) are also often employed.

4.11.2 Pulse-Width-Modulated Amplifiers

One of the major difficulties with the linear amplifiers described in the preceding section is that, very often, the output is only a fraction of the total supply voltage, for example, during the initial or final portions of a move or when the move is deliberately performed at low speeds. This is accomplished by operating the power transistors in their active (i.e., linear) regions, which means that the collector-to-emitter voltage drop V_{CE} of the transistor(s) that is (are) conducting is significant. Consequently, the power dissipated in the collector (i.e., the product of collector current and the collector-to-emitter voltage) can be large (on the order of tens of watts and sometimes as high as 100 W), so the transistors and heat sinks must be sized accordingly. Although it is certainly possible to obtain these large transistors with the technology currently available, the added cost incurred is not always warranted. Fortunately, it is now possible to use a different approach that is generally more cost-effective (i.e., pulse-width modulation, PWM).

With the advent of power transistors that can be switched at megahertz rates, the use of PWM amplifiers to drive servomotors in robotic applications, as well as other incremental motion applications, has become quite practical and attractive. The major advantage of a switched device over a linear device is that in the former, the power transistor is either "off" or in (or close to) saturation. In either case, the power dissipated in the collector is considerably less than in an equivalent linear amplifier. This is easily understood by recognizing that since little or no collector current flows when the transistor is turned off, the power dissipation is quite small. When current does flow, however, the transistor is in saturation, which means that the drop across its collector is only 1 or 2 V. Thus the dissipation is still quite small (i.e., under 12 W for a continuous armature current of 6 A). An equivalent linear device might dissipate 72 W (assuming a 12-V drop across the collector).

Just as with linear servo amplifiers, PWM devices can be of the H or T type and the same comments concerning the advantages and disadvantages of both are pertinent (see Figures 4.11.1 and 4.11.2). However, unlike the linear case, the output voltage of the T or H circuit will be almost equal to the *full value* of either the positive or negative dc supply voltage (see Figure 4.11.3).

How can these types of signals provide the required variation in armature voltage and hence rotor speed? The answer to this question is found by recognizing that the servomotor is a low-pass filter [e.g., see the transfer function in Eq. (4.3.5)]. With T_S defined as the period of the switching signal waveform, then if the radian switching frequency $\omega_S = 2\pi/T_S \gg \omega_E$, the electrical pole of the motor (i.e., $\omega_S > 100 \, \omega_E$), the filtering action of the motor will cause the effective armature voltage to be the "average value" of the waveforms in Figure 4.11.3.* Mathematically, this means that

$$(V_{\text{arm}})_{\text{ave}} = \frac{1}{T_S} \int_0^{T_S} V_{\text{arm}}(t) \, dt \qquad (4.11.1)$$

Thus applying Eq. (4.11.1) to the waveforms in Figure 4.11.3, it is seen by inspection that the motor will not move for the square wave in part (a) because $(V_{\text{arm}})_{\text{ave}} = 0$, whereas the nonzero average value of this quantity for the waveforms in (b) and (c) will produce rotor motion. It is important to understand that Eq. (4.11.1) will not be strictly correct if the switching frequency is too low. For example, if it is only about 10 times higher than the electrical pole of the motor, the effective armature voltage will be somewhat less than the average value and the armature current may exhibit significant ripple (see Problem 4.33).

In actual use, a PWM servomotor drive can be made to produce practically any type of acceleration, velocity, or position profile that might be required in a given application. For example, if it is desired to cause a servomotor to turn with a trapezoidal velocity profile (see Figure 4.6.7), this can be achieved by making the pulse width, T_p in Figure 4.11.3, vary trapezoidally with time (see Problem 4.34). In a robotic application the joint processor converts the velocity error samples into equivalent values of T_p. This is accomplished by causing the associated control logic to command the appropriate power transistor(s) in the PWM amplifier to turn on for T_p milliseconds. In view of the discussion of the preceding paragraph, faithful reproduction of the desired profile will occur only provided that the switching frequency is "high enough." This statement, in effect, implies that the frequency must be chosen so that the *sampling theorem* is satisfied.

Unlike the linear servo amplifier, there is another cause of power dissipation in a PWM device, and this places a practical upper limit on the switching frequency.

---

*Recall that a periodic waveform such as a square wave can be represented by a Fourier series:

$$V_{\text{arm}}(t) = V_{\text{dc}} + \sum_{n=1}^{\infty} [A_n \cos (n\omega_S t) + B_n \sin (n\omega_S t)]$$

If this signal is passed through a *low-pass* filter network with a cutoff frequency below ω_S (and hence $n\omega_S$), only the dc term will be transmitted, and the output will be V_{dc}.

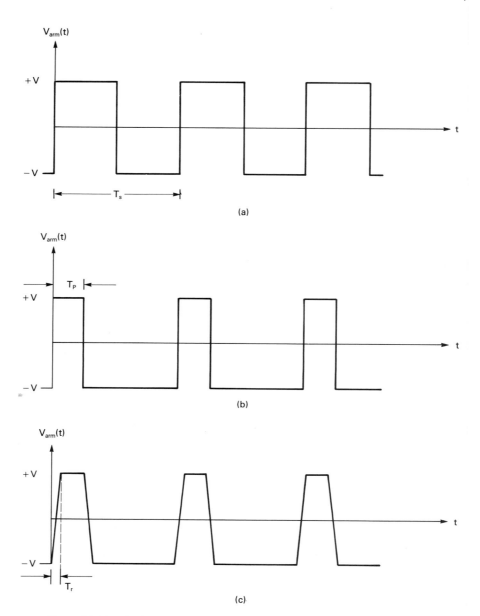

Figure 4.11.3. Typical PWM waveforms: (a) no load PWM output, ideal switch, $(V_{arm})_{ave} = 0$, motor does not turn; (b) loaded PWM output, ideal switch, $(V_{arm})_{ave} = -V/2$, motor turns CCW; (c) same as part (b), except nonideal switch and power transistors in active region during T_r.

Since switching cannot physically occur instantaneously but rather takes a finite time T_r, the power transistors spend a portion of the switching cycle in the active region (see Figure 4.11.3c). If the switching rate is extremely high, it is possible for this time to become a significant portion of the overall switching period, with the result that the overall power dissipation can be quite large, approaching that of the linear case. As a consequence, practical PWM servo amplifers usually work at switching rates of 1 to 15 KHz. (The lower limit is often determined by human factors, since a low-frequency switching rate can produce annoying and sometimes intolerable audible noise.)

4.11.3 Effects of Feedback in Servo Amplifiers

In this section the effect of using voltage, current, and combined voltage and current feedback with the power amplifier is considered. The reader should recall that the use of feedback can "stabilize" whatever quantity is being fed back. Thus, in the case of voltage feedback, an amplifier's output voltage is held constant regardless of changes in the load's impedance. This is sometimes referred to as a *voltage-stabilized* amplifier.

Figure 4.11.4 shows the voltage–current characteristics for an amplifier with voltage feedback and a corresponding op amp implementation. Note that the intersection of any single constant resistance line with a particular constant-voltage output curve defines an *operating point*. It is important to understand that regardless of the value of the load resistance, the voltage-stabilized amplifier will produce an output V_{out} which is proportional to both the input V_{in} and the gain factor V_{vf} (units of v/v).

As stated in Section 4.11.1, the use of voltage feedback can reduce the deadband in a T amplifier configuration. The feedback will cause the appropriate drive signals to be applied to the transistors so that the output voltage will be a linear function of the input voltage and the gain. Figure 4.11.5 indicates how a T power stage may be driven by an op amp having voltage feedback. Note that to reduce zero crossover distortion, the signal to be fed back must be obtained at the *output* of the transistors, not at their bases.

The advantage of using an amplifier with voltage feedback for motor control is that the voltage delivered to the motor terminals will be kept at the value commanded by the input. This is important because control strategies discussed so far adjust the applied voltage via the amplifier in an effort to control the position or velocity of the motor shaft. In fact, if power dissipation is not a problem, the voltage-stabilized amplifier used with an *unregulated power supply* (sized such that its loaded output voltage is still in excess of that required to produce the desired motor performance) can ensure that the voltage and current delivered to the servomotor are predictable despite fluctuations in the power supply (caused by line voltage variations, for example). In addition, for a multiple motor system (such as that utilized in a robotic manipulator) that employs a single supply, the gain of each motor's voltage feedback amplifier can be adjusted so that for the maximum

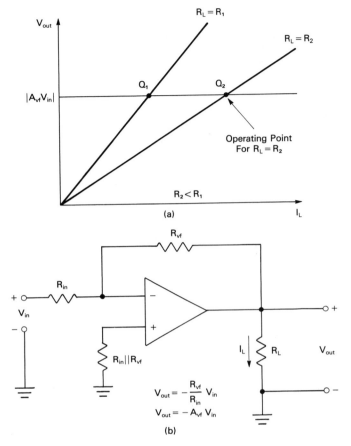

Figure 4.11.4. Voltage feedback amplifier: (a) *V-I* operating characteristic; (b) circuit diagram.

allowable input signal, the output voltage delivered does not exceed some pre-defined maximum value.

Figure 4.11.6 shows the operating characteristics for an amplifier that has a *stabilized current* output. In this configuration, the output of the amplifier is a constant current defined by the product of the amplifier's gain A_i (units of amperes/volt) and the input voltage. As the load varies, the output voltage changes in order to keep the output current constant. This type of amplifier is used when it is desired to adjust the current across the motor terminals (e.g., in a torque control application). One advantage of using such a device with dc servomotors is the fact that the current delivered will be the same regardless of changes in the motor's armature resistance (which, it will be recalled, is a function of the armature temperature). In addition, the voltage drops inherent in the wiring from the amplifier

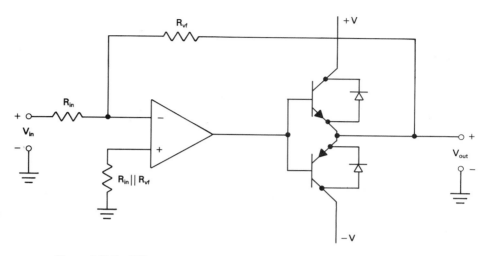

Figure 4.11.5   "T" type power stage driven by an op amp with voltage feedback.

to the motor will not affect the power delivered *to the motor*. This is clearly a benefit in high-performance systems.

If *both* current and voltage feedback are incorporated into the same amplifier, the operating characteristics are as shown in Figure 4.11.7. This particular configuration may also be viewed as a voltage-stabilized amplifier to which a fixed series resistor has been added. The resistance value is defined as (V_o/I_o) ohms, where V_o is the amplifier's open-circuit voltage and I_o the short-circuit current. A major advantage of combining the two types of feedback is that the power dissipated in the armature or the torque produced under stall conditions may be controlled. [In the case of a voltage-stabilized amplifier, the stall current is determined by the output voltage and the motor's armature resistance and produces a torque defined by Eq. (4.3.3).]

If it is desired to limit the torque at stall, possibly to prevent damage to a transmission or coupler attached to the motor's shaft, it is necessary to limit the output current. One way to accomplish this is to put a resistor in series with the motor's terminals, thus effectively increasing the armature resistance. It should be obvious to the reader that while such an approach would work, the size of the resistor could be physically quite large in order to accommodate the required heat dissipation. As will be seen below, the simultaneous use of voltage and current feedback with an amplifier can achieve the same result without the use of an external power resistor. Of course, power must be dissipated in the transistors of the power stage, but this is generally preferable to employing a large and possibly costly resistor.

We now demonstrate the current-limiting effect obtained by utilizing simultaneous voltage and current feedback. Using the component designations shown

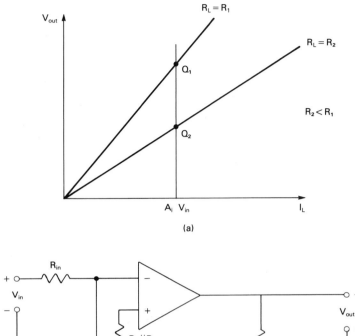

Figure 4.11.6. Current feedback amplifier: (a) *V-I* operating characteristic; (b) circuit diagram.

in Figure 4.11.7, the input–output relationship for such an amplifier is found to be

$$V_{\text{out}} = -V_{\text{in}} \frac{R_{vf}}{R_{\text{in}}} - I_L R_s \frac{R_{vf}}{R_{cf}} \qquad (4.11.2)$$

Note that the configuration is that of an *inverting summing amplifier*. Current limiting results because the two input voltages (V_{in} and $I_L R_s$) have *opposite* signs. Equation (4.11.2) is plotted in Figure 4.11.8 for three values of V_{in} (-1, -2, and

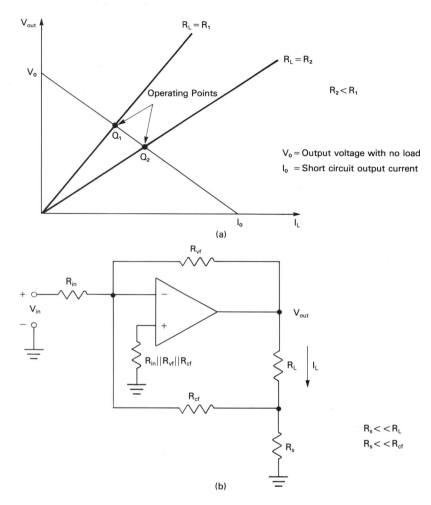

Figure 4.11.7.   Combined voltage and current feedback amplifier: (a) *V-I* operating characteristic; (b) circuit diagram.

-3 V) and three load resistances (0.1, 5, and 10 Ω). For the values used (see the table at the bottom of Figure 4.11.8), Eq. (4.11.2) reveals that the effective series resistor is 2.22 Ω. As stated above, this amplifier may be modeled as a voltage-stabilized amplifier whose load includes the series resistor equal to this value. To illustrate, consider the intersection of the constant-resistance line of 5 Ω with the $V_{in} = -2$-V line in Figure 4.11.8. At the operating point (Q_3), the output current is 1.94 A, with 9.7 V being supplied to the load. The same result can be obtained by using 14.0 V as the output of a voltage-stabilized amplifier having a total load of 7.22 Ω (5.0 + 2.22).

In any one of the configurations just discussed, there are practical limitations

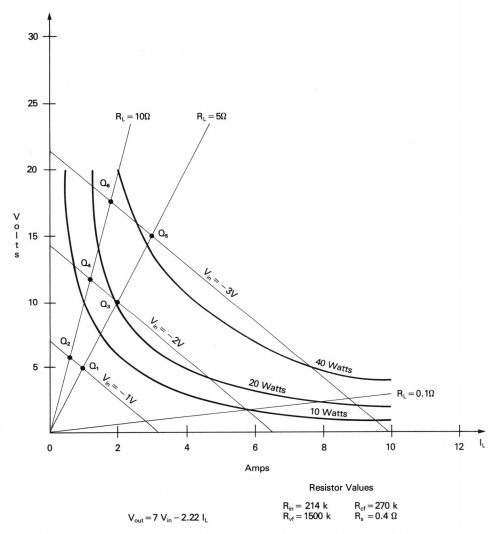

Figure 4.11.8. Load-line analysis for amplifier with combined current and voltage feedback.

$$V_{out} = 7\,V_{in} - 2.22\,I_L$$

Resistor Values

R_{in} = 214 k	R_{cf} = 270 k
R_{vf} = 1500 k	R_s = 0.4 Ω

that place bounds on the operating characteristics of the devices depicted in Figures 4.11.4, 4.11.6, and 4.11.7.   Such considerations as the ratings of the components and the available voltage and current from the power supply driving the amplifier may be used to define usable operating ranges for the two axes.   In addition, other circuitry may be added to prevent damage to the amplifier or other system elements. These circuits will usually operate only if a certain parameter is exceeded and, in general, can produce a variety of actions.   For example, a fuse will disconnect the amplifier and load.   Also, *hard limiting* can be implemented whereby the current is held at a fixed maximum value as the output voltage drops (i.e., short-circuit protection). Finally, it is possible to use *current foldback*, which produces a simultaneous reduction in the output current (once some predetermined maximum current has been reached) as the output voltage drops.

The remainder of this section is devoted to a discussion of the effects of each of the three types of amplifier configurations on the motor transfer function.

4.11.3.1 Voltage amplifier driving a servomotor

A dc servomotor driven by an amplifier utilizing voltage feedback is shown in Figure 4.11.9a. As discussed in previous sections, this amplifier usually consists of a power stage preceded by a preamplifier (normally implemented by an op amp). The characteristic of the device is that it has a low output impedance and is therefore termed a voltage amplifier. It is convenient to think of a system utilizing this group of components as producing a *voltage-controlled velocity*.

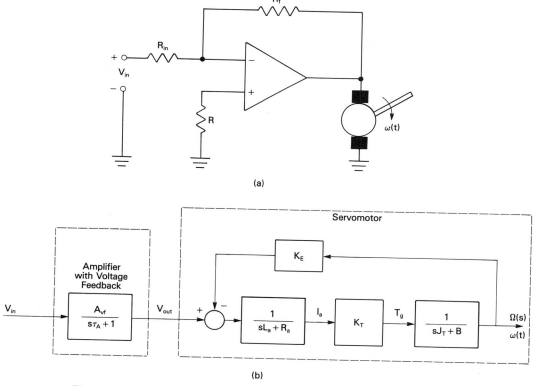

(a)

(b)

Figure 4.11.9.: (a) Servomotor driven by an amplifier utilizing voltage feedback; (b) block diagram representation of part (a). [Part (b) redrawn with permission of Electro-Craft Corp., Hopkins, MN.]

The voltage amplifier itself will have a finite bandwidth, and in most robot applications a single-pole model will adequately represent its frequency response (see, e.g., Section 4.4.2). Thus we will assume that its transfer function is given by

$$\frac{V_{\text{out}}(s)}{V_{\text{in}}(s)} = \frac{A_{vf}}{1 + s\tau_A} \tag{4.11.3}$$

where $1/\tau_A$ is the amplifier radian frequency bandwidth and A_{vf} is the magnitude of the gain, given by the relationship

$$|A_{vf}| = \frac{R_f}{R_{\text{in}}} \tag{4.11.4}$$

Figure 4.11.9b shows a block diagram representation of a dc servomotor driven by such an amplifier. The overall transfer function of this configuration is obtained by multiplying the transfer functions of Eqs. (4.3.5) and (4.11.3) and relates the motor shaft speed (transform) to the input voltage (transform). Thus

$$\frac{\Omega(s)}{V_{\text{in}}(s)} = \frac{(K_T/L_aJ_T)}{s^2 + [(R_aJ_T + L_aB)/L_aJ_T]s + (K_TK_E + R_aB)/L_aJ_T(1 + s\tau_A)} \cdot \frac{A_{vf}}{(1 + s\tau_A)} \tag{4.11.5}$$

The important point to note in this equation is that the use of voltage feedback has not affected the location of the motor poles. This, of course, assumes that no loading exists between the two devices, which is true for a *zero*-output impedance amplifier.

4.11.3.2 Current amplifier driving a servomotor

An alternative method of driving a servomotor is with a current amplifier (*high*-impedance source), as shown in Figure 4.11.10. The reader will recall that an amplifier with current feedback produces a constant output current for a given input voltage. Examination of the dynamic equations of a dc servomotor reveals that this implies that the terms associated with the electrical behavior (e.g., armature resistance and inductance together with the back EMF constant) do not influence the current actually being delivered to the motor.* By using the Laplace transforms of Eqs. (4.3.3) and (4.3.4) and ignoring the constant-torque terms (for the same reasons given in Section 4.3.1), the relationship between motor velocity and current is given as

$$K_TI_a(s) = (sJ_T + B)\Omega(s) \tag{4.11.6}$$

---

*This, of course, assumes an ideal current source capable of supplying any voltage necessary to produce the desired current.

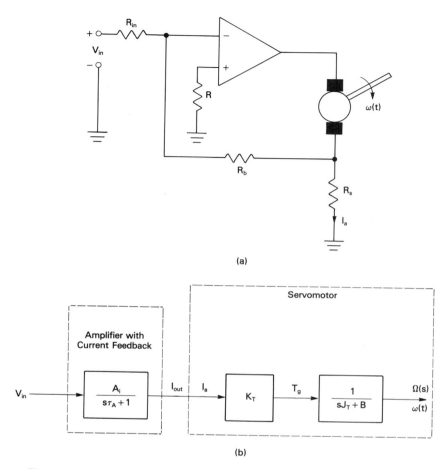

Figure 4.11.10.   (a) Servomotor driven by an amplifier with current feedback; (b) block diagram representation of part (a).   [Part (b) redrawn with permission of Electro-Craft Corp., Hopkins, MN.]

If Eq. (4.11.6) is rearranged to obtain a transfer function relating armature current to shaft velocity, we have

$$\frac{\Omega(s)}{I_a(s)} = \frac{K_T}{sJ_T + B} \qquad (4.11.7)$$

Equation (4.11.7) can be viewed as modeling a device that produces a *current-controlled velocity*.   Another way of looking at this result is to recognize that since torque is proportional to armature current, a current amplifier actually produces *torque control*.

In reality, the current amplifier has a finite bandwidth and can be modeled as

$$\frac{I_{\text{out}}(s)}{V_{\text{in}}(s)} = \frac{A_i}{1 + s\tau_A} \qquad (4.11.8)$$

where $1/\tau_A$ is again the amplifier bandwidth and A_i, the gain of the amplifier (units of amperes/volt), is a function of the input, feedback, and sense resistors. That is,

$$|A_i| = \frac{R_b}{R_{\text{in}}R_s} \qquad (4.11.9)$$

In Eq.(4.11.9) it is assumed that $R_s \ll R_b$.

The overall transfer function of a motor driven by a current amplifier is the product of the transfer functions in Eqs. (4.11.7) and (4.11.8), so that

$$\frac{\Omega(s)}{V_{\text{in}}(s)} = \frac{K_T A_i}{(sJ_T + B)(1 + s\tau_A)} \qquad (4.11.10)$$

Figure 4.11.10b shows a block diagram representation of this system. Comparison of Eqs. (4.11.5) and (4.11.10) indicates that the poles of the motor have been altered by the use of current feedback. In fact, the motor is seen to behave like a one-pole, *rather than a two-pole device*. Note that only the mechanical parameters of the system (i.e., the total inertia J_T and the viscous damping B have an effect on the behavior of the servo. It can be shown that the elimination of the pole (sometimes referred to as the *electrical pole*) due to the armature elements results in a larger velocity loop bandwidth (see Problem 4.42).

4.11.3.3 Current and voltage feedback amplifier driving a servomotor

Often, it is necessary (and desirable) to both limit the power that can be delivered to the motor and also increase the velocity loop bandwidth. This can be accomplished by combining the results of the two preceding sections to form an amplifier that utilizes both voltage *and* current feedback. Such a configuration is shown in Figure 4.11.11a. By examining the corresponding block diagram (shown in Figure 4.11.11b), it is seen that this system has a more complex structure than those previously discussed.

It can be shown that the transfer function relating the shaft velocity to the input voltage is given by (see Problem 4.38)

$$\frac{\Omega(s)}{V_{\text{in}}(s)} = \frac{A_{vf}K_T}{(1 + s\tau_A)\left[(sL_a + R_a)(sJ_T + B) + K_E K_T\right] + AbR_s(sJ + B)} \qquad (4.11.11)$$

where $b = (R_{\text{in}}/R_b)$ and $A_{vf} = (R_f/R_{\text{in}})$. A comparison of Eqs. (4.11.11) and (4.11.5) reveals the presence of an additional term in the denominator, which, of

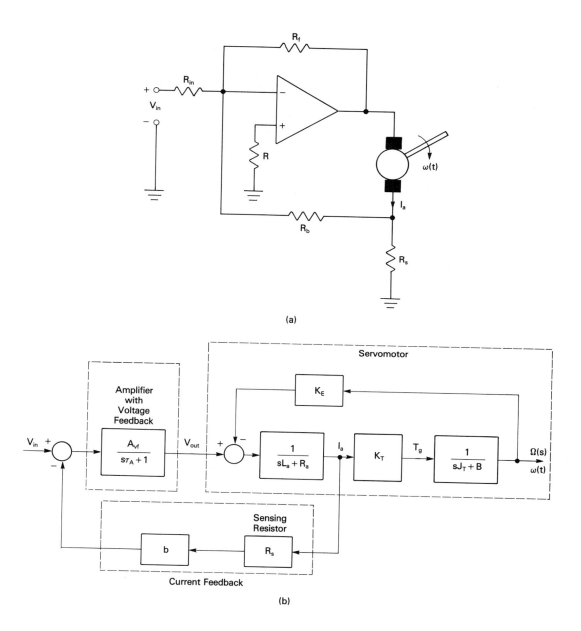

Figure 4.11.11. (a) Servomotor driven by amplifier employing both voltage and current feedback; (b) block diagram representation of part (a). [Part (b) redrawn with permission of Electro-Craft Corp., Hopkins, MN.]

course, can affect the pole locations. Note that the implementation shown in Figure 4.11.11a makes use of an inverting amplifier. If this circuit is used, the right-hand side of Eq.(4.11.11) would need a negative sign to indicate that the velocity and input voltage are 180° out of phase with each other. In practice, inversions occur throughout the circuits used to implement a loop and the negative signs tend to cancel. Thus they may be ignored in the analysis as long as the correct polarities of the signals exist in the loop.

To obtain a good understanding of how the motor poles are affected by using an amplifier with both current and voltage feedback, consider the following numerical example.

EXAMPLE 4.11.1

Let us find the poles of a motor–amplifier configuration such as that shown in Figure 4.11.11 for varying degrees of current feedback. The following are the assumed system parameters:

Motor parameters

$K_T = 11.9$ oz-in./A

$K_E = 0.084$ V/rad/s

$L_A = 0.0047$ H

$R_a = 2.24$ Ω

$J_M = 0.0054$ oz-in.-s^2

$B = 0.00288$ oz-in./rad/s

Resistor values

$R_{in} = 214$ kΩ

$R_{vf} = 1.5$ MΩ

$R_{cf} = 270$ kΩ

$R_s = 0.4$ Ω

Note that we have assumed that the load inertia is zero so that $J_T = J_M$.

Utilizing these resistor values in the amplifier, a Bode plot of the open-loop system consisting of the amplifier (with no current feedback) driving a resistive load indicates that there is a pole at about 7000 Hz. It is important to note that this pole is dependent on the particular op amp used in the implementation and the voltage gain chosen [5].

To see the effect of current feedback on the pole locations, we shall use

TABLE 4.11.1 SYSTEM POLES AS A FUNCTION OF THE CURRENT FEEDBACK

	$b = 0$	$b = 0.4$	$b = 0.79$	$b = 1.6$
p_1	-107.4	-60.9	-44.2	-28.6
p_2	-369.8	-658.0	-914.3	$-1,435.5$
p_3	$-44,000.0$	$-43,758.0$	$-43,518.6$	$-43,014.0$

four values of the parameter b (e.g., 0, 0.4, 0.79, and 1.6). The resultant system poles for each of these are summarized in Table 4.11.1. From these results it is observed that when $b = 0$, the current feedback is disabled and p_1, p_2, and p_3 are, respectively, the motor's mechanical and electrical poles and the pole of the amplifier. As current feedback is increased (the value of the parameter b is increased) the motor's mechanical pole moves toward the origin while its electrical pole moves in the opposite direction. In terms of Bode plots, current feedback increases the point where the 0-dB crossover occurs (i.e., the bandwidth). For example, if a velocity loop were to be implemented with a motor–amplifier and $b = 0$, the maximum bandwidth can be shown to be about 370 rad/s. However, when $b = 1.6$, the corresponding bandwidth is increased by almost a factor of 4. Note that the amplifier pole is relatively insensitive to variations in b.

4.12 SUMMARY

In this chapter we have presented a detailed discussion of the typical control structure used in each of the joints of a modern industrial robot. It has been demonstrated that we must ensure that the position servo bandwith is large enough to reproduce faithfully the profile of the desired position (thereby keeping the tracking error well within bounds) but not so large as to permit the servo to respond to the individual updates (i.e., the set points). It has also been found necessary to include an integrator as part of the servo loop's compensator due to the gravitational and friction disturbances that invariably act on most of the joints of a robot. Thus it is not surprising that a typical robot servo utilizes a PID controller. We have also found that most robots either employ a single position sensor (on each axis, e.g., an incremental encoder), and derive the velocity digitally from the information provided by this device, or else both an encoder and an analog tachometer are utilized to provide the required position and velocity data.

In addition to the material on the operation of the servo itself, we have presented detailed information on a variety of robotic actuators, including the dc servomotor, stepper (both rotary and linear) and brushless motors, and finally, a novel type of direct-drive motor that has already been incorporated into a commercial, high-performance robot. Besides these, hydraulic and pneumatic actuators, as they apply to robots, have been presented.

The last portion of the chapter has been concerned with the different types of servo amplifiers that can be employed, including linear and PWM devices. The advantages and disadvantages of both types have been discussed. In addition, the importance of utilizing flyback diodes to protect the power transistors from an "inductive kick" has been stressed. Finally, the effects on the performance of the servo (e.g., on closed-loop pole placement) when voltage, current, or combined voltage and current feedback amplifiers are employed has been demonstrated.

4.13 PROBLEMS

4.1 For the open-loop transfer function given in Eq.(4.2.1), sketch the root locus shown in Figure 4.2.3. In particular, obtain the expression for σ_A, the centroid of poles and zeros; σ_B the point where the root-locus "breaks away" from the real axis; K_B, the gain where this occurs; and the radian frequency where branches of the root-locus cross the imaginary axis. Also show that for $K > (1/\tau_A + 1/\tau_M)$ the closed-loop system in Figure 4.2.2 (with $K_g = 0$) is unstable.

4.2 Find the expression for the step response for the closed-loop system in Figure 4.2.2 (with $K_g = 0$). Thus show that if $K_B < K < (1/\tau_A + 1/\tau_M)$, this response will be similar to that shown in Figure 4.2.4.

4.3 Derive Eq. (4.2.4).

4.4 Derive Eq. (4.3.5) from Figure 4.3.3.

4.5 For $G_m(s)$ the transfer function of the servomotor given in Example 4.3.1:
   a. Find and plot the step response. That is, for $V_{arm}(t)$ a unit step voltage, find $\omega(t)$.
   b. Find and plot the armature current $I_a(t)$.
   c. Also find the maximum value of this current and the time at which it occurs.

4.6 Verify Eqs. (4.3.6) and (4.3.7).

4.7 Verify that if the position loop in Figure 4.3.5 is closed with unity gain and K_p is placed in the forward path (e.g., between the position and velocity loop summing points), the final value of the position to a step command of magnitude θ_d will be θ_d. Assume that the friction and gravitational torques are both zero. Also show that such a modification does not affect the dynamic response of the system. That is, show that the closed-loop poles of this system are identical to those in Eq. (4.3.7).

4.8 For the position servo with PID controller shown in Figure 4.3.7 and a gravitational torque disturbance given by Eq. (4.3.12), verify the results shown in Figure 4.3.13. Assume a step command (input) to the system of 1.57 rad and that $K_D = 0.02$ and $K_I = 75$. Two values of K_p should be used (i.e., 20 and 40). Note that some computational scheme that makes use of appropriate approximations will be required to do this problem since the disturbance input is a nonlinear function of the output.

4.9 For the open (position)-loop transfer function for the system in Figure 4.3.7, show that when $K_I = 0$, the system is type 0, whereas it is a type 1 when $K_I \neq 0$. Using these results, also show that if the disturbance input is a step function, the steady-state error will be zero only if $K_I \neq 0$.

4.10 For the tach loop represented by the open-loop transfer function in Eq. (4.4.1), obtain the Bode magnitude and phase plots shown in Figure 4.4.1. For the magnitude plot, use straight-line approximations to assist in obtaining the actual curve.

4.11 Applying the definitions in Eq. (4.4.5), show that the GM and PM for the tach loop in Figure 4.4.1 are 21.6 dB and 47°, respectively.

4.12 For the open tach loop in Eq. (4.4.1), suppose that $AK_g = 5$.
    a. Obtain the Bode plots for this new system and compare them with the ones in Figure 4.4.1.
    b. Find the GM and PM for this system.
    c. Obtain the *approximate* magnitude plot for the closed tach and compare it with that found in Figure 4.4.3. Show that the two curves differ initially by 14 dB and this decreases to 0 at $\omega = 1166.4$ rad/s.

4.13 For the system in Problem 4.12, obtain the approximate closed-loop tach transfer function.

4.14 For the closed tach loop:
    a. Show that the bandwidth for the approximate closed tach loop in Figure 4.4.3 is 336.7 rad/s.
    b. Show that the *actual* closed tach bandwidth is about 714 rad/s.

4.15 For the open position loop represented by the approximate transfer function in Eq. (4.4.8):
    a. Obtain the Bode plots for $K_p = 1$ and show that they are identical to those in Figure 4.4.5.
    b. Show that the GM and PM are 59.1 dB and 89.8°, respectively.
    c. Show that the 3-dB closed position loop bandwidth is about 1 rad/s.

4.16 For the open-loop transfer function in Eq. (4.4.8), let $K_p = 70$. Using the approximations for closing a loop, show that the GM and PM are 22.2 dB and 67.5°, respectively, and that the half-power bandwidth is about 67.5 rad/s.

4.17 Repeat Problem 4.16 if $K_p = 100$, thereby demonstrating that the gain and phase margins for the position loop are 19.1 dB and 68.2°, respectively, with the closed-loop bandwidth now about 94 rad/s.

4.18 For the lead compensator used in Section 4.4.5:
    a. Show that the tach GM and PM are 19.2 dB and 73.4°, respectively, with a closed-loop 3-dB bandwidth of 927 rad/s.
    b. If $K_p = 100$, demonstrate that the use of lead compensation in the tach loop causes the closed position loop bandwidth to increase to 98.6 rad/s.
    c. Moreover, show that with $K_p = 240$, this bandwidth can be more than doubled to 226 rad/s without causing the system to become unstable.

4.19 For the stepper motor in Figure 4.6.1, show that making the phase excitation sequence A, D, C, B, A will produce counterclockwise rotation of 60°.

4.20 Demonstrate that in order to be able to change direction, a stepper motor requires *at least* three phases.

4.21 a. Prove that for a stepper motor, the step angle $= 360° \, [|(N_s - N_r)|/N_s N_r]$, where N_s is the number of stator teeth and N_r is the number of rotor teeth.
    b. Also show that the number of steps/rev $= N_r N_s/|(N_s - N_r)|$.

4.22 For the motor in Figure 4.6.3, show that each step is 18° and that there are therefore 20 steps/rev.

4.23 Using the trapezoidal velocity profile shown in Figure 4.6.7 in Example 4.6.1 with ω_{pk} = 4000 steps/s and the acceleration and deceleration times still assumed to be equal, show that the acceleration torque is 5.5 oz-in., t_a = 160 ms, and the overall move time T = 410 ms.

4.24 For the motor having the force–velocity profile in Figure 4.6.11:
   a. Compare the time needed to accelerate a 2-lb load to a maximum velocity of 50 in./s by first utilizing the force available at 50 in./s (i.e., a constant acceleration) and then utilizing a three-segment piecewise linear velocity profile similar to that shown in Figure 4.6.12.
   b. Develop a general procedure for acceleration that will find the minimum time that a given load can be accelerated from 0 in./s to some maximum velocity V_{max} using the data from a force–velocity (F–V) curve and assuming that there are n regions of constant force. (Note that if n = 5, then five segments are to be used and the force that can be employed for that segment is the minimum for the corresponding region of the F–V profile.)

4.25 If a linear stepper has a pitch of 0.040 in. and microstepping is used to divide this distance into 400 microsteps (i.e., one microstep moves the motor 0.0001 in.), what pulse frequency must be applied to the motor to achieve velocities from 0 to 60 in./s?

4.26 Design a controller (i.e., the block diagram) capable of achieving the pulse rates obtained in Problem 4.25.

4.27 Investigate the use of "burst" in commercial stepper drivers. How is it implemented, and what performance improvement does it achieve?

4.28 Verify that for an eight-pole, three-phase BDCM, with three sensors placed every 120°, the sensor signals are those shown in Figure 4.7.3.

4.29 Show that if a constant current flows in each phase of the three-phase BDCM, T_{gen} will not be constant with respect to θ.

4.30 For the base drive signals shown in Figure 4.7.6, sketch the phase currents as a function of θ. Use these curves to verify that we get constant torque. Also indicate which coil pairs are energized.

4.31 Sketch the actual phase currents $I_{a\text{-}b}$, $I_{b\text{-}c}$, and $I_{a\text{-}c}$ as $f(\theta)$, assuming that rise and fall times are the same. Assume 10 and 20° to reach peak, and for both cases sketch actual torques produced by each phase and the overall torque T_{gen}. What happens if the fall time is essentially zero?

4.32 Determine the appropriate sequence of base drive signals that will cause the BDCM to rotate in a direction *opposite* to the one produced by the signals shown in Figure 4.7.6.

4.33 Show that for a PWM driving a servomotor, if the switching frequency is only about 10 times higher than the electrical pole of the motor, the armature current and angular velocity may exhibit significant ripple. Assume a two-pole model for the motor.

4.34 Show that a servomotor can be made to turn with a trapezoidal velocity profile (see Figure 4.6.7), by making the pulse width T_p (duty cycle) of the PWM signal to the motor terminals vary trapezoidally with time.

4.35 Derive Eq. (4.11.2).

4.36 Show that the equivalent series resistor in an amplifier with current and voltage feedback is given by $R_{equiv} = (R_{vf}R_s)/R_{cf}$.

4.37 Derive the characteristics of a voltage amplifier with a finite output resistance. Discuss how this affects the voltage and current delivered to a motor. Does the output resistance change the motor's transfer function? Explain.

4.38 Design a system consisting of five voltage-stabilized amplifiers (each driving a motor) being powered by a single unregulated power supply. The maximum input signal to each amplifier is ± 5 V. The maximum voltage and current requirements for each motor are as follows:

1. 47 V at 4.6 A
2. 45.5 V at 5.3 A
3. 42.5 V at 4.0 A
4. 34 V at 2.5 A
5. 26 V at 2.0 A

Specify the regulation of the power supply in terms of the supply voltage. Also discuss the power dissipated in each of the amplifiers.

4.39 Discuss how an amplifier with voltage and current feedback can be used to alter the transfer function of a motor.

4.40 Determine how much of the actual power is delivered to the load and how much is "dissipated" in the series resistor for an amplifier with voltage and current feedback. Is the percentage the same if the operating point changes? Explain.

4.41 If a voltage-stabilized amplifier is to drive a motor located a large distance from the amplifier, what precautions should be taken to ensure that the voltage *delivered* to the motor terminals is the value the designer wanted? Consider where the voltage feedback should be connected and compare this to the "cost" of using a larger-gage wire in terms of wire flex capability, wire insulation temperature, and utilizing three separate conductors versus two (for example).

4.42 Investigate the types of safety or shutdown circuits available as features in commercial power amplifiers. Discuss the advantages and disadvantages of each and how they can make a linear system behave in a nonlinear manner.

4.43 What is the effect of a finite output impedance of a voltage amplifier on the transfer function of Eq. (4.11.5)? That is, derive a result similar to Eq. (4.11.5) assuming that the amplifier is modeled by a Thévenin voltage given by Eq. (4.11.3) and a nonzero Thévenin resistance R_o.

4.44 Prove that eliminating the electrical pole in the motor model as shown by Eq. (4.11.10) can result in a higher velocity loop bandwidth. Do the analysis in terms of Bode plots and comment on the 0-dB crossover frequency. For the same motor driven by a voltage amplifier, what is the maximum bandwidth that can be achieved?

4.45 If a dc motor was to be used as a torque generator, define the transfer function. Discuss the advantage of using a current versus voltage amplifier to drive this motor.

4.46 Derive Eq. (4.11.11) starting with Figure 4.11.11b.

4.47 Assuming that $\tau_A = 0$ so that the amplifier pole is of no consequence,
   a. Derive an expression for the poles of Eq. (4.11.11) as the current feedback parameter b approaches infinity.

b. Compare this to the root-locus plot of Eq. (4.11.5) as the value of R_a (the motor's armature resistance) is varied from zero to infinity.

c. Compare the pole locations of Eq. (4.11.10) with those found in parts a & b.

4.48 Repeat Example 4.11.1 with an additional load inertia J_L of 0.02 oz-in.-s^2. Examine the trend in the poles as b increases. Are there any advantages of combined current and voltage feedback for driving a high inertial load? Explain.

4.49 (*optional*) Explain the use of combined current and voltage feedback in terms of state-variable feedback. What states are fed back? Can the transfer function be made into any second-order transfer function? Explain.

4.14 REFERENCES AND FURTHER READING

1. Dorf, Richard C., *Modern Control Systems*, 3rd ed. Reading, Mass., Addison-Wesley Publishing Company, Inc., 1983.

2. Lynch, Paul M., "Minimum Time, Sequential Axis Operation of a Cylindrical, Two Axis Manipulator," *Proceedings of the 1981 Joint Automatic Control Conference*, WP-2A, Vol. 1, Charlottesville, Va.

3. Nordquist, J. I., and Smit, P. M., "A Motion Control System for Linear (Stepper) Motors," *Proceedings of the Incremental Motion Society*, University of Illinois, Urbana, Ill., May 1985.

4. Welburn, R., "Ultra High Torque Motor System for Direct Drive Robotics," *Proceedings of the Robots 8 Conference—Applications for Today*, Detroit, Mich., June 4–7, 1984, Dearborn, Mich.: Society of Manufacturing Engineers, 1984, Vol. 2, pp. 19-64 to 19-71.

5. Chmielewski, Thomas, "A Note on the Effect of Amplifier Feedback on Closed Loop Pole Placement," in preparation.

The following are excellent texts on various aspects of control systems:

6. Kuo, Benjamin C., *Automatic Control Systems*, 4th ed. Englewood Cliffs, N.J.: Prentice-Hall, Inc., 1982. Covers classical and certain modern methods in analog control systems. It includes good material on Laplace transforms, Bode plots, PID controllers, state variables, and design.

7. Melsa, James L., and Schultz, Donald G., *Linear Control Systems*. New York: McGraw-Hill Book Company, 1969. Chapter 5 has an excellent treatment of Bode plots and minimum- and non-minimum-phase transfer functions.

8. Kuo, Benjamin C., *Digital Control Systems*. New York: Holt, Rinehart and Winston, 1980. An exhaustive treatment of digital control analysis and design.

9. Raven, Francis, H., *Automatic Control Engineering*, 2nd ed. New York: McGraw-Hill Book Company, 1968. Classical analog control from a mechanical engineering point of view. Includes a good discussion of hydraulic and pneumatic control systems.

10. Eveleigh, Virgil W., *Adaptive Control and Optimization Techniques*. New York: McGraw-Hill Book Company, 1967. A good solid introduction to optimal and adaptive techniques.

11. Athans, Michael, and Falb, Peter L., *Optimal Control.* New York: McGraw-Hill Book Company, 1966. A comprehensive, classic text in this field.

The following articles discuss various aspects of robotic controls:

12. Dubowsky, S., and DesForges, D. T., "The Application of Model-Reference Adaptive Control to Robotic Manipulators," *Journal of Dynamic Systems, Measurement, and Control*, ASME, Vol. 101, September 1979, pp. 193–200.

13. Luh, J. Y. S., Walker, M. W., and Paul, R. P. C., "On-Line Computational Scheme for Mechanical Manipulators," *Journal of Dynamic Systems, Measurement, and Control*, ASME, Vol. 102, June 1980, pp. 69–76.

14. Luh, J. Y. S., "An Anatomy of Industrial Robots and Their Controls," *IEEE Transactions on Automatic Control*, Vol. AC28, No. 2 (February), 1983, pp. 133–153.

15. Asada, H., Kanade, T., and Takeyama, I., "Control of a Direct Drive Arm," *Robotics Research and Advanced Applications, Proceedings of ASME Annual Meeting*, Dynamic Systems and Control Division of ASME, 1982, pp. 63–72.

16. Koivo, Antti J., and Ten-Huei Guo, "Adaptive Linear Controller for Robotic Manipulators," *IEEE Transactions on Automatic Control*, Vol. AC28, No. 2 (February), 1983, pp. 162–171.

Other references pertaining to the practical aspects of servo control.

17. *Moog Technical Bulletin 126.* A good reference on hydraulic servo applications.

18. Electrocraft Corporation, *DC Motors, Speed Controls, Servo Systems*, 1980. One of the best practical treatments of motors and servos.

5

Robotic Sensory Devices

5.0 OBJECTIVES

In this chapter we describe the operation of a variety of sensory devices that either are now used on robots or may be used in the future. In general, it is found that some are inherently digital devices, whereas others are essentially analog in nature. Sensors can be divided into two basic classes. The first, called *internal state sensors*, consists of devices used to measure position, velocity, or acceleration of robot joints and/or the end effector. Specifically, the following devices that fall into this class will be discussed:

- Potentiometers ("pots")
- Synchros
- Resolvers
- Linear inductive scales
- Differential transformers (i.e., LVDTs and RVDTs)
- Optical interrupters
- Optical encoders (absolute and incremental)
- Tachometers
- Accelerometers

The second class, called *external state sensors*, is used to monitor the robot's geometric and/or dynamic relation to its task, environment, or the objects that it is handling. Such devices can be of either the visual or nonvisual variety. The former group of sensors is treated in Chapter 6. In this chapter we discuss techniques that permit the monitoring of (1) distance from an object or an obstruction, (2) touch/slip, and (3) force/torque. Specifically, the following will be covered:

- Strain gages
- Pressure transducers
- Proximity devices
- Ultrasonic sensors
- Electromagnetic sensors
- Elastometric materials

The various aspects of the important topic of tactile sensing are presented in some detail. Welding sensors, devices that are probably the most advanced external robotic sensors, are also discussed.

5.1 MOTIVATION

As we have seen in Chapter 4, the successful control of most robots depends on being able to obtain information about the joint and/or end effector. It is therefore necessary to have devices (transducers) that provide such information and can be readily utilized in a robot for this purpose. In particular, position, velocity, and/ or acceleration (or at least analog or digital representations of these quantities) must be measured to ensure that the robotic manipulator moves in a desired manner (e.g., a straight line) with little or no oscillation (i.e., overshoot) at the final position. These so-called "internal state sensors" must not only permit the required degree of accuracy to be achieved, but they must also be cost-effective since *each* of the robot's axes will normally utilize such devices. As a consequence, the sensor selection and the decision to place it either on the load side or on the output of the joint actuator itself is influenced by such factors as overall sensor cost, power needs for a particular joint, maximum permissible size of the actuator, sensor resolution, and the need to monitor directly the actions of the joint. These ideas are discussed in this chapter along with the workings of the sensors themselves.

Although it is possible to utilize a robot without any external sensing whatsoever, more and varied applications require such devices. Thus, in addition to the control of the robotic manipulator itself, certain more sophisticated tasks require that a variety of quantities be monitored at the gripper. The data gathered by sensors placed on or near the gripper can then be utilized by the robot's controller to modify or adapt to a given situation. For example, if it is necessary to handle several different parts, some of which are rather fragile, it is important to measure the instantaneous gripping force being applied and adjust it to be sufficient to pick up an object without crushing it.   Of course, the particular application will influence the type, construction, and cost of such sensors.

Currently, the "state of the art" in external sensing elements, techniques, and materials, although fairly well advanced in some areas, is still in its infancy in others.   Many of the devices that have been developed are still only laboratory grade and are, therefore, not suitable for use in an industrial setting due to their inherent fragile nature or excessive cost.   This chapter will not only discuss some of the external sensor measurement techniques, devices, and materials that are now available but will include a brief overview of some of the recent research results and trends in this area.

5.2 NONOPTICAL-POSITION SENSORS

In this section we discuss the operation and applications of simple internal state sensors that can be used to monitor joint position.   Included are the potentiometer, synchro, resolver, and LVDT.   It will be seen that some of these devices are inherently analog and some are digital in nature.

5.2.1 Potentiometers

The simplest device that can be used to measure position is the potentiometer or "pot."   Applied to robots, such devices can be made to monitor either angular position of a revolute joint or linear position of a prismatic joint.   As shown in Figure 5.2.1, a pot can be constructed by winding a resistive element in a coil configuration.   By applying a dc voltage V_s across the entire resistance R, the voltage V_{out} is proportional to the linear or rotary distance of the sliding contact (or "wiper") from reference point a.   Mathematically, if the resistance of the coil between the wiper and the reference is r, then

$$V_{out} = \frac{r}{R} V_s \qquad (5.2.1)$$

For the pot to be a useful position sensor, it is important that the resistance r be *linearly* related to the angular distance traveled by the wiper shaft.   Although it is possible to obtain pots that are nominally linear, there is always some deviation from linearity as shown in Figure 5.2.2.   Generally, the nonlinearity of a pot (expressed as a percent) is defined as the maximum deviation ϵ from the ideal straight line compared to the full-scale output.   That is,

$$N.L. = 100 \frac{\epsilon}{V_{max}} \qquad (5.2.2)$$

The inevitable presence of this nonlinearity in any pot makes its use in systems where excellent accuracy measurement is required difficult and often impractical.

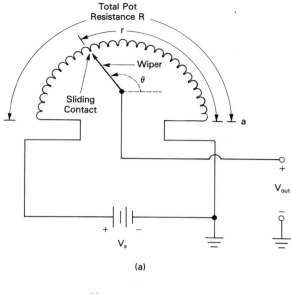

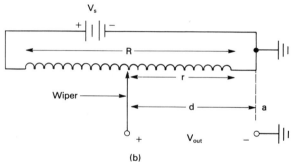

Figure 5.2.1.   Wire wound potentiometer ("pot"). Wiper makes physical contact with wires on the resistive coil. Note: point "a" corresponds to zero output (i.e., zero resistance): (a) rotary—output proportional to θ; (b) linear—output proportional to "d."

Thus except in the case of robots where extreme accuracy is not needed (such as in educational devices), the pot is not generally used as a *primary* position-monitoring sensor. In a later section of this chapter, it will be seen, however, that it is possible to utilize this type of device as one of the components in a position-measuring scheme.

Several other characteristics of a potentiometer limit its usefulness in high-accuracy position-measurement applications. The first has to do with the manner in which the device is constructed and hence how the voltage V_{out} is sensed. In particular, the wiper must physically contact the coiled resistive element. The problem with this arrangement is that with time, the contact will deteriorate due to wear and dirt/corrosion buildup. (The latter is especially true in a typical factory environment where a robot will be used.) This will cause the measured output voltage to deviate from the true value in an unpredictable manner. The so-called

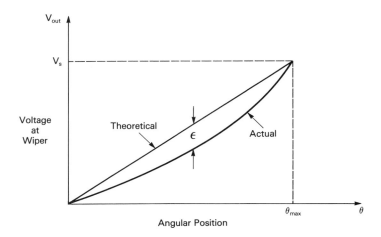

Figure 5.2.2. Output of a pot showing the theoretical linear and actual nonlinear characteristics as a function of angular position θ.

"electrical noise" introduced by the poor contact will cause apparent position errors that cannot be tolerated in a system that requires high-position accuracy.

A second problem with a wire-wound pot has to do with the smallest change in wiper position that can be sensed or its *resolution*. It is clear that with the wire-wound arrangement shown in Figure 5.2.1, a change in V_{out} will occur only when the sliding contact moves from one coil loop to the adjacent one. Thus if the pot is constructed with N turns of resistive wire, the smallest voltage change that can be observed will be V_s/N. The percent resolution can be defined as

$$\% \text{ resolution } = 100 \; \frac{V_s/N}{V_s} = 100/N \tag{5.2.3}$$

EXAMPLE 5.2.1

A wire-wound potentiometer is to be used to measure angular position. A 100-turn (i.e., 100 loops) resistive element is used and the wiper can rotate 300°; 10 V dc is applied to the pot. Determine the resolution of the device.

From the discussion above, the minimum voltage change will be

$$\Delta V = \frac{10}{100} = 0.1 \text{ V/turn}$$

The % resolution is, therefore,

$$\% \text{ resolution } = 100 \left(\frac{0.1}{10} \right) = 1.0$$

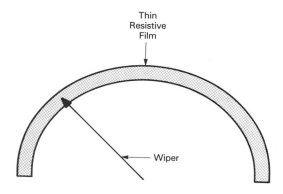

Thin
Resistive
Film

Wiper

Figure 5.2.3. Thin film resistive pot. This device has much better resolution than the wire wound version. Also, contact "noise" problems are reduced.

It is also observed that if the pot in Example 5.2.1 is connected directly to the shaft of a motor, for example, the position resolution will be 3° (300/100). Generally, in robot applications, this will not be good enough. That is, it will be necessary to "resolve" a smaller angular rotation. To accomplish this requires either a pot with more turns (for the same total angular rotation) or a modification of the motor drive and coupling scheme that permits more than 300° motor rotation in order to produce the same 300° of joint rotation. More will be said of the latter technique in a subsequent section. As far as the former "fix" is concerned, it may be possible to increase the number of turns of wire somewhat but, generally, not enough to improve the resolution significantly.

What, then, can be done either to improve the resolution of the pot or to reduce the noise problem due to wiper contact problems? The answer to this question is seen in Figure 5.2.3, in which a pot made from a thin resistive film is shown. It should be clear to the reader that using such a scheme will significantly reduce the smallest angular change that can be resolved. Also, the resistive film/wiper contact tends to be better "lubricated," so that the problem with noise is reduced. As the sensing mechanism still depends on a physical contact between the resistive element and the wiper, this technique does not, however, eliminate all contact problems.

The sensing devices described above produce a single-polarity voltage which is sometimes undesirable. The scheme shown in Figure 5.2.4a overcomes this problem, as seen in Figure 5.2.4b. The reader should note that the output voltage now swings from $-V_s$ to $+V_s$ over the entire range of travel of the pot's wiper.

Despite the problems described above, it is possible to use pots to provide a limited amount of feedback control in robots where high positional resolution and accuracy are not essential to the operation and where a certain amount of jitter caused by noisy contacts is tolerable. An example is shown in Figure 5.2.5, where a circuit for obtaining teachable end or stop points is given. Here the position of the particular joint is monitored by a bipolar pot and is fed back to the op amp feeding the power amplifier stage. The desired or *demand position* is set by

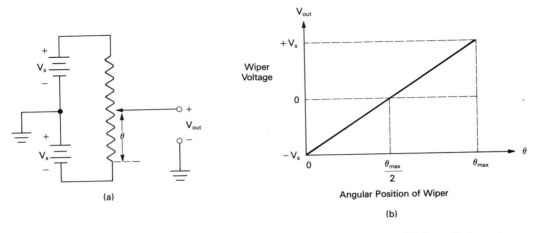

Figure 5.2.4. (a) Scheme for obtaining a bipolar voltage from a pot; (b) theoretical transfer characteristic for the circuit in part (a).

adjusting the voltage output of one of the input pots to the appropriate value (input position is thus proportional to voltage). Since the feedback is negative, the motor shaft driving the robot axis continues to rotate until the voltage across the position-sensing pot equals that coming from the demand-position pot. At that time, the robotic joint is at its programmed destination.

The analog multiplexer (MUX) is used to permit a number of different inputs to be selected under digital control. For example, for the three shown in Figure 5.2.5, a two-bit digital code could be used to select uniquely any one of the inputs. (If the fourth input was chosen, the joint would be commanded to go to its "home" or reference positon.) Programming of a robot utilizing such a control scheme would be accomplished by simply adjusting the input pots. A little thought should convince the reader that moves consisting of multiple points can be obtained simply by outputting a sequence of two-bit commands to the MUX (see Problem 5.2).

5.2.2 Synchro

As mentioned above, a significant practical problem with the pot is that it requires a physical contact in order to produce an output. There are, however, a variety of sensing devices and techniques that avoid this difficulty. The first one that we discuss is the *synchro*, a rotary transducer that converts angular displacement into an ac voltage or an ac voltage into an angular displacement. Historically, this device was used extensively during World War II, but technological innovations that produced other position-sensing elements caused it to fall from favor. In recent years, however, advances in solid-state technology have again made the synchro a possible alternative for certain types of systems, among them robots.

Normally, a synchro *system* is made up of a number of separate three-phase

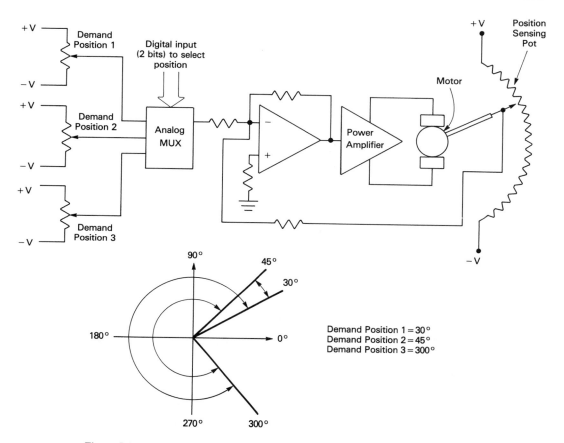

Figure 5.2.5. Teachable stop points using pots. The three pots are set to output voltages that will rotate the motor's shaft 30, 45, or 300° (all with respect to zero degrees) depending on the digital input to the analog MUX. Other moves are also possible with this configuration including 15° (from 30 to 45°) or 270° (30 to 300°).

components [e.g., the control transmitter (CX), control transformer (CT), and control differential transmitter (CDX)]. These elements all work on essentially the principle of the rotating transformer. Typically, two or three of the devices are used to measure angular position or the *difference* between this and a command position (i.e., the position error). For example, consider the two-element system shown in Figure 5.2.6. It is observed that an ac voltage is applied to the rotor of the CX and that the wye-configured stators of the CT and CX are connected in parallel. Using elementary transformer theory, it can be shown that the magnitude of the transformer rotor voltage $v_{out}(t)$ is dependent on the *relative* angle θ between the rotors of the CX and CT. In particular, this output voltage is

$$v_{out}(t) = V_m \sin \theta \sin \omega_{ac}t \qquad (5.2.4)$$

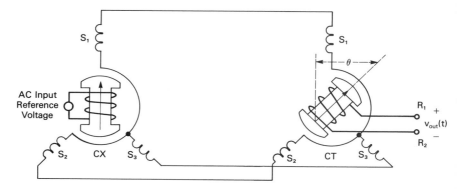

Figure 5.2.6. A two-element (control transmitter CX and control transformer CT) synchro system used to measure angular displacement. θ is the *relative* angle between the rotors of CX and CT.

where V_m and ω_{ac} are, respectively, the amplitude and radian frequency of the reference (or "carrier") ac voltage. Those readers familiar with elementary communications theory will recognize that Eq. (5.2.4) represents an *amplitude-modulated* function. The difference between the radio AM and synchro AM signals is, of course, that the modulation of the carrier in the latter case is due to the relative angular position θ of the CT rotor with respect to that of the CX rotor. In the former case, however, the modulation is achieved through the application of another voltage signal that varies with time.

From Eq. (5.2.4) and Figure 5.2.6 it is seen that the output voltage has its maximum magnitude when the two rotors are at right angles to one another and that it is zero when they are at either parallel or antiparallel. As a consequence, the CT is sometimes referred to as a "null detector." It is important to understand that in practice, the null is never exactly zero when the two rotors line up because of nonlinearities and electrical imbalances in the windings. These can produce "residual voltages" on the order of 60 mV (for a 115-V ac input). Due to the mathematical nature of a sine function, $v_{out}(t)$ will be approximately linearly related to θ if $-70° \le \theta \le 70°$. It is for this reason that where a linear relationship between output and angular position is important, the synchro must be used about an operating point of $\theta = 0°$.

Ideally, the ac signals from the CX are in phase with those produced at the CT. However, physical differences in the structures of the two devices that are inevitably present produce phase shifts that may be undesirable. A synchro control differential transmitter (CDX) is sometimes used to adjust the phase shift between the two synchro units. Such a device may also be used to produce a *variable* phase shift in applications where this is required. This is illustrated in Figure 5.2.7. Here the angular relationship between the master and slave rollers can be adjusted during the running of the process by rotating the shaft of the CDX.

The use of a two-element synchro in a "classical" position servo application

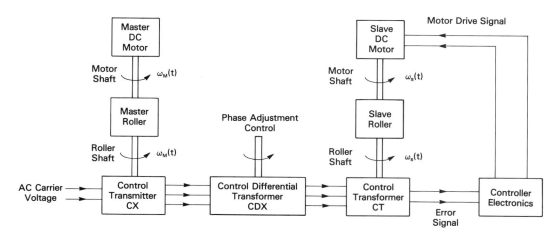

Figure 5.2.7. An example of a servo using a three element synchro system. To maintain a uniform product (e.g., steel sheets), the slave roller's speed $\omega_s(t)$ must be synchronized to that of the master, i.e., $\omega_M(t)$. The CDX is used to provide the desired angular relationship between the master and slave. The output signal of the CT is the difference between this desired and the actual master-to-slave angles, i.e., the error, and is used to provide the slave motor drive signal.

is illustrated in Figure 5.2.8. It is observed that the command or input (i.e., the angle θ_1) will produce a command voltage from the CX. The CT will then produce an error voltage in accordance with Eq. (5.2.4), where $\theta = \theta_1 - \theta_2$. This error signal is amplified and causes the servomotor to rotate until θ is again zero. In such an application, the two-element synchro provides a rugged, reliable, and cost-effective method of monitoring position error. However, the reader can readily appreciate that because of the need to convert the command position into a physical angular rotation of the CX rotor, such a system is not always practical in applications requiring the interfacing to digital devices. Thus, as mentioned above, it is not surprising that with the advent of microprocessor-controlled systems, synchros were quickly discarded in favor of other position-sensing methods more compatible with digital systems.

Recently, however, a number of advances in digital and hybrid technologies have produced a variety of devices that permit synchro systems to be easily interfaced with digital systems. For example, the digital-to-synchro (D/S) converter shown in Figure 5.2.9 replaces the CX in the position servo of Figure 5.2.8. A digital position command signal from a computer (e.g., the master) is transformed into a three-phase ac voltage by the D/S converter. (This voltage corresponds to that produced by the CX due to a physical rotation of θ_1.) The CT once again acts as a position error sensor and the system behaves in a manner that is identical to that of the one in Figure 5.2.8. The use of the D/S converter produces a position servo that is part digital and part analog.

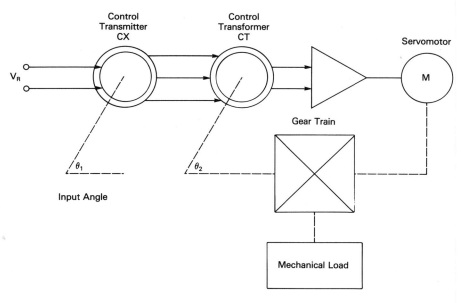

Figure 5.2.8. A synchro used in a position servo loop. The desired angular position is θ_1 whereas θ_2 is the actual angular position of the motor shaft. [V_R is the ac reference (carrier) voltage]. (Redrawn with permission of Walter Lewis, ILC Data Device Corp., Bohemia, NY. From the *Synchro Conversion Handbook*, p. 10, 3rd printing, 1982.)

In the next section we consider a device that is quite similar in operation to the synchro: the resolver.

5.2.3 Resolvers

The resolver is actually a form of synchro and for that reason is often called a "synchro resolver." One of the major differences between the two devices is that the stator and rotor windings of the resolver are displaced mechanically 90° to each other instead of 120° as is the case with the synchro. The most common form of resolver has a single rotor and two stator windings, as shown in Figure 5.2.10. With the rotor excited by an ac carrier voltage $B \sin \omega_{ac} t$, the two stator voltages become

$$v_{1\text{-}3}(t) = V \sin \theta \sin \omega_{ac} t \tag{5.2.5a}$$

$$v_{2\text{-}4}(t) = V \cos \theta \sin \omega_{ac} t \tag{5.2.5b}$$

where θ is the resolver shaft angle. It should be clear to the reader that such a device could, and often is, used in much the same way as the synchro CX to monitor shaft angle.

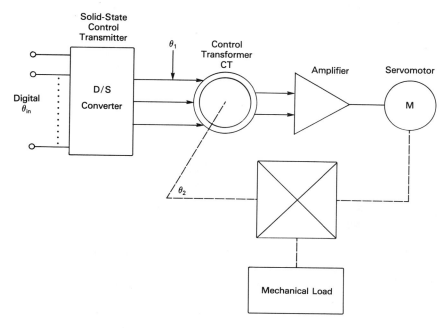

Figure 5.2.9.   The synchro control transmitter CX of the position servo shown in Fig. 5.2.8 is replaced by a D/S converter.   This scheme permits the desired input θ_{in} to be a digital quantity, i.e., makes the system microprocessor compatible. (Redrawn with the permission of Walter Lewis, ILC Data Device Corp., Bohemia, NY.   From the *Synchro Conversion Handbook*, p. 11, 3rd printing, 1982.)

An alternative form of a resolver has two stator and two rotor windings.   In actual use, the carrier voltage may be applied to any of these.   For example, if the former is used as an input, the unused stator winding is normally shorted.   The output voltages are identical to those given in Eqs. (5.2.5a) and (5.2.5b) and are monitored across the rotor windings.   Alternatively, one rotor winding can be used as the input with the two stator windings being used as the outputs.

To utilize a resolver in a servo system, it is usually necessary to employ two resolvers in much the same way as was done with the synchro system of Figure 5.2.7.   Figure 5.2.11 shows a resolver transmitter (RX) and resolver control transformer (RT) in a simple position servo.   Again, the reader should note that RX and RT are used to obtain the difference between the actual and desired angles (i.e., $\theta_1 - \theta_2$).   It is important to understand that although angular position can be monitored using a single resolver [see Eqs. (5.2.5a) and (5.2.5b)], this is usually not done in servo-controlled devices because of the need to utilize an error signal to drive the system actuator.

As in the case of the synchro, there has recently appeared a series of special-purpose chips that permit one of the elements of a resolver servo system to be eliminated.   For example, the Analog Devices Solid-State Resolver Control Trans-

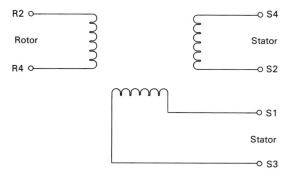

Figure 5.2.10. Electrical circuit of a simple resolver. With an ac carrier voltage input to the rotor, the output voltage amplitude of the two stator windings will be dependent on the sine or cosine of resolver shaft angle θ. (Redrawn with permission of Analog Devices, Inc., Norwood, MA. From *Synchro and Resolver Conversion*, Fig. 1-11, p. 7 © 1980, Memory Devices Ltd., Surrey, UK.)

former (RSCT 1621) shown in Figure 5.2.12 can be used in place of an RT. As can be seen, a 14-bit digital representation of a command input φ and the analog output of an RX, representing the actual angle θ, are input to the D/R converter. The output of this device is then an analog voltage that is proportional to θ − φ. This chip is a *hybrid* since it not only includes the digital and analog circuits necessary to process the two input angles but also has on board the appropriate input and output transformers. The only significant difference between a D/R and a D/S converter is in the transformer configurations.

A position servo utilizing such a chip is shown in Figure 5.2.13. Note that since the output of the D/R converter (or DRC) is an ac voltage, it is necessary

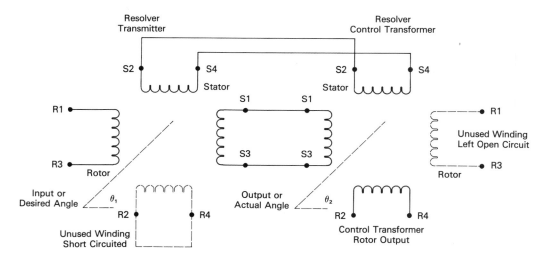

Figure 5.2.11. Resolver transmitter connected to a resolver control transformer. (Redrawn with permission of Analog Devices, Inc., Norwood, MA. From *Synchro and Resolver Conversion*, Fig. 1-13, p. 8, © 1980, Memory Devices, Ltd., Surrey, UK.)

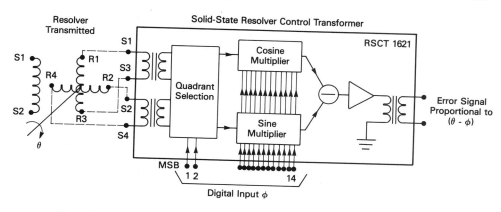

Figure 5.2.12.   Resolver transmitter and RSCT 1621 Solid-State Resolver Control Transformer functional diagram.   Use of this hybrid device permits elimination of separate input and output transformers.   θ is the measured or actual angular position and φ is the desired angular position.   (Redrawn with permission of Analog Devices, Inc., Norwood, MA.   From *Synchro and Resolver Conversion*, Fig. 5-11, p. 110, © 1980 Memory Devices Ltd., Surrey, UK.)

to use an ac amplifier, together with a phase-sensitive detector and integrator to obtain the appropriate drive signal to the servo amplifier.   As in the case of a comparable synchro system, this *servo* is functionally a hybrid since the command signal is digital, whereas the monitored position (and the error) is analog in nature.   (See Section 4.5.1 for further discussion of such a servo.)

It was seen in Chapter 4 that it is often convenient, in the control systems used in robots, to have a *digital* representation of the actual angular position of either the actuator shaft or the joint itself.   The tracking RDC shown in Figure 5.2.14 accomplishes this.   Here the RX is connected, either directly or through a gear train, to the shaft that is to be monitored.   The converter then "tracks" the shaft angle outputting a digitized version of it.   Thus it can be seen that the RDC takes the place of *both* an RT and an ADC.   Unlike the ADC, however, the tracking RDC *automatically* performs a conversion whenever the input voltage from the RX changes by a threshold value, as determined by the resolution of the RDC.   For example, if a 12-bit converter is used, a minimum angular change of $0.088°$ ($360/2^{12}$) in the resolver shaft will initiate a conversion.   Note that unlike many A/D converters, there is no need to trigger the R/D externally.

Tracking synchro-to-digital (S/D) converters are also now available.   The only difference between these devices and the RDC discussed above is that configuration of the input transformer on the chip is different since it must accept a three-phase rather than a two-phase voltage.   Insofar as the user is concerned, however, the devices are identical.

The reader may wonder whether there is an advantage to using a tracking R/D (or S/D) converter over a D/R (D/S) device since the decision seems to depend on the nature of the servo system configuration.   However, with reference to Figure

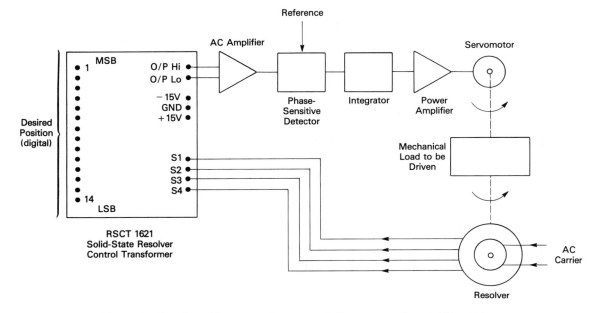

Figure 5.2.13. A position servo that uses a D/R converter (i.e., RSCT 1621). An ac amplifier is used because the D/R output is an ac voltage. (Redrawn with permission of Analog Devices, Inc., Norwood, MA. From *Synchro and Resolver Conversion*, Fig. 5-12, p. 110, © 1980 Memory Devices, Ltd., Surrey, UK.)

5.2.14, it is observed that the R/D converter has a velocity output. Available on many types of tracking R/Ds (or S/Ds), the analog voltage at this terminal represents the time rate of change of the input angle θ or angular velocity. Thus if the RX is connected to the shaft of a robot joint actuator, the R/D will provide both angular position and velocity information. As noted in Section 4.2, the position servo used in a robot joint requires velocity information in order to provide the desired amount of damping. Since an R/D (or S/D) converter with velocity output provides this information, it seems to represent a possible alternative to a system that utilizes separate sensors to monitor position and velocity or where a digital representation of velocity is found from the digital position (see Section 4.5.2). In fact, the velocity signal output of the R/D has better low-speed behavior and is more nearly linear than other velocity sensors and produces a better overall velocity signal than that *derived* from digital position sensors (e.g., an incremental encoder; see Section 5.3.2).

   Application of such a device in a position servo used to control one axis of a robot is shown in Figure 5.2.15. Note that this system corresponds to a hybrid servo described in Section 4.5.1 since the angular position command and error signals are digital, whereas the inputs to the velocity loop (i.e., the command and actual velocity signals) are analog voltages.

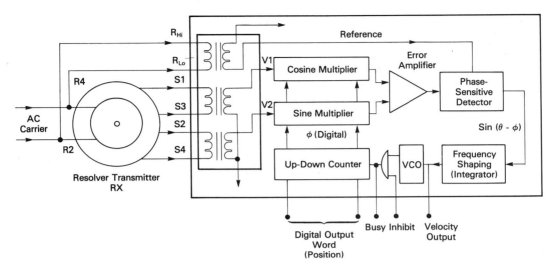

Figure 5.2.14. A tracking resolver-to-digital converter. The RX senses the actual position θ. The chip outputs a digitized version of this angle. The velocity is $d\theta/dt$ and is an analog quantity. (Redrawn with permission of Analog Devices, Inc., Norwood, MA. From *Synchro and Resolver Conversion*, Fig. 3-3, p. 46, © 1980 Memory Devices, Ltd., Surrey, UK.)

Theoretically, the development of R/D, S/D, D/R, and D/S chips to monitor angular position and velocity appears to make the resolver or synchro attractive for use in robotic systems. The rugged nature of the RX or CX is a particularly useful trait, and it is possible to obtain resolvers or synchros that have better angular resolution (i.e., the ability to sense smaller angular increments) than most other position sensors (e.g., incremental encoders or pots). However, there are a number of reasons why these devices are not commonly used in robots today. The first is cost since the converter chip and RX (or CX) combination is usually significantly higher than that of an optical encoder package (including the associated electronics) for the *same* resolution. A second is the potential problem of electromagnetic interference (EMI) due to the ac carrier signal. Although this problem can be overcome, careful shielding of certain critical subsystems of the robot controller is generally required to accomplish this and thus manufacturing costs may be increased. Finally, it is usually necessary to bring out a larger number of wires than with other position-monitoring techniques. This can be especially troublesome with the moving joints of a robot. All things considered, however, it seems logical that if the overall cost of resolver/synchro systems is reduced, such devices will indeed be used in robot servo systems.*

---

*In fact, this is already happening with a number of powerful RDC chips recently being introduced at prices that almost make the resolver system cost comparable to other more standard technologies (e.g., optical encoder).

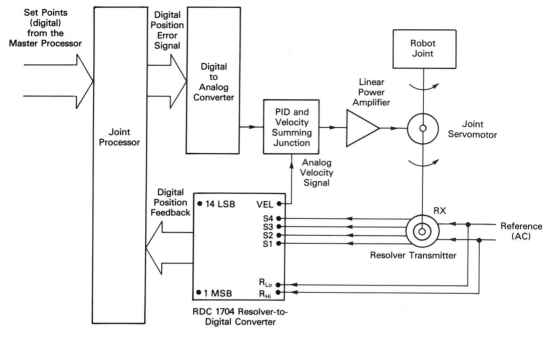

Figure 5.2.15. Tracking R/D converter with velocity-output used in the position servo of a single robotic joint. This is a hybrid servo since the position information is digital whereas the velocity information is analog. (Redrawn with permission of Analog Devices, Inc., Norwood, MA. From *Synchro and Resolver Conversion*, Fig. 3-42, p. 73, © 1980 Memory Devices, Ltd., Surrey, UK. Redrawn with additions.)

5.2.4 The Motornetics Resolver

As mentioned in Chapter 4, a new type of motor with the trade name *megatorque* was introduced in the early 1980s. Capable of producing the extremely large torques required by direct-drive robots, the motor would have been less attractive in this application without the concurrent development of a high-resolution position sensor. Fortunately, such a sensor was developed by Motornetics Corporation [1,2].

As shown in Figure 5.2.16a in schematic cross section and in Figure 5.2.16b as it actually appears when fabricated, this novel reluctance-based type of resolver has annular ring geometry and consists of a single multipole toothed stator with windings together with a toothed rotor without windings. In effect, the primary and secondary windings are combined so that all of the active magnetic area is utilized. This causes the sensor's accuracy to be improved and its signal level to be increased. In addition, it needs only a total of *four* wires, which is an extremely important benefit in robot applications.

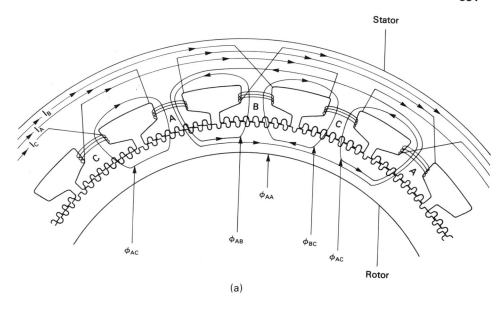

(a)

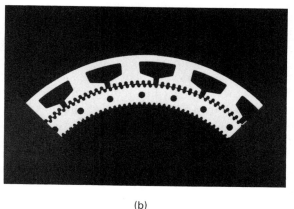

(b)

Figure 5.2.16. Motornetics resolver: (a) sketch showing current and flux paths. (Redrawn with permission of B. Powell and Motornetics Corp., a subsidiary of NSK, Santa Rosa, CA.   From I.B. Cushing, "A New High Accuracy Angular Position Transducer" Proc. of Motor-Con, April, 1984, p. 284, Fig. 1.); (b) portion of actual device.)

Although the stator and rotor of the Motornetics Resolver have the same number of teeth, tooth alignment varies in unison every third pole. This is accomplished by changing the mechanical phasing of the teeth of each pole (with respect to the immediate neighbors on either side of any tooth) by one-third of a tooth pitch. The reader can easily verify that such is the case from Figure 5.2.16a.

Electrically, every third winding is connected in series so that the self- and mutual inductances (with respect to the other two phases) vary cyclically. The cycle repeats each time the rotor moves one complete tooth pitch. In this way the mechanical angle is equal to the electrical angle divided by the number of rotor teeth N.* For example, if $N = 150$, the device can be thought of as behaving like a standard resolver placed on the input side of a 150:1 speed reducer since the electrical signal will go through 150 cycles for each mechanical revolution.

Although the Motornetics Resolver's three-phase nature makes it more closely resemble a synchro, electronic circuits are normally used to modify the signals so that more commonly available RDCs can be used to digitize the analog position information. A fairly inexpensive 10-bit RDC will produce an overall resolution of 153,600 (150 × 1024) "counts" per motor revolution. The corresponding number for a 12-bit RDC is 614,400. In either case, this is considerably greater than the resolution generally used by industrial robots of the mid-1980s (e.g., in the order of 40,000 to 60,000 counts/rev). However, as robot resolution requirements increase, it is clear that this sensor will be a candidate in certain applications.

It is important to understand that unlike the standard single-cycle resolvers described in the preceding section, the multiple-cycle device is an incremental position-sensing device rather than an absolute one. This means that when a robot utilizing such a sensor is powered up, the true position is *unknown* since the actual position is determined only within one cycle, but there is no way to know *which cycle, of the possible N, is being sensed*. The apparent difficulty is easily overcome by first causing the robot to execute a calibration procedure. For example, all joints may be driven (without regard to the position sensors' outputs) until they encounter mechanical end stops. Then the motors are reversed, causing the robot joints to "back away" a specified number of "counts" from these end stops. All digital position counters are then zeroed. To obtain absolute position information, it is only necessary for the hardware to keep track of both the count and the *cycle number*, which can easily be done.

5.2.5 The Inductosyn

A device that is used extensively in numerically controlled machine tools is the *Inductosyn*, a registered trademark of Farrand Controls, Inc., which developed it. Acknowledged to be one of the most accurate means of measuring position, it is capable of accuracies of 0.1 mil linear or 0.00042° rotary.

In actual operation, the Inductosyn is quite similar to the resolver. Regardless of whether the configuration is linear or rotary, there are always two magnetically coupled components, one of which moves relative to the other. For example, consider the linear Inductosyn shown in Figure 5.2.17. The fixed element

---

*Thus the resolver's design uses ferromagnetic teeth to multiply the number of poles of the sensor *without necessitating a large number of windings*. This permits manufacturing costs to be minimized and reduces failures due to breaks in the coils.

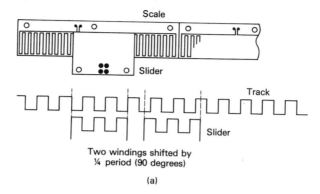

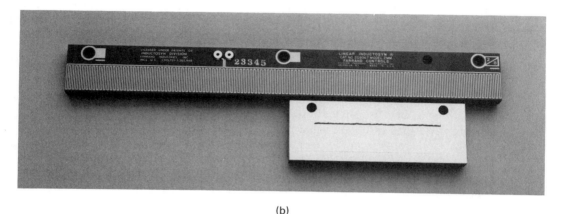

(b)

Figure 5.2.17.   Linear Inductosyn: (a) sketch of slider and scale with windings shown magnified. (Redrawn with permission of Farrand Controls, a division of Farrand Industries, Inc., Valhalla, NY.); (b) photo of actual device. (Courtesy of Farrand Controls, a division of Farrand Industries, Valhalla, NY.)

is referred to as a *scale* and the moving element as a *slider*.   Both of these are fabricated using printed-circuit technology, which is one of the major reasons for the high degree of accuracy that is achievable.   A rectangular-wave copper track having a cyclical pitch of 0.1, 0.2, or 2 mm is normally bonded to the substrate material.   The scale usually has one continuous track that may be many inches long (e.g., 10, 20, or longer).   The slider, on the other hand, is about 4 in. long and consists of two separate tracks of the same pitch as the scale but separated from one another by $\frac{1}{4}$ of a period (or 90°).   The slider is mechanically able to travel over the entire length of the scale, the gap between these two elements being about 5 mils.   (An electrostatic screen is placed between them to prevent accidental short circuits due to externally applied forces.)

As in the case of the resolver, an ac voltage $V \sin \omega_{ac}t$ is applied to the scale. Here, however, the carrier frequency ($\omega_{ac}/2\pi$) is in the range 5 to 10 kHz. The output at the two slider tracks is then

$$V_{s1} = V \sin \left(\frac{2\pi X}{S}\right) \sin \omega_{ac}t \tag{5.2.6a}$$

$$V_{s2} = V \cos \left(\frac{2\pi X}{S}\right) \sin \omega_{ac}t \tag{5.2.6b}$$

where X is the linear distance along the scale and S is the wave pitch. The amplitude of the sinusoidally varying input voltage is modulated spatially in much the same manner as the resolver [e.g., see Eqs. (5.2.5a) and (5.2.5b)]. Unlike the resolver, however, this spatial variation repeats *every* cycle of the scale track. Moreover, since Eqs. (5.2.6a) and (5.2.6b) represent the *average* voltage across a number of poles (i.e., cycles) of the scale, any variations in the pitch and/or conductor spacing are minimized, again contributing to the high degree of accuracy achievable with the device.

In its rotary form, shown in Figure 5.2.18, the stator (surprisingly) corresponds to the *slider* of the linear Inductosyn. Two separate rectangular track waveforms are placed radially on a circular disk. Again there are separate sine and cosine tracks, which, because they alternate physically, permit most of the error due to spacing variations to be averaged out. As a consequence, the rotary Inductosyn is probably the most accurate means currently available for monitoring position in commercial machine tools. As mentioned previously, typical accuracies are in the order of ± 0.42 millidegrees. Note that although laser devices are capable of giving considerably higher accuracies, their excessive cost makes them unattractive for this type of application.

The rotor of the rotary Inductosyn corresponds to the *scale* of the linear device in that it has a single, continuous, and almost rectangular printed track. Typically, there are anywhere from 128 to 1024 cycles (or 256 to 2048 "poles") on the disk. Because of the rotary configuration, however, the ac input voltage is applied to the *rotor* using brushes and slip rings. (A brushless configuration is also possible.) The output voltage of the device is monitored across the stator and has the same form as that shown in Eqs. (5.2.6a) and (5.2.6b) except that ($2\pi X/S$) is replaced by $N\theta/2$, where N is the number of poles of the rotor, and θ is the angle of rotation of the rotor with respect to the stator.

In actual operation, either form of Inductosyn can be used like a resolver. For example, one Inductosyn can act like a transmitter (RX) and the other like a receiver (RC) in a simple position servo. Alternatively, a resolver can be used as the RX and the Inductosyn as the RC. The advantage of the latter approach, however, is that one complete rotation of the resolver due to a position command signal will produce only a single cycle motion of the Inductosyn. Thus depending on the resolution of the latter device (i.e., the number of cycles per unit length over 360°), use of the Inductosyn would permit positioning of a machine tool to

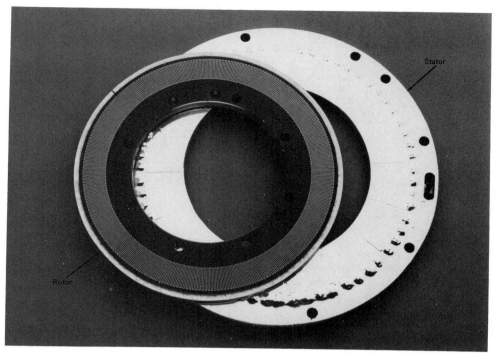

Figure 5.2.18.   Rotary Inductosyn®.   The stator corresponds to the slider of the linear Inductosyn®.   The ac carrier voltage is applied to the rotor. (Courtesy of Farrand Controls, a division of Farrand Industries, Valhalla, NY.)

close tolerances.   For example, a 1-mil linear resolution would not be unreasonable at all.

The configuration described above would be potentially attractive for use in either prismatic or rotary joints of robots.   However, gears or harmonic drives would still be required to obtain the torque multiplication from actuator to output. Thus the added cost of the Inductosyn, together with the additional electronics needed to digitize its output signals, would probably make the Inductosyn less attractive than other position-monitoring sensors.   However, if extremely high accuracies are required in the future, this device may someday be useful in the design of robots.

5.2.6 Linear Variable Differential Transformers

Another device that is both extremely rugged and capable of accurate position determination is the linear variable differential transformer (LVDT; see Figure 5.2.19).   It is observed from this figure that the LVDT consists of two parts, one of which is movable and the other fixed.   This electromechanical transducer is

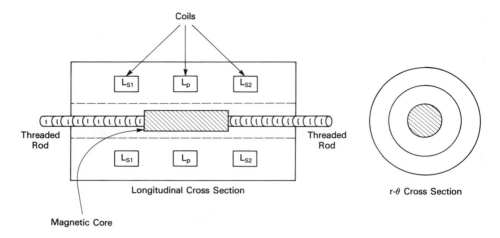

Figure 5.2.19. A linear variable differential transformer (LVDT) showing the single primary and the two sets of secondary coils. The magnetic core is generally the moving element of this sensor. (Redrawn with permission of Schaevitz Engineering, Pennsauken, NJ.)

capable of producing a voltage output that is proportional to the displacement of the movable member relative to the fixed one. Units having sensitivities on the order of 1 mV/mil with full-scale ranges of ± 25 mils to several inches are available. Because LVDTs are analog devices, they essentially have a resolution that is limited only by the external monitoring device (e.g., a voltmeter).

A common design of the LVDT has three equally spaced coils (L_p, L_{s1}, and L_{s2}) on a cylindrical coil form (see Figure 5.2.19). This is usually the stationary element. A rod-shaped magnetic core is also positioned axially inside the coil assembly and is free to slide back and forth. The purpose of this moving element is to provide a magnetic path for the flux linking the three coils.

To understand the operation of the LVDT, we consider the equivalent electrical circuit of the device shown in Figure 5.2.20. As can be seen, an ac voltage is applied to L_p, the primary side of the coil structure (this corresponds to the center coil in Figure 5.2.19). Since L_{s1} and L_{s2} on the secondary side are connected in *series opposing* (note the position of the dots on the windings), $v_{out}(t)$ will be zero if the coupling between the primary and each of the secondary windings is the same (i.e., the voltage induced in these coils will be the same). A little thought should convince the reader that this condition will exist when the magnetic core is positioned exactly in the center of the coil assembly.

If, however, the core is moved away from the central position, the coupling between L_{s1} and L_p will differ from that of L_{s2} and L_p. For example, the former will increase, whereas the latter will decrease. Consequently, the voltage induced in L_{s1} and L_{s2} will increase and decrease, respectively, with respect to their center core values. Thus $v_{out}(t)$ will be nonzero. With reference to Figure 5.2.19, it

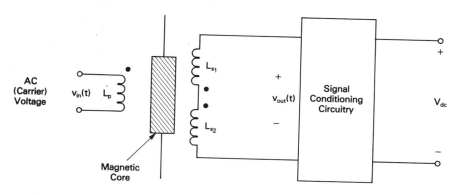

Figure 5.2.20. Electrical circuit of an LVDT showing the magnetic core. The secondary coils are connected in series opposing so that when the core is at or near the center of the LVDT $v_{out}(t)$ is zero. The signal conditioner is used to "demodulate" $v_{out}(t)$ and produces a dc voltage that is proportional to the core's linear distance away from the null (center) position. (Redrawn with permission of Schaevitz Engineering, Pennsauken, NJ.)

should be apparent that as the core is moved farther away from the central position, the coupling between the primary and one secondary coil will increase. There will be a corresponding decrease in the coupling between the primary and the other secondary winding. Thus the magnitude of $v_{out}(t)$ will be related to the linear displacement of the magnetic core. With proper design it is possible to make this output voltage *linearly* proportional to the displacement as shown in Figure 5.2.21. In normal operation, the object whose position is desired is attached to the rod-shaped magnetic core. A voltage that is proportional to the motion of the object results.

The signal conditioner shown in Figure 5.2.20 performs a number of functions. The first and perhaps the most obvious one is to convert the ac voltage $v_{out}(t)$ into a dc voltage V_{dc}. See Figure 5.2.21, where this was implied. The second task required of the device is that of determining the phase of $v_{out}(t)$ with respect to $v_{in}(t)$. This is necessary so that positive *and* negative displacements can be sensed.

A final operation performed by the signal conditioner is demodulation. The ac voltage applied to L_p may be thought of as a carrier signal. As the magnetic core is moved dynamically, the carrier is, in effect, modulated by the motion waveform $f(t)$. To be able to sense such dynamic behavior, the signal conditioner must eliminate the high-frequency ac component, that is, it must demodulate $v_{out}(t)$, thereby producing a voltage V_{dc} that is proportional to the physical motion waveform $f(t)$. It should be noted that carrier frequencies of 2 to 20 kHz are typical. Motions with a frequency content of approximately 1/10 of this value can be monitored (i.e., 200 to 2000 Hz).

LVDTs are often used to sense position where pure linear motion is performed (e.g., in a variety of machine tools, such as lathes). Rotary motion can be sensed with an RVDT (where R stands for "rotary"), a device that works on the same

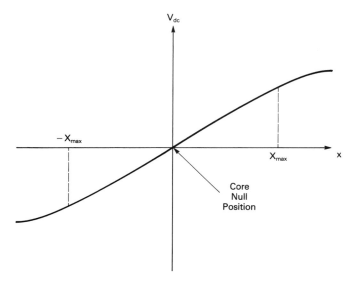

Figure 5.2.21. Output of an LVDT signal conditioner as a function of the core position x. For $|x| \leq X_{max}$, the characteristic is extremely linear. Outside of this range, however, linearity suffers. (Redrawn with permission of Schaevitz Engineering Pennsauken, NJ.)

principle as the LVDT but has a rotational configuration. When it comes to robots, however, neither device has been much used because of a number of problems.

The first difficulty with the LVDT in a robot application is that it is necessary to operate the device about its center point. Such an alignment is often quite difficult to perform and could create major manufacturing problems, thereby increasing costs. Difficulties with calibration of the robot on power-up could also be encountered. A second and associated problem with the LVDT is that the center or "null" position has a tendency to drift with time and temperature. Unless this is prevented (by the use of appropriate temperature compensation) a change in the robot's calibration would result, a totally unacceptable occurrence in most manufacturing environments. Possibly the major difficulty with using an LVDT and/or RVDT as a position sensor is that the joints of most robots do not move in pure straight lines or circular arcs. Motion is normally a complex combination of these trajectories. As a result, it is quite difficult, if not impossible, to configure the LVDT/RVDT so that the magnetic core is always collinear with the axis of the coils. Note that binding and probable damage to the unit will result if collinearity is not maintained. Finally, the RVDT has one additional limitation that makes its application to robots problematical. That is, it can only sense rotary motions of approximately $\pm 60°$. Since most robotic axes are required to move more than $120°$, this is most certainly a severe restriction.

Even if these difficulties were to be overcome, the fact that the LVDT is an analog device would make it inconvenient to utilize the device in a microprocessor-controlled servosystem.   In addition, the cost of the sensor, signal-conditioning circuitry, and A/D converter would be, no doubt, significantly higher than that of an equivalent optical incremental encoder and its circuitry.   Thus it seems likely that despite its good resolution capabilities, the LVDT will not soon be used on robots themselves as an internal position sensor.   It seems more realistic to expect that such devices will be utilized in equipment that is used in *conjunction* with a robot (e.g., in parts presentation).   Indeed, this is already a fact in some instances.

5.3 OPTICAL POSITION SENSORS

As we have seen, the sensors discussed in the previous sections can theoretically be used to determine the position of a robotic joint.   However, for one or more practical reasons, doing so is either not possible or often difficult and/or inconvenient.   Another class of sensor, utilizing optical hardware and techniques, can quite frequently be used to perform the position determination task with relative ease and surprising accuracy.   We now discuss such devices and their application to robotics.

5.3.1 Opto-Interrupters

It will be recalled that point-to-point-type robots require only that the beginning and end points be accurately determined (see Section 1.3.3.1).   The actual path between these points is not important, and hence little or no position information is utilized by the robot's control system except at the trajectory endpoint.   The actuators drive the joints of the robot until the final position is sensed, at which time the actuating signals are removed.   In effect, an open-loop control scheme is used.   "Programming" is accomplished by moving the endpoint sensors to different locations.

It might appear that a simple mechanical switch (or micro switch) is an ideal device for this application.   However, because of the need to interface the switch with a microprocessor, the inevitable contact bounce problem and the limited life expectancy make this approach relatively impractical for commercial robots.   (It *is* used in educational-type robots, however.)

An optical technique can be used to produce the required ability to sense "end of travel" without the problems associated with mechanical switches.   Called an *opto-interrupter*, its operation is quite easily understood.   Consider the arrangement shown in Figure 5.3.1.   A transparent disk with at least one dark sector is placed between a light emitter (e.g., an LED) and a light receiver or sensor (e.g., a phototransistor).   Light will reach the receiver until rotation of the disk causes the "black flag" to block it.   A binary or "on–off" signal can be generated and used to sense the endpoint of travel.   For example, the output (i.e., the collector) of the phototransistor will be *low* as long as light impinges on the transistor's base. On the other hand, the collector voltage will be *high* when there is no light.

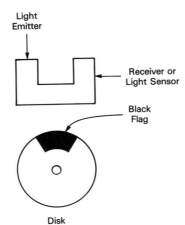

Figure 5.3.1. Simple opto-interrupter showing light emitter-receiver assembly and disk with "black flag."

The block diagram of a simple electronic circuit that makes use of such a sensor to drive a robot axis to the end of travel is shown in Figure 5.3.2. Here the system is actuated by momentarily closing the start switch. The motor will continue to rotate until the black flag on the disk prevents light from reaching the light sensor. When this occurs, the motor voltage is turned off and the axis coasts to a stop. (If desired, additional circuitry can be added to produce dynamic braking, thereby stopping the motor much more quickly.)

A possible realization of the logic and sensor electronics is shown in Figure 5.3.3. The waveforms of the digital signals S_1, S_2, and S_3 are shown in Figure 5.3.4. To understand the operation of this circuit, recall that the output of a NAND gate will be *low* (i.e., 0 volts or "logical zero") only when both inputs (S_1 and S_2 in this instance) are *high* (i.e., "logical 1" or for TTL logic circuits, 5 V). Any other combination of input signals will cause the output of the NAND gate to be *high*. Thus if the black flag on the disk is initially placed in the slot of the

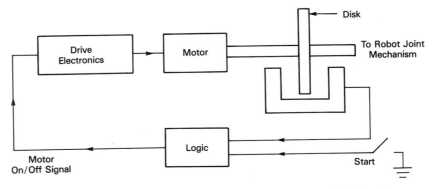

Figure 5.3.2. Block diagram of a simple unidirectional motor control circuit. The motor begins to rotate when the switch is closed.

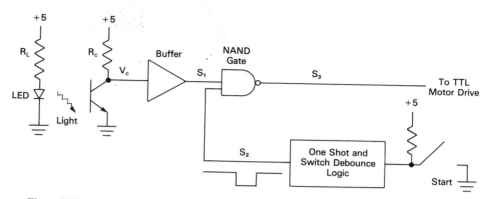

Figure 5.3.3. Possible realization of sensor and logic circuits for simple motor controller of Fig. 5.3.2.

opto-interrupter, the collector of the phototransistor will be about 5 V, so that S_1 will be high. In addition, if the one-shot and debounce circuit is designed so that its output is normally high and goes low only when the one-shot is triggered by the start switch being grounded, S_2 will normally be *high* also. Therefore, the signal to the motor drive circuitry is *low* and the motor does not turn.

As seen in Figure 5.3.4, when the start switch is depressed, S_2 goes *low*, which in turn causes S_3 to go *high*. The motor begins to rotate and will continue to do so until the black flag again interrupts the light, reaching the base of the photo-transistor.

It is important to note that this simple circuit permits only unidirectional rotation of the motor. Thus if it were used to actuate an axis of a simple robot, the manipulator would be limited to motion in one direction only. More complex circuitry would be required to produce bidirectional motion. In addition, as shown in this example, such a robot would be quite limited since there would be only a single endpoint. More endpoints could be obtained simply by utilizing more than one flag placed at appropriate places on the disk. In fact, "programming" such

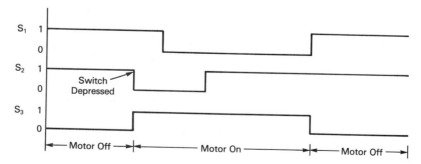

Figure 5.3.4. Timing diagram for the digital signals associated with the sensor and logic circuit of Fig. 5.3.3.

a robot axis would consist of producing a special disk with the correct number of flags at the proper locations (see Problem 5.11).

5.3.2 Optical Encoders

One of the most widely used position sensors is the optical encoder. Capable of resolutions that are more than adequate for robotic applications, these *noncontact* sensory devices come in two distinct classes: (1) absolute and (2) incremental. In the former case, the encoder is able to give the actual linear or rotational position even if power has just been applied to the electromechanical system using the sensor. Thus a robot joint equipped with an absolute encoder will not require any calibration cycle since the controller will immediately, upon power-up, know the actual joint position.

This is not so in the case of the incremental encoder, however. Such a sensor only provides positional information *relative to some reference point*. A robot utilizing an incremental encoder must, therefore, first execute a calibration sequence before "true" positional information can be obtained. The discussion at the end of Section 5.2.4 is pertinent in this context.

Although either linear or rotary encoders for both of the foregoing classes are available, the rotary device is almost exclusively used in robotic applications. One of the most important reasons for this is that revolute joints far outnumber prismatic ones in robots currently being manufactured. Even for joints that move in a linear fashion, as in the case of a spherical coordinate manipulator, the linear encoder is normally much more costly, and so rotary encoders are still employed. Therefore we restrict the discussion to the latter type, although much of what is said will apply directly to the linear sensor.

5.3.2.1 Rotary absolute encoders

As mentioned above, the absolute encoder is capable of giving the correct rotary position at all times even after power-up has occurred. The device produces a separate and *unique* coded word for each shaft position, and unlike the incremental encoder, every reading is independent of the preceding one. A major advantage of the absolute encoder is that even if system power is accidentally lost (due to a power outage or relay trip, for example) the device will "remember" where it is and will report this to the system as soon as power is restored. Calibration of machines using this type of encoder is, therefore, maintained *even if the position of the rotating member is moved when the power is off*.

Absolute encoders usually consist of three major elements:

1. A multiple-track (or channel) light source
2. A multiple-channel light receiver
3. A multiple-track rotary disk

Normally, light emanating from a linear, *N*-element light source (e.g., LEDs)

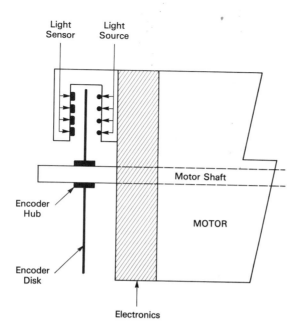

Figure 5.3.5. An absolute encoder mounted on a motor. Shown are the various components of the encoder including the disk, light sources, light detectors, and electronics.

is made to pass at right angles through the disk and is received (or collected) by a corresponding linear array of N light sensors (e.g., phototransistors) mounted on the opposite side of the disk (see Figure 5.3.5). The disk is divided into circumferential *tracks* and radial *sectors*. Absolute rotational information is obtained by utilizing one of several possible code formats. For example, Figure 5.3.6 shows a four-track 16-sector pure binary-coded disk. Other coding schemes that can be used include binary-coded decimal (BCD) and Gray code. It can be seen from Figure 5.3.6 that the resolution of the disk is 22.5° (360/16) since one complete disk revolution is 360° and there are 16 sectors. If the shaded areas are assumed to represent a binary "1" and the clear areas a binary "0," the outputs of each of the four light sensors will represent a 4-bit sequence of ones and zeros. For the binary code used in Figure 5.3.6, the decimal equivalent of this number is the actual sector number. As an example, if sector 11 is in the region of the LEDs, the output of the photo transistors will be 1011 or decimal 11. It is clear from this discussion that the absolute disk position is known simply by reading the photodetector outputs.

In practice, it is possible to produce absolute encoders with up to 13 separate channels (i.e., 13 bits) which means that resolutions of up to $360/2^{13} = 0.044°$ are possible for a single complete rotation of the disk. Often, however, it is necessary for the device being monitored by the encoder to undergo *many* rotations. Since it is clear that the coded binary sequence *repeats* for each complete disk cycle, something else is needed. In this case it is possible to use a *second* disk placed on the same shaft as the first but geared down so that a complete revolution of

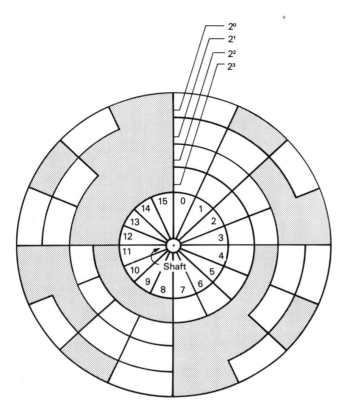

Figure 5.3.6. A four-track, sixteen-sector pure binary-coded disk used in an absolute encoder. (Redrawn with permission of Dynamics Research Corp., Wilmington, MA. From "Techniques for Digitizing Rotary and Linear Motion," Fig. 3-1, p. 3-2, 1976.)

the first moves the second only a distance of *one sector*. The first one is used for absolute positional information for *any* single shaft revolution, whereas the second disk gives the actual rotation number.

Although their ability to "remember" position is an extremely attractive feature for robot applications, absolute encoders are generally *not* used because of their excessive cost. For example, a 10-bit device can run several hundred dollars. A comparable incremental encoder, on the other hand, can be purchased for about $80 to $100. It is, therefore, not surprising that most robots utilize incremental devices, which we now discuss in some detail.

5.3.2.2 Optical incremental encoders

As mentioned above, optical incremental encoders are widely used to monitor joint position on robots. In addition, they are the sensor of choice in a variety of machine tools, including lathes, *x-y* tables, and electronic chip wire and hybrid die bonders. The major reason is that they are capable of producing excellent resolution at a significantly lower cost than a comparable absolute device. However, absolute position information can be obtained only by first having the robot or other machine tool perform a calibration operation.   This is usually not considered to be a major disadvantage, since such an operation generally has to be executed only after power has been applied.   It is important to understand that if power is accidentally lost during an operation, calibration must be performed again since the incremental encoder has no "memory."

Just as in the case of the absolute device, the incremental encoder in its simplest form consists of a disk, an LED light source, and a corresponding set of light receivers (e.g., phototransistors).   However, there are significant differences between the two.   For example, there are usually only a *single* LED and *four* photodetectors.   Also, the thin circulator disk (usually made of glass, Mylar, or metal) contains a *single* track consisting of *N* radial lines, as shown in Figure 5.3.7. The resolution of an encoder containing such a disk is normally defined as the *number of lines, N*.   This implies that the encoder can resolve an angular position equal to 360°/*N*.   Typically, encoders with resolutions of 100, 128, 200, 256, 500, 512, 1000, 1024, 2000, and 2048 lines are available, meaning that angular resolutions ranging from 3.6° down to 0.175° are achievable.   Generally, in robot applications, 200- to 1000-line disks are quite adequate, even where it is necessary to position a part or tool to within ±1 or 2 mils.   We will shortly see that it is possible to increase this resolution electronically.   However, before discussing this, let us describe how the incremental encoder produces positional information.

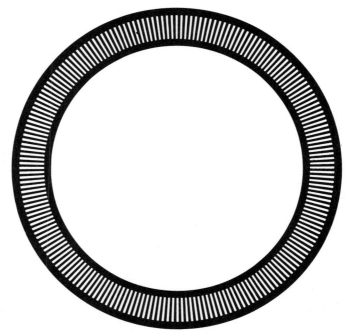

Figure 5.3.7.   An incremental encoder disk having 200 radial lines.   A resolution of 1.8° is possible with such a device.   (Courtesy of Dynamics Research Corp., Wilmington, MA.)

If the encoder disk is mounted on a rotating shaft (e.g., of a servomotor as shown in Figure 5.3.5), then as the disk turns, light to the photodetectors will be interrupted by any line on the disk that passes in front of the LED source. It can be shown that the detector's output will be a waveform that is approximately sinusoidal. Often, a comparator is used to convert these signals to TTL pulses, thereby making them more suitable for digital systems. There are two problems with this arrangement. The first is that although a single photodetector will produce a sequence of N TTL pulses per revolution, it should be clear that it will be impossible to determine the *direction of rotation* of the disk. A second difficulty arises due to variation or drift in light source and/or ambient light intensity. Since a comparator is used for TTL conversion, the width of the pulses will be quite sensitive to the amount of light collected by the detector (see Problem 5.12). This is an undesirable condition, especially in cases where the disk is spinning at a high rate of speed (e.g., more than 5000 rpm; see Problem 5.13).

Both of these problems can be overcome by employing *multiple* light sensors. For example, a second photodetector separated from the first by 90° (electrical) will produce a second, or B output channel which is identical to the first, or A channel, except that it yields TTL signals approximately 90° out of phase with the original ones. Clockwise or counterclockwise rotation of a motor shaft can be determined simply by noting whether A leads or lags B (see Figure 5.3.8).

The solution of the light-variation problem requires the use of additional photosensors. To understand this, consider the single-channel encoder (with only a small, magnified section of the disk indicated) shown in Figure 5.3.9. Here we have placed a stationary plate or *reticle* in front of the light sensor. This component consists of a number of optical "slits" (i.e., lines) and is used to direct light from about 20 lines on the encoder disk to the single photodetector. An overall im-

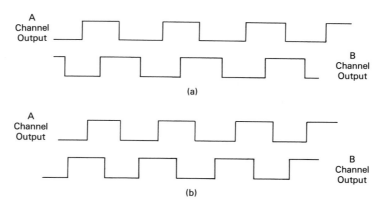

Figure 5.3.8. TTL outputs of the A and B channels of an incremental encoder: (a) A leads B when clockwise rotation occurs; (b) A lags B when counterclockwise rotation occurs.

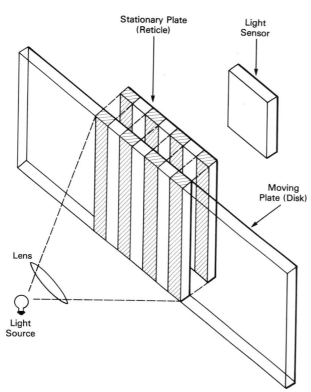

Stationary Plate
(Reticle)

Light
Sensor

Moving
Plate (Disk)

Lens

Light
Source

Figure 5.3.9. Section of a single channel encloder. (Redrawn with permission of Dynamics Research Corp., Wilmington, MA. From "Techniques for Digitizing Rotary and Linear Motion," Fig 4-1, p. 4-2, 1976.)

provement in performance is realized by reducing the encoder's sensitivity to both dirt and variation in line placement.

In actual operation, when the disk is rotating, the photosensor voltage output will vary theoretically in a triangular fashion, as shown in Figure 5.3.10. Actually, the waveform is more nearly sinusoidal, primarily due to the finite line widths in the shutter assembly (i.e., the disk and reticle). The maximum sensor output voltage E_{max} is proportional to the intensity of the LED. The minimum voltage E_{min} is not zero because light cannot be fully collimated by the shutter (i.e., there is always some light leakage). This value can be minimized, however, by reducing the clearance between the shutter and the light source (e.g., a 1- to 10-mil gap is typical). It is desirable to do this because the usable component of the sensor output is the peak-to-peak value E_1.

If a comparator is used to digitize the sensor output signal, a TTL pulse will be generated each time the voltage passes above the average value E_{ave}. This will theoretically produce a train of pulses with a 50% duty cycle provided that the disk is rotating at a constant velocity (see Figure 5.3.11a). However, if E_{ave} drifts due to LED and/or ambient light intensity variation or photodetector sensitivity

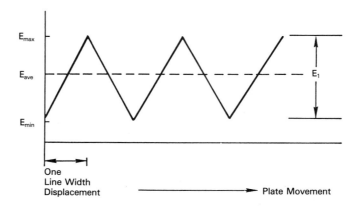

Figure 5.3.10. Output voltage of the sensor in Fig. 5.3.9 as the encoder disk moves relative to the reticle. (Courtesy of Dynamics Research Corp., Wilmington, MA. From "Techniques for Digitizing Rotary and Linear Motion," Fig. 4-2, p. 4-3, 1976.)

changes (caused by elevated temperature or high-frequency operation), the pulses will no longer have a 50% duty cycle, as shown in Figure 5.3.11b. Although at low speeds this is not a problem, high-speed applications will cause the pulses to be so narrow as to produce sensing errors (i.e., pulses may be missed).

This problem can be overcome by employing a second sensor (and reticle)

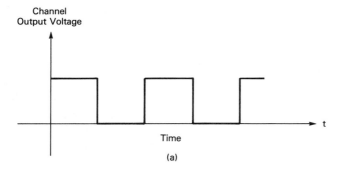

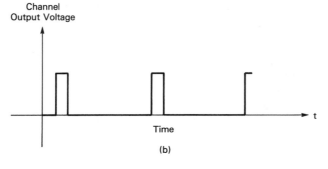

Figure 5.3.11. TTL output of a single encoder channel. The disk is assumed to be rotating with a constant angular velocity: (a) ideal 50% duty cycle; (b) the duty cycle is not 50% due to drift in the average value of the sensor voltage E_{ave}.

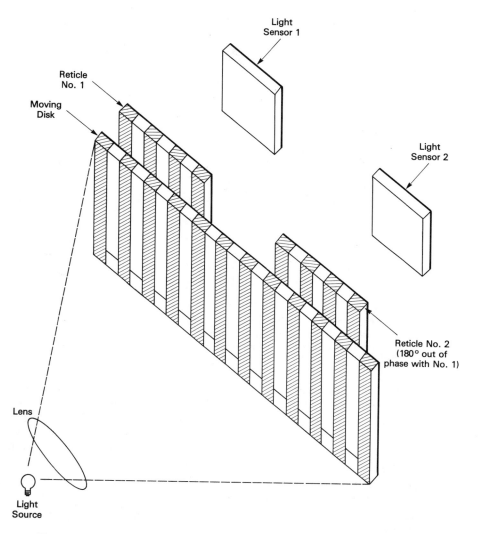

Figure 5.3.12.   Use of two sensors with the *same* light source to reduce the variation in the average value of the encoder's output, E_{ave}.   (Redrawn with permission of Dynamics Research Corp., Wilmington, MA.   From "Techniques for Digitizing Rotary and Linear Motion," Fig 4-4, p. 4-5, 1976.)

placed 180° out of phase with the first, as shown in Figure 5.3.12.   Note that the *same* light source is used to illuminate both sensors.   If the outputs of the two photodetectors are connected in "push-pull" so that the two signals are subtracted, a triangular waveform centered about zero and having approximately twice the peak-to-peak amplitude of either signal will be generated (see Figure 5.3.13).   In practice, differences in the two sensors cause the average value to differ somewhat

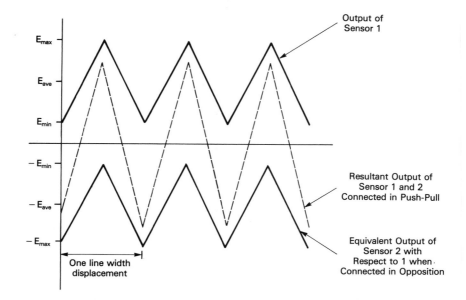

Figure 5.3.13. Push-pull output from two sensors. This arrangement significantly reduces the effect of average value drift on the TTL duty cycle. (Redrawn with permission of Dynamics Research Corp., Wilmington MA. From "Techniques for Digitizing Rotary and Linear Motion," Fig 4-6, p. 4-6, 1976.

from zero. However, this is a second-order effect and can easily be offset with a bias voltage applied directly to the difference amplifier.

The push-pull configuration has a number of advantages over a single sensor device. First and most important, the optical encoder is much less sensitive to variations in the average value of the photodetector output since the light sensors will be equally affected. As a direct consequence, the interpulse spacing variation (at constant velocity) is reduced to about one-half that found for a single sensor unit *for the same drift in average light intensity*. In addition, temperature and/or frequency effects are minimized because, once again, both sensors are affected to the same degree.

As mentioned above, the single sensor encoder cannot give any information about the direction of rotation. A little thought should convince the reader that the use of a second photodetector placed 180° out of phase with the first one does not alter this situation. The encoder obtained is still a *single-channel* device. To determine direction, a second set of photosensors, placed 90° out of phase with the first set, must be used as shown in Figure 5.3.14. Here the (push-pull) output of the first set becomes the A channel, whereas that of the second is channel B. A typical two-channel output is shown in Figure 5.3.15. Using the same convention as in Figure 5.3.8, the situation in Figure 5.3.15 would represent *clockwise* (CW) disk rotation. Note, however, that such an assignment is *arbitrary*, and therefore

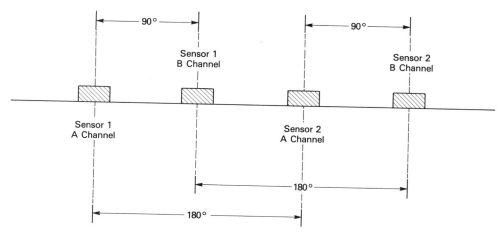

Figure 5.3.14. Two sets of sensors, placed 90° apart, are used to determine the direction of rotation.

one could just as easily consider "A leading B" to be a counterclockwise (CCW) rotation. Which definition is used is unimportant, but consistency must be maintained.

5.3.2.3 Increasing incremental encoder resolution electronically

Besides permitting the determination of (rotational) direction, the addition of a second channel to an optical incremental encoder has an added important benefit: increased resolution. One would normally expect that a disk containing N lines would have a resolution of $360°/N$. However, it is possible to utilize the information contained in both channels to reduce this number *electronically* by a factor of 2, 4, or even more (in fact, reductions as high as 20 have been obtained!). We now present methods for doubling and quadrupling the resolution of an optical incremental encoder.

Double Resolution. A single, two-input Exclusive OR gate can be used to produce a "$\times 2$" ("times two") circuit, thereby doubling the effective number of lines on the encoder. The explanation of this is left as an exercise for the reader (see Problem 5.15).

Another way to implement a $\times 2$ circuit that can detect direction at the same time is with a ROM used in a feedback configuration (see Figure 5.3.16). For the 16-by-4-bit ROM, the two most significant output (i.e., data) bits, D_2 and D_3, are fed back to the two most significant address lines, A_2 and A_3, respectively. The two least significant data bits are sent to the CU (count up) and CD (count down) inputs of an up/down counter. To understand how this circuit works, assume that a clockwise rotation produces a pulse on the CU line and a counterclockwise

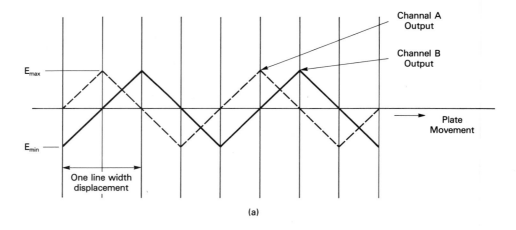

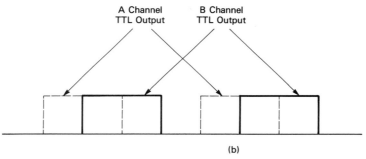

Figure 5.3.15. Two channel, incremental encoder outputs: (a) "raw," triangular signals; (b) digital signals. A channel leads B. [Part (a) redrawn with permission of Dynamics Research Corp., Wilmington, MA. From "Techniques for Digitizing Rotary and Linear Motion," Fig. 4-7, p. 4-7, 1976.]

rotation, a corresponding pulse on the CD line. Furthermore, suppose that a portion of the *ROM data table* is as shown in Table 5.3.1.

First assume that the encoder disk is at rest and that the A channel is outputting a 1 and the B channel a 0 so that $A_0 = 1$ and $A_1 = 0$. It can be seen that for row I of the ROM data table, the data at address 1010 is 0010. This is a *stable* situation since D_2 and D_3 are identical to A_2 and A_3, respectively. Moreover, both the CU and CD lines are 0, so that the up/down counter is not changing. Now let the disk begin to move in a clockwise manner. When $A_1 = 1$, the ROM address becomes 1110. If the data stored at this location are those shown on row II of the table, we will have an *unstable* situation since $D_3 = 1$ (i.e., the address "wants to be" 1111). As a result, the system will rapidly move to row III. However, before this can occur, the count-up line will have gone *high* for a short period of time, causing one to be added to the count. The reader can easily verify that

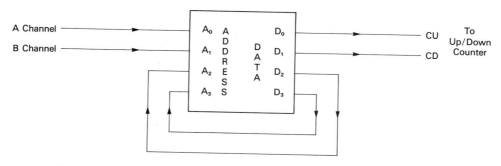

Figure 5.3.16.   A circuit used to double the resolution of an incremental encoder and sense direction of rotation.   A 16-by-4 bit ROM and feedback are utilized.

once row III is reached, a stable situation is again present and no further counts or shifts in addresses will occur unless the encoder disk continues to rotate.

Using the same logic as above, it can be seen that starting at row III in the data table and rotating the encoder disk CCW will produce a short-duration pulse on the CD line.   Row IV will represent an unstable situation and cause the address to change rapidly from 1011 to 1010 (row V).   It is possible to complete the remaining 12 data entries for the ROM, so that any CW or CCW rotation of the disk produces an appropriate up or down count (see Problem 5.16).

Quadruple Resolution.   The $\times 2$ circuits just described effectively recognized pulses from *both* the A and B channels of the encoder.   The reader may have already noted, however, that for each line on the disk, there are actually *four edges* (or transitions) produced (see Figure 5.3.8, for example).   This fact leads to the conclusion that simply by *counting these edges*, a fourfold increase in resolution should be possible.   A circuit that effectively does that is shown in Figure 5.3.17.

To understand the operation of this circuit, refer to the timing diagram for CW motion of the disk shown in Figure 5.3.18.   Suppose that the 16-bit counter (realized with four 4-bit binary up/down counter chips that respond to *negative edges* on the CU and CD inputs) is initially cleared (i.e., its count is 0, so the LSB + 2 through the MSB are low).   It can be seen from Figure 5.3.17 that the LSB of the count is obtained by exclusively OR'ing the A and B channel signals,

TABLE 5.3.1   ROM DATA TABLE

	A_0	A_1	A_2	A_3	D_0	D_1	D_2	D_3	Comment
I	1	0	1	0	0	0	1	0	Stable
II	1	1	1	0	1	0	1	1	Unstable
III	1	1	1	1	0	0	1	1	Stable
IV	1	0	1	1	0	1	1	0	Unstable
V	1	0	1	0	0	0	1	0	Stable

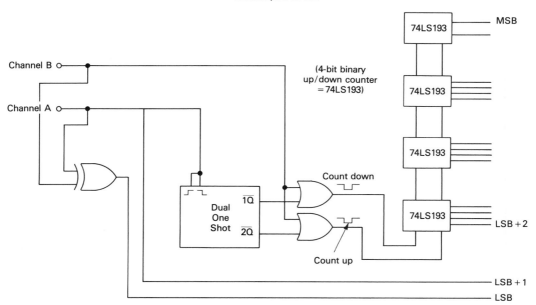

Figure 5.3.17. Possible circuit for increasing the encoder resolution by a factor of 4. Note that either a count-down or a count-up pulse will be generated depending on whether A leads or lags B.

whereas the LSB + 1 is just the A channel information itself. In addition, it is observed that a count-down pulse is generated for each encoder *line* sensed. (See $\overline{1Q}$ + B in Figure 5.3.18).

With respect to Figure 5.3.18, it can be seen that just before t_0, the position count is zero since A and B are both low and the most significant 14 bits have been assumed to be zero. On the leading edge of the first pulse on the A channel (at t_0), both the LSB and LSB + 1 become 1 and because a count-down pulse is also generated, the most significant 14 bits also change from 0 to 1. The count in hexadecimal has changed from 0000 to FFFF. The first B-channel positive edge (at t_1) causes A(XOR)B to go low, implying that the LSB = 0, LSB + 1 = 1, and the count is FFFE. The first negative edge on the A channel (at t_2) produces a 1 on the output of the XOR gate and the count is now FFFD. The first negative edge (at t_3) on the B channel results in both the LSB and LSB + 1 being 0. However, another count-down pulse is generated so that after the first encoder line, the count is FFFC. Clearly, the resolution has been increased fourfold.

As mentioned above, it is possible to increase the resolution of an optical incremental encoder by factors considerably greater than 4. To accomplish this, however, it is necessary to use the "raw" or (almost) sinusoidal waveforms from the photodetectors. Additional channels of information are generated from these signals by appropriately adding and subtracting them to obtain additional sets of

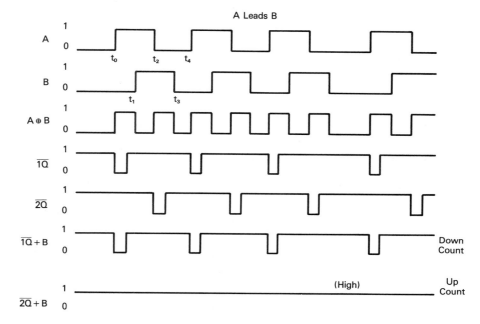

Figure 5.3.18.   Timing diagram for the $\times$ 4 circuit of Fig. 5.3.17.   Clockwise motion of the encoder disk causes A to lead B and results in a *decreasing* count.

signals having different phase relationships with respect to the original ones.   TTL conversion is then performed, and an increased number of counts for each encoder line sensed results when schemes similar to that shown in Figure 5.3.17 are utilized. In effect, the extra signals are used to produce "more edges" which can be counted with the appropriate digital circuitry.   To avoid sensing errors that cause counts to be "dropped" for this type of higher-resolution scheme, it is necessary to carefully maintain the quadrature relationship between the A and B channels.

5.4 ROBOT CALIBRATION USING AN OPTICAL INCREMENTAL ENCODER

In the preceding section we described the optical incremental encoder in considerable detail since for robots it is generally the position sensor "of choice."   One question that may have occurred to the reader is: How can an *incremental* device be used to obtain the *absolute* position information required by a robot?   Indeed, it appears on the surface that it cannot and that the need for knowledge about where a robot axis actually is within its workspace can be met only by an absolute device.   However, as we will now see, this is not the case and, in fact there are at least two distinct methods in current use which permit the incremental encoder to be utilized as an absolute device, with the resultant cost savings.   We now consider each in turn.

5.4.1 Zero Reference Channel

The difficulty with an incremental encoder is that it provides positional information (or the "count") only for a single rotation of the encoder disk. Since almost all robotic axes require that the actuator (e.g., the servomotor) must complete well over 100 turns in order to cause the joint itself to make one complete rotation, some method of keeping track of the rotation number must be included. Clearly, it is not difficult to do this. As an example, for a 300-line encoder, a "rotation counter" could be implemented so that every time a full 300 lines was counted, the rotation count would be incremented or decremented depending on the shaft rotation direction.

The major problem with this scheme, however, is that on a power-up condition, the digital system does not know which "turn" is being sensed. All information is *relative* to wherever the robot finds itself when it is first energized. Thus some method of *calibrating* or initializing the system is obviously required. A commonly used technique for accomplishing this is to add an additional channel of information to the encoder disk. Called a *zero reference* channel, it consists of exactly one line and produces a single *index pulse* for each encoder revolution. Clearly, counting these pulses immediately provides information about the revolution number. This is only one use of this channel, however.

A second and more important role for the zero reference channel is in the calibration of the robot axis. When power is applied to the robot, each joint is caused to move (often at a constant velocity, although this is not critical) in a predetermined direction toward a mechanical end stop on the axis. The actuator continues to turn until the end stop is encountered. The stoppage can most easily be detected by using the encoder and looking for a situation where over a (short) period of time, the count does not change. Note that it is not necessary to know the value absolutely. All that is required is that the current count be the same as that obtained, for example, 100 ms before. Once the system recognizes that the axis has reached its mechanical end of travel, the actuator is reversed and continues until the *first* index pulse is generated. At this point, the counter can be initialized to zero. All subsequent motions will be *relative* to this calibration point, and absolute position can be obtained simply by reading both the encoder count and the number of index pulses accumulated.

Note that it is not always necessary to utilize an index pulse to obtain absolute position information. As before, the calibration procedure begins by having each axis move until it reaches its mechanical end stop. At this point, the encoder count is noted and the actuator reverses direction. However, in this case, the motion ceases only after a *specified number of encoder counts have been accumulated* (relative to the end stop value). Finally, the count is initialized, thereby completing the calibration phase.

Although this calibration procedure is used on some robots today, there are some problems. For example, the calibration point is sensitive to both temperature and line voltage fluctuations. Yet this potential difficulty must be weighed against the cost reduction obtained by eliminating the zero reference channel. For some robots, this can make such a scheme attractive from a cost standpoint.

5.4.2 Absolute Position Using a Pot and an Incremental Encoder

A second technique exists for utilizing an incremental encoder in an absolute position application. In this case, a pot is used together with the encoder. The reader may be surprised at this in view of the discussion of Section 5.2.1, where it was stated that it is difficult to use pots as position sensors on robots because of severe reliability problems caused by electrical noise. However, it is possible to get around this problem by using the pot for "coarse" and the encoder for "fine" position information.

In actual operation, the pot is mounted on the same shaft as the optical encoder disk. It may be necessary either to use a multiturn device of an N:1 gear reduction so that N actuator turns produce a single rotation of the pot wiper. Regardless of which scheme is used, the pot voltage is divided into N *discrete* values, each corresponding to a specific actuator revolution number, stored in a table in the system memory. By comparing the actual pot voltage with the tabular data, the particular rotation number can be determined. This is the coarse position information. The encoder count can then be used in the normal manner to obtain the true (i.e., fine) position within any particular rotation.

This scheme avoids the major difficulty with using a pot—that is, position errors due to noise produced by the sliding contact of the wiper. The reason is that the pot is being used to monitor only a *set of discrete positions* (i.e., the actuator rotation number). No error will be made as long as the pot's output voltage is within a *range* of $\pm V/2N$ volts of the ideal value, where V is the applied dc voltage to the entire pot resistance. It can be seen from this discussion that a certain degree of noise immunity has been imparted to the system by utilizing this approach. Of course, when the noise level exceeds $\pm V/2N$, the pot must be replaced. It is for this reason that a thin-film device is often employed to increase the time between such replacements.

One of the interesting features of this *hybrid* scheme is that the robot can be calibrated in almost any position in the workspace. When power is first applied to the controller, the axis actuators are commanded to rotate until the first index pulse on each encoder is sensed. The pot voltages are then read and the particular actuator rotation number is determined by table lookup. At this point, the position counters are appropriately initialized, thus completing the calibration phase. As an example, if a pot voltage corresponds to the tenth actuator revolution, the rotation portion of the position count is set to 10. (The encoder count is, of course, initialized to zero since the index pulse is being read.)

Although the pot can be used to determine the rotation number after calibration has been completed, this is not usually done. The use of the index pulse for this purpose is essentially noise-free, whereas employing the pot increases the probability of an error. Thus the pot is only a factor during calibration, with its major role being to permit the initial absolute position to be obtained.

There are two major disadvantages of this system as compared to the pure digital/mechanical end stop technique. The first has been mentioned already: that is, the necessity of periodic replacement of the pot. Maintenance is, therefore, greater when the hybrid scheme is used. A second disadvantage is that there are positions within the workspace where the robot *may not be able to perform a calibration without striking one of its links*. To prevent this from occurring, the robot will not permit a calibration to be performed. In this case it may be necessary to power down and manually move the offending joint (or joints) to a "better" position within the work volume. Note that this step is never necessary when a mechanical end stop is an integral part of the calibration procedure, as in the case of the pure digital method described in the preceding section.

A final word is in order concerning the two methods described in this section. The major difference between the two techniques is that when power is first applied to the robot controller, the system utilizing the hybrid scheme "knows where it is" with an error of at most one actuator rotation. On the other hand, a system that employs the pure digital technique (with an end stop) only knows where it is once the calibration phase is completed. Before that it is truly "lost in space." Although this may appear to be a disadvantage, it really is not since uncalibrated robots are not very useful devices. Thus regardless of the scheme employed, calibration must first be performed before any useful work can be done.

5.5 INSTABILITY RESULTING FROM USING AN INCREMENTAL ENCODER

In the preceding section we saw how an incremental encoder could provide accurate and reliable position information. However, a potential difficulty with this device is that the mechanism that incorporates such a sensor in its servo may actually oscillate. Let us see how this situation can occur and also what steps can be taken to prevent it from being a problem.

5.5.1 Digital Jitter Problem

First suppose that a robot axis is required to *hold* a load horizontally. As there is no motion required, the desired and actual positions are the same and the error signal in the position servo is zero. However, this situation cannot continue indefinitely since the robot's joint will begin to rotate downward due to the influence of *gravity*. If for the moment we assume that no multiplying circuitry is utilized with the N-line incremental encoder that is on the joint actuator, the axis will continue to move until the *first* encoder line is counted. This means that *the servo will have no knowledge that the desired and actual positions are different until the actuator has rotated 360/N degrees*! At this time, an error signal will be generated and the actuator will return the joint to the desired horizontal position. The reader can readily understand that the entire cycle described above will repeat so that the load will not remain in the "home" position but will, instead, oscillate about it. This oscillation, which is characteristic of most digital-position servos, is a "limit cycle" and is often referred to as *digital jitter*. Clearly, such behavior is quite undesirable in a robot, and for that matter, any precision positioning device. What can be done about it?

One possible way to reduce the digital jitter problem is to employ multiplying circuitry with the incremental encoder. Based on what was said in the preceding section, a $\times M$ circuit will increase the resolution of the device M-fold so that the joint actuator will now rotate only $360/(N \times M)$ degrees before the error signal is generated. Also, an encoder with more lines can be utilized with the same result expected. It is clear that both of these techniques *reduce but do not entirely eliminate* the digital jitter problem.

Note that where a fair amount of *friction* is present, as in the case of some robot joint mechanisms, this may be all that is needed to *prevent* the limit cycle from occurring. Where the friction does not damp out the oscillation, however, other measures are required.

5.5.2 Analog Locking of a Position Servo

The digital jitter just described comes about due to the *discrete* or quantized nature of the error signal. Obviously, if this error was *continuous*, there would be no such problem. In this regard, one of the advantages of an *analog* position sensor, such as a pot, is that it does produce a continuous signal, and thus its use would prevent digital jitter from occurring. However, the reader will recall that the pot suffers from other problems that make it unsuitable for most precision positioning applications, including robotics. Fortunately, the encoder can again be used to "save the day."

As mentioned earlier, the output of the photodetectors used in an optical encoder is very nearly sinusoidal. It is this fact that can be put to good advantage to practically eliminate the digital jitter problem. Under normal operating conditions (i.e., during motion), the "raw" or sinusoidal information is still converted into TTL pulses, so that the digital nature of the encoder remains unchanged. However, when the final desired position has been reached, the raw encoder signal is utilized to provide the required analog information. The digital command (i.e., desired final or home position) input to the servo is switched out and is replaced by an *analog* voltage of zero, resulting in a scheme that is sometimes referred to as an *analog lock*.

To see how analog lock operates, consider the previously described situation of the loaded robotic joint being held horizontally. As soon as the axis begins to rotate away from the desired home position, an error signal is *immediately* sensed and a restoring torque generated, thereby causing the joint to return to "home." With reference to Figure 5.5.1, if gravity causes the joint to move to point A (or

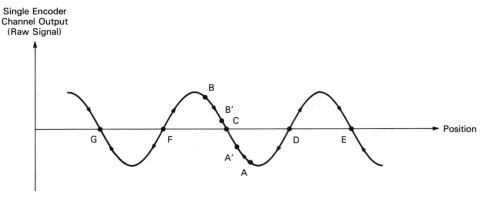

Figure 5.5.1. "Raw" signal for one channel of an encoder. When used in a position servo, points D and F are "unstable" whereas points C, E, and G are stable as indicated by the arrows.

B), a nonzero error signal is produced that is sufficiently large to drive the joint in a direction so as to minimize this error. It is clear that a very stiff servo can be produced if the voltage from the photodetector is followed by an amplifier with a large gain. In this case, a smaller angular rotation of the joint will produce the required error drive so that point A (or B) will now be A' (or B'). Thus the system will sense the error "more quickly," that is, closer to point C corresponding to encoder line K, the one nearest the desired final point. By increasing the gain of the amplifier, we can make A' almost coincident with C.

It is important to note that points D and F can never be "held" (i.e., they are unstable positions; see Problem 5.26). Also, the analog locking scheme can produce an error of one full encoder line if steps are not taken to prevent this from occurring (see Problem 5.27).

Currently, most robotic servos do not require an analog lock because the joint mechanism friction is sufficient to damp out, or inhibit, the limit cycle. However, as higher-performance robots become more commonplace, it is expected that friction will be reduced and some method of preventing digital jitter from occurring will be necessary. This is already the case in other high-precision positioning devices (e.g., high-speed wire bonders).

5.6 VELOCITY SENSORS

As noted in Chapter 4, a robotic servo must make use of *both position and velocity* signals to produce the desired manipulator performance. Up to this point, the monitoring of position has been discussed. The question of how one obtains velocity information is the topic of this section.

As we will learn, it is possible to determine the angular velocity of a rotating shaft in several different ways. For example, the *dc tachometer* has been used extensively for this purpose in many different control applications, including robotics. In addition to this *analog* device, however, it is possible to utilize an *optical encoder* and a *frequency-to-voltage converter* to obtain analog velocity. Alternatively, the optical encoder itself can be made to yield *digital* velocity information when combined with the appropriate software. We now discuss, in turn, these various techniques for measuring velocity.

5.6.1 DC Tachometers

It is well known that rotating the shaft of a dc motor will produce an analog voltage that increases (or decreases) with increasing (or decreasing) shaft angular velocity. In effect, the motor becomes a dc *generator* and can therefore be utilized to measure the shaft speed. Although it is possible to use almost any dc motor in this application,* dc tachometers are usually *specially designed* devices. There are a number of reasons why this is so.

The first and perhaps the most important one is that the tachometer ("tach") should produce a dc voltage that not only is *proportional to the shaft speed* but also has a voltage versus speed characteristic that is ideally *linear* over the entire operating range. (Some deviation from linearity is usually acceptable at speeds below 100 rpm, however; see Figure 5.6.1.) This permits the tach to be most easily used as a velocity sensor in control applications. Normally, the generated voltage produced by a dc motor will not possess the degree of linearity required in these cases.

A second reason for not using a motor in such an application is that the tach's output voltage should be relatively free of voltage ripple in the operating (i.e., speed) range of the device. Although a certain amount of ripple is permissible and can usually be handled with a low-pass filter, too much may produce unwanted jitter in the device being controlled. This would be particularly offensive in the case of a robotic manipulator. In general, a dc motor will produce too large a ripple for most control applications, so a specially designed device is preferable.

The final reason for not using a dc motor as a tach is that volume and/or weight is often an important system design consideration. As we mentioned before, this is certainly the case for the axes of an industrial robot, where the actuator must often be carried along in the joint itself. Since the tachometer supplies little if any current to the rest of the servo system, the output power requirement of the device is minimal. Thus it hardly makes sense to use a motor in this application, and a smaller device is quite satisfactory.

It is found that a permanent-magnet iron–copper armature tachometer will satisfy the above-mentioned characteristics. The speed–voltage curve of this *analog* device is quite similar to that shown in Figure 5.6.1. The underlying principle

---

*Geiger, Dana, "Regulating Servo Speed without a Tachometer," *Control Engineering*, October, 1979, pp. 73–74.

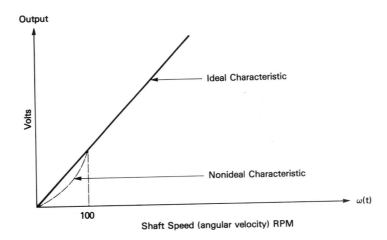

Figure 5.6.1. Output voltage versus speed transfer characteristic of an analog tachometer. At low speeds, the actual device (nonideal characteristic) is not linear.

of the tachometer can be understood by recalling that a wire moving in a magnetic field will induce a voltage across the wire that is proportional to its velocity and the sine of the angle between the magnetic field direction and the coil's plane. This angle is 90° when the wire's plane and the field are perpendicular to each other and results in the maximum voltage being developed.

In practice, the armature's copper (or aluminum) coils are wound longitudinally on a cylindrical piece of iron as shown in Figure 5.6.2. It can be seen that the ends of the coil are connected to a *commutator*, which is a segmented ring. Here only one coil is detailed, but normally there will be many (e.g., 11) spaced equally around the circular cross section. The corresponding commutator will then have twice as many segments as coils. The sliding electrical contact is usually obtained by a set of two or four carbon *brushes* which touch the various segments of the commutator.

Based on the above, the operation of the "rotary iron" dc tach can be understood. As any single coil rotates in the field of the permanent magnet, the induced voltage varies sinusoidally with angle. Thus at constant velocity, the voltage will also be sinusoidal in *time*. The brush/commutator assembly will act as a rectifying element by reversing the coil connection for each half of a complete revolution. In this manner, a *pulsating dc voltage is produced*. All other armature coils will also produce a sinusoidal voltage of differing phase with respect to the first one. Since the coils are evenly distributed around the armature's cross section, the *net* voltage output by the brushes is very nearly constant (i.e., dc). The small ac component of the voltage that is present is referred to as *ripple*. Tachometers currently being manufactured usually produce ripples of about 3 to 5% of the dc output.

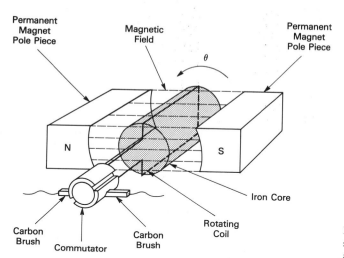

Figure 5.6.2. Analog tachometer showing one coil (of many) mounted on an iron core.

A more costly alternative to the rotary iron design described above is to use a *moving coil* for the armature. In this instance, a significant reduction in weight is achieved by employing a hollow "cup" whereby most, if not all, of the armature's iron is removed. This is accomplished by fabricating a rigid cylindrical shell out of the copper (or aluminum) coils or *skeins* using polymer resins and fiberglass. In addition, it is possible to utilize more coils (e.g., 19 to 23). By eliminating the armature's iron, the inductance of this type of tachometer is reduced, thereby permitting the ripple voltage to be quite a bit smaller than for a rotary iron device. Typical values are in the order of 1% of the dc output. Also, because the moving-coil design allows more coils to be utilized, the low-speed performance of the tachometer is improved over that obtained by the rotary iron version.

It should be clear that if an analog tachometer is used in a robotic application, the moving-coil version is quite probably the more attractive of the two designs because of the reduction in weight. On axes where the actuator is *not* carried and hence weight is not a consideration, the rotary iron design may be preferable due to the reduced cost. Despite the fact that in this case, the increased ripple can be handled with a low-pass filter, its low-speed performance may still be objectionable, so that the moving-coil device may still be the unit of choice.

As of this writing, the most common class of industrial robot that makes use of an analog tachometer is the SCARA (see Section 1.3.2.3). The primary reason is that the configuration of such a robot does not require the actuator to be lifted against gravity. Recall that the major axes of a SCARA move *perpendicular* to the gravitational field, thus the added weight of the tach does not present a significant additional burden (i.e., torque load) to either the servomotor or the mechanical structure of the manipulator. However, where the motor must be moved against gravity, it is usually preferable to employ a different technique for obtaining the velocity signals. We now discuss two such methods.

5.6.2 Velocity Measurement Using an Optical Encoder

As mentioned above, the added weight penalty that must be incurred when using a permanent-magnet tachometer is often unacceptable in robotic applications where the actuator must be moved with the particular manipulator link against gravity. In this instance, an alternative to the extra piece of hardware is required. Fortunately, the optical encoder described in an earlier section of this chapter, and already used for position determination, is available for monitoring shaft velocity.

Two techniques exist for doing this. The first utilizes both the encoder and a frequency to voltage converter (FVC) to provide an *analog* voltage that is proportional to shaft speed. As far as the user is concerned, it behaves very much like the dc tachometer described in the preceding section. The second technique makes use of the encoder and appropriate software to provide a *digital* representation of the shaft velocity; pure digital servos, as described in Chapter 4, would utilize this approach. In fact, most robots today do indeed use the optical encoder to produce digital position and velocity information. We briefly describe these two methods.

5.6.2.1 Encoder and frequency-to-voltage converter

An earlier section of this chapter showed how the TTL pulses produced by an optical incremental encoder could be used to monitor position. The question arises: How can these signals be processed so that velocity information is also obtained? The answer is found in the basic definition of velocity; that is, the *time rate of change of position*. Thus if the number of encoder pulses is observed (and counted) periodically and this number is converted to a dc level, the signal so produced will in fact be proportional to the shaft velocity. Clearly, we are approximating the derivative by $\Delta x / \Delta t$. Here, Δt is the "sampling" interval (or period) and Δx is the number of TTL pulses produced during this time interval.

A device that accomplishes the above is referred to as a frequency-to-voltage converter or FVC. This product of advanced integrated-circuit technology accepts both channels of the TTL encoder pulses and, using its own internally generated clock, counts these pulses during each clock cycle. The binary count is then output to an internal DAC which produces the desired dc voltage that is proportional to the encoder disk speed and hence the motor shaft speed. An example of an FVC is the Analog Devices AD 451 shown in Figure 5.6.3 in block diagram form. This unit will produce a 0- to 5-V output for pulse repetition rates of dc to 10 kHz. (The AD 453 will go to 100 kHz.)

How does the velocity signal produced by this device compare to that of an analog tachometer? First, the output of the FVC has *less* ripple than that of the tach, and in fact the nature of this ripple is totally different. The internal DAC produces a *piecewise constant output* which, depending on its conversion rate, will

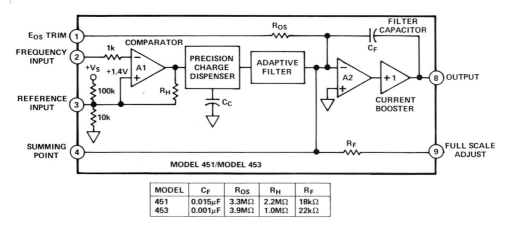

MODEL	C_F	R_{OS}	R_H	R_F
451	0.015μF	3.3MΩ	2.2MΩ	18kΩ
453	0.001μF	3.9MΩ	1.0MΩ	22kΩ

Figure 5.6.3. Block diagram of an Analog Devices 451/453 frequency-to-voltage converter. The 451 has a frequency range of dc to 10 kHz, whereas the model 453 can handle pulse repetition rates up to 100 kHz. (Courtesy of Analog Devices, Norwood, MA.)

have a period (i.e., an update rate) which is so small that it will cause the FVC's output to appear to be continuous in most applications. Thus, unlike the analog tach, no low-pass filter is needed when using the FVC. Second, the FVC will exhibit more time delay than the tach, the exact amount depending on the internal clock rate. In high-performance systems, such as semiconductor wire bonders that require servos having large bandwidths, this delay can create stability problems which must then be dealt with using additional compensation. However, in the case of the servos used to control robot joints, the extra phase lag created by the delay is usually not of any consequence due to the much smaller bandwidth requirements.

Why is it that we do not find the FVC being used extensively in robotic systems? One reason is that current prices for the devices are on the order of $50, making the FVC almost as costly as an analog tachometer. As we will see, there *is* a considerably less expensive way to use the encoder signals to obtain velocity information. Another reason for not using the FVC is that hybrid servos, involving digital position and analog velocity, are not as frequently utilized. More common in robot systems are pure digital servos.

5.6.2.2 Encoder and software

As indicated above, there is a way to obtain velocity information using an incremental encoder by processing the position data. One simple technique is to once again approximate the velocity by $\Delta x/\Delta t$. Rather than having a special-purpose chip perform the operation as in the case of the FVC, the following short computer algorithm can be utilized assuming that the position is updated (i.e., the encoder count is read) every T seconds:

1. Read and store the current encoder $P(kT)$.
2. Retrieve the previous encoder count $P((k - 1)T)$ (this was the position T seconds earlier).
3. The approximate velocity $V(kT)$ is given by

$$V(kT) = \frac{P(kT) - P((k - 1)T)}{T}$$

4. Increment k and repeat steps 1 through 3.

This algorithm is quite easy to code and requires little time to execute. However, it is based on a first-order approximation of the derivative which may be inadequate in some applications. In this instance, a second- or higher-order approximation must be used. Note that if this is done, the execution time will be increased and additional phase lag will be introduced into the digital servo loop, thereby necessitating extra compensation in order to stabilize the system.

Also, it is important to select carefully the update time T in the algorithm. Too large a value may cause the "sampling theorem" to be violated so that there will be a large error between the actual velocity and $V(kT)$. Too small a value will mean that low-speed performance of the algorithm will be poor since the position count will change very little, if at all, from update to update. This problem was discussed in Chapter 4 (see Section 4.5.2).

5.7 ACCELEROMETERS

Besides monitoring the position and velocity of a physical system, it is also possible to monitor its acceleration. Normally, linear acceleration is *measured*, whereas angular acceleration is most often *derived* from angular velocity by differentiation. Let us consider briefly how a device that can be used to obtain linear acceleration, and referred to as an *accelerometer*, operates.

From Figure 5.7.1, it can be seen that an accelerometer consists of three basic elements:

1. A mass M
2. Some type of linear displacement sensor (e.g., an LVDT)
3. A set of springs having an equivalent spring constant K

Based on one of Newton's laws (i.e., $F = Ma$, where F is the force needed to accelerate the mass (M) a linear units per second per second), it is easy to

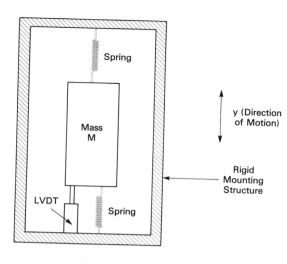

Figure 5.7.1. Basic elements of a linear accelerometer. The LVDT is used to monitor the relative displacement of the lower spring, a quantity that is proportional to the acceleration.

understand the operation of this device. Suppose that the entire accelerometer begins to move (i.e., is accelerating) in a downward direction. The force required to do this (e.g., *Ma*) will be opposed by the springs supporting *M*. As they bend *upward* a distance *y*, this force will be equal to *Ky*. Thus

$$Ma = Ky \tag{5.7.1}$$

Solving for the *linear acceleration a* gives

$$a = \frac{K}{My} \tag{5.7.2}$$

From this equation it is apparent that the acceleration of the mass is *proportional* to the distance. If an LVDT is used to determine linear position, as is often the case in commercial accelerometers, the output of that sensor will be proportional to the actual acceleration. Thus the signal-processing device used in an LVDT and described previously can be made to read acceleration directly.

Not mentioned in the discussion above is the fact that to be a useful sensor, the accelerometer must also include enough damping so that the spring–mass combination does not "ring" significantly (i.e., produce damped sinsoidal oscillations). Normally, it is desirable to have a small amount of overshoot of the final displacement position so that a damping constant of 0.6 or 0.7 is used. Under these conditions, an accelerometer can monitor motions having frequency components that are at least 2.5 times lower than the undamped (or "natural") frequency of the second-order mechanical system composed of the mass, spring, and damping. That is, the operating frequency f should obey the inequality

$$f \leq 0.4 \left(\frac{1}{2\pi} \right) \sqrt{\frac{K}{M}} \qquad (5.7.3)$$

Although commercial accelerometers can be obtained that will measure accelerations ranging from ± 5 to thousands of g's,* these devices have had limited use in robots currently being manufactured. One reason is that, as mentioned above, it is usually possible to determine only *linear* acceleration directly. Since most robot joints are revolute rather than prismatic in nature, it has been deemed more useful to measure position and/or velocity of the joint directly. If necessary, the acceleration signal can be *derived* from these data. Another reason for not using acceleration is that even if it were easy to measure this rotational quantity directly, it would still be necessary to process the information so as to get the desired position and velocity data to control the robot joint. The reader will recall (Chapter 4) that with robots, we are dealing with a *position servo* so that it seems to make more sense to monitor the position directly rather than indirectly (i.e., by performing two integrations on the acceleration signal). Certainly, one would expect fewer errors and/or uncertainties using the direct approach.

There have been a few experiments with accelerometers and robots, however. In most cases, the sensor has been used together with standard encoders to provide an estimate of the *actual motion* of the joint being controlled. It will be recalled that most position sensors are mounted on the actuator output before the rotary motion is geared down, in order to achieve the desired position resolution. The assumption is made that if the actuator's motion is controlled, the joint will respond in an identical manner. Clearly, this is not always true because mechanical linkages and/or couplings are not perfectly rigid. Generally, mechanical resonances that occur as a result cannot be compensated for by sensors so placed. However, an accelerometer mounted on the *joint structure* can provide information about what the joint is *actually* doing. These data, together with those from the actuator position sensor, can be possessed to compensate partially for the nonideal motion of a robot axis. The accelerometer is thus used to "observe" the actual joint behavior. In fact, at least one commercially manufactured robot utilizes a single axis accelerometer on its prismatic joint to reduce undesirable arm oscillations caused, in part, by imperfect mechanical linkages. Readers interested in learning more about the use of such sensors are encouraged to read the references at the end of this chapter, where linear optimal estimation theory is employed in an attempt to improve the performance of a robot [3, 4].

Recently the Pennwalt Corporation of King of Prussia, Pennsylvania, has begun to market a piezoelectric polymer (i.e., PVDF)-based accelerometer. This device has been shown to be far more rugged and sensitive than the more traditional

---

*A g represents the acceleration due to gravity; that is, 32.2 ft/s^2.

devices. Moreover, its cost is about an order of magnitude less than the industry standard unit *with virtually the identical frequency response.*

5.8 PROXIMITY SENSORS

Up to this point, we have discussed the behavior and application of sensors that were used to measure the position, velocity, or acceleration of robot joints (or more accurately, their actuators) and were called collectively *internal state sensors*. A second major class of robotic sensor is used to monitor the robot's geometric and/or dynamic relation to its task. Such sensors are sometimes referred to as *external state sensors*. Machine or robotic vision systems represent an important subclass of this group of devices and are treated separately in Chapter 6. The remainder of the current chapter is devoted to non-vision-type sensors that either can be or have already been used to make the robot more aware of its external environment. Although some of these may utilize *optical techniques* as part of their sensing system, they are not properly classified as visual sensors, so we describe them in this chapter. Also, as will be seen, many are still under development in various research facilities and are therefore not ready for use in an actual manufacturing environment. Since it is not possible to treat this subject in an exhaustive manner, readers desiring more information are referred to the references at the end of the chapter. Of particular note is an excellent summary report by D. J. Hall [5] of Carnegie–Mellon University's Robotic Institute which was quite helpful in preparing much of the remainder of this chapter.

In this section we describe a number of sensors used to tell the robot when it is near an object or obstruction. This can be done either by using a *contacting* or a *noncontacting* technique. Often, such sensors are called *proximity* devices, but the distinction between proximity and touch and/or slip is not clear-cut. That is, some proximity devices can also be used as touch (or *tactile*) sensors. We will consider only the proximity feature here, deferring the discussion of their application to touch and/or slip detection to a subsequent section.

5.8.1 Contact Proximity Sensors

The simplest type of proximity sensor is of the contacting variety. As Figure 5.8.1 shows, such a device consists of a rod that protrudes from one end and a switch or other linear position-monitoring element located within the body of the sensor. As the robotic manipulator moves, the sensor will become active only when the rod comes in contact with an object or an obstruction. When this occurs, the switch mounted inside the sensor will close (or open, if that is more convenient). The change of state of the switch, monitored through the robot's I/O interface, will cause an appropriate action to take place. Examples include an immediate (or emergency) halt if the device is used to sense obstacles or the branching to

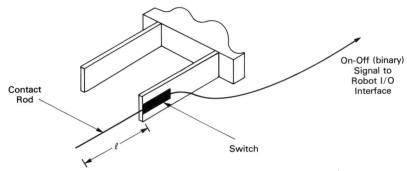

Figure 5.8.1. Simple contact rod proximity sensor mounted on one "finger" of a robotic gripper.

another part of the robot's program, thereby causing a particular operation to be performed (e.g., closing of the gripper). Such contact monitors can be placed anywhere on the robot's arm and/or wrist, and it is possible to utilize more than one. Thus simultaneous obstacle *and* object sensing is possible.

If the simple on–off switch is replaced by one of the linear position-sensing devices described in an earlier portion of this chapter, the "binary" contact proximity sensor becomes one that can detect actual position of the object (or obstacle). For example, a simple pot or an LVDT can be employed. Then once the protruding rod makes contact, further motion of the manipulator will push the rod into the sensor. If this rod is attached to the magnetic core of an LVDT or the wiper of a potentiometer, the motion will be converted into a voltage that will be proportional to the actual distance of the end of the rod (and hence the object) from some reference point on the robot (e.g., the end of the gripper). In addition, the approach velocity can be obtained from this information by performing either an analog or digital differentiation. Thus both distance and *approach velocity* can be monitored using such a contact sensor.

It is important to understand that a *single* contact proximity device cannot provide any information about the shape or nature of the object or obstacle (i.e., no object recognition capability is possible.* It will be seen in a later section, however, that some degree of object recognition can be obtained by utilizing *arrays* of these devices.

5.8.2 Noncontact Proximity Sensors

In contrast to the devices described above, a much larger class of proximity sensor does not require any physical contact at all in order to produce a signal that can be used by a robot to determine whether it is near an object or obstacle. These

---

*Some information about object shape *can* be obtained by having the robot execute a search, once the first contact has been detected. This will, most probably, require a good deal of time to perform, however (see Problem 5.42).

noncontact devices depend on a variety of operating principles in order to make the proximity determination. For example, reflected light, ultrasound, or variation in capacitance, inductance, or resistance have all been used. We now briefly describe a number of such sensors.

5.8.2.1 Reflected light sensors

One of the simplest types of proximity sensors that uses light reflected from an object and has been used experimentally on a robot gripper is shown in Figure 5.8.2a. The sensor consists of a source of light and a photodetector separated by about 8 mm and tilted symmetrically toward one another. This, together with lenses mounted in front of the assembly, produces focused incident and reflected beams. Figure 5.8.2b shows the photodetector voltage as a function of object distance from the (detector) lens. Figure 5.8.2c indicates that several of these sensors can be placed on a robotic gripper. In this way, proximity in several directions can be monitored simultaneously (e.g., ahead and below the robot's hand).

Although the maximum sensor output will occur when an object (or obstacle) is at the focal point, Figure 5.8.2b reveals a basic difficulty with this device. That is, two different object positions produce the *same* voltage except when the object is located exactly at the focal point. Since a one-to-one correspondence between position and detector voltage does not exist, additional logic or hardware is required to eliminate the ambiguity. For example, if the robot is moving and the sensor signal is *increasing*, it is clear that the object is on the *far* side of the focal point (i.e., has yet to reach this point, and so the output corresponds to the *larger* of the two position values). If, however, the signal is *decreasing*, the focal point has been passed and the smaller distance should be used. Several sensors placed at angles can also be utilized to eliminate this ambiguity. In addition, Jet Propulsion Labs (JPL) has embedded fiber optic filaments inside the fingers of the robot so that the effective voltage characteristic of the sensor is *monitonically decreasing with distance* (see Problem 5.32).

Besides this difficulty, other problems with the sensor exist. For example, ambient light will shift the curve in Figure 5.8.2b up or down depending on the intensity. The problem has been solved at JPL by *pulsing* the light source at a 6-kHz rate. However, a more difficult and perhaps impossible problem to overcome is that the sensor is sensitive to the reflectivity of the object or obstacle. A highly reflective surface will obviously produce a larger output voltage than one that is less reflective. Thus it might be necessary to "calibrate" the sensor to *each* object so that the maximum output voltage could be found. Then, knowing the characteristic of the detector, position could be determined *relative* to this maximum value. Alternatively, careful control and/or an a priori knowledge of the surface reflectivity would be required.

Even with these measures implemented, it would still be difficult to use the sensor for absolute position monitoring. The major reasons are that the device

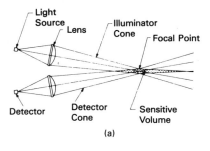

(a)

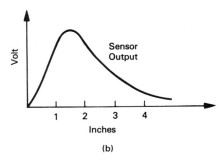

(b)

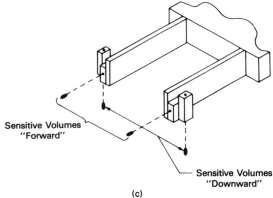

(c)

Figure 5.8.2. Reflected light sensor: (a) light source-detector assembly; (b) detector output voltage as a function of distance; (c) two-dimensional sensor array mounted on a parallel jaws gripper. (With permission of A.K. Bejczy and Jet Propulsion Laboratories, Pasadena, CA. From reference [6].)

would be quite sensitive to variations in light-source output, drift in the detector characteristics (due to ambient-temperature fluctuation), and environmentally caused changes in the reflectivity of the object. Thus all that could be reasonably expected would be to sense the proximity of an object to the robot's gripper within a *band of distance*. A threshold detecting circuit might be utilized to achieve this.

It is rather disheartening to realize that what at the outset appeared to be a simple and ideal noncontact proximity sensor has such great problems associated with it that its application to robots is not likely. It is for this reason that other devices are normally used to monitor proximity in a manufacturing environment.

5.8.2.2 Fiber optic scanning sensors

Fiber optics have been used to develop several different types of noncontact proximity sensors. As reported by Fayfield [7], there are at least three systems for utilizing this important technology in the robotics and/or the manufacturing fields. With regard to Figure 5.8.3, one employs transmitted light, whereas the other two make use of reflected light. It is important to understand that it is not possible, in all cases, to obtain reliable absolute position information. The devices can only tell whether or not a part is present.

In the *opposed* or beam break configuration (Figure 5.8.3a), the object is detected when it actually interrupts the beam of light. Such an optical interrupter depends on the object being opaque and is, obviously, not useful where parts are made of transparent or translucent materials. By employing high-gain amplifiers and noise-reduction schemes, these sensors can detect objects as close as a few mils and as far away as several inches. However, they are limited to informing the robot that something is or is not present. That is, absolute position information cannot be obtained. In addition to this, the receiver fiber bundle alignment is fairly critical. Thus, anything that would tend to misalign it from the emitter bundle would obviously affect the sensor's effectiveness. Finally, it would be necessary to use units with different gaps and/or lengths, depending on the type and size of the object to be sensed.

A second type of fiber optic proximity sensor is referred to as a *retroreflective* device since it employs a reflective target placed some distance from the body of the unit (see Figure 5.8.3b). An opaque object entering the area between the end of the fiber bundle and the target is sensed since reflected light reaching the receiver is considerably reduced in intensity. This is also true for parts made of translucent materials because the incident beam of light and that reflected from the target are both attenuated when they pass through such an object. The use of thresholding circuits on the receiver side of the sensor is important in both of these cases. The retroreflective scheme utilizes a bifurcated fiber bundle so that incident and reflected light is carried by the same set of fibers. Clearly, this eliminates potential alignment difficulties associated with the previous technique. However, the need for a separate target somewhat restricts the use of such a sensor to parts detection only. It is clear that unless an unforeseen (and unpredicted) obstacle happens to disrupt the light from the reflecting target, it will not be sensed by a retroreflective mode fiber optic sensor.

The last fiber optic proximity scheme is shown in Figure 5.8.3c. Here a bifurcated fiber bundle is again used, but there is no retroreflective target. The sensor actually can measure the amount of light reflected from an object up to a few inches away from the fiber bundle. Since most materials reflect some light, this "diffuse" device can be used to detect transparent and translucent objects. As in the case of the reflected light proximity detector described above, some degree of absolute position monitoring is possible under ideal conditions. However, all of the difficulties with that type of sensor that were described previously

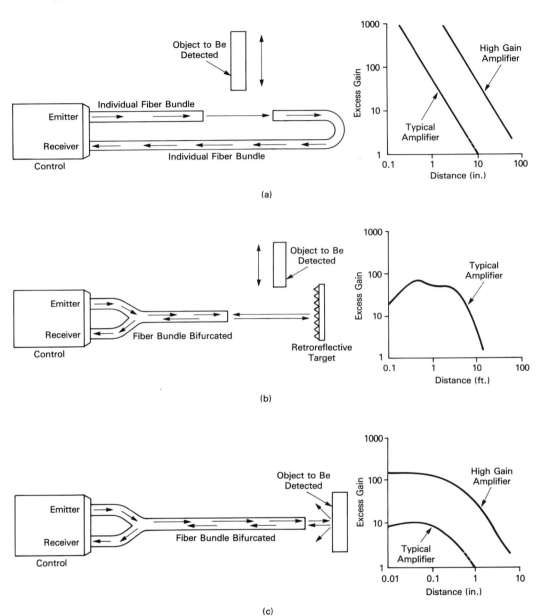

Figure 5.8.3. Three possible configurations of a fiberoptic scanning sensor: (a) opposed or beam break; (b) retroreflective; (c) diffuse. (Redrawn with permission of Production Engineering, Cleveland OH. From reference [7].)

are also present with the fiber optic version.   Nevertheless, this mode of fiber optic sensing is the most commonly used one since it is self-contained (i.e., no target is required), rugged, weighs very little, and is relatively inexpensive, making it ideally suited for a manufacturing environment.

5.8.2.3 Scanning laser sensors

A considerably more involved and costly proximity sensor is shown in Figure 5.8.4.   Consisting of a laser light source, two mirrors, one of which is rotated by an ac motor, and a lens-photo-receiver assembly, this scanning laser device has been used to permit an industrial robot to arc-weld curved objects [8].   The incident light beam from the laser (helium-neon) is "swept" across the object surface by the action of the motor-driven triangular mirror.   Note that this occurs three times for each motor revolution.   A lens mounted in front of a photodetector (e.g., a phototransistor) permits light reflected from only one point on the object's surface to be acquired.   Distance from the sensor to this point is determined by synchronizing the ac motor voltage with a high-frequency clock.   The number of clock pulses from the time this voltage is zero until the photodetector receives reflected light is a measure of the distance.   Tracking (in the case of the welding application, for example) is achieved by mounting the sensor on the end effector of the robot, thereby allowing the entire sensor to be moved to different locations in space. Black, transparent, or extremely shiny objects cause problems for this proximity technique.

5.8.2.4 Ultrasonic sensors

Ultrasonics has been used to provide ranging and imaging information for many years.   For example, naval vessels have used sonar sensing systems to detect submerged submarines since the early 1940s.   Also, since the late 1970s ultrasonic imaging has been used to provide "pictures" of various human organs without

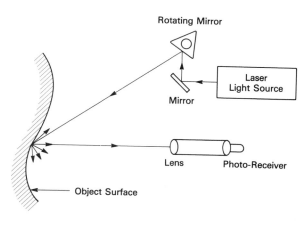

Figure 5.8.4.   Scanning laser proximity sensor.   Object scanning is accomplished by having a motor rotate a mirror at a constant angular velocity. (With permission of M. Ueda.   From reference [8].)

subjecting patients to more objectionable forms of radiation (e.g., x-rays). The advent of Polaroid Corporation's Sonar Sensing element brought ultrasonic ranging to the nonprofessional photographic market. As employed on their instant camera, the Polaroid sensor projects a low (energy)-level electrostatically generated ultrasonic pulse and measures the time for the reflected beam or "echo wave" to return to the sensor. This information is used to measure the distance to the object and then to automatically adjust the camera's focus accordingly. In recent years, this and other similar detectors have been adapted to robots.

Although there are several different sonar sensing techniques, robots normally utilize devices that produce short bursts of a sinusoidal waveform whose frequency is above the audio range (e.g., 40 kHz) [9]. From the block diagram in Figure 5.8.5, the operation of such a sensor can be understood. When an initiate signal is given, the transmit (sinusoidal) oscillator is enabled for 1 ms. This causes 40 cycles of energy at 40 kHz to be transmitted. In addition, a timing or sampling window enables the AND gate for a specific period of time, thereby permitting

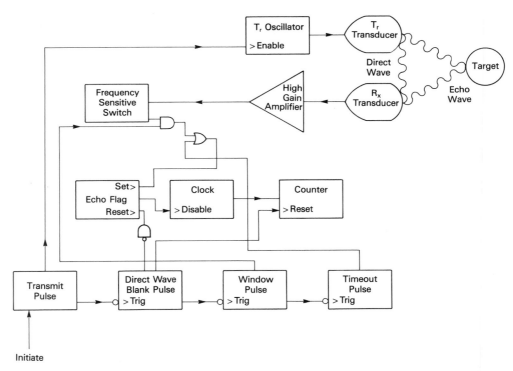

Figure 5.8.5. Block diagram of a pulse sonar, noncontact, proximity sensor. Note that T_r and R_x stand for transmit and receive, respectively. (Redrawn with permission of M.H.E. Larcombe. From reference [9].)

any reflected pulses (from the echo wave) to be counted. This count is proportional to the distance of the object from the transducer. Thus the maximum range of the sensor can be varied by adjusting the sampling window. Errors due to spurious signals are prevented by using a narrowband filter tuned to the frequency of the transmitted beam (e.g., 40 kHz). The direct wave from the transmitter is not counted due to the action of the blanking pulse, which keeps the counter disabled until after the transmit oscillator is off (i.e., disabled).

In the case of the Polaroid device, the burst actually consists of *four distinct high frequencies*: 50, 53, 57, and 60 kHz. This prevents the surface topology or material of the object being scanned from "looking" like a *matched termination* to the incident ultrasonic energy which would eliminate or "cancel" the echo wave, thereby rendering the object effectively "invisible" to the sensor.

The idea behind using a group of frequencies is that the matching effect is frequency sensitive. Thus *some* energy will always be reflected. The "penalty" one pays for this is that the sensing electronics used in conjunction with such a device will be more complex since they must be able to handle four different frequencies rather than a single one as described above. The accuracy of the Polaroid sensor is reported to be about 1% (of its range).

Arrays of Polaroid sonar sensors have been used by the National Bureau of Standards to create a safety "curtain" of sonar energy about an industrial robot [10]. If this curtain is broken by someone attempting to enter the workspace, the robot is immediately halted and must be *reset* before it can resume operation. This scheme prevents an accident even when an intruder is *entirely* within the work envelope. Note that permitting the robot to continue moving once the sonar shield is reestablished would not safely handle such a situation.

Polaroid and other ultrasonic sensors have also been used on a number of mobile robots to determine the distance from walls and obstacles. In one experimental device developed at Drexel University in 1983, the common problem of loss of position calibration due to wheel slippage was practically eliminated using a single sonar sensor rotated 180° by a stepper motor to determine the actual distance from walls and *known* obstacles [11]. The robot's path was taught and followed utilizing distance information from incremental encoders mounted on the drive-wheel servomotors. Recalibration of the system from time to time was achieved "on the fly" by comparing the actual distance from a particular "landmark" (e.g., a wall) reported by the sonar sensor with that given by the encoder. Position counters were then adjusted accordingly. A position error of about 0.5% was obtained using this procedure. In another mobile device, 14 narrow-angle ultrasonic sensors were placed at various locations on the French HILARE robot [12]. The transmission angle of these detectors was 30°, whereas the receiving angle was only 15°. The large number of sensors was required to eliminate "blind spots" around the perimeter of the robot that would otherwise be present due to the narrow-angle detectors used. Measured distances of approximately 2 m with an accuracy of about 0.5 cm were reported.

5.8.2.5 Eddy-current sensors

Another type of sensor used as a proximity switch and for determining the accuracy and repeatability of commercial robotic manipulators operates on the *eddy-current* principle. A typical device of this class utilizes a sensing coil to induce high-frequency (eddy) currents in a ferrous or nonferrous (e.g., aluminum) conductive target. The amplitude of the sensor's generated oscillation depends on the distance between the metal surface and the coil, and this in turn determines the amount of magnetic coupling in the overall circuit. Consequently, position can be obtained by monitoring the amplitude.

One technique for doing this employs a "killed oscillator," [13] as shown in Figure 5.8.6. The presence of a metal target near the coil in the sensor's probe causes the oscillator's amplitude to drop since the induced eddy currents represent a loss mechanism and thus produce damping or "killing" of the sinusoidal waveform. A demodulator, essentially an integrator, responds to such a change by producing a smaller dc output. In a *proximity switch* application, a thresholding circuit is used to detect when the level drops below some predetermined value, at which point the switch's state is changed. By adjusting the threshold value, a robot manipulator can be stopped at a desired distance from a part or object. This scheme can also be used to prevent an inadvertent collision between the end effector and another piece of machinery within the robot's workcell.

Alternatively, the thresholding circuit and switch can be eliminated and the actual dc level from the demodulator used to provide information about the absolute distance of the sensor from the object. This type of device has been employed by some manufacturers to determine both the repeatability and accuracy of their robots. In addition, it is clear that an eddy-current sensor mounted on a manipulator's end effector could be used to inform the robot of its distance from a metallic part. Another potential application would involve the use of such a sensor and a robot to perform various types of inspections on manufactured parts (e.g., shaft

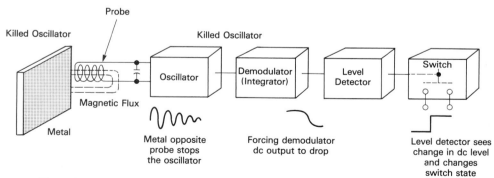

Figure 5.8.6. Killed oscillator, eddy current, noncontact, proximity sensor. By removing the level detector and switch, the device can monitor absolute distance. (Redrawn with permission of *Machine Design*. From reference [13].)

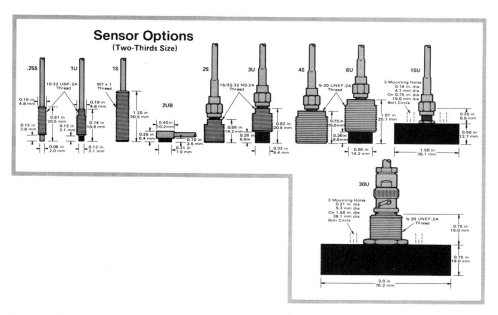

SYSTEM MODEL NO. / SYSTEM PERFORMANCE	KD-2310 -.25S	KD-2310 -1U	KD-2310 -1S	KD-2310 -2UB	KD-2310 -2S	KD-2310 -3U	KD-2310 -4S	KD-2310 -6U	KD-2310 -15U	KD-2350 -30U	Units
TARGET MATERIAL											
Non-Magnetic	Yes	Yes	Yes	Yes	Yes	Yes	Yes	Yes	Yes	Yes	
Magnetic	No	No	No	Yes	Yes	Yes	Yes	Yes	Yes	Yes	
MEASURING RANGE	10	40	40	80	80	120	160	240	600	1200	mil
(+20% Nominal Offset)	0.25	1	1	2	2	3	4	6	15	30	mm
NONLINEARITY	.05	.2	.2	.4	.4	.6	.8	1.2	3.0	6.0	± mil
(½% of Measuring Range)	0.00125	0.005	0.005	0.01	0.01	0.015	0.02	0.03	0.075	0.15	± mm
RESOLUTION – ≤ MID RANGE											
Analog	0.004	0.004	0.004	0.008	0.008	0.012	0.016	0.024	0.060	0.120	mil
	0.0001	0.0001	0.0001	0.0002	0.0002	0.0003	0.0004	0.0006	0.0015	0.003	mm

Figure 5.8.7.   Operational specifications and dimensions for a family of commercial eddy-current sensors.   (Courtesy of R. Dambman and Kaman Instrumentation Corp., Colorado Springs, CO.)

and commutator runout, plate and disk flatness, or the thickness of nonconductive materials).   Gaging of certain metal components would also be possible with this class of sensor.

Eddy-current sensors are ideally suited for the manufacturing environment because they are able to operate reliably in areas contaminated by oil and dirt and also where there is significant variation in temperature and humidity.   They also provide good linearity, with one manufacturer (e.g., Kaman Instrumentation Corp.) reporting a specification of 0.5% of the overall measuring range (see Figure 5.8.7). This same manufacturer also specs analog resolutions of from 0.004 to 0.12 mils, depending on the sensor's range. These sensors can be used to measure static and dynamic displacements and hence are well suited for determining manipulator position overshoot under a variety of load conditions. In addition, they can sense frequency variations of up to about 50 kHz, so that it is possible to use them to measure discontinuities on moving metallic objects.

The major disadvantage of these devices is that they must be calibrated for the type of metal used for the target (e.g., aluminum and steel will produce different sensor outputs). Thus in a robot application, the detector might have to be re-calibrated for each different type of part inspected. Also, the effective linear measuring range is determined by the size of the sensor, with larger distances implying physically larger probes. For example, as shown in Figure 5.8.7, a 1.5-in.-diameter unit is required to sense distances of up to 600 mils, whereas a 10-mil distance needs one that is only 80 mils in diameter.

5.8.2.6 Resistive sensing

A problem encountered in the robotic application of arc welding is keeping the welding tool (also referred to as a "torch" or "gun") tip at a specified constant distance from the seam that is to be welded. By doing this and also keeping constant the speed with which the gun is moved, the uniformity and strength of the weld can be controlled. In addition, a strong weld can be ensured by having the robot accurately "track" this seam. A technique that has been developed to meet these requirements is called *through-the-arc resistive sensing*. The fundamental principle underlying such a sensory technique is that for a constant voltage applied to the welding tool, the arc's resistance (or more correctly, the current) is a measure of the height of the torch tip above the surface that is to be welded. Inductive current monitoring is utilized because welding normally produces large currents (e.g., 100 to 200 A).

In the case of gas metal arc welding (GMAW), more commonly called the metal inert gas (MIG) technique, a blanket of inert gas (e.g., argon, helium, or carbon dioxide) protects the welding torch's electrode, as well as the material being welded, from exposure to the air and, hence, rapid oxidation. The electrode consists of a wire continuously supplied from a drum. The composition of this wire varies depending on the nature of the welding application because the electrode's metal is used as a filler in the MIG process. It is found that for this type of welding, the relationship between the arc current I and voltage V is given by

$$V = R (h - L)I$$

where R = average resistivity per unit length of the electrode wire
$\quad\quad L$ = arc length
$\quad\quad h$ = height of the tool tip from the metal surface.

Normally, V is a constant, so that I is an inverse function of h. Thus, by adjusting the robot's position, the arc current can be kept fairly constant. It has been found,

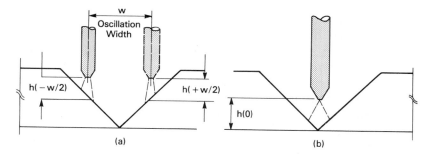

Figure 5.8.8. Through-the-arc "resistive" (position) sensing for automated seam tracking in an arc welding application. (Redrawn with permission of G.E. Cook. From reference [14].)

for example, that in the MIG process, I will vary by about 1 to 1.5% for each 40-mil change in h.

The problem of automated seam tracking has been solved by a modification of the foregoing method (see Figure 5.8.8). In this through-the-arc position sensing, the welding tool is deliberately moved back and forth a small distance across the seam between the two pieces of metal [14]. This action is often called *weaving* and is available as a selectable motion, often with a variable oscillation amplitude, on many robots (e.g., a Unimation PUMA 550). If the center of the seam is being properly tracked, the arc current at the maximum and minimum points on the weave (or oscillation) will be the same. However, if the gun has moved away from the seam's center, these two currents will differ, and this difference signal (or error) can be used to realign the manipulator. The vertical height h can also be controlled by sampling the arc current at the center of the torch oscillation and then comparing this current with some reference value, which has been determined in advance (i.e., off-line) and depends on the type of material being welded. Again, the difference between the reference and actual currents produces an error signal that can be used to readjust the robot's position above the welding surface.

A similar seam and height tracking procedure is possible for the other major type of arc welding [i.e., the tungsten inert gas (TIG) process, also called gas tungsten arc welding (GTAW)]. In addition, other weld tracking systems have been developed, but these are usually of the contact variety. We will discuss them in a subsequent section when tactile devices are described.

5.8.2.7 Other proximity sensors

Besides the noncontact proximity devices described above, a number of others either have potential for or have been used as external robotic sensors. These include:

1. *Capacitive probes*, which utilize a bridge or oscillator circuit to detect changes in capacitance produced by variation in the distance between two parallel plates [15].

2. *Constant charge probes*, which measure the height of a ferrite magnetic head above a rotating magnetic disk by monitoring the change in voltage across the capacitor formed by these two elements. The capacitor is assumed to have a constant charge [16].

3. *Air pressure sensors*, where proximity sensing is obtained by monitoring the change in back pressure produced when an object comes near a jet delivering high-pressure air [17].

4. *Hall effect sensors*, where the magnetic field intensity (e.g., produced by either permanent or electromagnets *mounted on an object*) determines the generated voltage output [18].

5.9 TOUCH AND SLIP SENSORS

Of all the senses that human beings possess, the one that is probably the most likely to be taken for granted is that of touch. It is only when a hand or arm is amputated that the ability to recognize objects and/or adaptively control the grasping force that comes from the human tactile sensory apparatus is truly appreciated. It is therefore not too surprising that in the attempt to imbue robots with some of the attributes of human beings, developments in robotic vision have outshadowed those in the area of touch and slip sensing.

In the last few years, however, as new and more sophisticated applications for robots have been conceived, tactile sensing has been recognized as an extremely important machine sense. In the area of parts handling, for example, it has become increasingly important to be able to detect any misalignment (i.e., the actual orientation) of the parts as they are presented to the robot. In addition, it is often necessary to know *where* a part is being grasped by the robotic gripper and whether or not it is *slipping*. Although vision has been used (or proposed to be used) in this respect, it appears that tactile sensing may be a less costly and faster (computationally) solution to the problem.* Also, a major advantage of tactile sensing over vision is that it can yield the desired information about part position and

---

*Hillis has given three reasons for this [19]:

1. "There are far fewer data to be analyzed than in a visual image." As a consequence, *real-time processing* is possible.

2. "Collection [of data] is more readily controlled." This is the equivalent of analyzing a visual scene where the background, illumination, and point of view are optimally adjusted (e.g., to provide ideal contrast between the part and the background).

3. "Properties that we actually measure are very close, in kind, to the properties that we wish to infer." For example, size and shape can be obtained directly for a tactile sensor but must be inferred if a visual sensor is used.

orientation *within the jaws of the gripper*. Moreover, there are many applications where the limited resolution/pattern recognition capabilities of a tactile device is more than adequate for the desired task. For these reasons, recently there has been a significant increase in research and development in this area both at universities and in industry (robotic and otherwise).

The term *tactile sensing* does not have a universally accepted meaning. For example, Harmon [20] has defined it to be "the continuous-variable sensing of forces in an array." The implication is that the sensor should possess *skin-like properties* and be capable of detecting differing levels of signals and *parallel patterns of touching*. In contrast to this, Harmon refers to *simple contact* or *force sensing*, whether binary or continuously variable at one, or at most a few, specific points as *simple touch sensing*. However, other researchers have chosen to describe as tactile sensors those devices that *do* produce a binary signal from the active elements in the array. That is, thresholding of the signal from an array element is used to determine whether contact has been made. It should be noted that just as in the case of binary machine vision (see Chapter 6), it is possible to obtain a good deal of object recognition information in this instance. Also, because there are usually significantly fewer data, processing time is reduced significantly. In this section we use this modification of Harmon's definition.

In Harmon's report, he presented the results of a survey of 47 researchers and manufacturers in which they were asked to give the desired attributes of an "ideal" tactile sensor. In summary, he found that what these people wanted was that:

1. Tactile sensors should be compliant and rugged (i.e., durable in the manufacturing environment).
2. Sensors (i.e., sensory arrays) should be "smart," meaning that they should process most of the information before communicating with the robot.
3. Sensor resolution should be on the order of 100 mils, although some applications could require a larger or smaller value.
4. The sensors should be able to detect forces as low as 5 to 10 gr, with a dynamic range of about 1000 to 1.
5. Sensor response should be stable (with time), monotonic (and preferably linear, although some nonlinearity would be acceptable), and most important, not exhibit hysteresis (i.e., be repeatable).

The prototype experimental and the (few) commercially available tactile sensors meet some but not all of these requirements. In fact, much research and development remains to be done before a device exists that will permit a robot to lift an egg without crushing it or a block of brass without dropping it when the objects are presented in *random sequence*.

In the first part of this section, we present a number of tactile sensing techniques that have been used as either simple touch or tactile sensors with varying degrees of success. The second portion of the section discusses several methods of slip sensing.

5.9.1 Tactile Sensors

A variety of techniques and materials have been used in an attempt to produce a tactile sensor that is sensitive, rugged, and reliable (i.e., meets the requirements listed above). As of this writing, none do this, although a few satisfy some of items on the list. We will briefly describe a number of devices that utilize different sensing principles. In particular, we discuss an extension of the simple contact rod proximity sensor (see Section 5.8.1) to produce a *three-dimensional* tactile sensor. Other devices covered make use of photodetectors, air pressure, conductive elastomers, or polymers as their sensing elements. The section concludes with a description of several tactile arc welding seam trackers.

5.9.1.1 Proximity rod tactile sensors

As mentioned earlier in this chapter, certain simple proximity-sensing techniques can be extended to produce a robotic tactile sensor. An example of this [21] is shown in Figure 5.9.1, where the single-contact rod proximity sensor has been replaced by an *array* of such sensors (i.e., $4 \times 4 = 16$). A possible mode of operation requires that the robot wrist on which the device is mounted be moved down toward and *parallel* to the table or other surface on which an object is resting (Figure 5.9.1c). Descent continues until the base of the sensor is at a distance approximately equal to the length of the sensing rods above the tabletop (Figure 5.9.1e). At this point, mechanical or electrical switches connected to each of the sensor rods are checked for closure (i.e., contact). In this manner, a two-dimensional or *binary* pattern of the object is obtained. Image processing techniques similar to those employed with binary vision systems can be used to provide object type, shape, and orientation information (see Chapter 6). An appropriate set of actions can then be performed by the robot, [e.g., reorientation of the gripper (if necessary) and closing of its jaws].

A major difficulty with this technique is that the robot must know exactly how far to descend toward the table surface. If it does not go far enough, it is possible that not all of the sensing rods will come in contact with the object. If it goes too far, the table will appear as part of the object. One method of overcoming this problem is to replace the (binary) switches with elements that measure actual distance (i.e., provide *gray-level information*). With such a modification, as the sensor moves toward the object, the rods are once again pushed back into the body of the device (Figure 5.9.1d). However, in this case, the robot stops its descent when *all* rods have moved a minimum (or threshold) distance, thereby indicating that the sensor's elements have come in contact with either the object or the tabletop (Figure 5.9.1e). Measuring the distance moved by each of these rods (relative to their *starting position*) yields a three-dimensional image of the

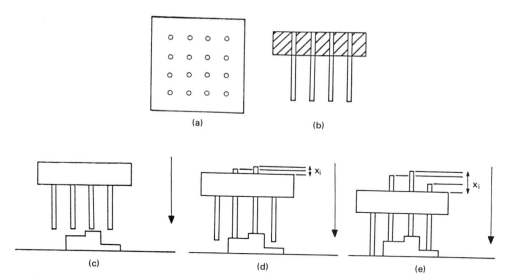

Figure 5.9.1. Three-dimensional proximity rod tactile sensor: (a) top view; (b) side view; (c) sensor descending; (d) sensor in partial contact with object; (e) sensor in full contact with object. (Courtesy of W.B. Heginbotham, Nottingham, England. From Page, C.J. and Pugh, A., "Novel Techniques for Tactile Sensing in a 3-D Environment," *Proc. of 6th Int'l Symp. on Industrial Robots*, University of Nottingham, England, March 24–26, 1976.)

object being "scanned." Gray-level image processing techniques similar to those employed with vision systems can be used for this purpose (see Chapter 6).

This procedure has a problem also. Since the rods must be able to move quite freely, it is possible that false deflections may be obtained. Spring-loading of the rods is possible, but a better solution suggested by the authors is to *vibrate* the tabletop. The robot will then continue to move toward the object until *all rods are vibrating*. At this point, the robot is commanded to stop, the relative rod deflections measured, and the object recognition algorithms used to process these data.

Besides the originally proposed switch sensors, a variety of linear measuring techniques can be used to obtain the relative rod deflections. For example, the authors used rods made of ferrous material. Magnetic detection methods were then used to sense distance (see Figure 5.9.2). This was accomplished by causing the robot to move vertically (using stepper motors) and looking for a rod to move the ferrite cylinder into or out of the sensing coil. (Such an action produced a significant change in voltage across a coil.) The travel distance of each rod could be deduced from the instant each one caused a switch. The state of all the matrix of switches was continually scanned to determine the appropriate switching pattern and length of rod travel. In this manner, the part *contour* was sensed. In a later version of the tactile sensor, it was suggested that each rod be connected to a pot [22]. All the comments of Section 5.2.1 concerning the problems of using a pot are pertinent in this respect. Obviously, many of the other position-sensing methods discussed in earlier sections of the chapter could also be used. However, the more expensive ones (e.g., the LVDT) would not be practical since each rod would require a separate position sensor.

5.9.1.2 Photodetector Tactile Sensors

Among the noncontact proximity sensors discussed in Section 5.8.2, it will be recalled that one used the "beam break principle." Over the last decade, several tactile sensors have been developed utilizing this technique [23]. In 1983, the Lord Corporation described a commercial device [24] shown in Figure 5.9.3. [Actually, only one sensor element is indicated. In reality the sensor would, of course, consist of an array of such elements, (e.g., 8×8 for the Lord LTS 100).] It can be seen that the portion of the sensor that comes in contact with the object to be sensed is covered with an elastomer (a rubber-like material). In addition, a piece of this material extends through the sensor structure. Mounted on the back of the body of the device is a photo emitter-detector assembly (see Figure 5.9.3a). When the object comes in contact with the touch surface, if the elastomer is compressed a minimum distance, the material extending through the body breaks the beam of the photosensor (see Figure 5.9.3b). Obviously, a thresholding circuit can be used to provide binary information about the object, that is, whether or not each element of a sensor composed of such devices is in contact with a part. In a manner similar to that described for the proximity rod contact sensor, two-dimensional information about size, shape, and part orientation can be obtained.

It is also possible to determine information about the relative deflection at each array point. In this respect the reader will recall (from the discussion in Section 5.3.2.2) that the voltage ouput from a photodetector varies with the incident light intensity. Thus, by monitoring the actual signal from the individual photodetectors, the voltage level can be related to distance traveled by the sensing element. Depth (e.g., three-dimensional) information is limited since the overall travel distance is quite small. For example, the elastomer used in the Lord LTS 100 will deflect a maximum of 2 mm. However, the voltage variation at each array point can be related to the *pressure* or *force* being applied by the robotic gripper. Clearly, this is a desirable attribute of such a sensor. The above-mentioned device will sense a force of 1 lb applied to any single sensing site at a full mechanical deflection of 2 mm.

At least two potential difficulties occur with such a sensor. The first has to do with mechanical hysteresis in the elastomer. This implies that the rubber will not return to its *original* position after it has been compressed. For a binary device, proper thresholding of the photodetector signal level will probably minimize the problem. However, this effect will create severe problems with a sensor that is supposed to provide absolute voltage-level information (as is the case with a device that also gives pressure or force data). The severity of the problem is reduced somewhat, however, by the small travel distance (e.g., 2 mm).

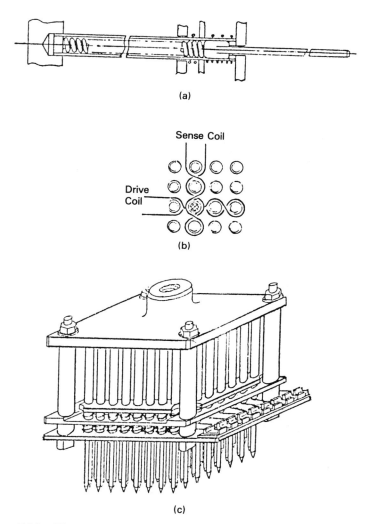

(a)

Sense Coil

Drive Coil

(b)

(c)

Figure 5.9.2. The sensing principle of a 64 (8 × 8) element proximity rod tactile sensor: (a) single spring loaded ferrite cylinder element; (b) top view of the sensing and drive coil configuration; (c) drawing of the actual sensor. (Courtesy of W.B. Heginbotham, Nottingham, England.)

The second problem with using this type of tactile sensor has to do with ruggedness in a manufacturing environment. Since the elastomeric surface must actually come in contact with the object being grasped by the robot, it is quite likely that significant wear will take place. This means that unless the rubber is carefully protected, it will have to be replaced quite frequently. Depending on the actual application, this may or may not be an acceptable solution.

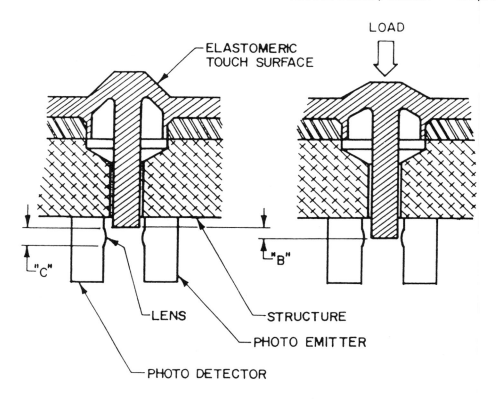

LOAD

ELASTOMERIC
TOUCH SURFACE

"C"   "B"

LENS

STRUCTURE

PHOTO EMITTER

PHOTO DETECTOR

ZERO DEFLECTION   MODERATE DEFLECTION
(a)   (b)

Figure 5.9.3. One element of a photodetector tactile sensor: (a) no load applied
to the sensor's surface; (b) a load results in partial obstruction of the light beam
from the photo emitter causing a change in the photodetector output. (Courtesy
of J. Rebman and the Lord Corp., Industrial Automation Division, Cary, NC.)

5.9.1.3 Conductive elastomer sensors

Elastomers provide varying degrees of compliance and are generally made
from either foams or rubber (natural or silicon based). The Lord sensor described
in the preceding section used one such elastomer as a *deflectometer*. However, it
is also possible to imbue many of these materials with interesting *electrical* properties
(e.g., increased conductivity) by impregnating them with silver or carbon. When
this is done, it is found that their resistance changes as they are deformed (e.g.,
compressed). A number of tactile sensing devices have been developed that utilize
this variation in resistance as their active elements. One such unit is shown in
Figure 5.9.4. It is observed that the device has a three-layered "sandwich" struc-

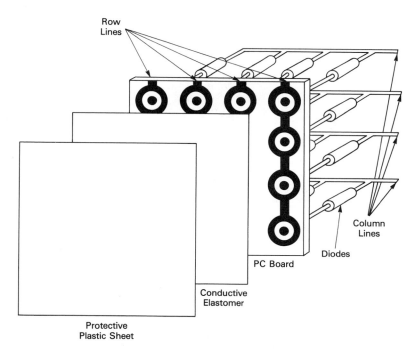

Figure 5.9.4. Conductive elastomer sensor. (Redrawn with permission of W.E. Snyder and reused with permission of the original copyright holder, The Institute of Electrical and Electronics Engineers, Inc., New York City, NY. From reference [25].)

ture consisting of a sheet of conducting elastomer covered with a protective plastic coating. These two parts are placed over a printed circuit board on which 16 pairs of concentric ring electrodes have been etched. Thus there are 16 sensing sites arranged in a 4 × 4 rectangular array. Groups of four outer ring electrodes are connected together (i.e., in parallel) to form the four *rows*. The columns are obtained by first connecting cathodes of the individual diodes to the inner electrodes. Groups of four diode anodes are then connected in parallel. A single fixed current-limiting resistor, connected to a 5-V dc supply, is placed in series with each of these groups, thereby completing the row structure. The elastomer provides a pressure-sensitive resistive path between the inner and outer electrodes. It is important for the reader to understand that the grouping described above for a sensor having $n \times m$ elements is a practical necessity in a robot application since only $(n + m)$ external wires are then needed for accessing individual array sites. This should be contrasted with the situation where *every* array element has an external lead and hence $(n \times m)$ wires must be carried in a wire harness that may be subject to severe twisting motions during normal robot operation.

The method of determining which element (or elements) in the sensing array

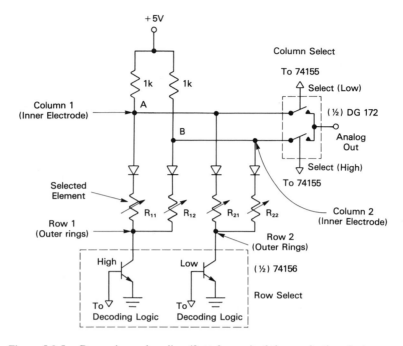

Figure 5.9.5. Row-column decoding (2 × 2 array) of the conductive elastomer sensor in Fig. 5.9.4. (Redrawn with permission of W.E. Snyder and reused with permission of the original copyright holder, The Institute of Electrical and Electronics Engineers, Inc., New York City, NY. From reference [25].)

is in contact with an object can be understood by considering the simplified 2 × 2 array row–column "decoding logic" circuit shown in Figure 5.9.5. The extension of this circuit to larger ones is straightforward. Suppose that only a single array point is being compressed. A row of electrodes is "selected" by causing one group of outer rings to be grounded. For example, the correct 2-bit code applied to the Select lines of a 74156 will cause the leftmost output transistor to be on and all the others off. Row 1 is enabled, and as a result current will flow from the inner to the outer ring for *each* of the two elements of this row (i.e., through R_{11} and R_{12}). In a similar fashion, the proper 2-bit code to the Select inputs of the 74155 causes either the voltage at point A (V_A) or B (V_B) to appear at the Analog Out terminal (i.e., column 1 or column 2 to be decoded). Note that the diodes are required to prevent "cross talk" between adjacent elements in the array from occurring.

If neither of the elements is the one in question, the voltages at points A and B will be almost equal. However, if one of these elements is being compressed (e.g., the one in row 1, column 1), R_{11} will be different from R_{12} and so V_A and V_B will also differ. For example, if the resistance of the elastomer decreases as the material is compressed, R_{11} will be smaller than R_{12} and V_A will be less than V_B. The use of appropriate thresholding circuitry will convert this information into a binary format which can be used for object recognition. However, force and/or pressure monitoring from such a device would require the utilization of the voltage *levels*. Note that in some respects the enabling of the rows and columns of this sensor is similar to decoding of a computer keyboard.

An experimental sensor that also uses a conductive silicon rubber but in a different way has been described by Hillis [19]. The device also has a sandwich structure. However, unlike the previous one, the printed circuit has etched on it a set of *parallel* line electrodes, implying that conduction is in only one direction. The active element used is anisotropically conductive silicon rubber (ACS), which is conductive only along one axis in the plane of the sheet. The PC board and elastomer sheet are mounted so that their individual conduction directions are at *right angles*. The intersection of each of the contact points of the two sets of conductors forms the separate sensing elements.

As shown in Figure 5.9.6, a separator (either made of a fine-mesh nylon stocking or formed from a nonconducting paint) is placed between the PC board and the ACS. With no pressure on the sensor, the ACS does not touch the gold electrodes on the PC board (Figure 5.9.6a). As the pressure is increased, the contact area between the ACS and the electrodes increases (Figures 5.9.6b and c). Since the contact resistance is inversely proportional to the contact area, the larger the pressure, the smaller this resistance. Row and column decoding is accomplished using a scheme similar to the one described above.

Several problems exist with these sensors. One of the major ones is that the elastomeric material has a limited life. It has been found that after a relatively small number of operations (e.g., hundreds), the material begins to deteriorate and cracks develop. In addition, attempting to grasp objects with sharp corners will also damage the elastomer. In either case, the electrical properties are significantly changed, and eventually the sensor is rendered useless. The only remedy is to replace the active material, which may not be acceptable (because of system downtime) in some processes.

Another major problem with many elastomers is that they exhibit significant hysteresis. This means that the array resistances (e.g., R_{11}, R_{12}, R_{21}, and R_{22} in Figure 5.9.5) do not return to the values before compression occurred. Unless a "calibration" is performed before each use, repeatability can be greatly affected by this situation. Even for those elastomers that do not exhibit a permanent hysteresis, these materials often require a long time to return to the uncompressed resistance value. Thus unless there is sufficient *time* permitted between uses, there will still be a hysteresis-like effect. As of this writing, the Barry Wright Company, which has produced a commercial touch sensor that utilizes a conductive elastomer, has stated that it is currently developing special compounds that will reduce or eliminate many of the problems associated with conductive rubber tactile sensors. If successful, such sensors will no doubt become commonplace in the manufacturing environment.

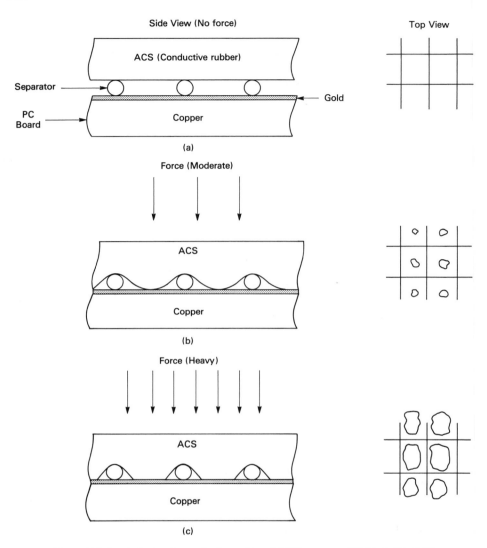

Figure 5.9.6. Anisotropic conductive rubber (ACS) sensor. Views at the right show that the contact area between the rubber and the gold electrodes increases with increasing force (pressure): (a) no applied force; (b) moderate force applied—contact resistance down somewhat; (c) large applied force—contact resistance down significantly. (Redrawn with permission of W.D. Hillis, Thinking Machines, Corp., Cambridge, MA.)

5.9.1.4 Pneumatic switch sensors

IBM has been interested in manufacturing advances for a number of years. As early as 1973, they announced the development of an experimental end effector equipped with 100 pneumatic switches on each of the gripper's "fingers" [26] (see

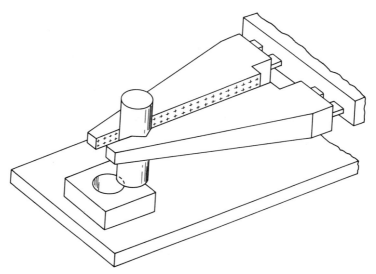

Figure 5.9.7.   IBM pneumatic switch sensor.   Each of the fingers of this experimental end effector has 100 pneumatic switches.   (Redrawn with permission of the IBM Corp., Purchase, NY.   From reference [26].)

Figure 5.9.7).   An exploded view of two of these switches is shown in Figure 5.9.8. The entire sensor is covered with a flexible "skin" whose purpose is to increase the friction between the walls of the gripper and the object being grasped (e.g., a peg).   In addition, it is important that this skin (or covering) electrically insulate the part and the gripper.   Some form of nonconducting flexible rubber or foam is suitable for the application.   Underneath the covering, a *thin* metal sheet (or diaphragm) is secured to the sensor's solid body.   With nothing inside the gripper's jaws, a source of pneumatic (or liquid) pressure keeps the diaphragm pressed against the skin. However, the gripping of the peg, for example, will cause a force (F_1 or F_2) to be applied to a number of switch locations.   If the force at any of the sites is large enough to overcome the applied pressure, the diaphragm will snap to the inward position as indicated by the dashed lines in Figure 5.9.8.   This will, in turn, allow the diaphragm to come into contact with the fixed electrode, thereby completing the electrical path (i.e., closing the switch at that particular site).   The metal sheet will return to its original position (switch open) when the external force is removed due to the internal pressure.

It is clear from this discussion that such a sensor will be strictly binary in nature.   However, the actual force exerted on any part can be varied somewhat by preprogramming the amount of internal pressure and stopping the motion of the gripper's jaws when any single switch is activated.   Sheet dimensions and material type also affect the force required to close the switches.

Although some binary *pattern recognition* is also possible with the sensor, the

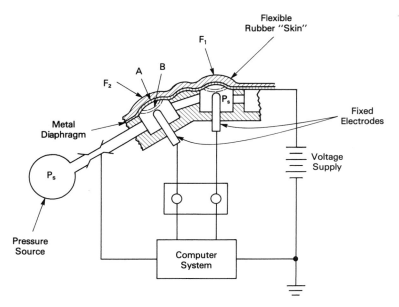

Figure 5.9.8.   Exploded view of two elements of the pneumatic switch sensor shown in Fig. 5.9.7.   The thin metal diaphragm is in position A if the force F_2 is below some threshold whereas it is in position B if the threshold is exceeded.   (Redrawn with permission of the IBM Corp., Purchase, NY.   From reference [26].)

experiment that was actually performed involved the attempted insertion of a peg into a hole.   The peg was placed within the gripper's jaws and moved to the approximate hole position.   The air pressure was set high enough so that the act of grasping the peg caused no switches to be closed.   The (binary) distribution of forces (switch closures) caused by any misalignment of the peg in the hole was sensed, processed by the computer, and this information used to modify the gripper's orientation so as to facilitate insertion.   It is important to understand that since the manipulator has little or no compliance, any misalignment of the peg with the hole would normally prevent insertion.

Despite the promising nature of the pneumatic sensor, IBM did not continue its development.   The most likely reason was its rather limited dynamic range (i.e., 0 to 50 gr).

5.9.1.5 Polymer tactile sensors

Piezoelectric materials have been considered as another possible active element in a number of experimental tactile sensors for robotic applications.   The property that makes them potentially attractive is that they *generate* a voltage when their dimensions are altered. Of the many such materials that are available, the polymer polyvinylidene fluoride, referred to as either PVF2 or PVDF, is one that has been most extensively used. The major reasons are that it is extremely rugged, even in the thin sheets used for the sensing application, and very sensitive, capable of producing large voltages for extremely small mechanical deformations.

A thin sheet of PVDF (e.g., 28 μm thick) can be considered as a capacitor when it has electrodes on both of its surfaces. If a pattern of such electrodes is deposited directly on the PVDF film, the *developed charge* at the compressed sensing point will flow through the cross section of the particular row and column electrodes which make up the capacitors. It is important to understand that the PVDF is capable of producing an output when it is either compressed (the "direct mode") or caused to bend (the "stretch mode"). For thin films, the latter mode has far better sensitivity (i.e., large response for small sensor deflection).

One characteristic of PVDF is that when bending takes place, the developed output is transient in nature. That is, the charge will "leak" off the capacitor, causing the voltage eventually to return to zero. Fortunately, the time constant of this behavior is on the order of many seconds so that detection of the maximum signal is not too difficult. Sensors utilizing this material obviously can only monitor *changes* in force (or pressure).

Another important property of PVDF is that is is *pyroelectric* in nature. That is, changes in temperature will produce a voltage response, and as a consequence extremely sensitive fire detectors have been built to take advantage of this behavior. It is possible that a robotic sensor could be constructed based on the pyroelectric effect, and in fact one has been proposed [27]. However, it is extremely likely that such a sensor would be of the proximity type because poor resolution would, no doubt, limit its ability to yield tactile information. Alternatively, it could be combined with another class of sensor to provide the desired tactile data. It is important to note that any sensor based on the piezoelectric effect would still be sensitive to changes in the temperature of the objects being grasped by a robotic gripper. In addition, changes in the ambient temperature itself will also cause PVDF pyro-generated voltages, although this is not a problem because these variations tend to occur over long periods of time.

As mentioned, most of the work toward developing a PVDF tactile sensor is experimental in nature. For example, a major effort is currently under way at Carnegie–Mellon University, where a 4 × 4 prototype has been constructed and tested. Also, at the Centro "E. Piaggio" at the University of Pisa, Italy, an array of 16 × 16 circular sensing elements has been used as a *binary* (i.e., "simple touch") sensor to yield information about position, orientation, and shape of a part [28]. The device depicted in Figure 5.9.9 uses a sheet of PVF2 that is 100 μm thick. The sensor is rugged, low in cost, and sensitive. However, there are difficulties with circuit reliability and EMI.

Another project, under the direction of one of the authors, involved the joint participation of U.S. Robots, Inc., Temple University, and Drexel University. The

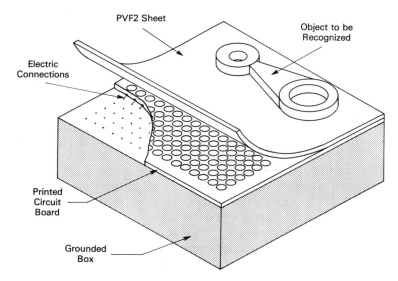

Figure 5.9.9. A 16 × 16 element PVF2 binary tactile sensor. (Redrawn with permission of P. Dario. From reference [28].)

design of a prototype sensor is shown in Figure 5.9.10. Here it can be seen that a multilayer sandwich structure is utilized. A protecting sheet (of nonconducting plastic, for example) is used to cover the film of PVDF. Metallization of the polymer is performed to produce the sensor's row electrodes. Layers of nonconducting silicon rubber are used to provide a certain amount of compliance and to prevent current from passing through any uncompressed sensing points. A PC board containing the column electrodes also provides mechanical stability to the overall package. At a compressed sensing location, the conductive rubber element contacts the column electrode so that any developed charge caused by deforming the PVDF can be read out.

A major concern with this design is the potential difficulty of efficiently manufacturing such a unit. Another problem (which is common to all such sensors) involves crosstalk, whereby uncompressed elements will appear to be activated because of a closed loop of capacitors. A possible solution is to utilize a set of multiplexing chips on board the sensor itself. The use of diodes does not appear to be feasible here (see Section 5.9.1.3), due to the large capacitance (compared to that of each sensing site) introduced by these elements. Even the multiplexers may not completely solve this problem since they also have a certain amount of capacitance (e.g., on the order of 5 picofarads) which could still permit some crosstalk to occur. Finally, one last difficulty with PVDF sensors is that it is quite important to control the *thickness* of the bonding material used (e.g., epoxy). Silkscreen techniques seem to be applicable in this case.

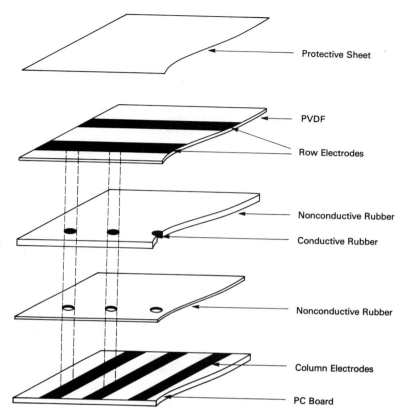

Figure 5.9.10. Tactile sensing device utilizing PVDF. A multilayered sandwich configuration also employs both conductive and nonconductive rubber. (Courtesy of U.S. Robots, Inc., Temple University, and Drexel University.)

5.9.1.6 A hybrid tactile sensor

The prototype sensor shown in Figure 5.9.10 actually employs both PVDF and a conductive rubber. It can therefore be considered to be a "hybrid" device. However, the conductive rubber is not used because its resistance varies with the applied pressure. Instead, this material provides a compliant contact that permits the developed charge to be sensed. There is, however, a true hybrid experimental sensor that has been developed at Bonneville Scientific. The prototype unit described by Grahn and Astle has 12 array elements with a 1-mm resolution [30]. It was reported to be able to detect pressures ranging from 0.15 to 300 psi (i.e., a 2000:1 dynamic range).

This device is essentially ultrasonic in nature, but it does utilize certain properties of both PVDF and silicon rubber in its design. To understand the operation of the sensor, consider the array construction shown in Figure 5.9.11. Here the

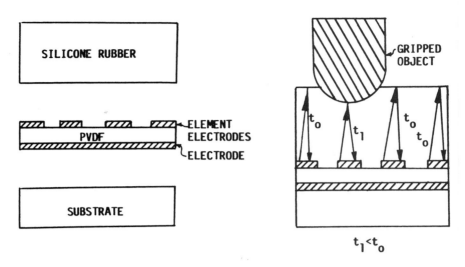

Figure 5.9.11. Array construction and principle of operation of an ultrasonic hybrid tactile sensor. The time for a reflected pulse to be sensed is related to the silicone rubber pad thickness and hence to the applied force. (Courtesy of L. Astle, Bonneville Scientific, Salt Lake City, UT. From reference [30].)

PVDF is used as an ultrasonic transducer that generates *pulses* of ultrasonic energy (see Section 5.8.2). This is accomplished by applying an electric field *to* the polymer, which causes it to mechanically deform. Nonconducting silicon rubber is placed over this transducer and used not only to protect it but also for compliance. However, the major role played by the elastomer pad is to provide a path for the incident and reflected acoustical (ultrasonic) wave. The pad thickness can be determined by monitoring the time for a reflected pulse to return to the receiver. If no object is being grasped, it will be t_0. However, the compression of the pad will reduce this time to t_1 ($<t_0$). An ($n \times m$)-element array of such sensors will provide some information about the size, shape, and orientation of the object using pattern-recognition techniques. In addition, if the relationship between the applied force as a function of the pad's compression is known a priori, it is also possible to obtain information about the actual force from the thickness data.

A similar prototype ultrasonic sensor utilizing PVDF was tested by Rossi et al [27]. Such a device was combined with the polymer tactile sensor mentioned above. Results of this work are quite preliminary.

Clearly, this class of sensor has a number of potential problems. The major ones involve the effect on object shape and applied force determination due to changes in the silicon rubber's *acoustic* properties. These result from a variety of causes, which include aging, temperature variations, fatigue, and improper use (i.e., sharp edges on objects being grasped). It is important to understand that anything that modifies the round-trip travel time of the sonar pulse in a *nonpredictable* manner will create an error.

5.9.1.7 VLSI-conductive elastomer sensor

One of the desired attributes of a commercial tactile sensor is that it be "smart" (i.e., that most of the signal/data processing be done "on board" before communication to the robot controller is made). One experimental device that attempts to do this has been developed by Raibert at Carnegie–Mellon University and is shown in Figure 5.9.12 [31]. In essence, the top portion of the VLSI chip has a thin layer of glass insulation into which an array of holes has been etched. As pressure is applied to the device, the conductive elastomer comes in contact with a set of aluminum electrodes that lie beneath the glass. Embedded still further under this sandwich of sensing elements are a set of "local processors" which perform the appropriate deformation and pressure determination. This is accomplished by comparing the sensed current flowing between the electrodes and the elastomer with a small "test current."

Recently, a 3 × 6 element device has been built that has a sensing area of about 1 mm². A major problem with the use of VLSI technology is that it is not now possible to produce such chips that are 100% defect free, that is, where all of the sensing sites are working. A possible solution to this dilemma is to utilize *redundant* design in the chip, thereby allowing for multiple computing elements at each sensing site. In addition to this difficulty, all of the problems with conducting elastomers that were described previously are still of concern.

5.9.1.8 Optical tactile sensors

As mentioned previously, Chapter 6 is devoted entirely to the important topic of robotic machine vision. However, several newly developed *tactile* devices utilize optical techniques. In fact, we have already described one device that utilizes an array of photodiodes and photodetectors to obtain tactile information (see Section 5.9.1.2). In this section we discuss two sensors, one based on light traveling through an acrylic plate and the other on arrays of optical fibers.

British Robotic Systems of London, England, with the assistance of the University College of Wales (Aberystwyth), has developed a high-resolution tactile sensor that utilizes light conduction through a light pipe and a charge-coupled-device (CCD) imaging system. As shown in Figure 5.9.13, the device consists of a compliant membrane which is separated by a small air gap from a light guide made from a piece of clear acrylic [32]. With the membrane undeformed, light entering from the left side undergoes almost total reflection at the air/acrylic boundary. As a result, little, if any, light reaches the CCD imager (see Figure 5.9.13a). However, if a part is placed on the sensor, as indicated in Figure 5.9.13b, the area where the membrane touches the acrylic will cause light to scatter. Some of the scattered light rays will incident the opposite surface of the acrylic at an angle that permits the rays to pass through the guide. They are then focused onto an array of CCD devices in a manner similar to that employed in solid-state video cameras. Standard optical image processing techniques can be used to determine the nature of the object. In addition, since the intensity of the reflected light (i.e., the degree

PHYSICAL LAYOUT

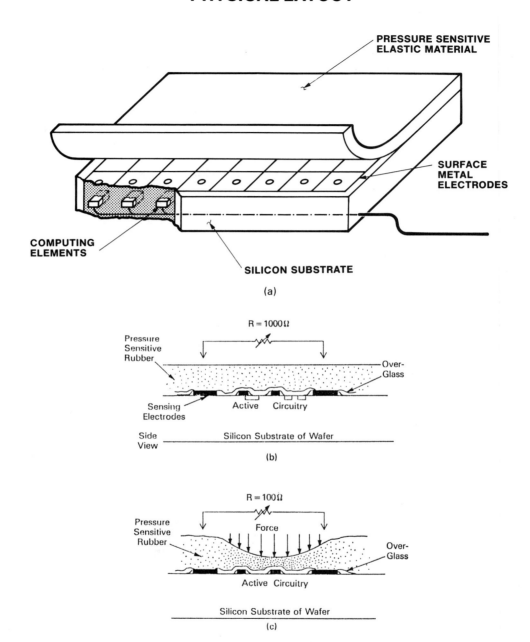

Figure 5.9.12. VLSI-conductive elastomer tactile sensor: (a) physical layout of the device; (b) side view—no applied force; (c) side view—an applied force reduces the resistance by an order of magnitude; (d) electronic sensing circuit for one element. (Courtesy of M.H. Raibert. From reference [31].)

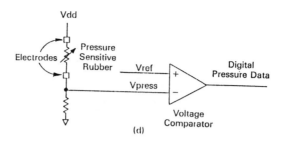

Figure 5.9.12. (continued)

of scattering) is approximately proportional to the pressure on the membrane, a sensor made in this way is capable of producing force information.

A prototype device has been constructed that measures about 0.64 in. by 1 in. Within this area, there are currently 64 × 64 (4096) individual sensing points, which provides extremely fine resolution (perhaps much more than is actually required, in fact). It is reported that an array of 256 × 256 points is also quite feasible. In the initial experiments, two such sensors have been placed on the fingers of a parallel-jaws gripper. Information provided from the devices has been processed and displayed on a video monitor. Eventually, however, it is to be expected that these data will be used in a detection and force feedback scheme. It is clear that this scheme offers the possibility of producing a rugged tactile sensor that can be used in a manufacturing environment.

The second optically oriented tactile device is currently being developed at MIT under the direction of T. B. Sheridan and J. L. Schneiter. The first prototype unit that they built utilized 34 layers, each containing 35 optical fibers, which were individually epoxied together. The fibers in every other layer acted as receivers, whereas those in the remaining ones were emitters. A white silicon rubber "skin" was used and provided compliance and even more important, a light-reflecting surface. With nothing pressing against the sensor, each of the receiving fiber optic bundles collects light from the various emitting layers. The output of these receivers is sensed by a standard TV camera and then digitized for subsequent processing. This represents the ambient or quiescent signal level from the sensor. When something presses on the rubber skin, reflected light picked up by the fiber bundles increases in those areas of the sensor that are in contact with the object. This occurs because the distance between the reflecting surface and the receiving layers is decreased, thereby reducing the amount of scattered light. Consequently, there is a change in the ambient signal level which can easily be determined digitally.

A second prototype based on the identical sensing principle utilizes the *same* fiber as both an emitter and a receiver. This is accomplished through the use of a beam splitter. In addition, there are only 25 layers, each having 26 optical fibers. In either unit, a major advantage of this approach is that it is possible to achieve high resolutions because of the ease with which fiber optic bundles can be "stacked." In fact, Sheridan has stated that densities of up to 2100 sensing points per square

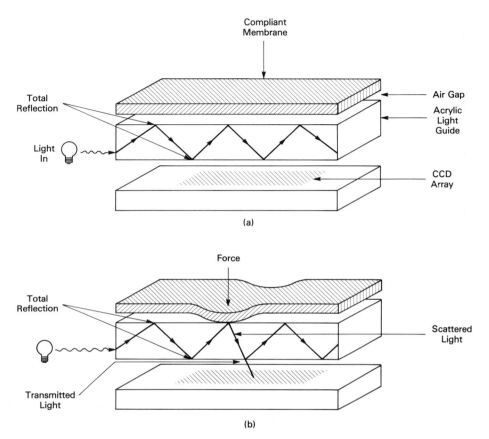

Figure 5.9.13. Acrylic light guide and CCD imaging sensor. The focusing arrangement is not shown: (a) input light undergoes total internal reflection when the compliant membrane is not deformed; (b) a force applied to the membrane causes light to scatter. This scattered light is focused and collected by the CCD array. [Reprinted and modified with permission of *Electronic Design*, Vol. 33, No. 3 (Feb. 7, 1985). Copyright 1985 Hayden Publishing Co., Inc. Hasbrook Heights, NJ.]

inch are possible. Note that the British device described above has an even higher resolution.

5.9.2 Robotic Arc Welding Sensors

An important robotic application is that of arc welding. It will be recalled that there exist a number of special-purpose robots that do nothing but this particular task (e.g., the Unimation Apprentice). However, it is also possible to utilize the continuous-path programming option that is available on other robots to perform such a job.

Regardless of which type of robot is employed, a real problem often encountered is that the programmed path may not conform *exactly* to the actual path required in order to produce a good weld. Such a situation can occur because of part misalignment or due to small dimensional variations in castings. Consequently, there is a need for a sensor that can be used to inform the robot when these paths differ, and hence permit some degree of repositioning of the welding tool to be implemented "on the fly." We have already described a noncontact technique whereby the seam was tracked by monitoring the arc current at the maximum and minimum points of tool-tip travel in a direction transverse to the actual weld curve (see Section 5.8.2).

Also, a number of *contact* sensing methods have been developed for this purpose. Although these devices do not provide much information about the nature of the object being welded, they do yield enough data to permit high-quality welds to be made. In addition, of all the external sensors in use today on robots, those utilized for seam tracking are probably the most advanced.*

In this section we consider a number of such devices. It is interesting to note that the contact techniques to be described here were developed in Yugoslavia, whereas the noncontact procedure introduced in Section 5.8.2 is American, having been developed by Cincinnati Milacron.

5.9.2.1 Simple active optical seam tracker

A technique of active seam tracking that was introduced in 1981 is shown in Figure 5.9.14 [33]. The device works in the following manner: The needle of the touch sensor is initially placed in the seam to be welded by commanding the robot to lower the tool-sensor assembly until contact is made with the metal surface. As welding commences, the manipulator begins to move along the preprogrammed path. Any discrepancy between this and the actual seam results in the sensor needle being displaced in a direction transverse to the seam. A special optical sensor located inside the body of the seam tracker detects this motion and provides the necessary error signal to the robot controller so as to permit modification of the programmed path.

The details of the special optical sensor, used in conjunction with the passive seam tracker, are shown in Figure 5.9.15. It can be seen that this sensor consists of a set of six photoemitters (LEDs) and photodetectors (photodiodes) together with a Gray code mask and slit assembly. When the needle is first placed in the seam, the narrow viewing window provided by the slit yields a particular 6-bit binary "word" which can be noted or stored in memory (i.e., calibration phase). By periodic sampling of the photosensor outputs (i.e., the binary word), the deviation from the initial position can be determined as the needle moves transverse to the seam. Since six bits are used, 64 different positions can be sensed using

---

*This will, no doubt, change in the next few years as more effort is expended on the development in other areas of robotic sensing.

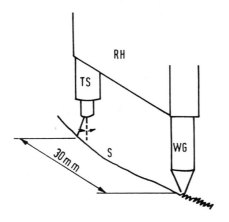

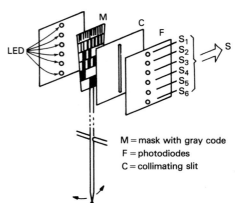

M = mask with gray code
F = photodiodes
C = collimating slit

Figure 5.9.14. A simple active optical arc welding seam tracking sensor. RH = robot hand, TS = tracking sensor, WG = welding gun, and S = seam to be welded. (Reproduced courtesy of S. Presern, University of E. Kardelj, Jozef Stefan Institute, Yugoslavia. From reference [33].)

Figure 5.9.15. Optical sensor used in the active optical seam tracker shown in Figure 5.9.14. (Reproduced courtesy of S. Presern, University of E. Kardelj, Jozef Stefan Institute, Yugoslavia. From reference [33].)

such a scheme. Positional resolution for the system is reported to be about 0.5 mm. It will be noted that this device is actually a form of "linear encoder" (see Section 5.3.2.2).

The choice of a Gray code rather than straight binary is based on the need to minimize detecting errors. The reader should understand that this code permits only a *single* bit change for consecutive binary numbers, and hence any situation that results in more than a 1-bit variation can be interpreted as an error condition. It is important to realize that because of the extremely high level of EMI present in the vicinity of a welding sensor [due to large arc currents (e.g., 100 to 200 A)], processing errors are possible. The use of an optical detection system rather than one based on induction principles, and Gray code, instead of a conventional binary code, minimizes the probability of sensing errors.

5.9.2.2 Passive seam tracking sensor

One of the difficulties with the previous seam tracking procedure is that it cannot be used to monitor path variations in the *z*-direction. That is, if the surface metal being welded has a different contour from that originally programmed, it is possible that the sensing needle will either move out of the seam or be driven down with enough force to damage the detector. Springs could, of course, be used to prevent this from occurring to some extent. However, where fairly large variations can occur, it is important to be able also to monitor the needle's actual vertical position.

A simple passive device that permits this to be done is shown in Figure 5.9.16. The sensor utilizes two LVDTs to detect variations in height (z) and one direction transverse to the seam (either $+x$ or $-x$ but not both). The needle is guided by the seam itself (i.e., "passively" with the springs helping in this respect) [34].

In actual use, the signals from the LVDTs are sampled periodically. When the needle is first placed in the seam, any change in their initial (or calibration) value is used as an error signal, which causes the robot to reposition the welding tool. A sensor utilizing these detectors has a positional resolution that is dependent on the resolution of the LVDT employed. Typically, this would be much finer than the one described in the preceding section.

This type of arc welding sensor has the advantage of being fast, accurate, fairly robust, and relatively simple. However, there are also a number of problems. The major one is that, as can be seen from Figure 5.9.16, the device is one-sided. Thus it is necessary to rotate it in order to sense deviations from the taught path that are on the opposite side of the seam. Since it is not always possible to predict which way such variations will occur, this can create difficulties. A second problem

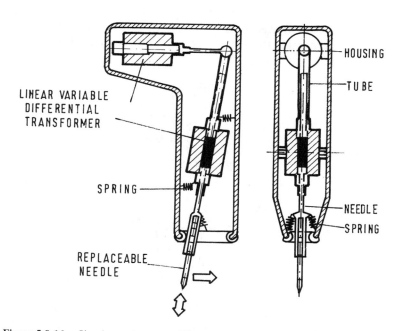

Figure 5.9.16. Simple passive arc welding seam tracking sensor with two degrees of freedom. (Reproduced courtesy of S. Presern, University of E. Kardelj, Jozef Stefan Institute, Yugoslavia. From reference [34].)

is with the LVDT itself. The reader will recall that one characteristic of these detectors is that they are sensitive to changes in temperature (see Section 5.2.6). Periodic recalibration of the unit might be required in order to prevent temperature-related calibration drift from causing position errors and hence poor welds. The last difficulty with such a sensor comes from the requirement that the needle must be in contact with the seam. This produces added drag (i.e., friction) and therefore increased torque must be supplied by the robotic actuators. Moreover, the added friction will also cause the tip of the needle to wear. A simple solution to this problem is to employ a small ball on the end of the needle.

5.9.2.3 Active nonoptical seam tracking sensors

Recognizing the problems associated with the passive seam tracker discussed in the preceding section, the same Yugoslavian group of researchers developed a sensor that overcame most of its difficulties [34]. With regard to Figure 5.9.17, the reader will observe that the device is *active* because it has the means by which the sensing needle is moved. That is, it does not depend on the seam itself to produce needle motion, unlike the two other detectors described above. A 9.8-W micromotor weighing 180 gr periodically lifts the needle out of the seam and then back into it, thereby reducing drag on the overall system. The rate at which this is done is about 10 Hz, although the frequency can be increased if a complex path is to be followed. A piezoelectric crystal is used to determine when the needle is contacting the seam. The signal from an LVDT is sampled at this instant

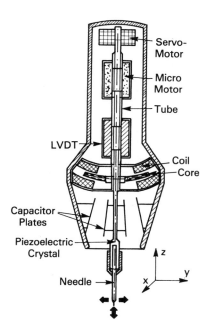

Figure 5.9.17. Nonoptical active arc welding seam tracking sensor with three degrees of freedom. Reproduced courtesy of S. Presern, University of E. Kardelj, Jozef Stefan Institute, Yugoslavia. From reference [34].)

and provides a measure of the z-position.   To compensate for unforeseen varations in the metal surface's height, a small dc servomotor is also employed to keep the *average* z-position of the sensor near enough to the seam so that the micromotor's action permits the needle always to touch the welding surface. X and y displacement of the sensor is accomplished by selectively energizing a set of coils, which in turn attracts the appropriate magnetic core.   (This action is similar to that of a solenoid-activated relay.)   Position is monitored using the capacitive plates shown in the figure.

The ability to drive the sensor element in the x- and y-directions is especially useful in cases where there are large seam gaps.   The data obtained in this instance can be used to increase the weave amplitude of the robot, thereby permitting the wider seam to be "filled."   Another advantage that this active seam tracker has over the other two welding sensors described previously is that it is less troubled by rough seams/surfaces.   This is due, primarily, to the *discrete* rather than *continuous* nature of the sensing method employed.

The major problem with the active, nonoptical, arc welding sensor is that in the sometimes less-than-ideal manufacturing environment, dirt tends to fill the space between the capacitor plates.   This can create x- and y-direction measurement errors.   One solution to the difficulty is to clean the sensor periodically. Another might be to seal it hermetically, although the authors do not suggest such a technique.

5.9.3 Slip Sensors

One of the capabilities of the human hand certainly taken for granted is its ability to determine when an object that is being grasped is slipping.   The biological control system associated with the hand utilizes the inputs from the appropriate slip receptors and causes the gripping force to be increased or decreased, as the case may be.   Machine determination of slippage of a part or object when in the grasp of a robot or other electromechanical "hand" is still in the experimental stage.   Of all the "external" robotic senses, slip detection is perhaps the least developed, and in fact, much of the research in the field has been oriented toward prosthetic applications.   It is the firm belief of the authors, however, that this situation will change.   Most certainly, in the next few years, as more complex and sophisticated assembly applications become commonplace, it will be necessary to detect slip rapidly and to adjust the gripping force "on the fly" to prevent the part from being damaged in a fall.

Perhaps the simplest way to determine if a part is slipping (or has not been properly grasped) is to use what is often termed the *lift-and-try technique* (see Figure 5.9.18).   This entails using the *motor current* of a particular joint or set of joints on a robot as a measure of whether or not a part is slipping.   In this respect, current monitoring can be performed either digitally or in an analog manner. Regardless of which technique is employed, the gripper is first oriented correctly, next placed over the particular part, and then a certain minimum grasping force

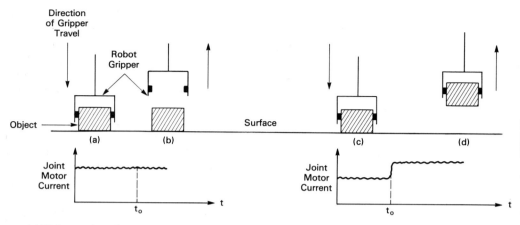

(a) Minimum gripper force applied @ $t = t_o$

(b) Part is not lifted since force is not large enough. Joint motor current does not change as gripper is raised above surface.

(c) Gripper descends again and force is incremented.

(d) Part is successfully lifted [force is high enough]. Joint motor current is seen to increase @ $t = t_o$ indicating that gripper is loaded.

Figure 5.9.18.   Lift and try technique for slip detection.

applied.* As the manipulator attempts to lift the object in question from the surface (e.g., a pallet, table, or conveyor) the motor current in one or more joints should *increase* due to the added load torque. If no increase is detected, the manipulator is commanded to return to the starting point. The force is then incremented by some predetermined amount and the robot "tries again." The procedure is repeated until the monitored joint current does increase, at which time it is assumed that the part is not slipping and is properly grasped.

   There are obvious difficulties with this technique. The first is that even if the part is successfully raised above the resting surface, there is no guarantee that it will not slip out of the gripper as the manipulator moves. If, in fact this occurs, the procedure outlined above will not detect the slippage *while the robot is in motion*. A second problem is that if a fragile part is to be lifted, the minimum applied gripping force should be small, to avoid crushing. On the other hand, a heavy, more robust part could easily handle a larger initial force. If the mix and order of parts to be lifted were not known a priori, it is possible that either damage could be done to some or that it would take far too long to acquire others. The last difficulty with the technique is that monitoring motor current is not always error free. Care would have to be taken to prevent spikes due to brush noise

---

*This assumes that the gripping force can be controlled. It is important to realize that as of this writing, few if any commercially available grippers have such an ability. However, it is expected that as more sophisticated sensors are developed, this situation will change.

from being mistaken as a current above the "lifting threshold." Obviously, the use of brushless motors would reduce this problem but would increase the cost, due to the need for electronic commutation.

In addition to the lift-and-try procedure, a number of experimental devices based on optical, magnetic, or conductive sensing techniques have been developed. We now describe briefly several of these slip detectors, which were proposed by groups working in Japan and Yugoslavia.

5.9.3.1 Forced oscillation slip sensors

In 1972 a group of researchers working at the Nagoya University in Japan reported on a number of experimental devices that were developed for detecting the slippage of an object being held by a robotic gripper. A variety of sensing techniques were employed. For example, "forced oscillation" was used whereby any translation of the part in a direction *tangential* to the surface of the gripper jaws (i.e., orthogonal to the direction of the applied gripping force) caused a short burst of voltage (i.e., a "spike") to be generated. One method of accomplishing this is shown in Figure 5.9.19 and is very much like that used to play back signals recorded on a (analog) phonograph record [35].

Here a sapphire needle protrudes from the surface of the sensor and is in contact with the object being grasped. If the part begins to slip, the needle will be displaced and will produce mechanical deformation in a piezoelectric crystal (e.g., Rochelle salt). The resultant generated voltage spike can be sensed using a threshold detector and the gripping force increased incrementally until the part stops slipping.

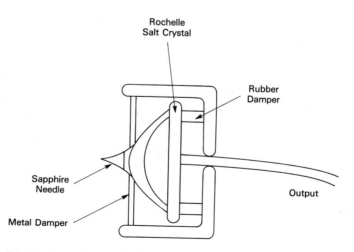

Figure 5.9.19. Prototype slip detector using a sapphire needle and piezoelectric (i.e., Rochelle salt) crystal to produce a forced oscillation. (Redrawn with permission of IIT Research Institute, Chicago, IL. From reference [35].)

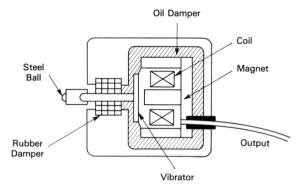

Figure 5.9.20. Improved forced oscillation slip sensor. (Redrawn with permission of IIT Research Institute, Chicago IL. From reference [35].)

There are a number of difficulties with this type of slip sensor. The major one is that, although the rubber damper shown in the figure makes the device less sensitive to non-slip-related motions, the unit still tends to respond to mechanical vibrations of the robot manipulator itself. It is important to note that such a detector must be able to sense slippage accurately *while a part is being moved*. Thus it is necessary to reduce as much as possible this sensitivity to non-slip-related motions. Another obvious disadvantage of the sensor described above is that the needle must move against the surface of a part in order to detect slip. This will cause eventual wear, meaning that periodic replacement of the needle will be required (just as in the case of the phonograph stylus). Finally, the entire device tends to be easily damaged if it is dropped or rapidly decelerated (also like a phonograph pickup cartridge).

The unit shown in Figure 5.9.20 incorporates a number of improvements which eliminate most of the difficulties described above [35]. In particular, the sapphire needle has been replaced by a more robust steel ball (0.5 mm in diameter). The fragile crystal transducer has also been eliminated and in its place a permanent magnet–coil arrangement is used. Any motion of a part along the gripper's jaws will cause a mechanical displacement of the coil which will, in turn, produce a voltage output. Again, the use of dampers (both rubber and oil) and thresholding circuitry will reduce the sensitivity of the sensor to manipulator vibrations.

5.9.3.2 Interrupter-type slip sensors

The same Japanese research group also developed additional prototype slip sensors using two technqiues borrowed from other types of sensors [e.g., optical (or magnetic) encoders or interrupters]. One such detector is shown in Figure 5.9.21. This device consists of a rubber roller that protrudes above the sensor's surface and contains a small permanent magnet. Any slip of the part in the gripper's jaws will cause roller rotation. As the permanent magnet passes over the magnetic sense head (e.g., one used in a tape recorder can be employed), a pulse is generated and the gripping force can be increased accordingly.

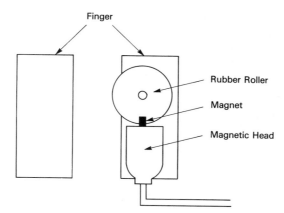

Figure 5.9.21. Magnetic roller type slip sensor. (Redrawn with permission of IIT Research Institute, Chicago, IL. From reference [35].

A major disadvantage of this technique is that the slip "resolution" is not good. It can be seen from Figure 5.9.21 that if the sensor is not placed initially in a "reset" position (e.g., magnetic section of the roller over the read head), it will take some fraction of an entire rotation before any slip is detected. Even if the reset position is used to begin with, so that slip will then be detected as soon as almost any rotation occurs, further slip will *not* be sensed until one additional roller revolution occurs.

A possible solution to the resolution problem would be to use a number of magnets rather than one. These elements might also be replaced by small pieces of ferromagnetic material symmetrically embedded around the circumference of the rubber roller. In addition, the "read" head would consist of a simple dc energized coil. In this case, part slippage would cause a change in reluctance (and a corresponding voltage pulse at the terminals of the coil) after only a small fraction of a full rotation. Continued slip would also be sensed in this manner without having to wait for a complete roller revolution.

A second interrupter-like device uses optical rather than magnetic detection (see Figure 5.9.22). In this instance the rubber roller of the previous sensor has

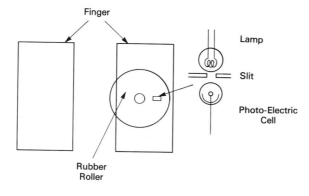

Figure 5.9.22. Optical interrupter type slip sensor. The photoemitter-detector assembly consisting of the lamp, slit, and photo-electric cell is shown in detail at the right. (Redrawn with permission of IIT Research Institute, Chicago, IL. From reference [35].)

a small slit placed in it.   A photoemitter–detector assembly is employed to detect any part motion.   Clearly, the device has the same resolution difficulty as described above.   Again, an obvious way to increase the resolution would be to place more slits in the roller.

5.9.3.3 Slip sensing "fingers"

In addition to the slip sensors described above, the same researchers also attempted to model the slip detection capabilities of the human hand by replacing neurons with pressure sensors (see Figure 5.9.23).   One type of pressure transducer utilized in experiments is shown in Figure 5.9.24.   If a slip occurs, the protruding needle will distort the hard rubber diaphragm.   The resultant pressure change will be detected by a standard pressure sensor and the generated electrical signal used to increment the grasping force until the part is stationary within the gripper's fingers.

A possible problem with this scheme is that the action of increasing the gripping force to stop slip may, itself, produce a change in pressure, which might then be interpreted as additional "slip."   The use of appropriate signal processing techniques would probably reduce or eliminate this situation since the frequency content of the slip response signal would, no doubt, differ considerably from that produced when the grasping force was increased.   Another obvious difficulty with the mechanical device shown in Figure 5.9.23 is that it is not possible to obtain the density of human hand receptors using such pressure sensors with present technology.   While prosthetic devices may need a large number of slip-sensing sites over a small surface area, manufacturing applications will generally require considerably fewer, so that this may not be a serious limitation for an industrial robot.

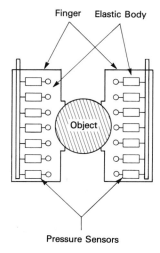

Figure 5.9.23.   Engineering model of human fingers.   (Redrawn with permission of IIT Research Institute, Chicago, IL.   From reference [35].)

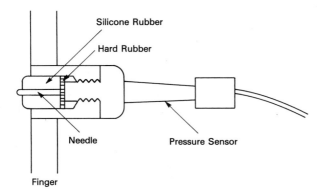

5.9.3.4 Belgrade hand slip sensors

As part of a research program aimed at developing a prosthesis that closely approximated the function and form of a human hand, a number of slip-sensing devices were introduced in the mid-1970s by workers in Belgrade, Yugoslavia. Although they differ from one another somewhat in form, all utilize the technique of forced oscillation to produce an electrical signal that is used to sense slip.

The first sensor developed is shown in Figure 5.9.25. As an object slips on

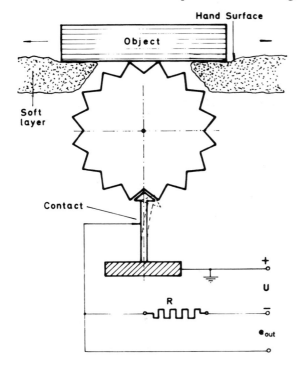

Figure 5.9.25. The first Belgrade hand slip sensor. A slipping object causes the knurled roller to make the reed (contact) vibrate thereby producing a modulated square wave at e_{out}. (Courtesy of Z. Stojiljkovic. From reference [36].)

the hand's "skin," a knurled roller protruding slightly above the surface revolves, causing a reed to vibrate against a grounded contact. This action produces pulse-width modulation in the square wave R, which results in the overall signal e_{out}. Any nonzero output with a duty cycle less than 50% indicates that slip is occurring.

The major problems with this transducer are that it is both difficult to miniaturize (important for a prosthetic device) and has a limited reaction angle (i.e., slippage force must act in the plane of the roller). Also, it can sense slip in only one direction.

A second slip sensor for the hand is shown in Figure 5.9.26. Here the knurled roller has been replaced by a needle. An object sliding along the surface of the hand causes the needle to vibrate against a grounded contact formed by part of the sensor structure. Once again, pulse-width modulation of a "carrier" square wave results. Besides the difficulties of the preceding device, this one will work only when small grasping forces are applied. (Large forces will cause the needle to "lockup," thereby preventing mechanical vibration.)

The final slip sensor, which overcomes many of the problems associated with the other two, is shown in Figure 5.9.27. It can be seen that the roller or needle has been replaced by a small conducting spherical ball that can rotate in *any direction* and is partially covered with nonconducting fields (i.e., the dark areas). A slip (in any direction) can be detected by monitoring the *differential* voltage across the terminals of the two contacts. Once again, a pulse-width-modulated signal is generated when the object slips.

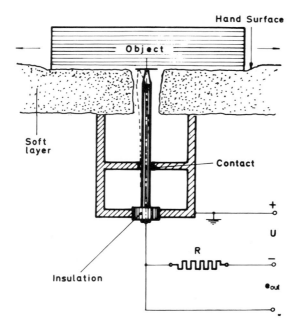

Figure 5.9.26. The second Belgrade hand slip sensor. The knurled roller in Fig. 5.9.25 is replaced by a needle. (Courtesy of Z. Stojiljkovic. From reference [36].)

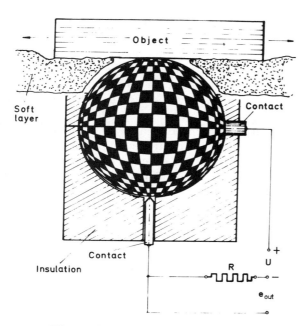

Figure 5.9.27. Final design of the Belgrade Hand slip sensor. Slip in any direction is sensed due to the motion of the small sphere. Note that the dark areas do not conduct. (Courtesy of Z. Stojiljkovic. From reference [36].)

The major advantages of this design over the other two are that slip can be sensed in any direction, the device is considerably less sensitive to external vibration (i.e., non-slip-related), and it is fairly easy to miniaturize the ball, thereby increasing the unit's ability to detect small tangential motions. In fact, the sensor was actually installed on the Belgrade hand and a servo system built. It is possible that such a design could be adapted for robotic applications, although this has yet to be done.

5.10 FORCE AND TORQUE SENSORS

In many manufacturing applications involving industrial robots, it is extremely important to be able to adjust and/or monitor the force and torque being applied to a part. A number of the tactile sensors described in the preceding section have the ability to provide information about how much force the jaws of a robotic gripper are exerting on an object. For example, the one produced by the Lord Corporation can be used as a force sensor in addition to its limited pattern-recognition capabilities. The reader is referred to Section 5.9 for further information.

There are, however, applications where only simple force/torque detection is required and where the inclusion of object recognition provided by an array of sensors may actually slow down the particular manufacturing process. In this section, we describe a number of techniques/devices that have been developed for monitoring these quantities.

5.10.1 Force Sensing by Motor Current Monitoring

In Section 5.9.3 we described briefly the lift-and-try method of acquiring a part by a robotic gripper. This procedure required that the motor current of a particular manipulator joint (or set of joints) be monitored. An increase in this quantity above some predetermined threshold value indicated that the object had been grasped successfully.

It is also possible to use motor current as a measure of the grasping force. If we assume that the gripper itself is servo controlled and that the actuator is a servomotor*, the armature current (monitored differentially across a small resistance placed in series with one of the motor leads) is proportional to the torque generated by the motor. That is,

$$I_a = \frac{T}{K_T} \tag{5.10.1}$$

where I_a = current

T = total generated torque

K_T = motor's torque constant

Through the action of a mechanical rotary-to-linear motion converter (e.g., a rack and pinion; see Section 3.4.2.2), this torque is converted into a force (see Figure 5.10.1). With a pinion gear of radius R, the transmitted force F is

$$F = \frac{T \times \eta}{R} \tag{5.10.2}$$

where η is the efficiency of the rack and pinion assembly (usually on the order of 90%, i.e., $\eta = 0.9$). Substituting Eq. (5.10.1) into Eq. (5.10.2) provides the desired relationship between the motor current and generated force. Thus

$$F = \frac{K_T I_a \eta}{R} \tag{5.10.3}$$

Other rotary-to-linear motion converters (e.g., a lead screw) will yield similar results.

From Eq. (5.10.3) it can be seen that when the robot is commanded to grasp a part, the motor current can be increased to a value that produces sufficient force to prevent slipping but is small enough to avoid crushing (or significant mechanical deformation that might cause permanent damage). These maximum and minimum values of I_a would have to be determined a priori (i.e., off-line) for *every part* to be handled and the information stored in the robot's controller. This is not a major disadvantage, however. It should also be noted that the actual force applied

---

*For example, such a gripper was developed by a group of students working under the guidance of one of the authors.

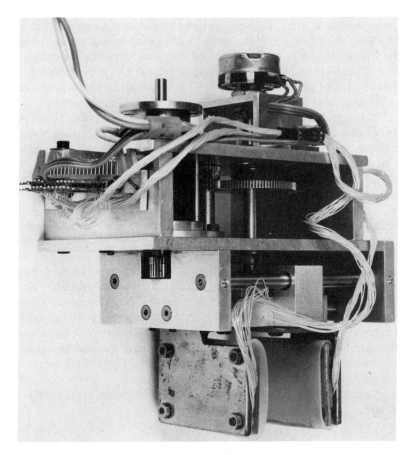

Figure 5.10.1. "Electrogrip," a servo-controlled gripper, developed by U.S. Robots and Drexel University, utilizes a rack and pinion and dc servomotor to drive the fingers of a parallel jaws robotic gripper.   Any position within the overall range of 2 in. can be obtained with an accuracy of 1 mil.   (Courtesy U.S. Robots, Inc. and Drexel University.)

to an object by such a gripper would, of course, be somewhat smaller than that indicated by Eq. (5.10.3) because of friction and gravity loads.   That is, some motor torque (current) would be required to overcome the friction in the various mechanical linkages used in a servo-controlled gripper.   The controller would have to make the appropriate corrections for such "loss terms."

There are some obvious problems with this force-controlling technique.   The major one has to do with the temperature sensitivity of both the servomotor torque constant K_T and its armature resistance R_a.   The reader will recall from the discussion in Chapter 4 that K_T decreases as the temperature rises, whereas R_a increases.   Therefore, Eq. (5.10.3) reveals that the gripping force will decrease with rising temperature. Thus to control carefully the force applied to an object within the jaws of a gripper using motor current, it would be necessary to monitor the temperature and have the robot's controller make the required corrections. This could be done using a table-lookup procedure to provide the necessary correction factor at any given temperature. Additional memory and somewhat increased computational time would be the penalty for such a feature.

Difficulties with the technique would also result from brush noise and variation in brush resistance, both of which would make accurate monitoring of the armature current a problem. However, as stated in the preceding section, the use of a *brushless* dc servomotor on the gripper would significantly reduce or even completely eliminate this problem. Such a gripper would, of course, be more costly, due to the additional cost of the motor and commutation electronics.

Understand that the procedure described above is not really a true *sensing* technique but is, rather, a method for maintaining a desired force. It is motor current and not force that is actually monitored, with the current being increased until a predetermined value is reached. The servo is used to achieve this. The actual force being applied to the object is *inferred* from the current but never measured. In some applications, this may be good enough. However, variations in the part weight or some electrical and/or mechanical parameter in the gripper itself will not always be sensed. In these cases, the object may either fail to be picked up or may be damaged. What is obviously needed is a sensor that provides direct information about the actual force being applied to a part.

One way of meeting this requirement is by making use of an LVDT. The reader will recall from Section 5.2.6 that such a device can accurately measure linear position. In addition, the force–displacement relationship for a linear spring is given by

$$F = Kx \qquad (5.10.4)$$

where K is the spring constant and x is the displacement about an equilibrium point. Obviously, the change in the spring length is a measure of the force causing the change, that is,

$$x = \frac{F}{K} \qquad (5.10.5)$$

In a *force transducer*, K is known and constant over some range of operation and the LVDT is used to measure the distance. Thus the output of the latter device is proportional to the applied force. Schaevitz Manufacturing produces a line of these sensors, which can measure forces from a few grams up to hundreds of pounds. (Obviously, different springs must be used as the force range is extended.)

In addition to the difficulties with using an LVDT already described in Section 5.2.6 (i.e., nonzero drift, high cost, need to zero-calibrate, and temperature sensitivity), such a sensor also suffers from additional temperature sensitivity because K varies with this quantity. In addition, its ability to measure force in only a single direction (usually parallel to the longitudinal axis of the LVDT) limits its usefulness in robotic applications, where one cannot guarantee that forces will always be orthogonal to the fingers of a gripper. Thus it is not surprising that these transducers have not been generally utilized on robots. Clearly, some other type of device is required.

5.10.2 Strain Gage Force Sensors

One of the simplest methods of sensing a force (or pressure) exerted on an object is to detect the deflection of the fingers of the robotic gripper in response to such an applied force. The strain gage provides a convenient and accurate means of doing this. We now describe briefly the operation of such a device. Readers interested in learning more are referred to an article by Mounteer and Perrin. [37].

The principle underlying the operation of a strain gage is that a mechanical deformation produces a change in resistance of the gage, which can then be related to the applied force. To better understand the operation of this device, consider a simple strain gage consisting of a plastic body (or some other flexible, nonconducting base) whose top surface is coated with a thin layer of a conducting material (e.g., aluminum or copper. See Figure 5.10.2). If the conductive coating is assumed to have a uniform cross sectional area A, then the resistance of the device is given by

$$R_g = \frac{L}{\sigma A} \qquad (5.10.6)$$

where σ is the conductivity of the conducting material and L is the length of the gage. When glued to an object, any deformation will cause the gage to bend either concave up or down (see Figure 5.10.2b and c). Since the conducting material thickness is assumed to be small, this bending produces little change in A. However, the gage length is either reduced (Figure 5.10.2b) or lengthened (Figure 5.10.2c). It is seen from Eq. (5.10.6) that such action causes a corresponding decrease or increase in the gage resistance.

There are four basic types of strain gages:

1. Unbonded wire
2. Bonded metal foil
3. Thin film
4. Semiconductor

The first does not utilize a nonconducting substrate. Here the sensing element usually consists only of an extremely fine platinum tungsten wire. In the second class, a thin copper or aluminum alloy foil is glued to either a nonconducting base (bonded) or directly to the object under study (nonbonded). With the thin-film gages, vacuum or sputter techniques are used to deposit resistors onto heat-

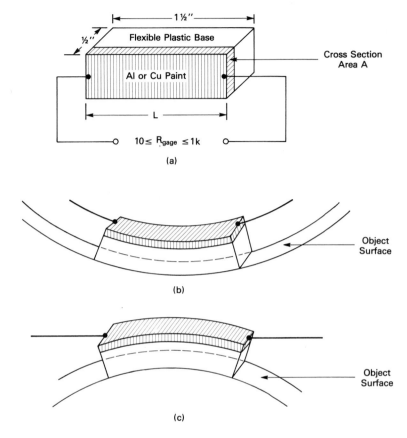

Figure 5.10.2. A simple strain gage: (a) undeformed device—gage length L = 1.5 in.; (b) concave up deflection of an object—L < 1.5 in. and R_g decreases; (c) concave down deflection of an object—L > 1.5 in. and R_g increases.

treated steel substrates. As such, these devices represent the "state of the art" in strain sensors. Probably the most sensitive of the four types, however, is the semiconductor gage, which is actually divided into two subclasses. One is made of silicon elements bonded to a cantilever beam or diaphragm and is called a "bonded bar semiconductor strain gage." The other borrows photolithographic and diffusion techniques from integrated-circuit technology to produce a "diffused-type semiconductor strain gage."

Regardless of what type is employed, some electronic circuit must be used to sense the change in gage resistance and produce a voltage output as a result. Such a circuit is shown in Figure 5.10.3. Here the gage is one arm of a Wheatstone bridge (R_g). Another arm is adjusted so that its resistance R_1 is approximately

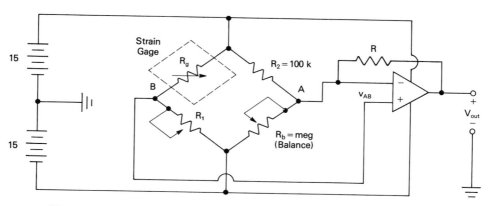

Figure 5.10.3.   Strain gage bridge circuit. R_1 is approximately equal to R_g when the gage is not deformed. R_b is used to balance the bridge before any deformation occurs. V_{AB} or V_{out} can be related to the applied force.

equal to that of the unstressed gage.   A 100 kΩ fixed resistor R_2 and a 1 MΩ balance pot R_b complete the circuit.   If the bridge is initially balanced, so that

$$R_g R_b = R_1 R_2 \qquad (5.10.7)$$

the output voltage V_{AB} will be zero.   Any change in R_g caused by a deformation of the gage will unbalance the bridge and cause V_{AB} to have a nonzero value.   By applying a known force to the object onto which this sensing device has been placed, the system can be calibrated (i.e., the value of V_{AB} can be related directly to force).

An obvious difficulty with this type of force sensor is that variations in ambient temperature tend to change the gage resistance, thereby causing the bridge to become unbalanced even when no force is applied.   It is possible to overcome such a problem by either (automatically) rebalancing the bridge periodically or by utilizing two gages (and two bridges) and using the *difference* of their outputs as the actual sensing signal.   The latter technique requires more circuitry but makes temperature drift a second-order effect.

Strain gages can be used to produce a robotic force sensing element.   For example, as shown in Figure 5.10.4, these devices can be placed on the back of the "fingers" of a parallel jaws gripper.   Then, as the fingers begin to grasp an object, the resultant deflection will be monitored by the gages.   If the deflection versus force characteristic for the gripper material and structure is known, the grasping force can be related to the output voltage.   For this technique to be successful, the objects being grasped must be made of a solid material that is significantly *less deformable* than the gripper's fingers.   Obviously, there will be problems with using such a sensor to control the output force while lifting a fragile object (e.g., an egg or one made from Styrofoam), and some other force-sensing device will be necessary.

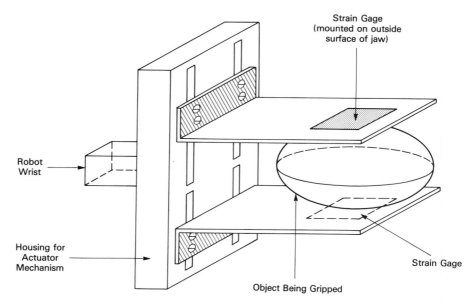

Figure 5.10.4. A robotic force sensor can be made by placing strain gages on the outer surfaces of the fingers of a parallel jaws gripper. These gages monitor the finger deflection when an object is grasped.

Another robotic force sensor making use of strain gages was developed in the mid-1970s at Stanford University by V. Scheinman as part of his graduate work [38]. A modified version was produced at SRI-NASA Ames for the Ames Anthropomorphic robot (see Figure 5.10.5). In reality, this device was a six-dimensional force and torque detector that made use of clever machining of a piece of aluminum tubing to create a series of elastic beams onto which were bonded *pairs* (for temperature compensation) of foil gages. The unit was 3.2-in. long and had a 3-in. O.D. and a 0.18-in. wall thickness.

As can be seen from the figure, there are eight narrow beams, with four oriented so that their long axes are in the z-direction (denoted by P_{x+}, P_{y+}, P_{x-}, and P_{y-}), and with the remaining four perpendicular to the z-direction (denoted by Q_{x+}, Q_{y+}, Q_{x-}, and Q_{y-}). The gage pairs are indicated as R_1 and R_2 and are oriented so that a vector from the center of the latter passes through the center of the former along the positive x, y, or z direction: for example, the gages on beams P_{x+} and P_{x-} are perpendicular to the y direction. The neck at one end of any beam has the effect of "amplifying" the strain at the gage positions while transmitting negligible bending torque.

If the output voltage from any pair of gages is given the same name as the beam, e.g., P_{x+} is the voltage due to R_1 and R_2 located on beam P_{x+}, it can be shown that the three forces F_x, F_y, and F_z and the three torques M_x, M_y, and M_z

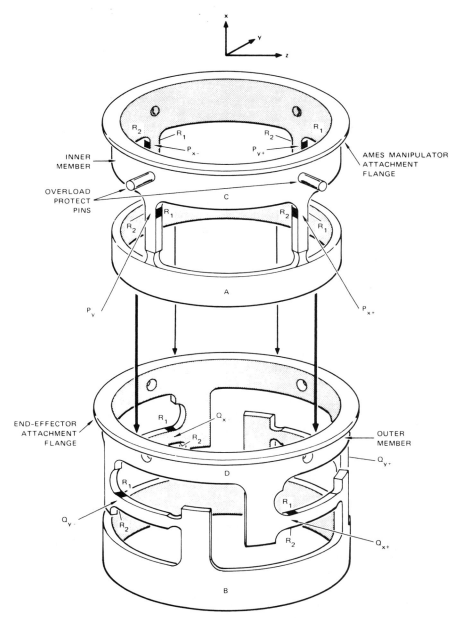

Figure 5.10.5.   The SRI-NASA Ames force-torque sensor, a modification of the work of V. Scheinman, was developed for the Ames Anthropomorphic robot. (Courtesy of A.K. Bejczy, Jet Propulsion Laboratory, Pasadena, CA.)

are proportional to

$$F_x \sim P_{y+} + P_{y-}$$

$$F_y \sim P_{x+} + P_{x-}$$

$$F_z \sim Q_{x+} + Q_{x-} + Q_{y+} + Q_{y-}$$

$$M_x \sim Q_{y+} - Q_{y-}$$ (5.10.8)

$$M_y \sim Q_{x-} - Q_{x+}$$

$$M_z \sim P_{x+} - P_{x-} - P_{y+} + P_{y-}$$

Rather than use individual bridges for each gage, the potentiometric arrangement shown in Figure 5.10.6 was utilized. Although not as sensitive as a Wheatstone bridge, this voltage-divider circuit obviously requires far fewer components to produce the desired temperature compensation effect. It should be noted that since aluminum is a good thermal conductor, any variation in the ambient temperature will cause the two gage resistances to change by the same amount. Thus the circuit output will not change.

To prevent failure due to excessive loading, the shear pins were included. These permitted the wrist to handle safely maximum x, y, and z forces of 70, 70, and 108 lb, respectively. The maximum safe x, y, and z torque loads were 72, 72, and 144 in.-lb, respectively.

A refined version of the foregoing sensor was subsequently built for the Jet Propulsion Laboratory under a contract with Scheinman's company, Vicarm, Inc. (later acquired by Unimation) [6]. As shown in Figure 5.10.7, this device replaced the elaborately machined cylinder with a "Maltese cross" configuration that was more easily fabricated out of a single piece of aluminum. Semiconductor strain gages, mounted on each of the four sides of the deflection bars of the cross, were used in place of the foil types to provide increased sensitivity. Once again, a potentiometric circuit was employed and produced the eight outputs W_1 to W_8, as indicated in Figure 5.10.7. A 6×8 transformation matrix was then used to convert these data into actual x, y, and z components of force and torque as indicated in this figure. The information derived from this sensor permitted the adjustment of the grasping force applied to several test objects. Bejczy reported that the results were "highly repeatable."

A commercial robotic force sensor that utilizes strain gages is currently available (B and B Machine and Engineering, Allston, Massachusetts). This device places the gages on the fingers of a gripper and can yield x, y, z force information. (Three twising moments can also be obtained from the force data and gripper opening.) Its sensitivity is 0.5 V/lb; it has a linear range of ± 20 lb with overloads of up to ± 40 lb permissible, and weighs about 2 oz.

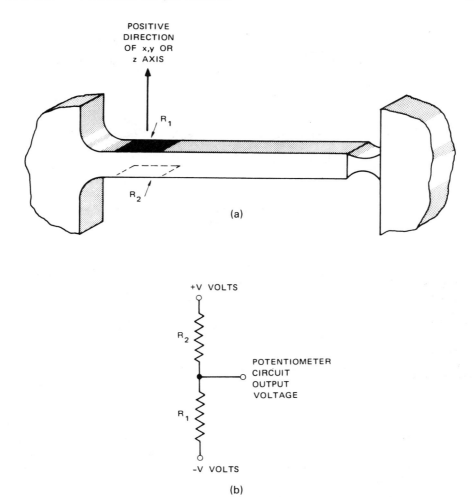

POSITIVE
DIRECTION
OF x,y OR
z AXIS

R_1

R_2

(a)

+V VOLTS

R_2

POTENTIOMETER
CIRCUIT
OUTPUT
VOLTAGE

R_1

−V VOLTS

(b)

Figure 5.10.6.   A single degree of freedom on the SRI-NASA Ames force-torque sensor of Fig. 5.10.5: (a) mechanical mounting of the strain gages; (b) potentiometric circuit.   (Courtesy of A.K. Bejczy, Jet Propulsion Laboratory, Pasadena, CA.)

5.10.3 Compliance and Assembly Operations

Up to now we have concentrated on sensors that provide information to a robot's controller, thereby permitting an object to be acquired without crushing or slipping. The signals from such sensors are used to adjust the gripping force, pressure, or torque.   Besides picking up parts, there are other operations that may require a different type of force adjustment, however.

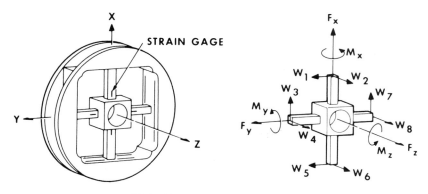

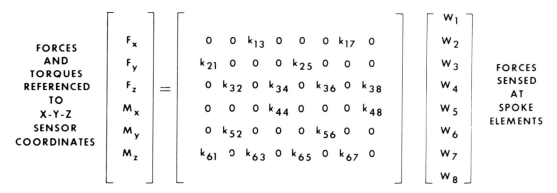

**TRANSFORMATION MATRIX
UNDER IDEAL CONDITIONS**

$$
\begin{bmatrix} \text{FORCES} \\ \text{AND} \\ \text{TORQUES} \\ \text{REFERENCED} \\ \text{TO} \\ X\text{-}Y\text{-}Z \\ \text{SENSOR} \\ \text{COORDINATES} \end{bmatrix}
\begin{bmatrix} F_x \\ F_y \\ F_z \\ M_x \\ M_y \\ M_z \end{bmatrix}
=
\begin{bmatrix}
0 & 0 & k_{13} & 0 & 0 & 0 & k_{17} & 0 \\
k_{21} & 0 & 0 & 0 & k_{25} & 0 & 0 & 0 \\
0 & k_{32} & 0 & k_{34} & 0 & k_{36} & 0 & k_{38} \\
0 & 0 & 0 & k_{44} & 0 & 0 & 0 & k_{48} \\
0 & k_{52} & 0 & 0 & 0 & k_{56} & 0 & 0 \\
k_{61} & 0 & k_{63} & 0 & k_{65} & 0 & k_{67} & 0
\end{bmatrix}
\begin{bmatrix} W_1 \\ W_2 \\ W_3 \\ W_4 \\ W_5 \\ W_6 \\ W_7 \\ W_8 \end{bmatrix}
\begin{matrix} \text{FORCES} \\ \text{SENSED} \\ \text{AT} \\ \text{SPOKE} \\ \text{ELEMENTS} \end{matrix}
$$

Figure 5.10.7. JPL "Maltese cross" version of the sensor shown in Fig. 5.10.5. The matrix transformation relating the strain gage outputs to the actual x, y, and z components of the forces and torques is also given. (Courtesy of A.K. Bejczy, Jet Propulsion Laboratory, Pasadena, CA.)

As an example, consider a common assembly operation which involves inserting a rod (or screw) into a hole (see Figure 5.10.8). Depending on the amount of clearance that exists between the rod and the side of the hole, any misalignment caused by either part movement or problems with the robot's repeatability will cause binding, thus preventing a successful insertion (see Figure 5.10.8b and c). Successful insertion will generally require that the gripper descend *along the hole's center line*, as shown in Figure 5.10.8a.

The major problem with such an operation is that it is not always possible to keep the robot positioned properly. Due to the mechanical stiffness of most industrial robots, such a situation will normally result in damage to either the part or the robot, or cause a joint timeout. What is obviously called for is "give" (or compliance) in the manipulator. As discussed in Chapter 1, SCARA-type robots

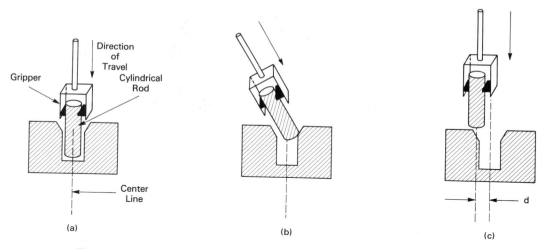

Figure 5.10.8. Insertion of a cylindrical rod into a hole: (a) successful insertion occurs if the robot's gripper descends along the hole's center line; (b) binding occurs when the gripper descends at an angle to the center line or; (c) parallel to the center line but displaced by a distance "d."

have some ability to do this electronically and also because of their jointed cylindrical structure. However, it is possible to provide almost any robot with the desired compliance using a *passive* device called a remote center compliance (RCC) placed between the gripper and the wrist flange* (see Figure 5.10.9). The Lord Corporation (Erie, Pennsylvania) manufactures a series of these units that combine high axial stiffness (i.e., along the direction of the insertion force) and low lateral and torsional stiffness (i.e., either perpendicular to or about the direction of insertion force). By varying the properties of the laminated rubber elastomer components, the compliance of a given RCC can be changed.

The terms "compliant center" or "elastic center" are used to describe a point about which rotation occurs when a moment is applied. Lateral translation will occur only when a lateral force is applied at the elastic center. The RCC permits both of these motions, which facilitate certain types of assembly operations. This can best be understood by considering the situation depicted in Figure 5.10.10. A position offset between the rod and hole causes a lateral force to be generated at the far end of the rod due to the chamfer (see Figure 5.10.10a). With an RCC, this force acts through the center of compliance, which causes the shaft to undergo a lateral translation, thus permitting it to be inserted more easily into the hole (see Figure 5.10.10b). In the case of a nonparallel misalignment such as that shown in Figure 5.10.11a, the binding of the rod against the side of the hole will produce a moment. This, in turn, will cause a rotation about the compliant center, thereby

---

*The RCC is also discussed in Section 3.7.2.

Figure 5.10.9. Family of remote center compliance (RCC) devices that combine high stiffness along the direction of the insertion force with low lateral and torsional stiffness (either perpendicular to or about the direction of the insertion force). (Courtesy of J. Rebman and the Lord Corp., Cary, NC.)

allowing the shaft to align properly with the hole (see Figure 5.10.11b). Understand that although the RCC is a passive device and does not actually monitor forces due to misalignments, its design is such that, in effect, it utilizes such forces to move the gripper, holding a part in a direction so as to *reduce* the binding force, thereby correcting for any misalignment.

By employing different types of displacement sensors on an RCC, it is possible to obtain electrical signals proportional to various torques and forces that result from any misalignment between two parts. This information can then be utilized in a force or torque feedback scheme. Called an IRCC, the I standing for "in-

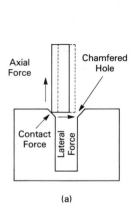

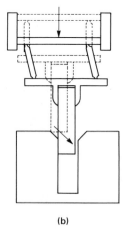

Figure 5.10.10. Position offset in an insertion operation can be corrected by an RCC: (a) example of a lateral force produced by such an offset; (b) an RCC permits the rod to translate laterally so that insertion can be accomplished. (Courtesy of J. Rebman and the Lord Corp., Cary, NC.)

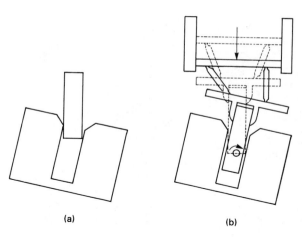

(a)   (b)

Figure 5.10.11. Nonparallel misalignment corrected by an RCC: (a) such a misalignment of the rod causes binding against the side of the hole; (b) the resultant moment causes a rotation about the compliant center of the RCC thereby permitting the center lines of the rod and hole to become coincident. (Courtesy of J. Rebman and the Lord Corp., Cary, NC.)

strumented," such sensors have been developed at the MIT Draper Laboratories. The first two prototype units each were able to measure x and y translational and rotational position for a coordinate system in which the z-axis pointed along the RCC and the tool. That is, they had four degrees of freedom (DOF). One unit used four Kaman Scientific proximity sensors (see Section 5.8) as the active element. The other one employed a Reticon linear diode array, thin metal shutter, and LED light source for each DOF. In the latter case, the number of illuminated photodiodes in the array gave the shutter position. Both sensors were developed primarily for laboratory (research) purposes and were not well suited for the manufacturing environment, due to difficulty in calibration and high cost.

A more practical device was also developed that had only three DOFs (x, y, and θz, i.e., z-axis rotation). The three positional sensors each consisted of an LED and dual-photodiode combination and were mounted on a Lord Corporation RCC. With the photodiodes connected differentially, any displacement of the RCC produced a signal that was proportional to the offset from the null or unforced position. In actual experiments this device was used on the wrist of a PUMA 600 robot and interfaced to the robot's controller through the I/O module utilizing the VAL robot programming language [39].

A number of applications were attempted, including chamferless insertion, edge following, and seam tracking. The first used the IRCC as a touch sensor to detect when a peg, tilted 10° from the hole axis, contacted the hole side (see Figure 5.10.12). The passive compliance of the RCC was then used to complete the insertion. In the case of edge following, the transducer provided force information which was used to adjust the y position so that a force of from 1 to 2 lb was exerted against the edge of the plywood material by a rotary cutting tool. The z-axis was perpendicular to the material surface and tool motion was along the seam in the tool x direction (see Figure 5.10.13). Inadequate sampling of the sensor signals,

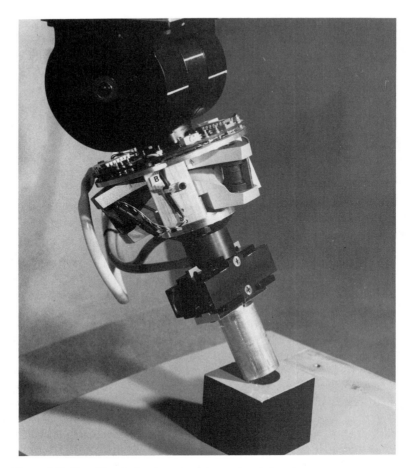

Figure 5.10.12.   Chamferless insertion of a peg in a hole using an IRCC. (Courtesy of D. S. Seltzer, the Charles Stark Draper Laboratory, Inc., Cambridge, MA.)

due in large part to using the PUMA's I/O lines, resulted in limited tracking speeds (e.g., about 12 in./min). The passive RCC has only a relatively small range of operation (e.g., ±0.1 in.), so that its ability to edge track is severely limited. The instrumented device, however, extends this range significantly.

With seam tracking, the IRCC was required to provided two-dimensional force information because of the need to maintain constant contact with two surfaces instead of one. (This would be a required task during an arc welding application). Once again, the low sampling rate limited the tracking speed (i.e., to about 11 in./ min). The sensor was also used in an experiment involving self-learning. Readers wishing to learn more about this are directed to Seltzer's paper [39].

Figure 5.10.13. Edge following using an IRCC. Force information from the sensor is used to adjust the y position of the rotary cutting tool so that a constant force of 1 to 2 lb. is maintained against the plywood. (Courtesy of D.S. Seltzer, the Charles Stark Draper Laboratory, Cambridge, MA. From reference [39].)

    Besides the MIT devices, a series of commercial six-axis IRCCs has been developed by the Barry Wright Corporation (Watertown, Massachusetts) (see Figure 5.10.14). Utilizing six foil-type, strain gages mounted on a standard RCC (produced by Astek Engineering, Inc., Watertown, Massachusetts), the Model FS6-120A provides electrical signals (obtained from six separate bridges) that are proportional to three forces (ranging from maximums of 25 to 150 lb) and three moments (ranging from maximums of 18 to 108 in.-lb). These sensors weigh about 1 lb and include in their on-board electronics package a microprocessor (TI 9995) to convert the information into a user-specified orthogonal coordinate frame. To accomplish this, the output from each of the strain gage bridges is buffered, low-

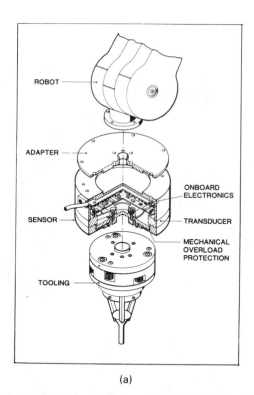

(a)

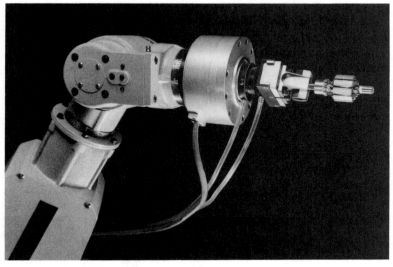

(b)

Figure 5.10.14. A commercial instrumented remote center compliance: (a) artist's conception of an IRCC attached to a robot; (b) actual IRCC mounted on a robot shown grasping a motor armature. (Courtesy of P. Cholakis, Barry Wright Corp., Watertown, MA.)

432

Figure 5.10.15. Successful insertion of a gear-shaft assembly into a bearing has been achieved by using a three-axis strain gage sensor mounted on the gripper's fingers to provide electronic compliance. (Courtesy of T.A. Brownell, General Electric Company, Corporate Research and Development, Schenectady, NY. From reference [40].)

pass filtered, and then input to a 6:1 multiplexer. A 12-bit DAC digitizes the data so that it can be handled by the microprocessor. Both analog and digital outputs are available for interfacing with the robot's controller.

Another technique for obtaining compliance has been demonstrated by Brownell at General Electric Research and Development [40]. In his experiments, a three-axis strain gage sensor developed at B & B Machine and Engineering (Allston, Massachusetts) was placed on each of the fingers of a parallel-jaws gripper and provided tool x, y, and z force components. These sensors were particularly well suited to robotic applications because of their light weight (2 oz), small size (1.5 in. $\times$ 2.5 in. $\times$ 0.84 in), and large force range (±20 lb with a ±40 lb overload capability). The information obtained by the B & B devices was then converted to joint coordinates by applying the appropriate matrix transformations for the Cartesian robot used. The resultant joint force feedback provided a means of varying the applied force (i.e., adjusting the compliance electronically).

Brownell found that with this scheme, it was possble to acquire fixed objects even when the robot's gripper was deliberately placed off-center. The same experiment failed when zero-force feedback was used. In addition, electronic compliance permitted successful insertion of a gear-shaft assembly into a set of bearings. In this application it was also necessary to mesh the gear teeth with those of a larger gear located nearby (see Figure 5.10.15).

5.11 SUMMARY

In this chapter we have treated extensively the topic of non-vision-based robotic sensors. These have been divided into two classes, those that provide *internal* information and those that provide *external* information. The former group of devices is generally used to keep track of the individual joint's instantaneous position, velocity, and/or acceleration. With the data from these sensors, the joints can then be controlled properly. Of all the sensors considered, the optical incremental encoder has been presented in great detail and many of the practical considerations necessary for its successful application to robots discussed.

The second group of sensors introduced in the chapter provides the robot with the information about its (external) environment. As discussed, most of these devices are still quite experimental in nature, with only a few commercial units available. In the future, it will be absolutely essential that robots performing complex manufacturing tasks possess the ability to apply just the right amount of force/pressure to an object. In addition, it will be important that these manipulators be able to determine what the object is from a tactile "image" provided by an array of sensors located in the gripper. One group of external sensors that are well developed are those used on welding robots. These units are currently often utilized to assist the manipulators in producing welds that are both accurately placed and of high quality.

5.12 PROBLEMS

5.1 For the position loop shown in Figure 5.2.5, the home position of a robot joint is assumed to be at zero degrees and occurs when both the input and sensing pots are outputting zero volts. Find the appropriate pot settings to produce demand positions DP_1, DP_2, and DP_3 of 30, 45, and 300°, respectively. Assume that the input and sensing pots are identical and that $\pm V$ corresponds to $\pm 180°$.

5.2 If the digital codes of 00, 01, 10, and 11 are used to select Home, DP_1, DP_2, and DP_3, respectively, determine the sequence of 2-bit numbers necessary to produce the following joint moves:
a. 0 to 30 to 300 to 45.
b. 0 to 45 to 30 to 300 to 0.
Discuss timing considerations in these moves.

5.3 The demand position pots in Figure 5.2.5 can be replaced by op amps which have feedback resistances that can be changed. Design a circuit that makes use of a set of digital switches to permit the selection of a number (e.g., eight) of resistances used in the feedback path of an op amp.

5.4 For an LVDT, it is necessary to convert the ac voltage across the two series opposing secondary coils into a dc voltage.
a. Show that a full-wave bridge will provide only magnitude information (i.e., direction cannot be obtained in this way).

b. Design a circuit that can be used to give *both* the magnitude and sign of a voltage that is proportional to the linear distance.

5.5 In Figure 5.3.3, design the one-shot and debounce circuit.

5.6 In Figure 5.3.3, why is it necessary to use a one-shot instead of a 555 timer, for example? Explain.

5.7 In Figure 5.3.3, why must the output of the one-shot be "short enough," and what happens if it isn't?

5.8 For Figure 5.3.2, design a circuit that permits the motor to turn in both directions.

5.9 Design a circuit for Figure 5.3.2 that allows the motor to be dynamically braked (as opposed to coasting to a stop).

5.10 Describe what might happen if the black flag located on the disk in Figure 5.3.2 was too small and the motor was permitted to coast to a stop.

5.11 Suppose that the motor in Figure 5.3.2 is driving a gear train (ratio n:1) and that the optical interrupter is on the *output side* of the gear train. Design the system so that a robot axis connected to the gear train will rotate to points located at 30°, 125°, 245°, and 325°.

5.12 Draw a comparator circuit (whose input is the triangular wave shown in Figure 5.3.10) and show that the TTL pulses in Figure 5.3.11 will vary in width as the light intensity changes (i.e., as E_{ave} in Figure 5.3.10 changes).

5.13 Discuss the reasons for a high-frequency limitation on an optical encoder. Consider the effect of narrow pulses due to drift in the average value of the output photosensor waveform.

5.14 Using a D-type flip-flop with clock input, show that it is possible to determine rotational direction represented by the two sets of waveforms shown in Figure 5.3.8.

5.15 Show that a two-input Exclusive OR (XOR) gate will double the effective resolution of an N-line encoder. Consider both CW and CCW rotations.

5.16 Complete all of the entries for the ROM of Table 5.3.1 so that all possible A- and B-channel transitions produce the appropriate up or down count for CW or CCW motion, respectively.

5.17 Discuss possible timing problems that might be encountered in utilizing the circuit of Figure 5.3.16. What might be a solution to such problems?

5.18 Consider CCW motion of an encoder disk. For the circuit of Figure 5.3.17, obtain a timing diagram similar to Figure 5.3.18.

5.19 For the situation in Problem 5.18 suppose that the counter is initialized at zero. Verify that the count does increase for each edge encountered on the A and B channels of the encoder. Write out the *binary count* for at least two encoder lines.

5.20 Consider the problem of ensuring that valid data exist on the output lines of the 16-bit position counter of Figure 5.3.17. Discuss the importance of valid data for a robot axis and devise a scheme for handling this problem.

5.21 For the $\times 4$ circuit of Figure 5.3.17, discuss any limitations on the dual one-shot period. Are there upper and lower limits?

5.22 Show that it is possible to obtain eight edges per encoder count by adding and subtracting the A- and B-channel signals.

5.23 Design a $\times 8$ digital circuit that utilizes the four TTL signals obtained from A, B, A − B, and A + B.

5.24 For the situation in Problem 5.23, discuss the importance of maintaining close tolerances on A- and B-channel quadrature (i.e., consider what happens to A − B and A + B if B is not exactly 90° out of phase with A).

5.25 Discuss the reasons why variation of temperature and line-voltage fluctuations can cause the calibration point to change when no encoder index pulse is used. How does the index pulse practically eliminate calibration sensitivity to such fluctuations?

5.26 With respect to Figure 5.5.1, use a plausibility argument to show that points D or F are unstable.

5.27 Show that if a robot joint is in the "home" position and its servo is operating in analog lock mode, any disturbance that causes the encoder to rotate beyond point D or F (Figure 5.5.1) will produce a position error of exactly one encoder line. Discuss how one might prevent this from happening.

5.28 Draw a block diagram of a servo that uses both all-digital-position information during motion and an analog locking scheme in the "home" position.

5.29 Derive the first-order velocity algorithm stated in Section 5.6.2.2.

5.30 If an encoder is read every T seconds so that the position is $P(kT)$, derive a second-order approximation for the velocity. Show that it will take longer to execute and that there will be more phase lag (i.e., time delay) introduced into the system.

5.31 Draw a computer flowchart that implements the scheme described in Section 5.8.2.1 (for a reflected light sensor) to eliminate the ambiguity in position due to the double-valued position versus voltage curve of the photosensor.

5.32 Discuss why recessing the sensor inside the gripper's fingers eliminates the position-ambiguity problem.

5.33 For the proximity rod tactile sensor described in Section 5.9.1.1, discuss the need for a switch closure threshold.

5.34 How would one convert the output of the Lord sensor described in Section 5.9.1.2 to monitor pressure or force?

5.35 For the Lord sensor, discuss the difficulties caused by rubber hysteresis.

5.36 Complete Figure 5.9.5, showing the correct inputs for the 74155 and 74156 decoders. Obtain a truth table for the appropriate inputs for row and column decoding.

5.37 Extend the circuit in Problem 5.36 to a 4 × 4 array.

5.38 Show the effect of crosstalk if the diodes are not used in Figure 5.9.5.

5.39 Discuss the problem of hysteresis with the Hillis sensor shown in Figure 5.9.6.

5.40 Draw a block diagram for a robot controller that incorporates the sensory information obtained form an arc welding seam tracker. What must be done to these data before the individual joints can be commanded properly?

5.41 Derive the relationship between force and motor current for a servo-controlled gripper. The servomotor is driving a gear train with a reduction ratio of $n:1$. The gear train, in turn, drives a lead screw that has a pitch P. Assume that the screw's efficiency is η_1 and that the gear train's efficiency is η_2.

5.42 Draw a flow chart for a robot's program that will permit shape information to be obtained if a single contact rod proximity sensor (like the one in Figure 5.8.1) is mounted on a robot's gripper.   Discuss any execution and/or time problems associated with such a process.

5.43 Show that for the voltage divider circuit shown in Figure 5.10.6, if R_1 and R_2 both change by either $+\Delta R$ or $-\Delta R$, the output voltage with respect to ground is unchanged.

5.13 REFERENCES AND FURTHER READINGS

1. Cushing, I. B., "A New High Accuracy Angular Position Transducer," *PCI/Motor-Con Proceedings*, Atlantic City, N.J. April 1984.

2. Welburn, R., "Ultra High Torque Motor System for Direct Drive Robotics," *Proceedings of the Robots 8 Conference—Applications for Today*, Vol. 2, Detroit, Mich., June 4–7, 1984. Dearborn, Mich.: Society of Manufacturing Engineers, 1984, pp. 19-63 to 19-71.

3. Luh, J. Y. S., Fisher, W. D., and Paul, R. P. C., "Joint Torque Control by a Direct Feedback for Industrial Robots," *IEEE Transactions on Automatic Control*, Vol. AC28, No. 2 (February 1983), pp. 153–161.

4. Szabo, Ernest, "Feedback of the Actual Position Utilizing the Existing Controller of a Robotic Arm," M.S. thesis, Drexel University, June 1984.

5. Hall, David J., "Robotic Sensing Devices," The Robotics Institute of Carnegie-Mellon, Report No. CMU-RI-TR-84-3, March 1984.

6. Bejczy, Antal K., "Smart Sensors for Smart Hands," *Proceedings of AAIA/NASA Conference on "Smart Hands,"* November 1978.

7. Fayfield, R. W., "Controlling with Bendable Bundles of Light," *Production Engineering*, Vol. 28, No. 7 (April 1981).

8. Ueda, M., et al., "One Trial to Use a Simple Visual Sensing System for an Industrial Robot," *Proceedings of the 6th International Symposium on Industrial Robots*, University of Nottingham, England, March 24–26, 1976.

9. Larcombe, M. H. E., "Tactile Sensors, Sonar Sensors, and Parallax Sensors for Robot Applications," *Proceedings of the 6th International Symposium on Industrial Robots*, University of Nottingham, England, March 24–26, 1976.

10. Kilmer, R. D., "Safety Sensor Systems," in *Robot Safety*, ed. M. C. Bonney and Y. F. Yong. Berlin: Springer-Verlag, UK: IFS (Publications) Ltd., 1985, pp. 223–235.

11. Einbinder, S., Kirshtein, J., et al., "Electronic Droid with Intelligence (EDWIN): An Autonomous Mobile Robot Prototype," *Final Report, Senior Design Project ECE 3*, Drexel University, Philadelphia, May 1983.

12. Bauzil, G., Briot, M., and Ribes, R., "A Navigation Sub-system Using Ultrasonic Sensors for the Mobile Robot HILARE," *Proceedings of the First Annual Conference on Robot Vision and Sensory Control*, Stratford-Upon-Avon, England, April 1981.

13. "Transducers," *Machine Design*, Vol. 54, No. 11 (May 13, 1982), Section 5.

14. Cook, G. E., Wells, A. M., Jr., and Eassa, H. "Microcomputer Control of an Adaptive Positioning System for Robot Arc Welding," *IECI Proceedings of Applications of Mini and Microcomputers*, November 1981.

15. Starke, L., "Limit Sensors and Proximity Switches and Control Techniques," *Elektro-*

meister and Deutschers Elektrohandwerk, Vol. 55, No. 18. September 1980 [untranslated].

16. Cupp, J. C., "Capacitive Probe and Constant Q Circuit for Measuring Head Flying Height," *IBM Technical Disclosure Bulletin*, Vol. 22, No. 1 (June 1979).

17. Belforte, G., D'Alfio, N., Quaglorotti, F., and Romti, A., "Identification through Air Jet Sensing," Proceedings of the 1st International Conference on Robot Vision and Sensory Controls," Stratford-Upon-Avon, England, April 1981.

18. "Hall-effect Transducer Improves Sensing Capability," *Automotive Engineering*, Vol. 89, No. 8 (August 1981).

19. Hillis, W. Daniel, "A High Resolution Imaging Touch Sensor," *International Journal of Robotics Research*, Vol. 1, No. 2 (Summer 1982), pp. 33–44.

20. Harmon, L. D., "Automated Tactile Sensing," *International Journal of Robotics Research*, Vol. 1, No. 2 (Summer 1982), pp. 1–32.

21. Page, C. J., Pugh, A., and Heginbotham, W. B., "New Techniques for Tactile Imaging," *The Radio and Electronic Engineer*, Vol. 46, No. 11 (November 1976), pp. 519–526.

22. Nobuaki, S., Heginbotham, W. B., and Pugh, A., "A Method for 3-D Part Identification by a Tactile Transducer," *Proceedings of the 7th International Symposium on Industrial Robots*, Tokyo, October 1977.

23. Hill, J. W., and Sword, A. J., "Manipulation Based on Sensor-Directed Control: An Integrated End Effector and Touch Sensing System," *Proceedings of the 17th Annual Human Factors Society Convention*, Washington, D.C., October 1973.

24. Rebman, J., and Trull, N. W., "A Robust Tactile Sensor for Robot Applications," *Lord Corporation Technical Article*, 9/83-250, 1983.

25. Snyder, W. E., and St. Clair, J., "Conductive Elastomers as Sensor for Industrial Parts Handling Equipment," *IEEE Transactions on Instrumentation and Measurement*, Vol. IM27, No. 1 (March 1978), pp. 94–99.

26. Garrison, R. L., and Wang, S., "A Pneumatic Touch Sensor," *IBM Technical Disclosure Bulletin*, Vol. 16, No. 6 (November 1973).

27. De Rossi, D., and Dario, P., "Multiple Sensing Polymeric Transducers for Object Recognition through Active Touch Exploration," *Proceedings of Workshop on Robotics Research: The Next Five Years and Beyond*, Paper MS84-492, Lehigh University, Bethlehem, Pa., August 1984.

28. Dario, P., Domenici, C., Bardelli, R., De Rossi, D., and Pinotti, P. C., "Piezoelectric Polymers: New Sensor Materials for Robotic Applications," *Proceedings of the 13th International Symposium on Industrial Robots and Robots 7*, Vol. 2, Chicago, April 1983, pp. 14-34 to 14-49.

29. Klafter, R. D., and Park, K., "A High Resolution Tactile Sensor," Progress report of work done at Drexel University, Temple University, and United States Robots for the Ben Franklin Partnership, Advanced Technology Center of Southeastern Pennsylvania, September 1984.

30. Grahn, A. R., and Astle, L., "Robotic Ultrasonic Force Sensor Arrays," *Proceedings of the Robots 8 Conference—Applications for Today*, Vol. 2, Detroit, Mich., June 4–7, 1984. Dearborn, Mich.: Society of Manufacturing Engineers, 1984, pp. 21-1 to 21-17.

31. Raibert, Marc H., and Tanner, J. E., "Design and Implementation of a VLSI Tactile

Sensing Computer," *International Journal of Robotics Research*, Vol. 1, No. 3 (Fall 1982), MIT Press, pp. 3–17.

32. *Electronic Design*, February 7, 1985, pp. 69–70.

33. Presern, S., Spegel, M., and Ozimek, I., "Tactile Sensing System with Sensory Feed-Back Control for Industrial Arc Welding Robots," *Proceedings of the 1st International Conference on Robot Vision and Sensory Controls*, Stratford-Upon-Avon, England, April 1981, pp. 206–213.

34. Presern, S., Ducar, F., and Spegel, M., "Design of Three Active Degrees of Freedom Tactile Sensors for Industrial Arc Welding Robots," *Proceedings of the 4th British Robot Association Annual Conference*, Brighton, England, May 1981.

35. Ueda, M., and Iwata, K., "Tactile Sensors for an Industrial Robot to Detect a Slip," *Proceedings of the 2nd International Symposium on Industrial Robots*, IIT Research Institute, Chicago, May 1972, pp. 63–76.

36. Tomovic, R., and Stojiljkovic, Z. "Multifunctional Terminal Device with Adaptive Grasping Force," in *Automatica*, Vol. 1, No. 6, Elmsford, N.Y.: Pergamon Press, 1975, pp. 567–570.

37. Mounteer, C., and Perrin T., Jr., "Pellicloid Pressure Sensor," *Sensors*, Vol. 1, No. 10 (October 1984), pp. 23–27.

38. Scheinman, V. D., "Design of a Computer Manipulator," *Artificial Intelligence Laboratory Memo AIM-92*, Stanford University, Palo Alto, Calif., 1969.

39. Seltzer, D. S., "Tactile Sensory Feedback for Difficult Robot Tasks," *Proceedings of the Robots 6 Conference*, Detroit, Mich., March 2–4, 1982. Dearborn, Mich.: Society of Manufacturing Engineers, 1982, pp. 467–478.

40. Brownell, T. A., and Ringwall, C. G., "Tactile Sensing for Manipulation," General Electric Company Report No. 84CRD003, February 1984.

6

Computer Vision for Robotic Systems: A Functional Approach

6.0 OBJECTIVES

The didactic material covered in this chapter is intended to expose the reader to many considerations required in providing computer vision for robotic systems:

- Computer vision components
- Computer vision task complexity
- General approaches to computer vision architecture
- Problem types amenable to solution by computer vision techniques

To reach these objectives, not all topics in computer vision can be treated, since complete coverage of this subject would easily require several volumes. Before the specific issues of computer vision systems are addressed, the functional requirements of machine vision should be presented. Let us consider certain visual tasks, requiring only an intelligent observer. Basically, these visual tasks concerning handling or acquisition of a part can be thought of as a series of questions that an expert viewer (or a robot) must answer:

- Where is it?
- Is it the one I am looking for?
- Is it defective or is it OK?
- How far away is it?
- Is it right side up?

440

- Is it being interfered with by another object of the same type? or a different type?
- What is the angle of the object relative to my hand?
- What color is it?

These types of questions lead us to consider the human visual intelligence system and its components:

- Eye(s)
- Neural connections (sensory and motor)
- Visual cortex

Physiologic subsystems can (very) loosely model the architecture of current (machine) vision systems. These include:

- Sensors
- Data paths
- Computer or processor

These major subsystems have numerous subtopics, but it is clear that the functional issues suggested by the questions above are addressed generally by the sensory $\longrightarrow$ data path $\longrightarrow$ computer configuration. This configuration is far too general and must be further subdivided. With this in mind, the topics that will be covered in this chapter are:

- Imaging components
- Image representation
- Hardware considerations
- Picture coding
- Object recognition and categorization
- Software considerations
- Need for vision training and adaptations
- Review of existing systems

6.1 MOTIVATION

"Computer vision" is a somewhat glamorous phrase that describes a wide class of computer image analysis topics. The use of the word "vision" is in many ways a poor choice, because it imbues a "sense" or "intelligence" to the robot that is undeserved and nonexistent. As used in this text and in most applied fields, "computer vision" is simply the analysis of photometric noncontact measurements obtained from a closed-circuit television (CCTV) camera or other photo-sensing element, such as a linear photoarray. Although there are a wide variety of photoimaging sensors, we will generally use the term CCTV in reference to the standard television camera, and not to the generic class of photosensors. We will often refer to the sensor as being a CCTV, but the reader should extend the meaning to other types of photosensors. The electrical voltage output from the CCTV is generally converted to a digital form, where processing of the image information by a digital computer may be accomplished.

The type of information to be produced by such a camera–computer couplet is:

- Location of a stationary object
- Object tracking as a function of time
- Object identification
- Object orientation
- Defect recognition and inspection of an object

These types of tasks may appear somewhat mundane rather than exotic, but they are far from trivial as far as computer implementation is concerned. Consider that most CCTV images are usually broken down into rectangular matrices of roughly 320×240 elements, with each element corresponding to an intensity from the scene being imaged. Consider the problems of searching through 76,800 (320×240) elements of an array at television rates (50 or 60 cycles/second), generally about 4.5 million searches per second. Scanning intelligently through this amount of data at these rates is no mean feat, and in fact it is the root of most of the problems facing today's computer vision systems.

Why, then, bother with "computer vision" systems if they require such high-speed, high-data volume solutions? The answer is simple: *noncontact measurement*! The advantages of noncontact location, detection, and identification are often so overwhelming that one is often forced to "computer vision" solutions. It is often the case that vision is the only sense capable of doing the job required, and human vision is often selected. In those cases where human interactive vision has proven necessary, it is reasonable to try to utilize computer vision toward the same end.

In situations where there is no alternative to vision, the function must be performed by either a human being or a machine. For example, in the reading of bar codes or OCR codes (optical character reader codes), vision processing must be performed, and the time penalty becomes irrelevant. One measure of the effect of the additional time penalty is the UPH or "units per hour" factor. For two sequential processes, such as "assembly" and "visions," one can define this factor as follows:

$$\text{UPH} = \cfrac{1}{\cfrac{1}{\text{assembly UPH}} + \cfrac{1}{\text{vision UPH}}}$$

6.2 IMAGING COMPONENTS

The imaging component, the "eye" or the sensor, is the first link in the vision chain. Numerous sensors may be used to observe the world. All the vision components have the property that they are "remote sensing" or "noncontact" measurement devices.

Vision sensors can be categorized in many different ways. For convenience, we will categorize them according to their dimensionality although they could also be classified by their wavelength sensitivity (i.e., do they respond to shades of black and white, or colors, or infrared, x-ray, ultraviolet, or the normal spectrum of human vision?). Vision sensors may be conveniently divided into the following dimensional categories or classes:

- Point sensors
- Line sensors
- Planar sensors
- Volume sensors

6.2.1 Point Sensors

The point sensors may be similar to "electric eyes," being either some type of photomultiplier, or more commonly, a phototransistor (see section 5.8.2.1). In either case, the sensor is capable of measuring the light only at a single point in space. For this reason they are referred to as "point sensors." These sensors may be coupled with a light source (e.g., a light-emitting diode) and used as a noncontact "feeler," as shown in Figure 6.2.1.

The "feeler" essentially monitors the light in a small "acceptance aperture." If an object falls in this acceptance aperture, light will be reflected from the object's surface and will be received by the sensor. If the acceptance aperture is clear, no light will be reflected into the sensor and it will not "feel" anything.

The point sensor may be used to create a higher-dimensional set of vision information by scanning across a field of view by employing some ancillary mechanism. For example, an orthogonal set of scanning mirrors (see Figure 6.2.2) or an x-y table can be employed to execute the scanning of the scene.

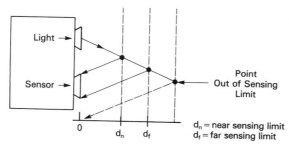

Figure 6.2.1. Noncontact feeler-point (i.e., proximity) sensor. The object is sensed only if its location falls between d_n and d_f. Points located beyond d_f are out of the sensing limit of the device.

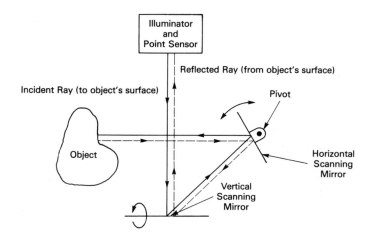

Figure 6.2.2. Image scanning using a point sensor and oscillating deflecting mirrors.

6.2.2 Line Sensors

Line sensors are one-dimensional devices and may be used to collect vision information from a scene in the real world. The sensor most frequently used is a "line array" of photodiodes or charge-coupled-device components. These devices are similar in operation, both being the equivalent of "analog shift registers" that produce a sequential, synchronized output of electrical signals corresponding to the light intensity falling on an integrating light-collecting cell. See Figure 6.2.3 for a schematic representation. The light output from these arrays is available "sequentially" (i.e., the individual cell outputs are not available in parallel or even on demand). The consequence of this is that the light intensity from the scene is available only in an ordered sequence and not at random on demand by the user. This has some consequences with regard to the time required for accessing a desired point intensity. The arrays may also be obtained in other than straight lines (e.g., circular arrays or crossed arrays are available; see Figure 6.2.4).

By proper scanning, line arrays may be used to image a scene. For example, by fixing the position of a straight-line sensor and moving an object orthogonally to the orientation of the array, one may scan the entire object of interest. Figure 6.2.5 is an example of such an application in a robot system.

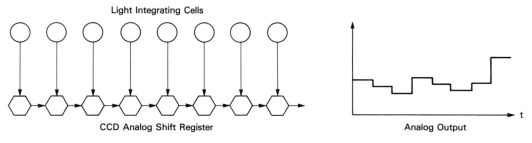

Figure 6.2.3. Schematic representation of line scanning arrays. The analog output signal is a sequential representation of the intensity of the light collected by the integrating cells.

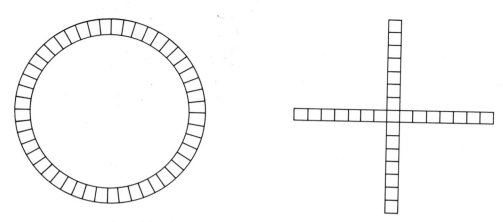

Figure 6.2.4.   Circular and crossed configurations of light sensors.

6.2.3 Planar Sensors

Planar sensors are an extension of the line-scan concept to a two-dimensional configuration.   Two generic types of these sensors are generally in use today: scanning photomultipliers and solid-state sensors.

The *scanning photomultipliers* are represented by television cameras.   The most common type of television camera used is the *vidicon tube* which is essentially an optical-to-electrical converter.   The photoelectric target "boils off" electrons

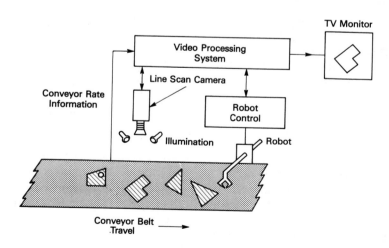

Figure 6.2.5.   An automated robot sorting system using a line scan camera to generate two-dimensional images.

when struck by photons in the visible spectrum. The electrons boiled off in this process are collected via the current in a scanning electron beam. This beam is scanned back and forth across the photoelectric target in a so-called *raster* fashion (see Figure 6.2.6). The rate and pattern of the raster scan differs throughout the world with a variety of existing standards.

A more detailed description of the operation of this tube follows. An optical image is focused on a photoconductive target layer which lies behind a transparent plate. When light strikes this layer, it becomes conductive and transfers electrons to the positive signal plate. The transfer causes a charge to build up on the face of the target. Its reverse face is scanned by a low velocity electron beam that tends to stabilize the target at the cathode's potential. This scanning causes a current to be generated which is indicative of the light intensity pattern on the photoconductive layer.

It is important to note that each time the target is scanned, it again begins transferring charge to the plate depending on the intensity of the incident light. In fact, the amount of charge may be thought of as being proportional to the integral of the light intensity. Since the raster scan is periodic in time, if a constant amount of light is incident on a particular point, the same charge will be present at each pass of the scan. From this description, it can be inferred that it is not possible to expose a vidicon to light and randomly read the data or expect consistent results by executing a single raster scan.

As in any physical system, the vidicon family has an upper limit to the amount of intensity that can be accepted by the photoconductive layer. If too much light is incident on a certain portion of the sensor, additional electrons (i.e., those above and beyond what would be removed by the normal incident light) from nearby areas may also be removed. This phenomenon is called *blooming* and is evidenced as an area of maximum intensity in the image.

Another important concept of the vidicon is that when the electron beam resets the charge, it may not reset it completely. Any remaining charge then decays exponentially. The term given to this phenomenon is called *lag* and is related to the speed of response of the camera. The physical manifestation of lag is that when viewing a moving object it may seem transparent. Additionally, if one observes an object as it moves into a camera's field of view and then stops, it may be necessary to wait for the motion to cease (i.e., 60 to 200 ms) before an accurate representation of the object can be obtained.

The RS-170 standard (described more completely below) used in North America specifies a 525-line 60-Hz physical format. Like line-scan devices, raster-scan devices provide serial or sequential access to the converted optical data, so that information about a given point in the field of view may be periodically inspected at a certain rate, but may not be accessed at random. In the case of a standard television camera, each intensity point from the camera may be inspected once every 33.3 ms (i.e., at 30-Hz = 1/33.3 ms). In many applications this time penalty is not severe, but in others it is almost intolerable. In most robotic applications, cycle times of 1 to 10 s are commonplace, and the additional time needed for robot vision processing would probably be acceptable. Most currently available vision systems provide about five vision cycles per second, so the extra 200 ms is not significant. However, in a small-parts assembly where the manipulator portion of the assembly cycle may be on the order of 1 s, the additional 200 ms for vision processing begins to have a significant effect. In an application such as semiconductor assembly, a 200-ms penalty would be prohibitive, since the additional time would reduce the process yield to unprofitable levels.

Although a vidicon camera sensor is not inherently a raster-scan device, the raster-scan format creates economies in manufacture (i.e., inexpensive cameras on the order of $200) and of course provides a simple mechanism for viewing the scene using ordinary television monitors, also relatively inexpensive.

Random-access scanning photomultipliers (image dissectors) are also available, but the photoelectric device is more expensive to manufacture (approximately $1000 for the tube itself) and the control circuitry is more complex because of the random-access requirements. Since this tube does not rely on the conventional raster scan, interesting variations such as spiral scans or radial scans may be implemented. Viewing the output of such a device would require a more costly monitor that is capable of accepting both raster and random-access inputs, with some type of mode switching required.

In addition to vidicon transducers, several types of *solid-state cameras* are available (e.g., photodiode and charge-coupled-device cameras). The solid-state camera is manufactured in a fashion similar to large-scale integrated circuits. The sensor elements themselves are very different from the photosensitive elements in a vidicon camera, but the arrays are still accessed in a serial or raster fashion. For this reason the solid-state cameras have no access-time advantage over the vidicon tube cameras. The solid-state arrays are inherently less noisy than the vidicon cameras, but are also considerably more expensive. This price/performance trade-off between the camera types must be carefully considered before final selection of a photo-optical transducer is made for a particular application. Many applications require the solid-state sensors because of weight and noise factors. This would be particularly important if it were necessary to mount the camera near or on the end effector of a robot.

Two-dimensional arrays (similar to line arrays) may be formed using either CCD (charge coupled device) or CID (charge injected device) technology. Both of these sensors are based on MOS (metal oxide semiconductor) transistor technology. Since they are discrete in nature, these devices will have a finite number of cells in both the horizontal and vertical direction. The solid-state array sensor is also an integrating detector and thus it is apparent that its sensitivity is proportional to exposure time.

It is important to understand how video information from the two-dimensional array is acquired. In the case of the CCD the most popular topology used is the frame transfer (FT) structure. FT technology makes use of an imaging area which

TABLE 6.2.1 BRIEF COMPARISON OF CAMERA
TECHNOLOGIES

Feature/specification	Vidicon	CCD	CID
Resolution	1	2	2
Sensitivity (light levels)	1	2	3
Speed	3	2	1
Bloom	3	2	1
Size	2	1	1
Reliability	2	1	1
Current cost	1	2	3
Future cost	2	1	3

Source: Tech Tran Consultants, Inc., Lake Geneva, Wisconsin from their report, "Machine Vision—A Summary and Forecast."

is exposed to light and generates charges proportional to the integral of the light intensity. There is also a storage area having the same number of cells as the imaging area. During the "frame time" when the image area is exposed, charge is accumulated at various cell locations in the array. After proper exposure, but before the next frame, this charge is clocked up to the corresponding cell in the storage area. (Note that the storage area is shielded from light.) A transfer register that permits each row of data to be moved in a serial manner is utilized in the operation. While the imaging area is being exposed for acquisition of the next frame of data, the charge pattern of the previous frame can be read out from the storage area by means of the readout register that operates in a manner similar to the transfer register.

In contrast to the CCD device, the CID camera may be thought to consist of a matrix of photosensitive cells arranged in rows and columns. As opposed to the CCD technology, each of the cells can be addressed randomly although commercial implementation relies on scanning compatible with raster techniques. After each cell is read, it is reset and begins integrating light for the next frame. Since the CID does not utilize the frame transfer structure of the CCD array, there is no delay between the electrical signal representing the image and the readout of the accumulated charge of a CID device.

Solid-state imagers also exhibit blooming. In this case, the charge carriers generated from an extremely bright part of an image spread to nearby elements because those elements located in the bright area are saturated. This charge travels to nearby locations thereby causing false information to be delivered to the vision system. In general, CID sensors are less sensitive to blooming than CCD-type sensors.

Table 6.2.1 is a brief comparison of the three major technologies used to implement cameras for image acquisition. Note that the ranking is from 1 to 3 with 1 representing the best performance.

6.2.3.1 Camera transfer characteristic

The transfer characteristic of a television camera (tube or solid state) can be defined by the parameter γ given by:

$$\gamma = \log(I/I_w)/\log(E/E_w)$$

where I denotes the signal current from the sensor, E the illumination on the photoconductive device, and I_w and E_w are respectively the value to which the signal and illumination are referenced.

For non-unity gamma, the contrast in the darker part of the picture is compressed while the lighter portion is exaggerated. Most television cameras are designed with $\gamma = 0.45$. This is the result of the natural characteristic of the CCTV monitors used to view the output. Normally, such monitors have a gamma ≥ 2. Thus if the camera is adjusted for the inverse (i.e., $\gamma = 0.45$), the picture on the monitor is pleasing to the eye. Unfortunately, while the image may appear pleasing to a human being, the signal provided to a computer will not contain the correct information. This is why it is extremely important to use a camera of unity gamma for imaging a scene for a vision application.

6.2.3.2 Raster scan

Previously, it was mentioned that the electron beam in a vidicon is scanned across the photosensitive element in a raster fashion. Figure 6.2.6 illustrates how

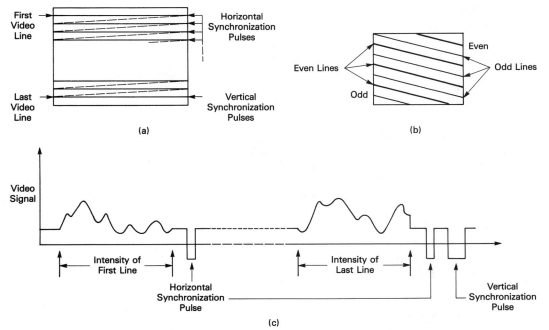

Figure 6.2.6. (a) Raster scan process on a television monitor; (b) magnified view of interlaced raster scan showing odd and even field lines; (c) two lines of analog video data.

TABLE 6.2.2 UNITED STATES
MONOCHROME TELEVISION STANDARD
RS-170 SPECIFICATIONS FOR RASTER
SCAN

Parameter	Value
Aspect ratio	
(width to height)	4/3
Lines per frame	525
Line frequency	15.75 kHz
Line time	63.5 μs
Horizontal retrace time	10 μs
Field frequency	60 Hz
Vertical retrace	20 lines per field

the raster scan would look on a television monitor. Examination of the figure shows that the beam starts at the top left and is moved across the face of the tube at a slight downward angle until it reaches the right side. Once at this location, it quickly moves back to the left and once again travels toward the right.

The process just described defines how a picture is recreated on a monitor or television screen. The picture consists of a frame that is made up of two fields: the odd and the even. This type of raster scan is referred to as interlaced scanning, where the even lines are first traced out, then the odd ones, then the even, and so on. The interlacing of the two fields is used as a method to permit pictures with moving objects to appear with minimum flicker.

Certain standards exist to define the signal associated with the output of a television camera and its subsequent reproduction on a monitor or television screen. As mentioned previously, the RS-170 specification is used in North America. Parameters used in a raster scan process are summarized in Table 6.2.2. Essentially, this standard specifies a 525-line, 60-Hz format for the video signal. Recall that raster scan devices produce serial data that repeat with a given period. Thus information about a particular point in the field of view may be inspected periodically at a certain rate but may not be accessed randomly.

Figure 6.2.6 also shows a representation for two lines of video. A complete frame consists of a more complicated waveform with additional pulses for synchronization of the fields and blanking pulses to ensure that the beam is turned off (blanked) during the time it is retracing. At this point, it is important to note that whatever type of imager is used (vidicon family, CID, or CCD) the electrical signal corresponding to lines in a frame (odd and even fields) is essentially that shown in Figure 6.2.6 with the parameters defined in Table 6.2.2.

6.2.3.3 Image capture time

An important parameter associated with image sensors is the actual image capture time. This can be defined as the time from when an object to be viewed is motionless to the time when the RS-170 video signal representing the scene contains both the odd and even fields (with valid data).   To this must be added the time that it takes to store the image in whatever type of vision system is being used.

Image capture time is dependent on the type of sensor and lighting.   Consider a CCD camera fixed in space and looking at an area into which an object is moved by means of some mechanical device (such as an *x-y* table or conveyor).   The camera, by its nature (integration of light levels) must be continually scanning. Since there will be a settling time for all mechanical motions, the instant when the motion of the object ceases will not be synchronized with the start of a field. Additionally, and as described previously, a frame transfer CCD camera transfers its image to a storage area which can be read out only when the next image area is scanned.   Referring to Figure 6.2.7, it can be seen that the worst-case time for image capture with a CCD is 83.3 ms.   It should also be apparent that if the object is motionless for at least three field times (enough to read out any invalid video

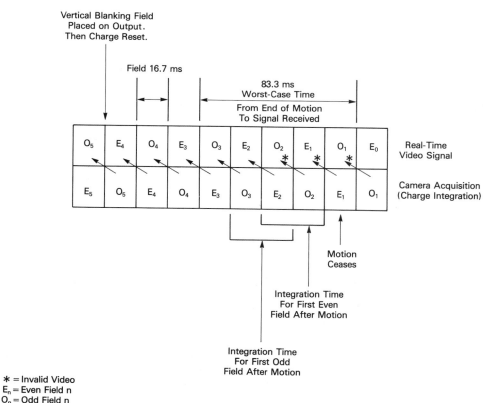

Figure 6.2.7.   Image capture for a CCD camera at the time when an object stops moving. Note: 33.3 ms must be allowed for integration of each field.

data from the frame transfer system), then the image can be captured in two field times (or 33.3 ms). That is, the data coming from the camera will be valid after three field times and the capture time is that needed to obtain both an odd and even field.

The same type of analysis can be performed for a vidicon where the lag time must be taken into account. The CID camera does not have a frame transfer architecture and therefore once the image is stationary, the video is available immediately. Its worst-case image capture time is 3 fields or 50.1 ms.

6.2.4 Volume Sensors

Volume sensors, providing general three-dimensional information, are not yet currently available on the market as a standard item. There are mechanisms that may be used to measure three-dimensional shape and orientation properties of solid objects. Stereo imaging using multiple two-dimensional arrays to image the object may be used. Three-dimensional information from solid objects may also

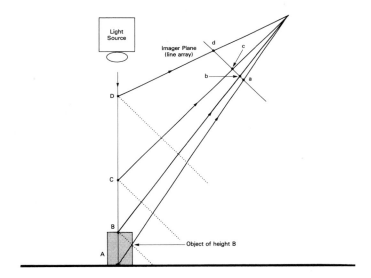

Figure 6.2.8. Schematic representation of a triangulation range finder. A point or line source of light illuminates objects directly below it, while the camera (in this case a line array) is set at an angle with respect to the light source. The camera "sees" the ray reflected from an object directly beneath the light. When no object is present, ray Aa intersects one end of the line array, while an object of height D causes ray Dd to intersect cell d at the opposite end of the array. If data relating the actual physical heights A, C, and D are known along with the address of the corresponding illuminated cells on the line array, then the following relationship may be used to solve for the unknown height B, given b (the line array cell illuminated by the light reflected from the object of height B): (AC/AD)/(BC/BD) = (ac/ad)/(bc/bd).

be obtained by use of directional lasers or acoustic range finders, and these techniques are becoming practical (see Figure 6.2.8).   The methodology for resolution of three-dimensional features is not yet solved in general, and no practical, general-purpose stereo vision for robots is currently available.

Structured light may be used for the location of various surfaces of objects. More discussion of this topic is given in the following section when illumination is discussed.

6.3 IMAGE REPRESENTATION

Currently, most vision for robotic systems is not stereoscopic, so we will restrict our discussion to representation of images as one observes on a television monitor. This is best described as an image intensity function of two variables, $I(x, y)$, a real function of two space variables (x, y), which corresponds to the light intensity falling on the photoelectric target.   The intensity function is used to represent an image, such as shown in Figure 6.3.1.

$I(x, y)$ is an idealized image representation and does not account for commonly occurring natural distortions, such as lens geometric aberrations, beat frequencies between source and scanner, and noise in the scene illumination.   A more complete image description is

$$V(x, y) = i(x, y)R(x, y) \qquad (6.3.1)$$

In this representation, $V(x, y)$ is the received visual or video information, for example, at the face of a vidicon target.   The function $i(x, y)$, in this case, represents the illumination function that corresponds to physical illumination by some incident energy source (e.g., visible light). $R(x, y)$ can represent a reflectance function if we are imaging by reflectance, or a transmittance function if the object is being

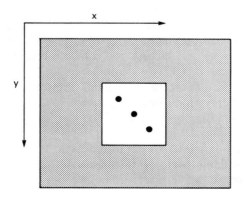

Figure 6.3.1.   Example of an intensity function $I(x, y)$ for an image.   One side of a cube (e.g., the three side off a die) is shown.

transilluminated (e.g., as a slide or transparency would be projected).   In general, $R(x, y)$ contains the information-bearing portion that will be of interest in the robot vision system.   Frequently, the illumination source is designed to be constant [i.e., $i(x, y) = k$], so that the measured signal $V(x, y)$ in Eq. (6.3.1) closely matches the desired image function $R(x, y)$.   The image illumination $i(x, y)$ may be manipulated so as to take advantage of some natural property of the scene.   For example, one may illuminate the field with red light so as to bring out certain "red" portions of a scene and to subdue other "less red" features.   One can accomplish a similar effect by placing a red filter in front of an imaging lens.   Other illumination variations may be designed to light a portion of the scene, for example, with a line of light, so that only certain edges or aspects of a field being viewed are imaged.   Illustrations of this are the structured vision system developed for use with robots by the National Bureau of Standards (see Figure 6.3.2), the General Motors Consight system, as well as several commercial products (see Figure 6.9.1).

The image function $V(x, y)$ in Eq. (6.3.1) represents a continuous measurement of intensity, as a function of the horizontal and vertical space variables x and y.   In Figure 6.3.1, the x axis was oriented from left to right, and the y axis from top to bottom.   This convention is taken from the normal raster scan of television, where the scan progresses from upper left to lower right.   The television type of image scanner naturally digitizes the y axis, since the raster scan generates only individual horizontal scan lines.   In the American Standard, there are 525 individual horizontal lines.   Of these, approximately 40 do not contain active video information from the scene.   These 40 are used for timing and synchronization and consequently only about 485 lines of valid image information are available.   The natural quantization of the y axis can be extended to the x axis, so that the intensity elements of a field will be defined at discrete points in space.   This has an obvious advantage if the image is to be processed digitally, as is the case with almost all robot vision systems.   Finally, one may digitize the intensity amplitude in addition to the x and y coordinates, so that the entire picture is digital in nature.   The picture in this form is said to be made up of individual picture elements, or *pixels*.   Each pixel's intensity is said to be its *gray level*, since the intensities have a value from "black" to "white."   Figure 6.3.3 shows a representative digitization process for the picture elements for a rectangular letter "O".

The question of sampling density must be raised since there are considerations of the sampling rates that must be made.   The other physical components of the imaging system (e.g., the lens and target) may cause the image formed on the target to be imperfectly focused.   This will cause the scanning system to collect information from a finite-sized aperture.   This and other effects limit the spatial resolution, usually to a degree that does not cause problems with regard to undersampling the image components (i.e., does not violate the Nyquist sampling theorem).

The amount of digital data that is used to represent the image varies widely.   Although 525 lines are available from the raster-scanned image, the x and y axes are typically digitized from 64 to 512 pixels.   The number used for (x, y) sizing is

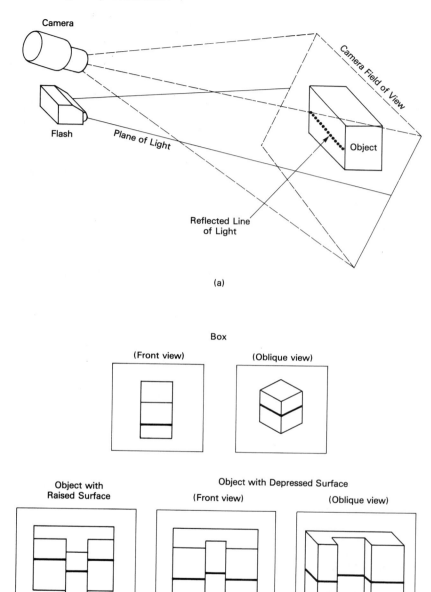

Figure 6.3.2. (a) Structured light imaging for the NBS robot vision system; (b) example objects and the line segment pattern formed by a plane of light for a box and an object with both a raised and depressed surface. In all instances, a dark line represents reflected light and is the only image "seen" by the camera.

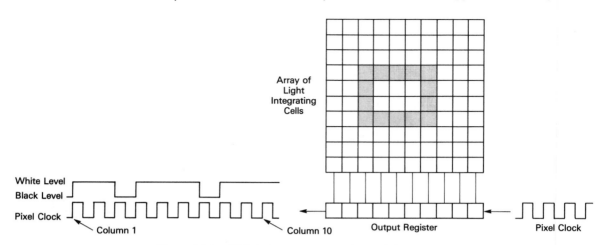

Figure 6.3.3. Digital picture representation and data readout. The signal from the output register is shown for row 5 or 6.

often chosen to be an integral power of 2, and often a representation consisting of less than the full resolution is adequate. These decisions are judgmental and require analysis of each specific application. This would then require a matrix of pixels from 64×64 to 512×512 (i.e., from 4096 to 262,144 picture elements). The intensity quantization is usually from 1 to 8 bits, so the amount of memory required for storing images typically ranges from 4096 to 2,097,152 bits. Translating this into 8-bit bytes means that from 512 to 262,144 bytes must be available for storing images. In most of the image systems available, the images are stored in "video RAM" (random-access memory) on the order of 128×128 to 512×512 pixels. Each of these pixels may be from 1 to 8-bits. Since these data are usually acquired in 16.67 ms, the data rates are on the order of 2 megabytes/s (up to 7.8 megabytes/s for a 512×512 image at 30 frames/s). (A *frame* consists of the 525 lines scanned by the vidicon camera. Every frame is composed of two interlaced *fields*, each consisting of 262.5 lines.)

6.4 HARDWARE CONSIDERATIONS

An exhaustive treatment of hardware for vision systems would require hundreds of pages, so our attention will be focused on the major considerations. Once again the consequences arising from the use of standard television specifications will direct our primary discussions. We will also briefly treat some nonstandardized situations.

Since commercial television cameras are serial access devices, and since most vision system processing will require random access to the picture information, one commonly encountered problem is the storage of pixels for future reference. This generally requires an analog-to-digital (A/D) converter in a configuration similar to that shown in Figure 6.4.1.

The pixel information must be sampled and digitized so that the information may be made available to a computer or microprocessor. The A/D subsystem essentially matches the rate at which video information is available to the rate at which a computer can acquire it. If we are dealing with 262.5 lines per field, the sampling period required is approximately 240 to 300 nanoseconds (one line/256 pixels = 62.5 μs/256). Unfortunately, ordinary microprocessor components cannot directly keep up with this data rate since typical instruction times are in the range of 0.1 to 5 μs. This mismatch in speed may be compensated for by sampling more coarsely, and taking more frame times to accumulate the picture information. For example, one may use components with a long time constant and sampling aperture and sample only one pixel every other frame (assuming that the raster is interlaced), the digitizer must return to the same field to collect the data from a uniform grid of spatial points. If this approach is taken and a 256 × 256 matrix of image points is required, the time for image acquisition is 256 × 256/30 = 2184 s or 36 min., which is obviously unacceptable. One may take a single pixel per line and capture 256 pixels per field; the time required for acquisition would then be 256/30 = 8.5 s, which is much more acceptable but still too slow for practical robotic applications. Even if one took 10 samples per line in order to match the instruction times of a typical microprocessor, we would still need to return to the same line 25 times, so it would require about 25/30 = 0.83 s to acquire an image, which is still too long for most practical applications. The generally applied solution to the time problem is to use a frame buffer that is capable of storing an entire image. These frame buffers, which usually have a built-in digitizer to convert the data to digital form, are available off the shelf in a variety of configurations that match the bus specifications of almost any minicomputer or microprocessor. Using such a device then permits higher-level languages and general-purpose image analysis algorithms to be applied to what is essentially a large array of data points.

In many robotic applications where a manipulator is to grasp an object for placement elsewhere, the silhouette of the part is very often sufficient to permit orientation of the manipulator. In these situations, the image required for use is said to be *binary*; that is, the pixels describing the object need only one bit to

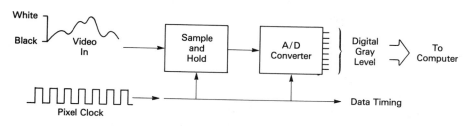

Figure 6.4.1.   A/D subsystem for converting analog video signals to digital gray levels.

describe the presence or absence of light intensity with respect to the portion of the object being viewed at that instant. Since it is inefficient to use an 8-bit byte to store binary pixels, binary images are often stored in a packed format. Packing allows 8 pixels to be stored into each byte, therefore requiring only 8192 bytes of memory for each field. (Verification of this is left as an exercise for the reader.) For gray-scale images one may require 32 kilobytes to 64 kilobytes of storage for 4-bit (16 gray levels) or 8-bit (256 gray levels) images. Even with the use of frame storage buffers, image-processing algorithms for robot vision must still be kept very simple if software techniques are to be used to process the images because so much data must be processed. Even moderately complex algorithms require considerable amounts of hardware to achieve sufficient preprocessing for general-purpose software to be applied. In some cases the hardware algorithm enhancements are so sophisticated that there actually is no need to store the image itself, but rather only a set of reduced picture parameters is accessed by the computer.

In addition to the considerations for image acquisition, the illumination of the scene itself influences the hardware considerations. In the simple cases, the lighting may be matched to the color or surface reflectivity of the objects under consideration. It may be necessary to use either colored light or colored filters to bring out certain features to match the color sensitivity of the camera. This often permits certain simple processing to be accomplished at the speed of light before the data ever reach the photosensitive target. One should remember that the speed of light is finite, and that light has a speed of about 1 ft/ns. By using oblique as well as normal perpendicular lighting, one may take advantage of shadows and surface anomalies, or one may choose to eliminate these effects. It is also common to use shutters and/or stroboscopic sources to "freeze" motion. Since the vidicon target is scanned at 60 Hz, motion blurring of moving objects is common, and a strobed light source can often be of great use. The principle in use is related to the fact that the vidicon will store the electrical equivalent of an image for a short time while the scanning is being completed. The strobe light essentially freezes the object's position while the camera stores its reflected light pattern. The scanning circuitry then transfers the electrical signal a short time later.

The appropriate use of transmitted and reflected illumination must be made to more easily satisfy the picture-processing requirements by the computing components, or there may be no feasible solution to the vision-processing task at hand in the limited time available. Again, the structured-light approach of the National Bureau of Standards is a good illustration of this principle in a robotics application (see Section 6.9.5.2).

6.5 PICTURE CODING

The representation of pictures has been briefly discussed previously. In this section we treat the topic more extensively and present the following coding concepts that represent the most commonly used methods in present practice:

- Gray-scale images
- Binary images
- Run-length coding
- Differential-delta modulation

6.5.1 Gray-Scale Images

Perhaps the best place to begin the discussion of picture coding is with the simplest distributional representation scheme, the gray-level histogram, which is a one-dimensional array containing the distribution of intensities from the image. Figure 6.5.1 shows a picture and its gray-level distribution. Assuming an 8-bit gray scale, the number of gray levels will be 256. One can see that the gray-level histogram destroys all geometrical information. This is illustrated simply by noting that if the image is rotated by any arbitrary angle, the gray-level distribution will remain the same. In a sense, the gray-level histogram is really an image transformation of a very simple type, so it is often very useful in evaluating imagery because of the enormous concomitant data reduction. For example, the original image may require 65 kilobytes of storage, while the histogram would require only 256 × 16 bits or 512 bytes, a saving of over 99% (512/65,536).

The gray-level histogram is related to the probability of occurrence of gray-level information. Using this interpretation, there are many useful ideas that evolve naturally. The average gray level, in conjunction with the minimum and maximum gray levels, can be inspected to decide whether or not the picture has an adequate contrast throughout the scene. This could, of course, be done automatically and the illumination increased or the sensitivity of the camera increased by opening the aperture setting on the camera.

Other picture properties may be deduced by measuring properties of the histogram, such as variance. The variance of a histogram is a measure of the "spread" of the distribution of gray values. The gray-level histogram function may also be used to determine a contrast enhancement function, so that the overall image quality is improved. For example, a simple linear contrast enhancement may be specified by amplifying the video signal and altering the offset. The mathematical expression for linear contrast enhancement is given as follows:

$$\text{new gray level} = K(\text{old gray level}) + B$$

where

$$K = \text{amplification or attenuation factor}$$
$$B = \text{bias or offset gray level}$$

The results of linear contrast enhancement are illustrated in Figures 6.5.1 and 6.5.2. Figure 6.5.1(a) shows an original poor contrast image with the corresponding histogram shown in Figure 6.5.1(b). The results of linear contrast enhancement and the corresponding histogram are indicated in Figure 6.5.2(a) and (b), respectively.

If one inspects the equation for linear contrast enhancement, it is easy to see

(a)

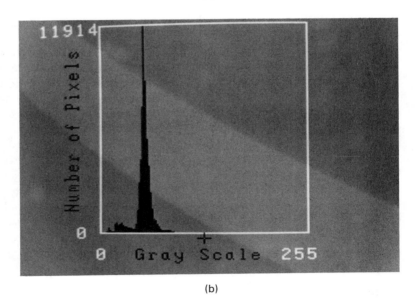

(b)

Figure 6.5.1. Original image and its corresponding gray-level histogram: (a) integrated circuit (original image); (b) gray-level histogram of part (a).

that if $K = -1$ and $B = 0$, the contrast-enhancement process becomes one of image inversion or image negativity (as in a photographic negative).

More sophisticated enhancement techniques may be derived from the histogram. For example, a procedure known as equalization may be applied to an image. Looking at the histogram in Figure 6.5.1, one can see that some gray

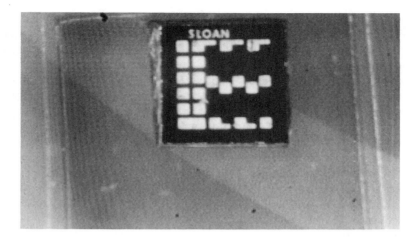

(a)

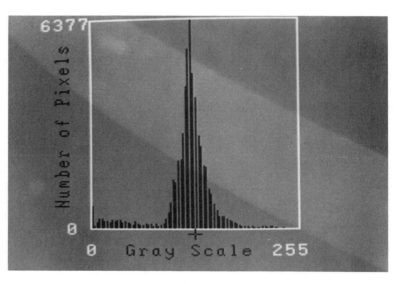

(b)

Figure 6.5.2. Linear contrast enhanced picture and its corresponding gray-level histogram: (a) contrast enhanced image; (b) gray-level histogram of part (a).

levels have a relatively low probability of occurring. If we redistribute the gray level to those bits that are relatively unused, we can increase the contrast in those gray level ranges where there is crowding of picture information. For instance, if the region where the most pixels occur is selectively enhanced at the expense of pixel intensities where few pixels occur, we obtain a more equalized distribution of information. In essence, this is increasing the use of the available display device. This is similar to the effect of an automatic gain control system (AGC) that one would find in a CCTV camera, whereby a special-purpose circuit measures the size of the input signal and automatically adjusts the system gain to maximize the amplitude of the electrical signal being output by the camera. It should be noted that a robot vision system (i.e., the processing circuitry) would not be affected by such an image alteration (since the processing circuitry is a quantitative system not subject to the contrast limitations of the human vision system), but the human being who usually sets up the system will have an easier task. Contrast enhancement features are often provided in robotic vision systems, so that the human operator will have more-well-balanced imagery to interact with during the training phases during setup of the vision system.

Other information may also be obtained by evaluating the histogram. For example, picture saturation may easily be detected by using the minimum and maximum gray levels found. The lighting may be controlled automatically to compensate for this distortion. The overall focus property of the scene may be determined as well, by looking at the range or the variance of the gray level as a function of lens focus position. In fact, an automatic focus procedure can be derived using a computer-controlled lens adjustment driven to the maximum range, provided that image saturation has not occurred.

The digital coding of images in gray-scale format is in many ways the simplest method, since this coding requires very few intelligent choices to be made. Basically, if one prevents the saturation of the picture intensity relative to the digitizer and the video amplification chain, the picture representation will reflect the scene being imaged. We shall see in the next section that binary image representation is very efficient with respect to memory storage, but the selection of a binarization threshold is often not a trivial process.

Gray-level coding requires anywhere from 2 to 8 bits, yielding from 4 to 256 gray levels. Standard television is specified so that only 10 gray levels can be perceived by a human observer, and most commercial television systems perform on that order. This means that 3 to 4 bits of digitization is sufficient to permit the maintenance of ordinary video standards. As far as the human visual system is concerned, a 4-bit picture is visually pleasing, but the discrete gray levels due to the digitizing may be noticeable. Under the best circumstances, human observers are able to distinguish 6 bits (64 gray levels) of intensity. Figure 6.5.3 shows examples of ordinary scenes digitized with 1, 2, 4, and 6-bits of information. Figure 6.5.4 shows computer-generated gray wedges to illustrate similar effects.

Gray-scale images may be digitized with little regard to image setup. However, the memory storage requirements are greater than for binary images, and the algorithms for processing gray-scale images are generally much more complicated and more time consuming than those used for binary image processing. This will become clearer in the following sections.

6.5.2 Binary Images .

Binary image coding requires that each pixel in the original image be coded into one bit: therefore the term *binary image*. In its simplest form, a fixed threshold may be applied over the entire scene. Figure 6.5.5 shows an example of such a process. Although this binary image appears visually pleasing, and a human being is easily able to recognize its content, in point of fact the image itself has been poorly digitized since there is an uneven or disproportionate distribution of black-and-white regions. In this figure the original subject matter was illuminated from one side, so that a significant shadow appears in the image and the selected binary image threshold is inadequate. In the case of a more realistic scene for robotic vision, this effect causes part of the object to extend past its actual limits, and part of the image to be eroded.

In the case of the binary portrait, it is important to appreciate the fact that no single binary threshold would be adequate under the lighting conditions used to present the image to the camera. Given this illumination, one could attempt to use an adaptive threshold that would adapt to the region locally surrounding the pixel to be binarized. One such adaptation technique is to use the local average intensity as the threshold. Figure 6.5.6 shows such a method applied to an image.

Because of the reduced memory requirements for binary image coding, as well as the reduced arithmetic requirements when dealing with a 1-bit pixel, many of the commercial vision products now manufactured frequently use binary images and have, to date, not generally used gray-scale imagery for object identification, location, and so on. This implies that shape and geometry factors, rather than gray-level textural parameters, are the most used in present-day robotic vision systems.

6.5.3 Run-Length Coding

Gray-scale and binary coding of images are direct methods for image coding, in that both systems maintain a map of the (x, y) coordinates and the corresponding intensity information. In the simplest form, this might be an array of intensity values, the array being as long as the number of pixels in the image. The term "data structure" refers to a representation of data in a structured manner useful for implementation and management by a computer system. The data structure for gray-level and binary images will be a continuous array in memory, whose index value *and* contents are directly related to a pixel location and intensity value. Figure 6.5.7 shows such a data structure for an 8-bit intensity mapped image. For this data structure, the location of the pixel under consideration is used to compute the index in memory associated with that pixel. In this example, the *entry index* is computed by taking the row number of the pixel and adding 256 multiplied by the column number of the pixel. (The row index ranges from 0 to 255, while the column index ranges from 1 to 256.) The actual *entry value* in the array is the

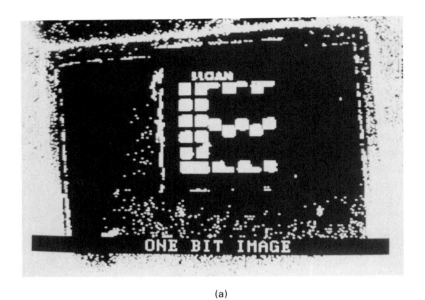

(a)

(b)

Figure 6.5.4. Computer-generated gray wedges (scales): (a) 1-bit gray-scale resolution; (b) 2-bit gray-scale resolution; (c) 4-bit-gray scale resolution; (d) 6-bit gray-scale resolution.

(c)

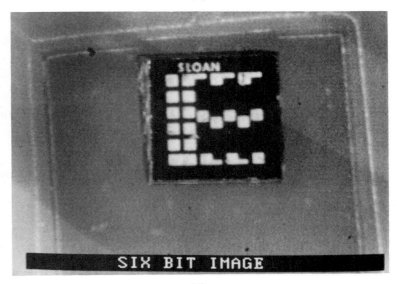

(d)

Figure 6.5.3. Continued

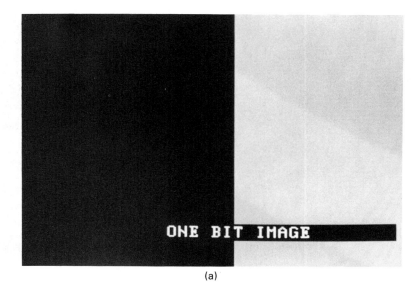

(a)

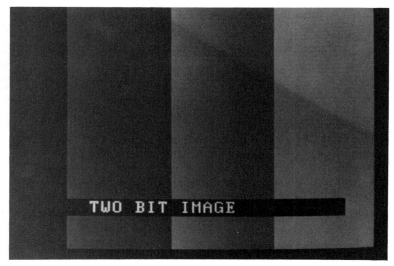

(b)

Figure 6.5.4. Computer-generated gray wedges (scales): (a) 1-bit gray-scale resolution; (b) 2-bit gray-scale resolution; (c) 4-bit-gray scale resolution; (d) 6-bit gray-scale resolution.

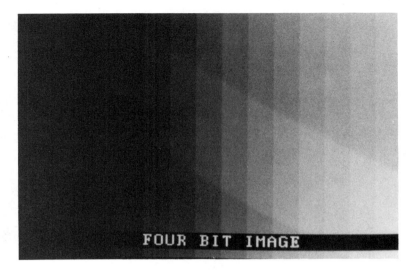

(c)

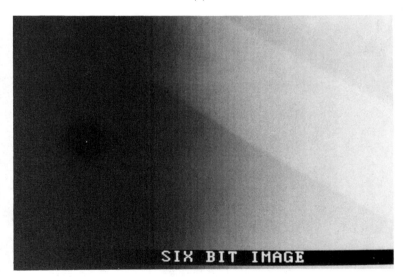

(d)

Figure 6.5.4. Continued

Figure 6.5.5.   Binary portrait image.

(a)

Figure 6.5.6.   Binarized images using the local average as a binary threshold: (a) original gray-level image; (b) original image with a 3 × 3 pixel local average subtracted; (c) binarized image with gray values of 124 to 132 mapped to black (all other values are white).

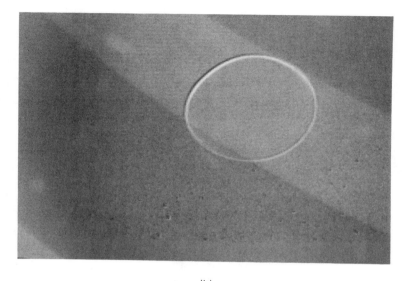

(b)

(c)

Figure 6.5.6.   Continued

digitized CCTV electrical value, which will range from 0 to 255.   For binary imagery, the data structure will require only 1-bit per pixel instead of 8-bits per pixel.

These direct coding and simple packing schemes are quite simple but do not take advantage of image structure.   If we look at the data structures above, es-

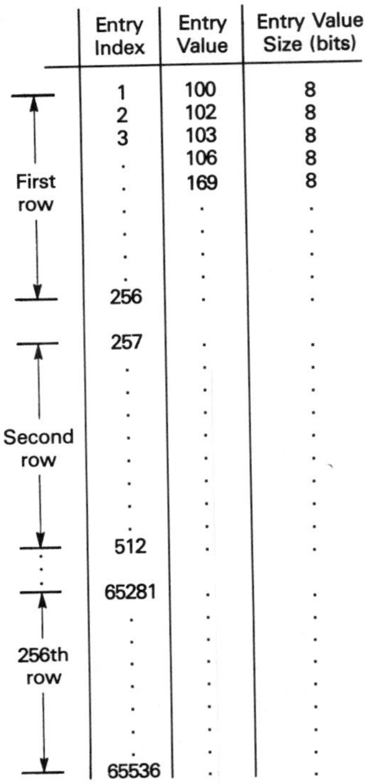

Entry Index	Entry Value	Entry Value Size (bits)
1	100	8
2	102	8
3	103	8
.	106	8
.	169	8
.	.	.
.	.	.
256	.	.
257	.	.
.	.	.
.	.	.
512	.	.
65281	.	.
.	.	.
.	.	.
.	.	.
65536	.	.

First row — (entries 1 to 256)
Second row — (entries 257 to 512)
256th row — (entries 65281 to 65536)

Figure 6.5.7. Simple data structure for an 8-bit intensity mapped image.

pecially for binary images, it is clear that there will generally exist long strings of pixels of the same binary value. In this case, a separate entry for each pixel is a waste of computer memory. If one simply stores the transition points and the string length, a (potentially) more efficient data structure can be applied to the images. Figure 6.5.8 illustrates such a "run-length coded" data structure.

For certain types of images that have a lot of "blobs," the memory requirements may be considerably reduced by the use of run-length coding. However, images with numerous small features may require more memory storage than the direct method. In many robotic vision applications, the use of run-length coding does offer a considerable saving. Gray-scale images may also be run-length coded, using a data structure similar to that shown above. Ordinarily, however, the run lengths are not as long, and memory savings are usually not as great as for binary images.

6.5.4 Differential-Delta Coding

Differential-delta coding (DDC) may also be used to code gray-scale images more efficiently. This coding technique uses the difference between the intensity of a pixel and the previous pixel. In ordinary scenes, the difference between successive pixels is not very large (on the order of 25% of the range or less), so the number

Entry Index	Entry Value	Entry Value Size (bits)
1	100	8
2	2	6
3	1	6
.	3	6
.	63	6
.		
256		
257		8
258		6
.		
512		
.		
.		
65536		

Entry Index (Transition point)	String Length	Value
1	100	0 (= black)
101	2	1 (= white)
103	50	0
153	3	1
154	25	0
180	.	.
.	.	.
.	.	.

Figure 6.5.8. Binary run-length coded data structure.

Figure 6.5.9. Data structure for DDC. The *first* entry value size of each row is 8-bits. All others require 6-bits.

of bits to encode the difference is two less than that required to encode the entire number representing the intensity. The data structure for this encoding is given in Figure 6.5.9. Figure 6.5.7 provides the data for Figure 6.5.9.

6.6 OBJECT RECOGNITION AND CATEGORIZATION

The preceding section dealt with the coding of images in a blind fashion, without intelligently addressing the issues of analysis of imagery from robotic applications. Although coding techniques are important, it is more important to understand the need for partitioning or segmenting the imagery from the real world. As was stated in the beginning of this chapter, several basic tasks are commonly encountered in robotic vision systems such as tracking, identification, and inspection. All these tasks require that the particular portion of the scene to be operated on can be extracted from extraneous picture information and can be analyzed efficiently. This naturally leads us to the two topics that will make the realization of these goals more likely:

- Dimensionality reduction
- Segmentation of images

6.6.1 Dimensionality Reduction

The direct coding of images using binary or gray-scale intensity coding results in storage requirements generally fixed and somewhat large (from 8 to 65 kilobytes and more).   Run-length coding and DDC generally reduce those figures when used for appropriate imagery.   These types of storage are still rather direct, in that they are coding the pixel values as they are found.   There is no basic intelligence embedded into the numbers stored to represent the picture.   In a sense, the images have been memorized, a photographic memory if you will.   We all know that those with photographic memories may not possess the wisdom to use the information in an intelligent fashion.   In a sense, the coding scheme makes up for the lack of intelligent coding, by brute force.   Once all the data for the image have been acquired, use must be made of the pixels to represent the content of the image rather than just the arrangement of the pixels.   The descriptions of the content of an image will generally yield a simpler description of the objects within the image field of view.   For example, one may describe an egg's orientation on a table as:

"egg is located at (x_0, y_0)

egg is oriented at an angle θ

egg short diameter is d_0"

This structured description of the egg's location is obviously much more efficient than the pixel description, and represents a significantly reduced dimensionality description.   However, these efficient descriptions would have little meaning out of the context of the example, whereas the pixel description would be more universally understood.

In a sense, this new data structure has been designed for the problem or product under consideration.   General methods for dimensionality reduction do not exist that are useful in all circumstances, but there are some methods that will be useful in many problems.   This is discussed in the next two sections.

6.6.2 Segmentation of Images

The dimensionality-reduction concept implies that the image must be processed in the context of specific situations.   The development of a self-teaching general-purpose vision system is a very remote dream.   It may someday exist, but for the present we must be satisfied with systems of a rather limited scope, albeit with high performance.

To develop an understanding of such high-performance systems, it must be understood that the reduction of dimensionality of images is an integral part of creating high-performance vision systems.   If these systems were required to process all the pixels in a sophisticated fashion, the amount of computation time and/or special-purpose hardware would be prohibitive.   By reducing the dimensionality of an image to a few salient features, we can then afford to spend a relatively long time evaluating those features.   To assist in this dimensionality reduction, we must develop mechanisms to isolate "regions of interest" (ROIs).   Isolation of the ROIs is generally referred to as image segmentation.

Segmentation can be attempted by many different techniques. We will treat the following methods:

- Color or gray level
- Edge detection
- Texture
- Regionization and connectivity

6.6.2.1 Color or gray level

Frequently, the color of an object is useful in separating it from other objects for the purposes of analysis. With proper lighting and proper filters in the optical path of the image sensor, one can often highlight the desired portion of the scene in order to accomplish segmentation.

Gray-level segmentation is also frequently used for image segmentation. For instance, the reflected light intensity from the surface of an egg can be used to define a ROI (region of interest) that separates the egg from the rest of the scene. Some of the more common techniques used in robot vision systems are silhouetting by backlighting of opaque objects, and floodlighting a scene to be analyzed so that the object is clearly well separated from the background.

6.6.2.2 Edge detection

Another popular and effective method of image segmentation is edge detection. In general, edges are portions of the image that have a high spatial variation of gray level. Not all such areas of high spatial variation contain edges. For example, in a scene with "salt and pepper" noise contamination, the spatial variation of gray levels will be high, yet there may not be a real corresponding edge in the scene. Assuming, however, that the imaging system has been set up with care, the use of the "edge emphasis" function may generally be useful in defining edges, and consequently used to isolate an object of interest.

There are numerous approaches to edge detection [5], but we shall consider some of the simpler algorithms only. The procedures to be specifically considered are:

- First difference/one-dimensional methods
- Sobel operator
- Contrast operator

Each of the operators requires consideration of at least two pixels; thus the local subimage shown in Figure 6.6.1 will be used. The local subimage matrix is interpreted as follows, A through I are intensity values [i.e., $I(x, y)$ values of the local image]. "E" represents the (x, y) location of the image point under consideration and A through D and F through I are E's neighbors.

```
A   B   C
D   E   F
G   H   I
```
Figure 6.6.1.   Local sub-image definition.

The first difference/one-dimensional operators are of the following generic type:

$$\text{Edge1} = |I - E| + |E - A|$$

$$\text{Edge2} = |F - E| + |E - D|$$

$$\text{Edge3} = |H - E| + |E - B|$$

$$\text{Edge4} = |G - E| + |E - C|$$

Edge1 computes a first difference function in the "northwest:southeast" direction. This particular edge function is most sensitive to edges perpendicular to the direction or orientation of computation and essentially returns a magnitude of edge in only one spatial direction. The functions Edge2 through Edge4 compute a similar type of one-directional edge property. Figure 6.6.2 shows the effect of the directionality biases of these operators.

The operators as shown above also incorporate some image smoothing, since they accumulate two measurements of edge contrast and then add them together. This type of operator will diminish the effect of an errant intensity point.

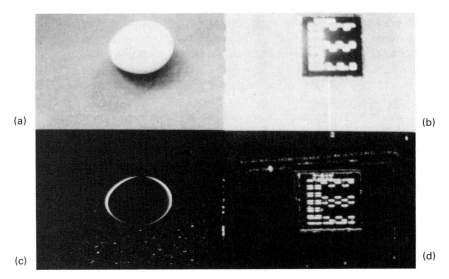

(a)   (b)   (c)   (d)

Figure 6.6.2.   Examples of one-dimensional edge operators on two images: (a) original egg; (b); original integrated circuit; (c) Edge2 horizontal difference; (d); Edge3 vertical difference.

Figure 6.6.3. Examples of Sobel and modified Sobel operators: (a) image of an egg; (b) Sobel difference operator applied to the image in part (a); (c) modified Sobel operator applied to the image in part (a); (d) difference between the images in parts (b) and (c).

The Sobel operator [5] incorporates edge information in two directions as follows:

$$\text{Edge} = (\langle A + 2B + C - G - 2H - I\rangle^2 + \langle A + 2D + G - C - 2F - I\rangle^2)^{1/2}$$

This operator computes a weighted-average intensity function along the borders of the subimage, and then forms two edge measurements perpendicular to each other. The two edge properties are then combined in a "quadrature" measurement. The Sobel operator generally enhances edges in an acceptable fashion. Note that it also has built-in smoothing of the local region. The squaring and square-root operations will be very time consuming, so one normally would modify this operator by replacing the squares with absolute values, and eliminating the square root in toto. This is referred to as the modified Sobel operator. Applications of the Sobel and modified Sobel operators are shown in Figure 6.6.3.

6.6.2.3 Contrast operator

Another approach to edge detection is the use of contrast differences. Consider the following:

$$\text{Edge} = E - \frac{A + B + C + D + F + G + H + I}{8}$$

This operator compares the intensity of the central pixel to the pixel's surrounding. Here, only the difference in contrast between the central pixel and its neighbors

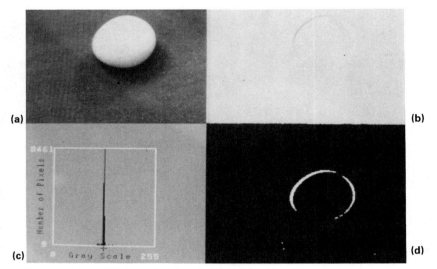

Figure 6.6.4. (a) Original 8-bit image; (b) contrast operator applied to the original image that was offset by 128 to avoid negative numbers; (c) gray-scale histogram of part (b); (d) binary image of part (b) with gray values of 124 to 132 mapped to black (all other values are white).

is considered, regardless of the distribution of the neighboring pixel intensities (see Figure 6.6.4 for examples).

The use of edge operators, as described above, can be used to expedite segmentation of images. For example, given that edge information is available, the extent of the object of interest may then be defined. The edge detection functions defined above have the property that they may be computed sequentially as the image is scanned (i.e., their need for random access to the image contents is minimal). The disadvantage to these procedures is that they characterize only edge points, and do not characterize edges. This is an important distinction because the edge points that are located are done so independently of each other, and there is no explicit or implicit connection among them. These techniques are useful, however, as implied above. For example, if one is scanning with a raster scan, the detection of an edge point may be used to activate another procedure for bounding the object of interest. In many cases, simply knowing the extent of the object in two dimensions is a good first step toward reducing the dimensionality of the image to be analyzed.

As there are serial- or sequential-access edge algorithms, there are also random-access edge tracking procedures that may be used for segmentation. One such procedure is a contour-following procedure.

6.6.2.4 Contour following

In contour following it is first assumed that we can determine whether or not a test point is interior to a region or exterior to it. A simple binary threshold technique may be appropriate for segmenting an object from the background, or alternatively, an edge detection function as described previously may be used to define a band of pixels surrounding the object. Another possibility would be the use of an adaptive threshold, as described in Section 6.5.2.

Contour following utilizes the following technique:

Find an edge point (between two pixels), and consider two test pixels diagonally ahead (with respect to the direction approached) of the current position. These pixels will be used to test if a candidate edge point is inside or outside of the region.

If the current point is inside the region, i.e., both test pixels are inside, turn left until the region is exited.

If the current point is outside the region, i.e., both test pixels are outside, turn right until the region is entered.

If the current point is indeterminant, i.e., one test pixel is inside and the other is outside the region, go straight.

This procedure is graphically illustrated below. (Note that "x" marks pixels interior to the region and "o" marks pixels exterior to the region. Also, the "x" and the "o" are the only two pixels under consideration at the present time.)

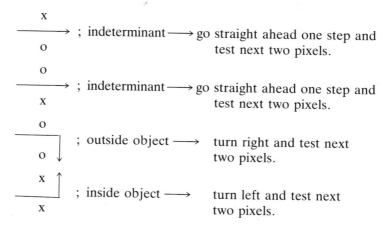

Figure 6.6.5 shows an example of contour following. In this example, the starting direction is toward the right at coordinate location (3, 9). The dotted line then indicates the contour path followed.

This contour following method is capable of following closed regions well, but will also follow any nook and cranny that exists in the contour. It must be emphasized again that proper original scene illumination can make the difference in any algorithm used to perform the required task.

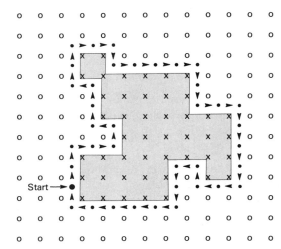

Figure 6.6.5. Example of contour following. "X" represents an interior pixel and "O" an exterior pixel. The shaded area is the binary image of the object.

6.6.3 Object Description, Categorization, and Recognition

Although the types of segmentation described in the preceding section are not the only types, they are often used for the purpose of isolating objects. Once objects have been segmented from the extraneous information in the scene, numerous methods can be used for object description, categorization, and recognition. There are two very common description techniques for objects. If the object is one that requires shades of gray to characterize it, the object will simply be described by a pixel map, with sufficiently high pixel resolutions in both space and light intensity. This description is not especially efficient but it is useful for comparing a known object with a test object.

The more usual situation is with objects that do not require gray-level description (i.e., a binary description is sufficient). In this case the outline of the object, along with interior holes, is sufficient. If an object has different mechanically stable states, several different outlines will be necessary to describe the object properly. It must be made clear that the outline that may be used to describe a binary object must have a data structure that is amenable to storage in digital form for later reference. One especially simple data structure is an ordered list of the (x, y) coordinates of the outline, but this is not very efficient or convenient.

Another more useful data structure for encoding the outline of an object is the chain code. Here a sequence of unit steps is utilized in one of the four or eight directions that may be traversed in reaching the next point on the outline. Such a coding scheme is illustrated in Figure 6.6.6 for a simple object. In this example, eight directions are used, and the "*" represents a point on the outline. Note that starting at different points on the outline gives different chain codes, but they can be reconciled with software techniques that are beyond the scope of this presentation.

```
Start at  ------> +*******
the "+"            *      *
and proceed        *      *
clockwise          *******

Chain code = <e,e,e,e,e,e,se,s,s,
              w,w,w,w,w,w,w,nw,n,n>

where      e = east

           w = west

           n = north

           s = south

           ne = north-east

           se = south-east

           nw = north-west

           sw = south-west
```

Figure 6.6.6. Chain code example. Starting at the "+", the sequence of pixel moves that defines the outline of an object is represented by the direction abbreviations shown within the $< >$, i.e., 7 east, 1 southeast, 2 south, 7 west, 1 northwest, and 2 north.

The use of gray-level or chain code object memorization is a simplistic approach to object description that is used in practice. However, one must recognize that description is only part of the problem. Given that a "master drawing" of the objects of interest is available, the real problem is to accomplish recognition of a new object. For example, assume that a known good part is available for characterization. The part would have to have some properties measured and stored. The outline and gray scale pattern of each mechanically stable state would also have to be memorized.

It can be shown that these simple data structures for object description require a significant amount of computer time for implementation (see Problem 6.7). The solution of the rotational and translational problems requires a three-dimensional search and comparison with the master drawing. To reduce the complexity of comparisons, it is very common practice to use less than complete geometrical descriptors or "features" for object characterization.

The most common techniques simply use parameters such as:

- Size or area
- Range of object projected onto x and y axes
- Ratio of the extent of the object in the x and y directions
- Center of gravity of gray scale or binary rendition of object
- Geometrical moment description
- Number of holes in the object

Figure 6.6.7 illustrates some shape and geometry features commonly used in object description. These features clearly reduce the dimensionality required to characterize the objects and will reduce the computation time for object identification. Since the dimensionality of the descriptions is reduced, we should expect to trade-off something in return. In fact, we will find that these features do not uniquely characterize the object, so there exist an unlimited number of different

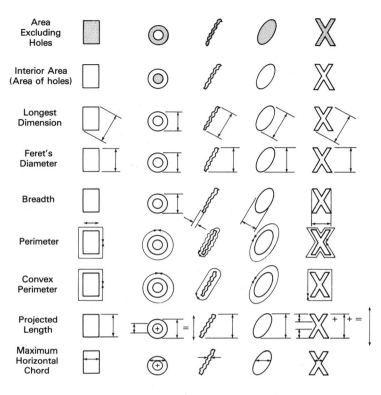

Examples of Derived Size Measurements

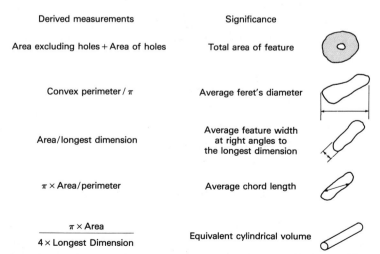

Derived measurements	Significance
Area excluding holes + Area of holes	Total area of feature
Convex perimeter / π	Average feret's diameter
Area/longest dimension	Average feature width at right angles to the longest dimension
$\pi \times$ Area/perimeter	Average chord length
$\dfrac{\pi \times \text{Area}}{4 \times \text{Longest Dimension}}$	Equivalent cylindrical volume

Figure 6.6.7. Commonly used shape and geometry features.   Redrawn with permission of Dynztech Laboratories, Inc., Imaging Products.

objects with the same parameters. Fortunately, in a real robotics application, we will know if these cases exist and add additional features to remove ambiguity.

One good method for categorizing an object is to derive a feature vector (i.e., a set of measurements describing the object) and then to compare that set with the known good set. The differences or error between the two vectors may be used to decide if the test object is in the same category as the object used for training. Obviously, this would have to be done for each stable state. Such a process is known formally as the "nearest-neighbor technique" [4] and is one of many statistical pattern-recognition techniques used for categorization. Unlike many statistical pattern-recognition problems, most robotic vision problems are not usually very statistical in nature, since the system designer can exert control over many processes (i.e., the engineer is often able to "design for automation").

6.6.3.1 Image comparison

In many applications, a reference image or sub-image is available. The reference or standard sub-image may represent the object to be located in subsequent test images. The sub-image is frequently used as a mask or memory engram for comparison with incoming test imagery. This search task may generally be described as:

$$O(x_0, y_0) = G\{T(x - x_0, y - y_0), t(x, y)\}$$

where

$$T(x, y) = \text{reference sub-image}$$

$$t(x, y) = \text{test image}$$

$$(x_0, y_0) = \text{spatial translation offset}$$

$$O(x_0, y_0) = \text{output response}$$

$G\{\cdot\}$ is referred to as the comparison operator.

Frequently, $G\{\cdot\}$ is chosen as a linear, two-dimensional convolution operator. Alternately, it can be defined as

$$G\{\cdot\} = \sum |T(x - x_0, y - y_0) - t(x, y)|$$

thereby representing the sum of the absolute differences of the reference and test image as a function of the spatial location of the reference (x_0, y_0) with respect to the test image. This technique is also known as a "nearest neighbor classifier" since the nearest neighbor to $T(x, y)$ in $t(x, y)$ will have the lowest score and will be the closest "relative." The summation is carried out over the image region in which the sub-image is coincident with a portion of the test image. In general, the sub-image is displaced spatially (in both x and y) so that it is placed over the entire test image.

Other operators may also be created. One of the simplest is called *binary correlation*. As expected, both the image and reference are binary in nature. This process may be defined as follows:

- A template is defined. For binary correlation, this is usually part of an image. It is important to note that it may take the form of a line, a square, a rectangle, or even a disjoint set of pixels (that maintain a fixed spatial relationship to each other).
- The template is overlaid on the test image. Some location on the template is chosen as a reference point (e.g., the top left corner). The corresponding point on the test image defines the coordinate where the results of the operation will be placed.
- The number (or fraction) of pixels on the template that match the portion of the test image over which it is placed are computed. A typical method of doing this is to sum the results of the complementary XOR function performed between each corresponding template and test image pixel.
- The value computed above is placed in a resultant image at the location corresponding to the template's reference point.
- The procedure is repeated until the template has been overlaid on each pixel location in the entire image.

Figure 6.6.8 shows an original binary image, a template, and the resultant image after the binary correlation process has been applied. As can be seen in part (c) of this figure, the locations where the best match occurs have the highest value. Various methods can be used to provide more image data in cases when the template extends beyond the image. One possibility is to add additional zeros. This was done in the example shown in Figure 6.6.8. Two columns of zeros were added (but not shown) to the right of the image and two rows of zeros were added at the bottom. Thus the top left edge of the (3×3) template could be placed on each pixel of the image.

While binary template matching is quite simple to implement (in hardware as well as software), some problems exist. Recall that generally information about an image is lost when it is binarized and pixels that constitute an edge must either be white or black. Thus accurate positional information may be lost. Additionally, since both the image and template are binary, it is important to have approximately the same threshold and lighting conditions when performing correlation as was used to define the template. This is because changes in scene illumination (and threshold) may cause the test image to differ from the image used to define the template, thereby resulting in incorrect matches or no match at all.

6.6.3.2 Template matching

As a matter of interest, the same procedure used for binary correlation can be used to perform edge detection or other mathematical operations on the image. When used this way, the process is referred to as template matching. In this case, the template (or model) is used to detect some property of the image such as an

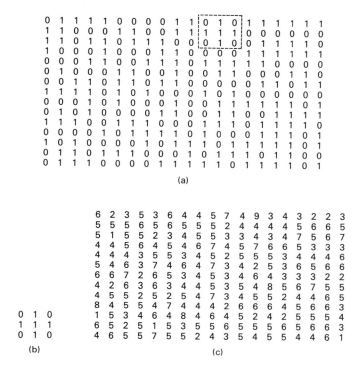

Figure 6.6.8. Example of template matching applied to a binary image: (a) binary image array (15 × 20); (b) binary template; (c) output of binary correlation (14 × 19). The dashed box in part (a) indicates a perfect template match has been achieved.

edge. For example, if both templates shown in Figure 6.6.9 are applied to an image and the square root of the sum of the squares of both process outputs at each (image) pixel taken, the edges of the resultant image are enhanced. These templates perform the same function as the Sobel operator described previously. In this case the reference location of the template would be the center pixel.

6.6.3.3 Correlation for gray-level images

The same concept used for binary template matching can be extended to a gray-level image and model (or reference sub-image). However, in this case it should be apparent that a high correlation could occur for features in the image that look nothing like the model due to the fact that they may be brighter than the pixels that match the model. Thus the method of cross correlation may not be adequate to uniquely define the feature (part of image that best matches the reference) that is being sought.

If the Fisher statistical correlation coefficient is used, the best match of a given gray-level model in a gray-level image can be found, independent of variations

1	2	1
0	0	0
−1	−2	−1

First Operator

1	0	−1
2	0	−2
1	0	−1

Second Operator

Figure 6.6.9. Templates used to implement the Sobel operator (described in Section 6.6.2.2) for edge detection.

in illumination. Essentially, we are able to pick out the feature based on the maximum value of the correlation of the model with the image. By definition the correlation coefficient is given by:

$$\rho(x_0, y_0) = \frac{N \, \Sigma I_i M - (\Sigma I_i)(\Sigma M)}{\sqrt{[N \, \Sigma I_i^2 - (\Sigma I_i)^2][N \, \Sigma M^2 - (\Sigma M)^2]}}$$

where

I_i = a sub-image of the test image which is dimensionally the same as the model (or template)

M = the model or reference sub-image

N = the total number of pixels in the model

(x_0, y_0) = the spatial location of the model with respect to the test image

Note that the model can be any size $(q \times r)$ with the provision that it is smaller than the test image.

The template or model is moved across the image as described previously. However in this case, besides multiplying the corresponding gray levels of the template and model, both the average value of the sub-image and the model (and the variance of each) are needed. Note that the average value of the model and its variance are constant. Variations on the equation above can simplify computation. For example, the correlation coefficient can be squared to remove the burden of taking the square root in the denominator.

A few interesting conclusions can be inferred from the use of the correlation coefficient. First, a value of 1 indicates a perfect match and if values less than zero are ignored, the maximum value obtained over the image defines the location of the best match. Of course, the higher the correlation coefficient the better the match. Second, if a feature in the image is partially degraded, the correlation coefficient will still generally identify it and its location. Third, all the locations in an image that matched the model could be found by simply looking for all the correlation coefficients above some specified threshold. Finally and as stated previously, the correlation coefficient is independent of linear changes in brightness. That is, assuming all the pixels of the image (or the model for that matter) are modified by the function

$$I(x, y)_{new} = a \, I(x, y) + b$$

for $a > 0$ and any b, the correlation coefficient is unchanged. The parameters a and b can be considered as gain and offset for either the image or the acquisition device.

If one chooses to make a template that defines some specific feature of an object such as a corner or edge, it is possible to use the gray-scale correlation technique just described to find these specific features. In this case, imagine a $(q \times r)$ model with the left half, $q \times r/2$, entirely white and the right half black. Using this template (or model) with gray-level correlation enables one to find all the areas of the image that resemble an edge having a light to dark transition.

6.6.3.4 Morphological image processing

In morphological processing, an output image is considered as the next generation of an input image. Essentially the processing may be carried out many times (i.e., the output becomes the input, the input is operated on, and becomes the next generation output) to cause certain characteristics of an image to be enhanced, removed, or otherwise changed. The operations on an input image consist of a set of rules that operate on the value of a spatially specified input pixel and its neighboring pixels. The rules generate a value for the pixel in the output image having the same spatial location as the input pixel. While morphological processing is commonly applied to binary images, it can be applied equally to gray-level images. However, there are no equivalent gray operators for each of the binary operators.

To briefly illustrate this type of processing, two of the most common operators, erosion and dilation, will be examined. Erosion in its simplest form, replaces a pixel by the local minimum of its neighborhood (e.g., a 3×3 local region). For this operator, bright objects will decrease in apparent "size" thereby becoming "eroded." (Of course, dark objects will increase in apparent size.) Dilation in its simplest form, replaces a pixel by its local maximum, thereby increasing the size of bright objects and decreasing the size of dark ones.

We will illustrate the morphological image procedure by considering a 3×3 kernel. In this case, imagine that each pixel surrounding the center one (as well as the center pixel) is the input to a transform rule. This rule generates a value of the output pixel which is placed in the output image at the same spatial location as the center pixel. Once again we may think of moving the 3×3 kernel area all over the input image to cover it completely. If we chose the logical AND of all the pixels in the 3×3 area as the transform rule, then the output will be a 1 only if a white pixel is surrounded completely by white pixels. This performs erosion since its effect is to reduce the size of white regions. Note that if erosion is applied enough times, it can completely eliminate all the white pixels of an image. As mentioned above, dilation is the opposite of erosion and corresponds to using the logical OR function for the pixel mapping rule. When it is applied to a binary image, the size of a white area is increased. Figure 6.6.10(a) shows an unprocessed binary image. Figure 6.6.10(b) shows the results of the white dilation operation performed on Figure 6.6.10(a) while Figure 6.6.10(c) shows the results of a white erosion (or black dilation) performed on the image of Figure 6.6.10(b).

If one performs erosion operations followed by dilations, small bridges between white objects will be broken. This cascaded operation is called an *opening*.

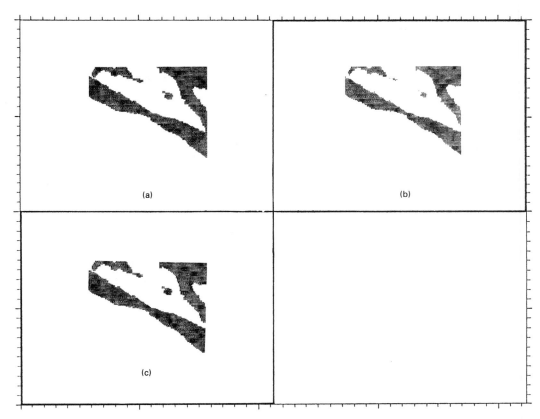

Figure 6.6.10.   Example of morphological image processing: (a) original binary image; (b) result of white dilation of the image in part (a); (c) result of white erosion of the image in part (a).

The operation of dilation followed by erosion has the opposite effect and closes up the spaces between adjacent regions.   This cascaded operation is called a *closing*.

Many other operators exist which allow filling in partially missing edges to restore the original shape of the image, measuring the extent of an image, and actually isolating objects and reducing their dimensionality to count specifically shaped and sized objects in a scene.

6.7 SOFTWARE CONSIDERATIONS

The techniques used for vision in robotic applications may rarely, if ever, be implemented totally in software.   Virtually all vision systems implement some algorithms in hardware.   The software considerations lie mainly in the ease with which the vision systems may be used.   Most of the present vision systems are, in essence, peripheral devices to the main robot controller and are invoked through a command/data structure that is rather simple. Basically, the peripheral (vision) device is given a string or stack of commands, and the peripheral returns data to the main processor. Some vision systems have their own user languages that range from cumbersome to friendly. For the most part, then, language and software considerations lie mainly in *control* of the vision peripheral, not in the actual implementation of the algorithms.

6.8 NEED FOR VISION TRAINING AND ADAPTATIONS

Although one might initially have believed that the definition of a prototypical imaged part is trivial, by now the reader should be aware that the amount of data required to define an object may indeed be huge in comparison to other digital data-processing applications. When considering the variety of degrees of freedom required to describe an object fully, it is evident that vision system training will be needed so that a reasonable amount of data may be retained by the vision system.

It is for these reasons that dimensionality reduction, as mentioned earlier, is so important. Specifically, it allows for the efficient representation of the visual data, usually in an independent manner. Another important consideration has to do with the potential for dealing with objects and parts that may be changing over time. For example, a conveyor belt carrying a part may not operate at a carefully controlled constant speed. As a consequence, parts will not arrive at known or precise intervals of time. In such an instance, tracking of the "trend" regarding parts arrival may be very useful in efficiently acquiring images and in processing the data.

Another case where this is true is in semiconductor assembly, where a die may be attached to a substrate by a die attach machine. If this machine has a slight but consistent drift, the imaging of the die for later bonding of wires to the lead frame and chip may be subject to placement errors, due to variable placement of the die. This type of adaptive updating of object positions is often necessary for efficient part handling.

6.9 REVIEW OF EXISTING SYSTEMS

In this section we review the major types of commercial vision systems currently available. The discussion of techniques used in these systems will be restricted somewhat, since robotic vision requirements may be very diverse. The major systems can be classified into the following type of general categories:

- Binary vision systems (utilizing either preprocessing or classification algorithms)
- Gray-level vision systems

- Structured light systems
- Character recognition systems
- Ad hoc special-purpose systems

6.9.1 Binary Vision Systems

Binary vision systems are those that use only two levels of image information. They are so-called silhouette systems, since very controlled lighting must be used to image objects reliably. Backlighting of parts is frequently selected so that the objects to be inspected stand apart from the background. The binary vision systems are used primarily for:

- Parts recognition
- Parts location
- Parts inspection

From a visual perspective, a binary vision device may be thought of as being able to operate on a part as if an inspector had picked up the part and held it up to a light source for backlighted inspection. One can see that there is a limited but useful class of information that one can glean from this procedure.

The SRI collection of algorithms is an example of a binary vision system. As typically implemented, it will permit arbitrary angular alignment of the part. A run-length-coded image is often produced because of the speed enhancements possible (see Section 6.9.5.3).

For objects where angular alignment is not an issue because of some prior orientation stage, but where the translational position is unknown, binary correlation techniques are often used. Binary correlation permits the object to be located with the value of correlation at the best match point often used as a measure of part quality. For arbitrary parts orientation this technique is not practical yet, because of the need to perform correlation in three dimensions (two translational, one rotational). Other binary vision systems frequently use so-called "pixel counting" for inspection of parts. These systems usually require spatial windowing of data and then counting pixels in those windows. This scheme typically requires significant application effort to choose the correct windows, and then requires a large degree of ad hoc adjustment to determine the significance of the pixel counts.

6.9.2 Gray-Level Vision Systems

Gray-level systems generally capture 4-, 6-, or 8-bit images, and then apply very tailored algorithms designed for a specific application. Gray-level template-matching techniques, for example, may be used to locate parts in nonsilhouetted environments. In many instances, highly controlled lighting may not be permitted, or the surface of the object has variable reflectivities that are useful for inspecting the object. Gray-level template comparisons may be used to locate objects that are angularly aligned, with the amount of template differences being used as a measure of object similarity, relative to a known "ideal" prototype.

6.9.3 Structured-Light Systems

The structured-light approach in general has proven very successful in numerous applications (e.g., see Section 6.9.5.2). We have already seen where backlighted objects are easy to analyze for certain types of object characterizations. Structured light carries that further by characterizing objects with slits of light, and then observing new samples in the same lighting environment. As a slit or plane of light falls on an object, various distortions and path deviations of the illumination may be seen and may be used to characterize location, orientation, and surface details. In addition to slits of light, one may also use "grids" of light and look for distortions in the grid pattern to characterize objects.

6.9.4 Character-Recognition Systems

It is often desirable to read labels or characters from parts, packages, and so on. Where bar codes may be placed on the parts to be identified, identification may be accomplished by simple bar-code readers. Alphanumeric codes are a different matter entirely, since recognition of arbitrary character sets has until recently been a very difficult image-processing task. Several systems available today are able to read a wide variety of character sets (after initial training) at high speeds (15 to 30 characters/s).

6.9.5 Examples of Early Robotic Vision Systems

A number of vision systems were developed specifically for use with robots prior to 1980. These include the GM Consight System, the one developed by the National Bureau of Standards, and also a system developed by SRI. We consider each of these in turn.

6.9.5.1 The GM Consight I system [12]

In the late 1970s, General Motors Research Laboratory demonstrated a robotic vision system that was capable of operating in the visually noisy environment often found in manufacturing installations. Called Consight I, this system was capable of determining the type, position, and orientation of a part on a conveyor belt without the need for enhanced (and often impractical) contrast techniques such as utilization of fluorescent paint on the belt's surface.

A block diagram of the system is shown in Figure 6.9.1. As can be seen from this figure, a minicomputer is used to process the scene detected by the solid state linear array camera (e.g., a Reticon RL 256L) that is mounted upstream from

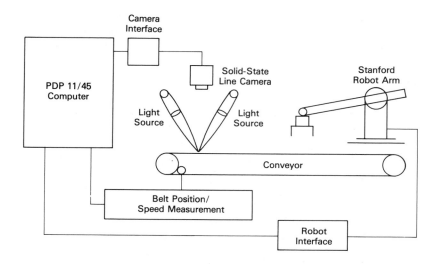

Figure 6.9.1.   GM Consight I system.

the robot's work station.   In addition, information about the conveyor's speed is sent to the computer.   In the time it takes a part to move from the vision system's location to that of the robot, the computer utilizes the visual and speed data to determine the location, orientation, and type of part and sends this information to the robot controller via an interface.   With this knowledge, the robot is able to successfully approach and pick up the part while the latter is still moving on the belt.   It is important to note from Figure 6.9.1 that the vision system is not mounted on the robot itself and, therefore, does not reduce the robot's useful payload.

It is necessary to monitor the conveyor's velocity continuously for several reasons.   First, typical moving belts that are found in factories generally do not have velocity servos controlling their speed.   Thus it is expected that the speed will fluctuate due to a variety of causes including load changes, line voltage variations, and wear of rotating parts (i.e., increased friction).   Since the robot must accurately know when the part arrives at its work station, the instantaneous belt speed must be available to the computer.   Second, keeping track of any belt speed variations has to do with the method used to acquire the two-dimensional visual scene.   This is discussed next.

As can be seen in Figure 6.9.2, the linear array camera scans the belt in a one-dimensional manner (e.g., in the y direction) and this is perpendicular to the conveyor's motion (e.g., the x direction).   Note that the camera will record 128 equally spaced points across the width of the belt.   The two-dimensional image is formed by instructing the camera to wait until the part has moved a specified distance down the conveyor before recording the next line image.   It should be clear that fluctuations in belt speed can produce distortion in the recorded image. This undesirable phenomenon is avoided by speed monitoring which permits the

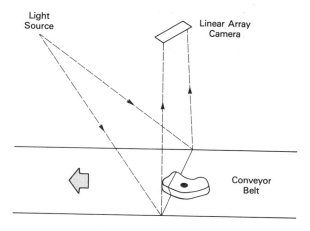

Light
Source

Linear Array
Camera

Conveyor
Belt

Figure 6.9.2.   Camera and light source configuration in the Consight I system. The basic light principle is illustrated.

time interval between acquiring two successive line images to be varied in order to compensate for any non-uniformity in the belt speed.

How does the Consight System avoid the problem of poor lighting conditions without resorting to contrast enhancing techniques? How does the system "know" if a part is present or not? The answer to both of these questions is through the use of *structured light*. With respect to Figures 6.9.3 and 6.9.4, it is observed that a light source, consisting of a long slender tungsten filament bulb and a cylindrical lens, which projects a linear (and fairly intense) beam across the belt's width is positioned downstream of the camera which is placed in a position so that it can sense this line of light. When no object is within the field of the camera, an unbroken line of light results. See Figure 6.9.4(a). However, when a part is present, the three-dimensional nature of the object causes a portion of the light beam to be intercepted before it reaches the camera position. When viewed from

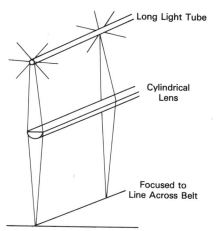

Long Light Tube

Cylindrical
Lens

Focused to
Line Across Belt

Figure 6.9.3.   Structured (linear) light source used in the Consight I system.

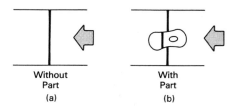

Without
Part
(a)

With
Part
(b)

Figure 6.9.4. Computer (line camera) view of a part: (a) with nothing on the conveyor, an unbroken line of light is seen by the camera; (b) a part causes an interrupted line of light to appear across the width of the conveyor.

above by the camera, this part of the line of light that is deflected by the object appears to be displaced (downstream) as shown in Figure 6.9.4(b). Thus the camera will sense a black image wherever there is an object and will record a light region where there is no part. As the part moves down the conveyor, the region(s) of black will change in length (i.e., y). The two-dimensional binary image recorded by the camera will, therefore, consist of regions of black (wherever there is a part) and white where there is none.

One potential problem with this procedure is shadowing, as illustrated by the dotted lines in Figure 6.9.5. Here, it is observed that the system will "detect" the leading edge of the part before it actually arrives at the camera position. This problem can be solved through the use of two or more linear light sources focused at the camera location on the belt as shown in Figure 6.9.5. The reader will observe that the second light beam prevents this position on the conveyor from becoming dark until the part is actually at the location.

The Consight System uses a *run-length coding* scheme for storing the two-dimensional binary image. Since the camera has 128 elements, only 7 bits are needed for this purpose. The remaining bit (usually the most significant) in any run length "word" is used to indicate whether the transition is light to dark or vice

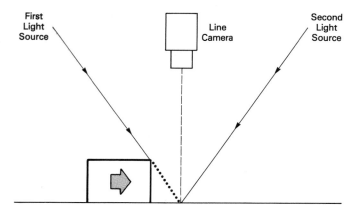

First
Light
Source

Line
Camera

Second
Light
Source

Figure 6.9.5. Use of two light sources prevents shadowing.

versa.   As mentioned in an earlier section of this chapter, it should be clear that unless the object being viewed has many holes, a considerable data compression will result from run-length coding and the memory required to store the processed data is far smaller than would be needed otherwise.   Moreover, it is found that processing time in subsequent steps is also decreased when this technique is employed.   Edge detection is accomplished utilizing a 6-connected region algorithm whereby connectivity is permitted along four sides of a pixel and along one of the diagonals.

Once the object outline has been found, the part must be classified and its position and orientation (relative to its leading edge) determined.   To do this, a small number of numerical descriptors (i.e., *features*) are computed or *extracted*. Some of the descriptors Consight employs are:

- Center of area (centroid)
- Axis of the least moment of inertia of the part silhouette
- Maximum radius point measured from the centroid to the image boundary

For a given object, comparing these and other simply computed features with those stored in the computer (for the entire "world" of permitted objects) allows the part to be recognized.

Orientation, a descriptor that is usually part specific, can be found in a number of ways including selecting the moment axis direction that points nearest to the maximum radius point measured from the centroid to the boundary.   This descriptor and the belt speed are then used to inform the robot when the object is within the workspace and where and how to grasp the part.   In some instances, it may be necessary to stop the conveyor for a period of time to permit the manipulator to acquire the object.

A major problem with the Consight system is that it cannot handle parts that are touching one another.   If such a situation occurs, the number of scan lines will usually be greater than for any single part.   Alternatively, no match between all the features of the two touching parts and those stored in the vision system's memory will occur (e.g., the overall area will generally be larger for the "compound object").   In either instance, the objects that cannot be "identified" are permitted to run off the end of the conveyor and into a reject bin where they can be recycled. It is important to understand that any object that can assume more than one stable position on the conveyor will require *a separate set of features* to be stored for each one.

Although the GM Consight was developed over ten years ago, it is still used commercially in a somewhat modified form today by the GMF Corporation and also by the Adept Corporation under a licensing agreement.

Having considered the various features of a vision system that was developed by a large private company, we next consider a robotic vision system developed with Federal money at the National Bureau of Standards.

6.9.5.2 National Bureau of Standards vision system [19]

In the 1970s, the United States Congress charged the National Bureau of Standards (NBS) with developing a fully automated machine shop by the latter part of the 1980s. As part of their effort to achieve coordinated control over robots and other less sophisticated machine tools (e.g., lathes, punch presses, milling machines, etc.), a need for a vision system that could be used in such an environment and could be interfaced with robots was perceived.

The research effort was undertaken by Dr. James Albus and his colleagues and produced a workable system in the late 1970s. The NBS Vision System, as it was first introduced, was able to process picture information in less than 100 ms and was estimated to cost about $8000. Since then, the system response time has been improved and some of the hardware has been modified. However, the basic operation technique has not changed appreciably as described next.

The major hardware elements of the NBS Vision System consist of three components: (1) a solid state camera capable of producing 16K pixels (128 × 128); (2) an electronic stroboscopic light that emits a *plane of light* and whose flash intensity can be modified digitally; and (3) a "picture processing" unit. To understand the operation of the system, consider Figure 6.9.6. [The reader should understand that the robot manipulator (in particular, its wrist) is not shown here. In actual operation, the camera and (structured) light source would be mounted on the robot's wrist whereas the processing unit would be in or near the robot's

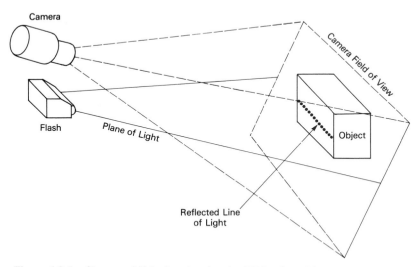

Figure 6.9.6. Structured light imaging for the NBS robot vision system. The camera and the flash (strobe) unit are mounted on opposite sides of the robot's wrist (not shown) such that the plane of light is parallel to the fingers of the gripper. The presence of an object causes one or more line segments to be seen by the camera. With permission of the National Institute of Science and Technology (formerly National Bureau of Standards).

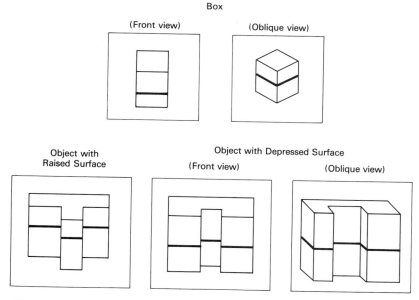

Figure 6.9.7. Example objects and the line segment patterns formed by the plane of light as seen by the camera for the NBS vision system. With permission of the National Institute of Science and Technology (formerly National Bureau of Standards).

own controller.] The strobe unit produces a plane of light that is projected *parallel to the wrist plane* (determined by the approach vector and *y*). The camera is mounted above the light source and is tilted down (i.e., so as to intersect the light plane). Its 36-degree field of view covers the region extending from inside the fingers of the gripper out to a distance of one meter. If the projected light strikes an object in this region, a pattern of *line segments* is formed on the object. See Figure 6.9.7. As the robot gripper moves closer to the object, these line segments will grow in size and will move *down* in the camera's field. Qualitatively, the reader should be able to conclude that the nearer the bottom of the image, the closer the object being scanned will be to the robot's end effector. However, how does this system provide quantitative information that will permit the robot to acquire the object?

To answer this question, consider the calibration chart (derived from simple geometric considerations) shown in Figure 6.9.8. Observe that the top and right axes are calibrated in pixels whereas the bottom and left axes are calibrated in centimeters representing the *x* and *y* distances between the camera and object respectively. For example, if the camera viewed the horizontal line shown in Figure 6.9.9 extending from pixel element (32,64) to element (96,64), the object producing this line would be located about 13 centimeters from the gripper and be about 10 centimeters in width.

The information contained in Figure 6.9.8 is actually stored in the vision

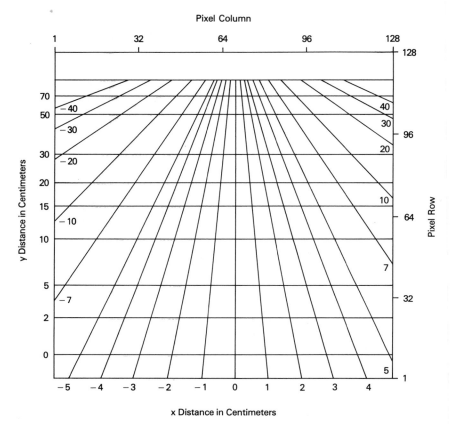

Figure 6.9.8. The calibration chart for the NBS vision system. The x and y distances are measured in the coordinate system of the fingers. The x axis passes through the two finger tips and the y axis is parallel to the wrist axis. The slight tilt in the figure is due to a misalignment of the chip in the camera. With permission of the National Institute of Science and Technology (formerly National Bureau of Standards).

system's processing unit and is used to determine the range and azimuth of each point of the reflected line segments. In particular, the interpretation algorithms use triangulation to extract range data, the slope of lines to indicate object orientation, and line endpoints to provide information on which edges of the object should be grasped by the robot. By studying the calibration chart in Figure 6.9.9, it should be evident that resolution of the system is coarse when the object is located far from the camera (i.e., about 1 meter) and is quite fine at short distances.

In actual application (e.g., a materials-handling problem) the task of acquiring an object that has been randomly placed within the one-square meter system workspace (e.g., a table top) is divided into a three-step sequence as described below:

1. The robot is commanded to go to a "home" position which is located at one

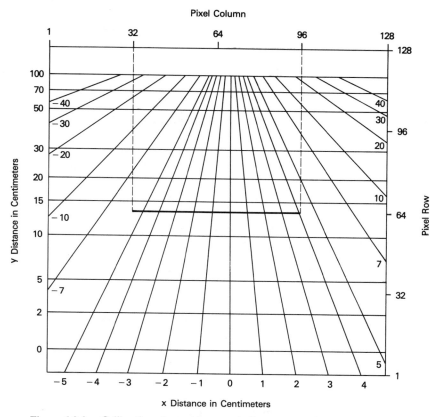

Figure 6.9.9. Calibration chart with example line segment (image) from (32, 64) to (96, 64) shown. With permission of the National Institute of Science and Technology (formerly National Bureau of Standards).

of the corners of the workspace. The table is then scanned (by firing the strobe unit) in a plane that is approximately parallel to its surface. The illuminated object appears in the image as a series of line segments. See, for example, Figure 6.9.7. Generally, this step in the process yields *coarse* range information.

2. The estimate obtained from step 1 is then used to move the robot's arm closer to the object. The flash unit is again fired and more accurate range information is obtained (recall that the resolution improves as the camera comes closer to the object because we are now operating in a higher resolution part of the calibration chart).

3. Based on the better range estimate obtained in step 2, the arm is moved above (or in front of) the object. The strobe is triggered a third time and the system makes fine positional and orientational corrections. The robot can then be commanded to grasp the object.

In the earlier versions of the system, there was a perceptible pause at each of the illumination points. A later version eliminated this delay thereby producing an extremely smooth motion. In addition, the processing speed was now so rapid that it was actually possible for the system to track a moving object.

It should be noted that the NBS system is fundamentally quite different from other machine vision devices because it is not "looking" all the time. Consequently, it is much more time efficient since only three picture scenes need to be processed (i.e., for the line segment information) during an object acquisition sequence.

Besides ranging and orientation data, it is possble to extract information on the *structure* of the object. This is also illustrated in Figure 6.9.7. For example, it is observed that an object with a raised portion of its surface will produce three disconnected horizontal line segments. This happens because the segment associated with the elevated section will be lower than the other two. In a similar manner, objects with depressed surfaces can be detected by the nature and number of disconnected line segments. The figure also indicates that obliquely viewed objects produce line segments that are connected but have different slopes (i.e., there is a cusp at their intersection). In addition, it can be shown that objects with cylindrical surfaces will produce *curved line segments* when illuminated by a plane of light.

Besides speed, another advantage of the NBS system over others is that the contrast problem is eased, or even eliminated, by the use of stroboscopic illumination. Even if the surface of the object is rather dull, it is possible to compensate for the "threshold" problem by increasing the flash duration under computer control. With the system used by NBS, the strobe time can be as short as 6.4 µs and as long as 1.6 ms. (This range is divided into 256 values.) The ambient light problem is solved by frame-to-frame differencing whereby data in the flash frame is compared with that from a nonflash frame at the same location. This does, however, require a "frame buffer" and hence additional memory.

At each step in the illumination process, a one-pass line following algorithm that looks for corners and gaps is utilized. For each image scan line produced*, the system calculates (in hardware) a run length of 8-bits and an intensity of 8-bits. (Note that gray-level information is utilized here unlike the GM system described in the previous section.) Based on this information, the possible run length of the next scan line is predicted (from the slope and curvature information of the "working" line). If the actual run length (RL) is within a specified ε, the point is added to the working line and the next run length is predicted. If, however, the RL $>$ ε, there are three cases:

1. RL = 128 which implies that a gap exists. The system will then begin to compute the gap width.

---

* In practice, the camera is made to read by columns first (bottom to top) thereby yielding only 128 data points/read.   Once the system senses a line (or a point on the line), it reads an appropriate row.

2. $\varepsilon < \mathrm{RL} < \Gamma$ (specified) which implies that the previous point corresponds to a corner (i.e., a cusp). The previous line has therefore ended and it is necessary to start a new one that has a different slope.

3. $\mathrm{RL} > \Gamma$ which implies that there is a *"depth discontinuity."* The system will terminate the previous line and compute the discontinuity.

If no line segments are detected or many faint segments are observed (based on the intensity information), the flash can be repeated at a higher intensity. On the other hand, if there is too much brightness in the image, which results in smearing, the flash intensity can be lowered. It should be clear that this adjustment in illumination intensity can be handled easily by the computer.

Although modified somewhat (e.g., the computer systems have undergone considerable change), the NBS vision system is still being utilized at the National Bureau of Standards in its automated machine shop project. At this time, however, it has not yet been commercialized.

Now that the reader has a good feel for two robotic vision systems and can see some of the similarities and differences between them, we next consider the one developed at SRI which is today being utilized by at least one commercial organization.

6.9.5.3 SRI industrial vision system [1]

A third robotic vision system was the one developed in the 1970s by SRI and had three objectives:

- Classification of objects
- Materials and/or parts handling by a robot
- Visual inspection of parts

In addition, these objectives were to be done for parts moving on a conveyor belt.

In the following sections, we describe four different aspects of the system that permit these objectives to be realized. They are:

- Lighting and imaging techniques
- Imaging hardware
- Feature extraction
- Automated parts recognition

Lighting and Imaging Techniques

It has been found that images that clearly show features of interest (e.g., holes), result in faster and easier image processing. To implement this "rule," the SRI system works only with *binary images* so that the pixels contain either black or white information. In addition, special lighting is utilized so that it is possible to employ simple thresholding techniques that can clearly indicate the relevant features of an object. This so-called *image contrast enhancement* is achieved in three different ways.

Contrasting Backgrounds. The first way to improve the contrast between the part and the background is to paint the conveyor belt with a red fluorescent. When the belt is illuminated with an ultraviolet light source, any nonfluorescent object appears *dark* against a *white background*. Note that another method of improving contrast is to utilize *backlighting*. One way to accomplish this is to place the object on a translucent plate and then view it from above. Although it is especially useful for hole detection, obviously such a technique is not applicable to the case of a conveyor belt.

Color Filters. Another technique that can be used to enhance the image is to mount a color filter in front of the camera lens. Clearly, this is only useful when the colors of either the object, the background, or *both* are known. In particular, when such a technique is combined with the fluorescent system mentioned above, a *red filter* (matched to the spectral response of the fluorescent belt) further enhances the image.

Special Lighting Arrangement. By illuminating an object in different ways, it is possible to control shadows and highlights. For example, *directional lighting* can be used to enhance shadows whereas multidirectional lighting can be employed to reduce them. Highlights (or reflections from shiny surfaces) can be enhanced by placing the illuminating source near the camera. Conversely, oblique lighting tends to reduce such highlights. Such techniques are generally referred to as *structured lighting* and are an important component of the SRI System.

Imaging Hardware

As mentioned above, the SRI Vision System is primarily aimed at applications involving parts that are being transported on a moving conveyor belt. Based on what was stated previously, the belt is painted with a red fluorescent and a red filter is placed over the camera lens. In addition, the belt is illuminated with an ultraviolet source. Thus the contrast between the background and any part is increased thereby improving the identification process.

Although it is possible to utilize a standard vidicon camera to acquire the binary image of the part on the conveyor, this is not done in the SRI system. The reason for this is that although, at the time of the experiments, it was possible to employ a high speed A/D converter to quantize the TV image to a raster of 120 × 120 points with 32 possible gray levels (i.e., 5-bits of brightness), the relatively large amount of memory (i.e., over 14K) needed to store the *entire* image being scanned by the camera made this fairly impractical*. Therefore, a linear array

---

* The reader should understand that with memory so inexpensive now, this would not pose much of a problem. Also, it is important to note that storing the entire image has the advantage of having it available for processing *at any time*. Thus features can be extracted anywhere in the process thereby permitting considerable flexibility in the image recognition algorithms.

camera consisting of 128 light sensitive diodes was used instead. With this device, one line of data, consisting of 128 one-bit samples was obtained each time the conveyor belt moved 0.05 inches. (Note the similarity between this approach and the one used in the GM Consight System.) During the time that the belt is moving to the next location, the raw data are processed (i.e., compressed) by converting it into *run-length format* (only 8-bits of information are needed since this is a binary system). At this point, the actual recognition process can be initiated.

Feature Extraction

To determine or *recognize* the part on the conveyor that is being scanned by the diode array camera, it is necessary to *extract* (compute) a set of *features* that permit the recognition to be performed. Before this can be done, however, it is necessary to obtain the outline of the object. In the SRI system, a *connectivity analyzer* is used to examine overlaps between lines in successive rows of the image, which in turn, permits connected components to be ascertained. In addition, holes located within the boundaries of the object are also found in this manner. In both cases, standard *edge detection algorithms* are utilized.

Once the part outline is determined, the initial or simple features are obtained. In the SRI system, these include:

- Area (the first *and* second moments are used)
- CG—Center of Gravity (area)
- Axis of the least moment of inertia
- Perimeter length (obtained from the list of perimeter points found by the connectivity analyzer)
- Extent of the object determined by the smallest enclosing rectangle
- Center of the extent (i.e., the extent rectangle's center)

With the above features now available, other features can then be determined. For example, the SRI system computes a *radius function*. That is, for each point on the perimeter of the object, the square of the distance from that point to the center of gravity (or extent) is found. Then, the maximum, minimum, and average radii become the features of interest.

Localizing features can also be used to determine the angular orientation of the object being viewed. They also permit attention to be directed to a specific portion of the part's outline. For this purpose, one can utilize:

- The maximum and minimum radius as determined in the manner described above
- Corners or notches (in relation to the maximum or minimum radius function)
- Holes

In the case of the water pump shown in Figure 6.9.10, the angle AOB between

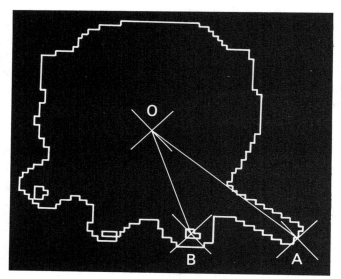

Figure 6.9.10.   Binary image of a water pump showing the hole B and the longest radius OA.

OA, the longest radius vector and OB, the vector formed from the CG to the nearest hole B, permits measurement of the position of the *movable* pump handle.

In the actual experiments conducted by SRI, there were 200 lines required to delineate the pump and with a PDP 11-40, image processing took about one second.   It should be clear that with the faster computers now available, this time could probably be reduced by quite a bit.

Automatic Part Recognition

One of the goals of the SRI Vision System is to be able to permit a robot to locate and acquire parts that are moving on a conveyor belt.   Besides having a complete *dictionary of features* for each of the "world" of objects that are to be distinguished, there are also other factors that affect the ability such a vision system has to reliably recognize an object*.

For any vision system, the designer must also take into account that variations in an image of a class of patterns are due to some or all of the following:

- Object rotation
- Object translation
- Lighting variations

---

* One refers to a system that can reliably operate under a wide range of conditions as "robust."

- Camera noise
- Quantization errors

Rotation and translation problems can be minimized by measuring (and using) features that are theoretically independent of those features. For example, area, total area of all holes located within the boundaries of the object, and the various radius statistics fall into this category.

For the remaining variations above, the SRI System makes certain important assumptions concerning the *statistics* of the features obtained for various objects being viewed. For example, it is assumed that the *variance* of these features depends almost entirely on measurement errors. That is, this statistical quantity is *class* (i.e., object) independent. While such an assumption is not 100% correct, acceptable results are obtained. On the other hand, the system also makes the assumption that the *class conditional* (i.e., the object dependent) distribution of any given feature is *normal*. This too is only approximately correct but once again yields acceptable results. In addition to the above assumptions, the system normalizes each feature's probability density by the standard deviation (SD) (i.e., it divides by the SD) thereby permitting a wide range of objects to be handled.

In actual implementation, a *sequential recognition* process is employed whereby a *binary decision tree* approach is utilized. Before generating such a tree, several *off-line* operations must be performed. Thus for *each object* included in the "world" of objects to be recognized, we must:

- Measure the distribution of N previously agreed upon features X_1, X_2, $\ldots$ X_N
- Compute $\mu_1(j)$, the mean value of the ith feature X_i for the jth class (object). See Figure 6.9.11.
- For each feature X_i, determine a *decision threshold*-ϕ in the middle of the *gap* that separates all the objects into two distinct groups (i.e., those with mean values above the threshold and those below it). With respect to Figure 6.9.11, it is seen that for the particular single feature X_i, the five objects being considered (i.e., classes $C1$ through $C5$) are divided into two groups ($C1$, $C2$, $C3$) and ($C4$,$C5$). Thus if the actual (i.e., measured) value of this feature for an unknown part is such that it is less than ϕ the part is probably either $C1$, $C2$, or $C3$. On the other hand, if the feature is greater than ϕ then the part is most likely either $C4$ or $C5$.
- Rank the N features by values of the largest gaps. It should be clear that the larger the gap, the more reliable the binary subdivisions will be and hence the object recognition.
- In the actual detection scheme, select for measurement the feature for which the largest gap is a maximum.

Once the operations above are performed, the decision tree can be constructed

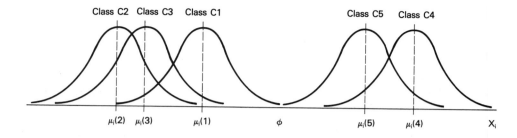

Figure 6.9.11. The Gaussian shaped distributions labeled Class C1 through C5 show the statistical distributions of the ith feature (e.g., perimeter) for a set of five objects. θ is the threshold that separates sets of classes.

and the actual detection process can be initiated. As an example, we consider a "world" of seven objects that are to be recognized consisting of four automobile parts (castings). Note that three of these have two stable configurations, e.g., a cylinder head (referred to in Figure 6.9.12 as Head 1 and Head 2), a piston sleeve (Sleeve 1 and Sleeve 2), and finally a disk brake caliper (Caliper 1 and Caliper 2). The connecting rod (Conrod) shown in the figure only has one stable state. Thus

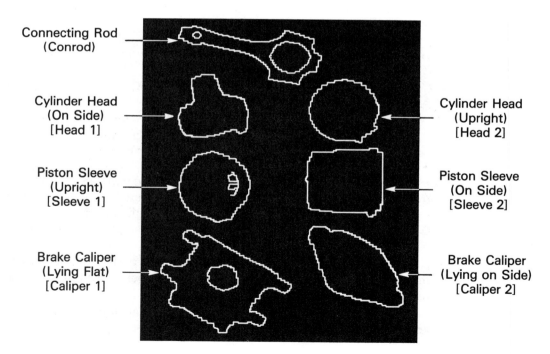

Figure 6.9.12. Seven outlines of foundry castings.

as far as the system is concerned, there are seven possible objects to identify. Moreover, for this example, the following seven features are used:

- X_1 = Perimeter of the figure
- X_2 = Square root of the area
- X_3 = Total hole area
- X_4 = Minimum radius
- X_5 = Maximum radius
- X_6 = Average radius
- X_7 = Compactness = X_1/X_2

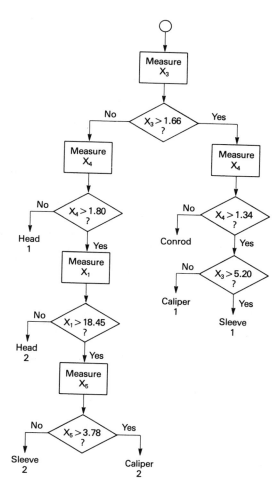

Figure 6.9.13. Binary decision tree for the seven images shown in Fig. 6.9.12.

Utilizing these features and based on the results of the "off-line" measurements described above, the decision tree for the seven objects results. As shown in Figure 6.9.13, the tree indicates that once the image of the object being viewed is stored in run length format, the system should first find X_3 (the total hole area) followed by X_4 (the minimum radius) and so on. It is interesting to note that for this rather small number of objects, only four of the features must be computed to accurately determine which object is on the conveyor. In fact, in the experiments conducted at SRI, an average of 2.7 features were required and recognition was accomplished in less than 1 second. As might be expected, most of the errors were found to be produced by poor binary images.

Although the SRI Vision System was developed over ten years ago, it is still being used in somewhat modified and expanded form by Automatix, Inc. under the trade name AUTOVISION*.

6.10 SUMMARY

This chapter illustrated many of the major aspects of machine vision, especially as applied to robots. The concepts presented permit the general understanding of the components, hardware, software, and algorithms that are often required in a vision or remote sensing application.

The reader should understand that the techniques presented here, while useful, will very often need to be modified before practical implementation is achieved. These modifications may be in the nature of using selected regions of interest, using computational approximations that provide for efficient implementation, or other similar techniques that permit the transfer of the theoretical or academic techniques into the real world of engineering and manufacturing implementation. Furthermore, the techniques presented in this chapter should be used as "vectors" that will point the implementer toward a good direction but will not give all the details required. For example, one may choose a horizontal edge operator for enhancing edges, but one also has to select the coarseness of the operator as well. The algorithm itself may be obtained from the literature, but the specific implementation parameters must be selected by the user.

Another issue of interest is the fact that the same problem may be solved in more than one way, and the method for selection of the most appropriate method will not be found in any textbook. The authors have sought to present many of the important techniques, and must regrettably leave the selection of the appropriate technique in any given environment to the user. The example in Chapter Nine serves to illustrate some of the problems associated with vision systems in robotics.

---

* For example, Automatix has developed and incorporated a special-purpose language called "RAIL," which permits the vision system to be more readily interfaced to the robot. In addition, modern 32-bit microprocessors have been utilized, which allow both binary and gray-level data to be processed.

6.11 PROBLEMS

6.1 a. Assume that an image has been digitized in x and y by increments of x_0 and y_0. To reconstruct the original continuous image, what is the maximum spatial bandwidth that the original image must have had?

b. What size scanning aperture would be necessary to prevent aliasing of the digital image in Part A with x_0 and y_0 as defined previously?

6.2 Assume that you have available a RAM of size 16,384 8-bit bytes. Organize a storage allocation map for this RAM so that you can store image representations from 32 × 32 to 1024 × 1024 pixels. Discuss and describe in detail the specific size of the pixel maps, with regard to x and y and gray level.

6.3 Draw a graph representing the general input–output relationship of linear contrast enhancement. Assume that an 8-bit gray-scale representation is to be used to encode each pixel. Draw graphs showing the transfer relationships for image inversion and image binarization.

6.4 For the data shown in Figure 6.5.7, design a generalized direct coding data structure, assuming that the intensity values are to be binarized at a threshold value of 106. Assume that the binary values from the direct coded binary data structure are to be redefined into a packed data structure, whereby each 8-bit word will now hold 8 pixels. Generate this new data structure and show an example.

6.5 Explain why the DDC method would not be useful for binary images.

6.6 Using square grid paper, construct a binary image of an ordinary object by taking a photograph from a magazine of an automobile, a wrench, or some other "blob"-type object. Construct the binary image by tracing the object on the graph paper and then marking each "pixel" that is covered by your object. Using the algorithm described in Section 6.6.2.4 for contour following, generate edge contours for the object by starting at the center of the leftmost boundary of your graph paper. Repeat this process by starting at the right-hand border, at the top, and at the bottom. Are all four contours identical? If not, explain why. If they are identical, explain the outcome. Are the edge contours always going to be identical for a general binary image? Explain what would happen if the object that you chose to binarize is placed at random locations (both translationally and rotationally) on your graph paper.

6.7 a. Find a coffee mug and compose drawings of all different stable states. Two states of the same object are said to be different if a simple rotation or translation of the object cannot reduce their graphic descriptions to congruency.

b. Consider a "master drawing" that is available for the coffee cup. How would you use this master drawing as a "template" for comparison of all new unknown cases?

c. Construct a flowchart or structured program to compare the master drawing to an unknown test case. Make certain that you have included a specification of the problem before you give the solution.

d. Assuming that you have available a 256 × 256 digital rendition of the master drawing and the test case, use the procedure you have detailed from Part C to estimate the number of computations required to identify the test object, and estimate the amount of computer time required if the calculations were to be done on an ordinary mini- or microcomputer.

6.8 a. Using the drawings you have constructed for Problem 6.7, derive a set of descriptors for the coffee cup.

b. Research the set of rotationally and translationally invariant moments given in [9], Section 7.2.2. Compute these moments for the coffee cup.

c. Draw a flowchart or compose a structured program for using this reduced dimensionality feature set for learning an object and then for categorizing an unknown object.

6.9 a. If each pixel in an image is modified by a linear change in brightness, i.e.,

$$I_{NEW}(x, y) = a\, I(x, y) + b$$

show (mathematically) that the Fisher statistical correlation coefficient (see Section 6.6.3.3) is unaffected by this transformation.

b. Repeat part (a) if the model M is affected by a different gain a' and offset b' [i.e., $M_{NEW}(x, y) = a'\, M(x, y) + b'$]

c. Show that the Fisher statistical correlation coefficient does not change if both model and image are modified by a different gain and offset value simultaneously.

6.10 Make a $10 \times 10 \times 4$ image consisting of a central square 6×6 pixels in size which all have a gray-level value of 15. The pixels surrounding the inner box should be filled in somewhat randomly with gray levels 0 through 14.

a. Define an Edge Template/model for the top, bottom, left, and right edges of your image. For example a left-edge template could be

$$0\ 0\ 0\ 1\ 1\ 1$$

$$0\ 0\ 0\ 1\ 1\ 1$$

Now search for each edge of your image by using the Fisher statistical correlation method.

b. Define a template for each corner, for example

$$0\ 0\ 0$$

$$0\ 1\ 1 \quad \text{(top left corner model)}$$

$$0\ 1\ 1$$

and use the Fisher statistical correlation method to find each corner.

6.11 For the image defined in Problem 6.10, find the edges by:

a. Sobel operator

b. Contrast operator

c. All first difference one-dimensional operator.

Compare the results obtained for each of the operators above and comment on the characteristics of each process.

6.12 REFERENCES AND FURTHER READING

1. Agin, G. J., and R. O. Duda, "SRI Vision Research for Advanced Industrial Automation," Second USA-JAPAN Computer Conference, Proceedings, pp. 113–117, Tokyo, Japan, August, 1975.

2. Baxes, Gregory A., *Digital Image Processing: A Practical Primer*, Englewood Cliffs, New Jersey, Prentice-Hall, 1984.

3. Castleman, Kenneth R., *Digital Image Processing*, Englewood Cliffs, New Jersey, Prentice-Hall, 1979.

4. Clarke, R.J., *Transform Coding of Images*, New York, Academic Press, 1985.

5. Duda, Richard O., and Peter E. Hart, *Pattern Classification and Scene Analysis*, Palo Alto, California, SRI, 1973.

6. Ekstrom, Michael P., editor, *Digital Image Processing Techniques*, New York, Academic Press, 1984.

7. King Sun Fu, *Syntactic Pattern Recognition*, Englewood Cliffs, New Jersey, Prentice-Hall, 1982.

8. Gonzalez, Rafael C. and Tou, Julius T., *Pattern Recognition Principles*, Reading, Massachusetts, Addison-Wesley, 1974

9. Gonzalez, Rafael C. and Wintz, Paul, *Digital Image Processing*, Reading, Massachusetts, Addison-Wesley, 1977.

10. Green, William B., *Digital Image Processing*, New York, Van Nostrand Reinhold, 1983.

11. Hall, Ernest L., *Computer Image Processing and Recognition*, New York, Academic Press, 1979.

12. Holland, S.W., Rossol, L. and Ward, M.R., "Consight-I: A Vision-Controlled Robot System for Transferring Parts from Belt Conveyors," Computer Vision and Sensor-Based Robots, Plenum, pp. 81–100, 1979.

13. Herman, Gabor T., *Image Reconstruction from Projections: The Fundamentals of Computerized Tomography*, New York, Academic Press, 1980.

14. Newman, W.M. and Sproull, R.F., *Principles of Interactive Computer Graphics*, New York, McGraw-Hill, 1979.

15. Pavladis, Theo, *Algorithms for Graphics and Image Processing*, Rockville, Maryland, Computer Science Press, 1982.

16. Pratt, William K., *Digital Image Processing*, New York, John Wiley & Sons, 1978.

17. Rosenfeld, A., editor, *Image Modeling*, New York, Academic Press, 1981.

18. Rosenfeld, A. and Kak, A.C., *Digital Picture Processing*, Second Edition, Volumes 1 and 2, New York, Academic Press, 1982.

19. VanderBrug, G.J., Albus, J.S., and Barkmeyer, E., "A Vision System for Real Time Control of Robots." Proceedings of the 9th International Symposium on Industrial Robots, pp. 213–231, Washington, D.C., March 13–15, 1979.

7

Computer Considerations
for Robotic Systems

7.0 OBJECTIVES

The purpose of this chapter is to provide an understanding of the computer architecture of robotic systems. The reader will gain an appreciation of the practical considerations that comprise the selection of a computer system from both the hardware and software point of view.

The topics to be treated in this chapter are:

- Architectural considerations (operating systems, multitasking, distributed processing, multiprocessors, bus structures, robotic considerations)
- Role of computational elements in robotic applications (communication functionality, calculation functionality, coordination functionality)
- Real-time processes (event-driven processes, sensor information processing)
- Robot programming languages
- Robot programming methods
- Artificial intelligence
- Path planning
- Robot's computer system

7.1 MOTIVATION

The use of computers and computational elements in robotic systems is essential, just as the presence of the brain in an intelligent animal is essential. In this chapter we present broad but by no means totally comprehensive coverage of some of the more important relevant topics. The material is intended as an introduction to the major topics of importance, and readers are advised to use other detailed textbooks and industry articles to round out their background.

One usually thinks of a computer as a device used for computation. In fact, Webster's *New World Dictionary* defines a computer as "an electronic machine that performs rapid complex calculations or compiles or correlates data." This simple concept is woefully inadequate when dealing with robotic systems, since many of the uses of computational elements are for tasks other than traditional computing. In particular, when one considers that most robotic manipulators utilize numerous microprocessors, the computational concept traditionally defined no longer spans the complete usage of computer components in a robotic system. In many cases, the microprocessors are much more "control" elements rather than "computational" elements.

Elsewhere in the controller, the computer components may be used for communicating to both the outside world and among other components in the robot controller. Additionally, a computer may be linked to a display unit (e.g., a color graphics display or CRT terminal) that is used to program the robot or to monitor its activities.

The more traditional computer tasks of language translation into instructions usable by the robot controller and path planning* are also performed by the computers in the robot controller.

7.2 ARCHITECTURAL CONSIDERATIONS

As will become evident, computers have a variety of roles to play in robotic systems. The efficient use of computational elements relies heavily on the use of more-or-less standardized or "off-the-shelf" devices, with custom units being kept to a minimum. The benefits of flexible automation through robotics can best be achieved by using standard programmable elements available from a wide variety of vendors and suppliers, reserving the custom engineering and task-specific activities for those tasks that require them. For example, a peripheral interface adapter (PIA) is a general-purpose programmable interface device that may be configured in hundreds of ways *under software control*. If one were to hardwire such a function every time it is needed, the cost of such interface units would be prohibitive. The availability of these programmable devices brings the unit cost of robotic flexible automation within reason. Other examples of such devices are stepper motor controllers and communications protocol interface devices.

---

*Path planning is a method whereby the path or trajectory of the end effector is computed from information about its current position, where it is supposed to go, how it is supposed to get there (e.g., in a straight line), its speed, and other criteria defined by either the user or external sensors.

A descriptive discussion of operating systems, multitasking, distributed processing, multiprocessors, bus structures, and robotic considerations will be presented in this section.

7.2.1 Operating Systems

A computer operating system handles most of the details of management of files, resources, program utilities, peripheral devices, communications among software developers, debugging, and documentation tools, such as code change tracking and automatic backup file creation. Even in the most inexpensive personal computers, an operating system removes much of the menial drudgery of software development, so that the programmer can be dedicated to developing the application software. Without operating systems managing the computer resources available, there is no question that almost no software would ever be developed. The point here is that the operating system makes it possible to develop software, since it manages the low-level details far better than any human being could ever hope to do.

Prior to the existence of operating systems, the human programmer was required to keep track of where programs were physically located (e.g., which bin of paper tape), which versions were the most recent, what changes were recently made, and so on. The operating system has improved the efficiency of software development and has permitted less skilled programmers to develop useful software.

Several standard computer operating systems have evolved and been developed recently. Two of these are listed below, with some of their attributes.

MSDOS: developed for the IBM family of personal computers and compatible products; provides a development and execution environment for a single user; provides the support and availability of numerous languages, compilers, assemblers; provides numerous utilities for file manipulation, directory manipulation, networking.

UNIX: initially developed by AT&T primarily for use within its corporate structure, it was soon made available for a wide variety of computing environments from very large computer systems to very small microcomputers. The system supports single- and multiuser environments with a very wide range of system utilities, languages, and communications support. UNIX has been adapted to execute on a very large number of different computers manufactured by many companies. It also provides facilities for transporting software among many different computer environments. Generally, UNIX is not suited to real-time applications, but there are variants, so-called real-time UNIX, that are suitable for real-time operation.

Operating systems provide the "hooks" for both programmers and programming languages to simplify the laborious task of program development. With these hooks, programmers do not have to be intimately familiar with the details of the hardware. For example, by utilizing a standard interface to an output device (such as a printer), the programmer may print messages and data by using a link to a subroutine that takes care of all device-dependent peculiarities, so that the problem and not the mechanics of printing may be focused on.

Using a commercially available operating system in a robot controller, one can speed up the development process and reduce the learning curve of potential users, since features such as file management, batch file generation, and on-line debugging tools are available.

Initially, since special microprocessor architectures were designed for use with each particular application, it was difficult to introduce the neophyte to the advantages of robotic technology. Recently, robots that utilize personal computers as their master controllers have become commercially available. These systems facilitate the implementation of the robot in the industrial or educational environment, since the time required for learning may be greatly reduced due to the general familiarity with the personal computer operating system and hardware. An example of this type of robot is the RTX, a SCARA-type robot arm. This robot can be interfaced directly with an IBM PC/XT or compatible. Installation is simple, requiring only a cable connection from an RS232-C port of the PC to the control port of the robot. "Programming" of the robot is equally simple since it may be manually driven using the cursor control keys as a teach pendant, or it may be programmed through software using standard programming languages. A library interface exists for the high-level language, PASCAL.

There are other proprietary operating systems available, too numerous to list, but in their own environments have many, if not more than the features discussed above.

7.2.2 Multitasking

Multitasking is an attribute of operating systems that permits the execution and management of several (many) processes in the same time frame. This does not mean that the programs are executing at the same time, since a single CPU (central processing unit) or MPU (microcomputer processing unit) can be performing only one instruction at a given time. Multitasking permits numerous users *each* to be executing several programs at the same time transparently to the other users, and each program operates transparently to the other programs. Since only a single CPU is present, multitasking will slow the execution speed of any single program, although this may or may not be perceptible to the user. For example, if one program is sampling a process through an analog-to-digital converter at 100 samples/s, and another program is printing a program listing, there may be no perceptible difference in program execution or performance. If, however, two programs are both using a floating-point processor to perform complex calculations, (e.g., as could be required in path planning algorithms), both programs may be severely affected relative to execution speed. It is also possible to permit different programs to have different *priorities* in such a way that one program's execution may be compromised (e.g., slowed down), so that another may be allocated more processing time.

Multitasking should not be confused with *concurrency*, which allows execution of co-processes that may share information. A multitasking operating system does not necessarily allow concurrency, and concurrency does not necessarily allow multitasking.

Multitasking-Robotic Considerations.   In a robot controller, many events can occur asynchronously. For example, if the controller is servicing a terminal in order to get commands from the user, it has no way of knowing exactly when a key may be depressed or which key may be depressed. To complicate matters, let us assume that besides servicing the keyboard, the controller is waiting for signals from each joint processor to tell it that the joint has reached its desired set point. Additionally, the controller must monitor the state of a digital input line that informs it if an intruder has entered the workspace and it must be precomputing the trajectory for the next move it is required to make. These and a host of other tasks warrant the use of a multitasking operating system to implement parts or all of a robot controller. By using a multitasking scheme, the controller can perform a certain task such as precomputing the next trajectory at real-time speeds and service other tasks when the appropriate interrupt occurs. For example, each time a key is depressed on the terminal, an interrupt may be generated that stops the current task, reads the character, places it with any other characters in a buffer, and checks for a termination character (such as a line feed). If no line feed occurs, the previous task continues; otherwise, a new task is started which interprets the command just received and may spawn other tasks to accomplish the directive.

7.2.3 Distributed Processors

Most computer systems used in nonrobotic application have a single processor or a collection of processors that act, from the user's perspective, as if there were only a single processor. However, in robotic systems, there are often distributed processors dedicated to specific tasks. For example, it is often the case that each axis will have a dedicated processor, and all of these are then controlled by a single master processor. This is done so that each axis can respond quickly enough to control (e.g., actuate, sense, and modify) some element outside the computer (e.g., the actuator, gripper, sensor, etc.).

The use of distributed processors typically permits a simpler program structure, and more computational power (and therefore controllability) in most applications. Of course, one of the disadvantages is that information must be passed among processors so that their activities can be synchronized.

7.2.4 Multiprocessors

Multiprocessors are a special case of distributed processors. Multiprocessors may be utilized to share computational load for the same task, for providing redundancy in computation, or for sharing the multiaxis controllability load in a system. For example, in space vehicles, multiprocessors execute the same task and the results of all the processors are compared in order to maximize safe operation of the spacecraft. To understand the redundancy issue, one need only remember space launches that were scrubbed because of computers that did not agree.

Shared processing can be visualized by considering weather prediction computer systems that would require 24 hours of computer time to predict weather 24 hours from the present. One might as well wait the 24 hours and walk outside to observe the weather instead of waiting 24 hours for the computer results. These tasks are so complex that multiprocessors are the only reasonable solution with today's technology to solve the problems. Even with multiprocessors, weather prediction is still not real time (or very accurate).

In robots, to date, such systems have not been utilized. However, it is possible that as the need for higher-performance manipulators and more sophisticated controls (e.g., optimal and/or adaptive) grows, multiprocessor techniques will become more important.

7.2.5 Bus Structures

A bus is a vehicle for transportation. In computer parlance, a bus is a vehicle for transportation of information. There are a multitude of bus structures that are used and have been standardized so that many standard products are available. Some of the standard buses are:

Type	Originator
IBM PC bus	IBM
Multibus	Intel
Multibus II	Intel
VME bus	Motorola
STD bus	
IEEE 488 bus	Hewlett-Packard
Q-bus	Digital Equipment Corp.
Unibus	Digital Equipment Corp.

In addition to these commonly used bus structures, there are proprietary buses designed by manufacturers that have not become industry standards but are used for a single company's products.

Attributes that may be associated with buses are numerous. These include:

- Bus width (i.e., how many lines are on the bus?)
- Functionality (i.e., are the lines handling addresses, data, control, or power?)
- Speed (i.e., what bandwidth of signal transmission is permitted?)

- Multiplexing (i.e., do all lines have the same function all the time, or do they have different functions depending on cycle location or periodicity?)
- Purpose (i.e., do the lines have specific cyclic functions, or are they programmable?)
- User adaptability (i.e., may the user define certain lines differently and permanently for different applications?)

In addition to these attributes, there are physical and mechanical issues, such as size of the bus, the type of connector, the electro-magnetic radiation properties, ruggedness, ability to withstand shock vibration, thermal shock, radiation, and so on. The variety of considerations is too large to cover fully in this text, but it should be understood that the issue of bus structure is as dynamic as the evolution of computer architecture itself, and the lifetime of a bus structure affects choices of components as well as basic system design considerations.

Currently, there are no standardized bus structures for robots, for this reason it is virtually impossible to interface one manufacturer's hardware with another. This should be contrasted to modern computer manufacturing where interface standards exist and are used to interconnect different vendors' hardware and software. The lack of a standard has hampered the growth of the robotics industry. Standards have been proposed by the SME and the IEEE, which if adopted, will begin to rectify the situation.

7.3 HARDWARE CONSIDERATIONS

Virtually every robot manufactured today has at least one computer within it. The simplest robot relies on the ability to control data flow and formats (protocols) to some degree and therefore has some sort of logical processing unit. This is generally considered to be a computer, and when so integrated may be a special-purpose hardware computational or control device. Nevertheless, it is still considered a computer, even though it may have no programmable features. Despite the fact that a computer was used to implement the functions required, its programming remains essentially fixed and consequently cannot be changed. Although certain options for its operation may be selected by the user by setting some switches or from a terminal, the sequences of operations remain as the designer originally chose them. The flexibility will not be compromised, however, within the design envelope, as one can program a computer to perform complex calculations even though the programmer cannot change the basic set of machine instructions.

More powerful robots must have the ability to perform coordinate transformations and/or straight-line coordinated motions. As a consequence, the computational tasks become significant and the computational power required increases correspondingly. Also, as sensor inputs and real-time signal processing are required of the robot in the future, the computational burden will increase even more drastically and the addition of more computational elements may well become expensive. In this instance, special-purpose hardware may become cost-effective in handling the external interface problems. At a certain level of complexity, there becomes no other way to solve the problems, except by designing special-purpose hardware with built-in sophisticated functionality. This is especially true for high-speed operations such as are encountered in assembly operations. An example is the use of special-purpose arithmetic-logic units (ALUs) customized for machine vision applications. The problem of parts alignment is often handled by the use of binary correlation. This is a process that is time consuming and requires several nested loops of high order. Performing these using computer language instructions, even with very fast 32-bit microprocessors still requires at least a second for parts having a moderate area. A specialized ALU may require only a few hundred milliseconds and may cost only a few thousand dollars.

7.4 COMPUTATIONAL ELEMENTS IN ROBOTIC APPLICATIONS

The use of computational elements in robotics covers a very broad spectrum of applications, including classical computational roles such as complex calculations, operating system and language functions, procedure definitions, and so on. In addition to these classical roles, computers are also used to control the actual joint servos and interface with external sensors to coordinate communications among elements, interface with factory host computers, and coordinate workcell activities. In the following few paragraphs, some of these roles will be detailed and will serve to illustrate the multifaceted role of these computational elements.

Figure 2.2.2 showed the details of a model of a robot controller. Recall that the controller's purpose was to provide the intelligence to cause the manipulator to perform in the manner described by its trainer. To accomplish this functionality, seven subsystems were defined (see Section 2.2), each with a specific task that taken together could provide relatively complicated robot control.

It is important to note that computational elements play an extensive part in the implementation of these subsystems of a robotic controller. Additionally, one should note that there may be many possible implementations for the controller. Factors such as cost, available technology, and required functionality and designer's choice contribute to the final architecture.

The following sections describe some of the roles of computational components as related to robotic control. The reader is encouraged to correlate the concepts presented in these sections with the general system overview of a robotic controller as presented in Sections 2.2 and 2.4 in order to gain an understanding of the internal functions of the controller.

7.4.1 Control Functionality

The simplest role of the computational element in a robotic application is that of a simple digital control unit, for example, activating or deactivating electromechanical relays or electrical switching elements (such as a transistor) in order to turn motors on or off with a so-called "bang-bang" strategy. The use of the computational element in this form is similar to a traffic light controller, whereby simple activation or deactivation of specific valid pathways for data or mechanical activation systems is effected.

Slightly more complex roles for control elements may be envisioned as programmable algorithm-level controllers for use in driving servos or displays. For example, a digital-to-analog (D/A) converter is used to convert a multibit signal in the internal domain to an analog, or continuous signal, in the external world. The result of a computational algorithm for control may result in a desired drive signal of 3.7 V. In the internal computer world, this signal will have a multibit representation in a unitless, abstract sense. Conversion of this unitless number to a measurable quantity is achieved through the use of the D/A converter. This output signal could, therefore, be used to drive controllers to illuminate a lamp to a certain intensity or to modulate an alarm sensor with a variable "chirp" signal (a "chirp" is a fixed amplitude, increasing frequency signal) or in fact may be used to produce a special output signal for communication between the computer and a human being. An important point to be made in conceptualizing computers as control elements is that these are rather nonstandard, nonclassical applications of computational devices.

Although all these examples clearly require computational elements to create the necessary signals, the output of this information through control actuators is often overlooked as a major functional role of computational elements.

Computational elements may also be used to interrupt sensor data and to plan and control actuation functions: for example, the use of analog-to-digital (A/D) converters to sense the environment or to sense conditions in the external world and to format those properly so that a computational program may then massage those data for appropriate external manipulation. Using these components as input (A/D) and output (D/A) devices creates the scenario whereby closed-loop control of actuation systems may be achieved.

Often, the individual joints of a servo-controlled robot are controlled by individual microprocessors. These are sometimes called slave or joint processors, and perform a number of control tasks, including acting as a digital summing junction which compares the set point from the master processor with the actual position data obtained from the encoder or other position sensor (see Chapter 4 and Appendix C). Other functions performed include linear interpolation of the set points to produce smoother joint motion, and finally implementation of a variety of digital computation schemes, including, for example, a PID controller.

7.4.2 Communication Functionality

Another role of computational elements is to provide communication (i.e., exchange of information between and among components). The controlled exchange of information permits processes to proceed according to the designer's plans, even though the processes may be asynchronous, random, or concurrent.

In a crude sense, the control components represent communication between the internal quantized computer world and the external world. They communicate in the sense that information is passed between these two environments. In a more general sense, however, there are no secure, formal mechanisms for guaranteeing data validity between the computer and its input/output devices. For example, if an A/D converter is used, the computer receives a number when it queries the device. However, the computer has no knowledge of the utility of the signal, or its validity.

The next higher level of complexity is that of a secure communication procedure, a so-called protocol by which elements communicate. This is perhaps best illustrated by the telephone as a communication element. It is clear that the telephone is used to pass information between at least two system elements. However, without the use of protocol, telephone communication would have much less information content. This need for protocol is illustrated by the following example. Imagine the difficulties a listener would encounter if a caller started to speak whatever thoughts or phrases came to mind without first identifying himself or herself. At an even lower level, one can imagine picking up the phone handset, punching in a number, and speaking without even waiting for the call to be answered. The point of this absurd example is to indicate the importance of protocol that permits confirmation and verification that the recipient of the information is prepared to receive it and that the packaging of the information is correct.

These concepts have been formalized in great detail, and standards have evolved for a variety of these so-called protocols. The following section will introduce some of the basic concepts but will not delve into great detail with regard to a standard definition of communication. Several national committees and organizations, such as the American National Standards Institute (ANSI), American Society of Mechanical Engineers (ASME), Electronic Industries Association (EIA), Institute of Electrical and Electronics Engineers (IEEE), National Electric Manufacturers Association (NEMA), Robotics International of the Society of Manufacturing Engineers (RI/SME) and the Semiconductor Equipment and Materials Institute (SEMI) have generally cooperated to define certain standards on these issues.

One of the simpler forms of communication between computers is a binary input/output (I/O). In a primitive sense, binary I/O is simply the passage of bit-organized information (i.e., a string of bits each able to take on a value of 0 or 1, true or false, on or off). The interpretation of the status of these bits resides with the user, and generally there is no security in this type of communication. This is the type of controlled communication scheme discussed in Section 7.4.1. It relies on other devices or elements to be ready and prepared to accept or deliver data when requested, in a so-called "asynchronous" mode. Asynchrony in this discussion implies that each element has the ability to arbitrarily (in time) assert its data.

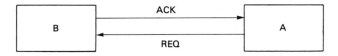

Figure 7.4.1. ACK/NACK pairing of states.

It is obvious that without a proper communication protocol, asynchronous data transfer can create havoc with the information flow. A simple mechanism to overcome the potential difficulties associated with asynchronous data transmission is known as *handshaking*. The simplest form of handshaking requires the dedication of two binary lines (ACK/REQ) in a so-called ACK/NACK pairing of states. This is illustrated in Figure 7.4.1. Using two such data lines to request and/or acknowledge readiness from a device requires careful selection of valid states for either the transmitter or receiver to occupy. These states are shown in Figure 7.4.2, whereby device A is requesting the attention of device B. Generally, 2 bits are required to guarantee absolute security of state preparedness on the part of two potentially independent processes.

The use of these bits is similar in principle to a flag system on a rural mailbox. In this situation, if the postal patron requests a pickup from an RFD mailbox, the patron will raise the red flag on the box. The mail carrier will observe the red flag, stop the vehicle, remove the material, and drop the flag. Confusion can, of course, result from this simple system if the same red flag is used to inform the postal patron that the mail carrier has deposited mail in the RFD box. Therefore, more than simple intelligence is required to understand the meaning of the red flag, and generally the context of the situation dictates its interpretation.

For example, on rural mail routes, the carrier typically arrives in a prescribed time window during a 24-hour time period. If the red flag is still raised after that typical time period, either the carrier was unable to keep to a regular schedule, or

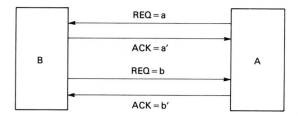

State Sequence

	1	2	3	4	5	6	7	8
a	0	1	1	0	0	0	0	0
a'	0	0	1	1	0	0	0	0
b	0	0	0	0	1	1	0	0
b'	0	0	0	0	0	1	1	0

Figure 7.4.2. Example of secure handshaking. The columns in the table indicate the system state. The numbers 1 through 8 represent the state sequence.

in fact the outgoing mail was picked up and mail deposited in the box at the same time. Clearly, the use of a red and blue flag in a so-called ACK/NACK paired scheme would eliminate this confusion (as is illustrated in Figure 7.4.2) since the carrier would raise the blue flag if mail was deposited, and would drop the red flag to indicate that the mail carrier had been there.

This type of handshaking is simple and secure, but also requires a great deal of overhead, since every communication must utilize the concept of flags. Techniques for exchanging information *packets* reduces some of this overhead, at the expense of not guaranteeing that every bit of information is sent and/or received only when the receiver/transmitter is certain to be ready. Packeted information transfer allows for error checking on a relatively infrequent basis and detects certain types of errors in transmission so that at the very least, errors can be logged and appropriate action taken. Such action may be a request for retransmission, or the data could even be ignored.

Figure 7.4.3 illustrates the packeting of information using a simple scheme to transmit a message of N bytes of information. This is known by many different terms, but the SECS1 (Semiconductor Equipment Communication Standard) protocol designates this technique as the "Data Link Protocol." The idea is simple in that first, the receiver is informed as to the length of the message (i.e., how many bytes will be transmitted). This is then followed by the message itself, and finally, a quantity called the checksum is transmitted. With such a format, the integrity of the message is preserved and errors in transmission detected.

Clearly, the key to the protocol is the checksum (which is also known as longitudinal redundancy check, or LRC). Normally, the checksum is the negative of the sum of the binary-coded numeric values of the message and is usually truncated to one or two bytes. As long as the structure of the message is known by the receiver, an error in transmission can be detected by computing a local checksum and then comparing it to the transmitted value. This technique will detect the occurrence of single-byte transmission errors.

It is important to understand that it is possible, although unlikely, to have multiple byte errors that will still produce the same checksum. If one wants to prevent this situation from occurring, more complicated error-checking techniques must be used, such as CRC (cyclic redundancy checks) that can detect multiple-byte errors. However, these require the evaluation of polynomial error formulas (as compared to the simple linear sum of the LRC) and are more time consuming. Moreover, it is usually not found to be necessary for most applications, except where there is the distinct possibility of noise-corrupted transmission.

In robotics applications, communication between slave and master processors would probably use LRCs to ensure accurate information up and down the information chain. In this respect, robotic communications parallels other multi-processing applications, requiring secure communications and needs no extraordinary techniques for implementation.

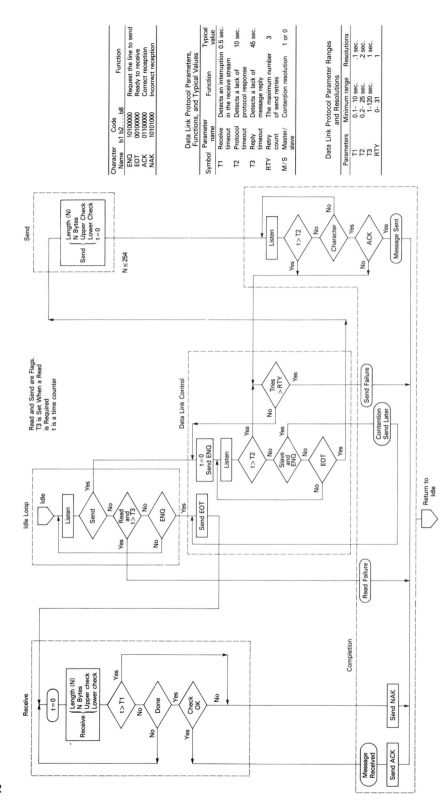

Figure 7.4.3. SEMI SECS1 data link protocol.

7.4.3 Calculation Functionality

In addition to the roles just described, more classical calculation roles may be attributed or assigned to computer components in a robotics system. One is that of performing a variety of *coordinate transformations* as will be developed mathematically in Chapter 8. Such transformations are necessary to develop drive signals for the control portions of the robot. For example, moving a gripper or manipulator from one point to another typically starts with specification of motion in a rectilinear or Cartesian coordinate system. However, to achieve the desired motion, these coordinates must be transformed into the specific joint space of the robot (e.g., Cartesian, cylindrical, spherical).

Usually, these transformations are mathematically complicated and require transcendental function evaluations. Consequently, some type of relatively sophisticated mathematical processing is required. This can be accomplished in a number of ways, including the use of software routines, hardware evaluation utilizing a floating-point processor, or employing software lookup tables. In a robot, the decision as to which technique to use is tied to the final system cost, speed, implementation, expandability, and generality.

Another major area where classical computer calculation-type functions are involved in a robot is in signal processing, e.g., noise removal from a distorted signal is a common requirement in sensor data analysis. Signal processing may be accomplished in both the analog and the digital worlds, and may be multidimensional. That is, there may be multiple lines of data coming in from the outside world in the form of binary input/output or in the form of analog or continuous signal input and output. An example of this is the processing of an ultrasonic acoustic signal from the outside world to determine position information, or perhaps to monitor acoustic emissions from a variety of electromechanical components such as motors, gears, and so on. In addition, metal surfaces scraping against other metal surfaces may produce acoustic emissions that are detectable and may be useful to the robot controller for preventative maintenance scheduling.

Another example of complex computational needs is in the field of vision, whereby one may be inspecting an outside world environment relative to the robot for the purpose of alignment: for example, in palletizing objects, alignment of integrated circuit chips or in the alignment of surface-mount components on printed circuit boards. These tasks are relatively computationally heavy and will generally require a dedicated processor for implementing these functions tin a timely and efficient manner. For example, one may need to direct the robot to position a camera so that it may "see" the environment. This information may then be passed to a vision processor which calculates the alignment offsets, passes that information over to the robot perhaps through a controller, and then to the transformational pathways or internal routines of the robotic computer. This permits the information to be translated into specific velocity and acceleration control signals.

Another possibility for using a vision system to augment the robot's sense of the environment would be to accept or reject parts, or to characterize or grade them. An example of this might be in a microchip dicing system, whereby one is inspecting a matrix of semiconductor components on a diced wafer. These wafers are often marked with ink dots to indicate rejection, and the robot may simply pick and place the good components into acceptance bins or packages. The poorer quality components or those that have been rejected may be either left on the wafer carrier or may, in fact, be taken off the carrier and deposited into a reject bin. A refined classification of this application would permit multiple ink dots or multiple coding of the surface of the chip so that one might grade parts into a variety of different categories. In this manner, one could fabricate variable quality assemblies by inspection, ranging from those with the highest down to those with the lowest.

In addition to the vision and coordinate transformation tasks, calculation is required in the area of direct axis (or joint) control. For example, if one is using a servomotor to drive a robotic axis, there are a variety of ways to accomplish this task. As discussed in Chapter 4, a digital-to-analog (D/A) converter could be used to drive the servo amplifier directly. In general, the calculation of the required drive signal is not trivial, and in fact the output will usually have to be shaped rather precisely in order to produce the desired robot performance (i.e., smooth, vibration-free motion). The so-called "on/off" or "bang-bang" control systems, whereby the input to a servomotor is a step of a known value, is a relatively straightforward control procedure. However, step signals, as explained in Chapters 3 and 4, will introduce high values of derivatives of position, velocity, and acceleration creating untoward effects in the output (e.g., excessive mechanical vibration). The obvious need to profile motions throughout space and to control axes simultaneously makes the problems associated with axis control computationally intensive.

In addition to the direct output requirements, position and/or velocity feedback information must be acquired and utilized. The acquisition of real-time information may be computer resource intensive, since some of those signals may have to be filtered digitally. Additionally, making use of the feedback signals to compute new positions, velocities, and accelerations may also present a large computational burden. Although not currently done, in the not too distant future external sensory data will be fed back to the master processor and will be used to modify the set points sent out to the joint processors. In effect, the system will then be recomputing the axis transformations to produce the desired manipulator motions. The role of the computer in this environment is therefore more or less traditional. Appendix C shows the computational algorithms necessary to accomplish these tasks.

Additional computational complexity is introduced by requiring coordinated motion control whereby all of the robot's joints must start and stop at the same time. A further level of computational difficulty results when the required motion must be in a straight line in three-dimensional space. There will invariably be velocity constraints, or there may be the requirement to use "via" (by-way-of) points whereby the robot must move through these "via" points to provide the desired path.

All of the examples above are computationally substantial tasks that may require floating-point processing, matrix calculations, and transcendental function evaluations, and as such serve to demonstrate the need for a powerful set of processors for robot control and implementation.

7.4.4 Coordination Functionality

There are other examples of nontraditional use of a computer element in a robot. For example, the control and coordination of multiple robots used to execute the same tasks, possibly with other material-handling equipment, is an important application. As discussed in Chapter 2, the concept of the coordination requires a cell controller to control the entire operation. This device must be able to coordinate the robotic manipulators, all sensory systems, as well as other material-handling systems, and of course must be able to keep track of the work in process and the location of individual completed subassemblies. The cell controller may also be required to report all of this to a factory host computer, which can interrupt or modify the plan of the specific cell controller.

The cell controller may either be located within the robot's controller or else may be a separate entity. For simple applications, the robot controller may perform the functions of the cell controller. However, in more complicated situations a separate unit may be required. The robot may not be aware that a cell controller is being used, but instead may just be following a preprogrammed path which is activated or deactivated remotely. In other situations, complex communication protocols between the cell controller and the robot may be required. This is especially true when multiple manipulators are being used.

As suggested earlier, the cell controller may be nothing more than a microprocessor with very simple processing capabilities, or may be a complicated minicomputer controlling a variety of components. One good example would be in a hybrid circuit manufacturing facility, where parts may be shuttled in wafer form mounted on sticky-tape frames (frames with a film and an adhesive surface) or waffle packs (plastic carriers with indented pockets for holding individual integrated-circuit dice in their own receptacles). These parts may then need to be assembled onto a small (roughly 2 in. × 2 in.) ceramic substrate which will have chips bonded to them (either glued or soldered). Wires must be attached internally between the substrate and the chips. Full testing of the chips may be required, and then they may be graded and placed into bins. This application requires a fairly complicated cell controller with use of common protocol with common languages throughout the system (See Figure 7.4.4.).

In addition to these types of coordination, the robot computer or computing elements may need to communicate with CAD, CAM, or CAE data bases. That is, one may have designed and simulated a specific assembly process on a host

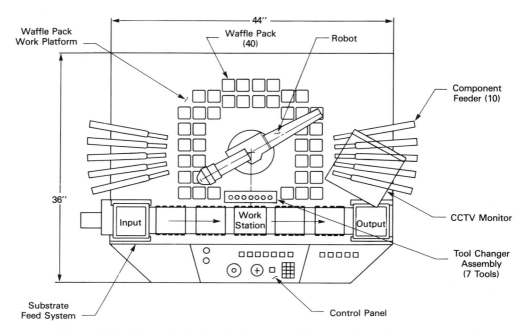

Figure 7.4.4. Schematic of a hybrid circuit assembly system that utilizes a robotic manipulator. The system utilizes a common protocol and language and requires a complicated cell controller.

computer. This assembly program may have been downloaded to a variety of computers, including the cell controller, for example, which may then be required to direct the assembly operation, including all the robotic manipulators so that the product is assembled. Note that in this case the robots were never taught directly, but obtained their "programs" electronically. Although this is not yet a widespread practice with robotic systems, it is one area in which to expect developments to be made.

In addition to the roles of the computers described above, there is the concept of coordinated path planning for the robot motions. For example, path planning might require that the work be moved in a prescribed manner through a specific set of workstations. This might require the integration of a number of robotic manipulators, perhaps incorporating information from a vision system as well as from other sensors.

As an example, we may look at the production of a wiring harness, which requires the stringing of wires of various lengths throughout a specific geometric pattern in space. Normally, wires will have to be routed around pins located on the harness board. Thus one must coordinate the stringing of a specific wire based on previous ones that the robot has installed. Furthermore, the work that the robot has completed may not be stable after installation. For example, when stringing a wire, there may be a curl or misposition of the wire after the robot releases the end of the wire. Then, when the robot goes back to place the next one, it must have some way to guarantee or to measure the placement of these previously laid wires so that the harness is produced in a reliable fashion.

7.5 REAL-TIME CONSIDERATIONS

In this section we discuss two important topics of real time event-driven processes and sensor information handling. The concept of "real time" is best thought of as "needed now." This needed-now concept gives the idea of urgency to the topics in this section, since either type of processing is so time-critical that if either process cannot be served *as soon as possible*, the robot and its environment may subsequently be uncontrollable, probably with catastrophic consequences (e.g., the robot may become unstable).

7.5.1 Event-Driven Processes

In many software applications, a computer must respond to input from the "outside world." Two methods of achieving this are:

- Program driven
- Event driven

Program-driven response implies that the input occurrence is expected in some sense. Entry from a keyboard is usually of this type, whereby a program is waiting for an input via a keystroke. Although the program does not know what the response will be, it does know that if there is a response, it will occur at a specific location in the program. Another example of such a response is that caused by a switch closing, indicating that a robotic gripper has successfully acquired a part. Here, the robot is expecting the closure of the gripper at a specific point in its program sequence. This is similar to a program waiting for an input of data prior to executing a calculation.

The above should be contrasted to an event-driven process, where the timing of the response as well as the type may be totally unpredictable. As an example, consider a pedestrian walking up to a busy street that has a pushbutton-activated street light. If the button is pushed, the traffic light controller will respond in time by changing the light to a yellow-then-red condition for traffic, and eventually to green for the pedestrian. If the pedestrian never pushes the button, the light will never change, and the internal "brain" will perform its normal tasks, keeping the light green for traffic, checking whether all the lamps are functional (by checking current through the filaments), and calling the traffic department to replace a bulb if necessary.

A similar event-driven response would be required if someone enters the workspace of the robot. If instrumented properly, the work envelope can be monitored by ultrasonic sensors, photo-optical interrupters, or pressure mats that will interrupt the robot's controller, and stop the robot motion activity so that the intruder will not be struck and possibly injured by the manipulator arm.

This concept of responding to random external events is not only useful for protecting an intruder or a pedestrian, but is a feature that all robotic computers must have in order to control and interact with their environment. The control of motors for coordinated motion, the coordination of assembling parts from various feeders, and an endless variety of robotic assembly tasks require the ability to respond to random events, because not all robotic sequences are deterministic by nature. Even though the specific global actions desired may be deterministic, the specific joint actuation sequences may be random, due to external perturbations from unknown or variable loads and/or because parts may arrive at pickup points randomly. The robot controller must be capable of properly handling these situations.

The response to the external events may be required in as short a period as several microseconds, so the computer's architecture must be such that this is possible. In some instances, where operating system overhead must be contended with, an interrupt latency may be experienced. Interrupt latency is the time from when the external interrupt occurred to the time when the interrupt is serviced or handled by a software interrupt service routine. If this latency becomes too long, it may be necessary to bypass or disable the operating system temporarily or permanently, and compose special software so that the interrupt may be handled in real time (i.e., rapidly enough so that the event requiring attention is handled in an appropriate and timely fashion). In virtually all modern-day computers, the ability to respond to interrupts is present, as is the ability to prioritize, queue up, and process hundreds or thousands of these interrupt requests.

In many parts of the robot, it is important for certain events to occur at known times. For example, in the digital control of a motor, it is extremely important for the position sensors (the encoders) to be sampled at a uniform rate (e.g., every 0.5 ms). The control signal must then be output to a D/A converter (DAC) at the same rate. To accomplish this, a real-time clock that generates a signal every 0.5 ms can be used to trigger an interrupt line. When the interrupt occurs, the sensor is sampled and the computations necessary to generate the control signal are performed. The control signal is transferred to the DAC at a known time after the input was sampled and the sequence repeats.

Other uses of real-time clocks are to generate time and date stamps and to time events. Also, in many applications, it is necessary to inhibit a robot from moving to the next taught point until mechanical settling occurs. This can be achieved by delaying the operation using an accurate timing program.

Very often, the operating system is able to assist in managing event-driven requests by providing utilities to service these requests, and by providing software tools to assist in the development of appropriate software (e.g., interrupt service routines).

7.5.2 Sensor Information Processing

Consider what happens if object acquisition is attempted without an external sensor. In other words, there are no sensory mechanisms required or available to confirm target location or that the robot has properly grasped an object. If it can be guaranteed that the object can be positioned within the tolerance of the pickup mechanism, acquisition can be achieved. However, when an object's position is not absolutely predictable, one must have some type of sensing mechanism. A variety of sensory devices that can be utilized in this respect have already been described in Chapters 5 and 6. In the case of vision-based sensors, computational requirements are very severe. In general, these high-rate external sensors must provide integral data compression so that the signals delivered to the robot are low-rate and the robot controller will not be overwhelmed by massive data handling requirements. For example, the vision-based sensor should reduce the imagery from the order of megabits to a few bits (e.g., the centroid location of the object to be grasped).

The important thing to understand, however, is that these tasks require complex sensory information processing and tend to be computationally large. For example, if one uses a moderate bandwidth sensor of only 50 to 100 Hz and attempts to use a high-speed, off-the-shelf processor to perform signal processing, it is readily discovered that the computer rapidly runs out of computing power. This means that very frequently, special-purpose hardware processors must be developed in order to handle the incoming data rates, or data must be simplified immensely so that the computer can make relatively simple decisions.

EXAMPLE 7.5.1

To illustrate the ideas above, consider a simple filtering operation for noise reduction. The following equation represents a simple single-pole low-pass filter:

$$G(n) = AG(n - 1) + (1 - A)F(n)$$

where $F(n)$ = input sequence from an A/D converter
$G(n)$ = output sequence
A = filter weight $(0 \leq A \leq 1)$

If $A = 0$, the input will pass directly to the output, and thus the filter will behave as an all-pass device. As A approaches unity, the filter properties will approach those of an ideal integrator. For intermediate values of A, low-pass filter characteristics result.

Now assume that the following times are valid for a hypothetical, high-speed microprocessor (e.g., a Motorola 68000 with a 12.5-MHz clock rate).

$$\text{data conversion} = 10.0 \text{ } \mu s$$

$$\text{add time} = 1.0 \text{ } \mu s$$

$$\text{multiply} = 6.0 \text{ } \mu s$$

$$\text{memory access} = 0.5 \text{ } \mu s$$

Assuming that A and $1 - A$ are precomputed and stored in memory, then for each new computation, one data conversion, five memory accesses, two multiplies, and one add will be required to filter the data. This corresponds to 25.5 μs or a data rate of 39,215 Hz. Assuming a 10 input sensor base, one can process these data at a rate of less than 4,000 Hz per sensor. Further assuming a 5-sample/cycle sampling rate, one can handle signals with frequency content up to 784 Hz with this high-speed processor.* It should be noted that slower rate processors would be at least proportionally poorer in performance. For example, a Motorola 6800 operating at 1 MHz would be approximately 12.5 times slower, which would yield a per sensor date rate of about 62 Hz. This 62-Hz rate may well be marginal in high speed applications, especially when other interfering processes, such as linear distortions, exist which require digital control system compensation.

Other types of sensory considerations have to do with nominal versus extraordinary conditions. This requires the ability to plan for exceptions to normal circumstances. For example, the robot manipulator may be programmed to go through a variety of movements. However, if an extraordinary situation occurs (e.g., the gripper is empty when it was not supposed to be), the robot must have some plan to handle this deviation from normal behavior. For example, where the part has not been properly acquired, the robot must then have the ability either to reacquire it or to inform its controller to execute an emergency stop sequence.

To accomplish the above, we may have a force sensor that produces a rather simple binary signal that indicates to the robot that either "Yes, the nominal condition is present and although there may be variations, nothing out of the ordinary has occurred" or "No, the contrapositive." Without this planning and use of even simple sensors, it is obvious that the exceptions to normal behavior that often occur will not be taken care of properly.

Exception handling is application-dependent and is usually left to the application programmer. The action to be taken is highly dependent on the nature of

---

* The theoretical sampling rate is two samples per cycle, but in practice one needs to sample at least five times per cycle.

the application, and what may be acceptable in one circumstance may be unacceptable in another.

A robot language is usually the medium by which exception handling is accomplished. For example, once the application is programmed, a subroutine can be added to test for the part being present once the gripper has been commanded to close. If the part is missing, the action dictated by the subroutine (signal for an operator, retry, continue without the part, etc.) can be carried out.

There are also some errors that are sensed by the system in all cases. For example, if the manipulator is commanded to move, a check could be made to ensure that the motion is occurring (e.g., by monitoring the error signals in each of the joint servos). If no motion occurs, it is possible that the robot arm has collided with another object or that one or more of the error signals may have exceeded a predetermined band. In the event that this type of error is detected, the arm could be stopped and the servo gains reduced so that the arm becomes "mushy." (This procedure prevents possible damage to motors and mechanical components of the robot.) It is also possible to monitor the control signal for each servo when the arm is not moving. In the event that this signal is too large, one may be able to conclude that the payload is too big and take appropriate action.

In addition to these considerations, there is the concept of self-adaptation to the environment. For instance, many robots are designed so that they may respond to a variety of inertial loads. For example, different payload weights should not affect the path that a robot takes in general. Often this is accomplished by changing servo gains to compensate for variations in the axes loads that would create undesirable deviations from the proper path. This would be equivalent to a young child picking up a lightweight toy and moving it from point A to point B in space. If however that lightweight toy was filled with lead shot, and the child attempted to follow the original path, difficulty might be experienced in overcoming this additional inertial load even if more muscle power was employed. That is, the child might have only limited ability to compensate. Although the computational algorithm might be there in the child's brain, the ability to handle that level of load might not exist.

The idea of inertial compensation can be built into the algorithmic control processes so that when the weight or inertia of the load changes (within limits), the robot may still move the load over the same path if it is instructed to do so. This idea of self-adaptation can be mathematically modeled and included in the robot's internal program.

Another situation that can be detected by proper monitoring of position and/or current sensors (in each joint) occurs when the robot strikes an object that was previously not known to be there. In this instance, there will be an increased amount of resistance to the arm's motion, which results in an unusual increase in motor current and/or a large position error. It is possible to program the computer to sense these conditions and take corrective action, such as stopping the motion or reducing servo gains. For example, if one visualizes a robot picking up the lead-weighted toy and moving the load from A to B, and one puts a chair in the way, it would be clearly desirable to have the sensory ability to detect that something out of the ordinary had occurred and take appropriate action. This self-adaptation concept more or less fits in well with the previous nominal versus extraordinary discussion.

As more external state sensors (see Chapter 5) are employed with robots, the information they provide will be used to modify the original program in real time. For example, tactile sensors placed on a robotic gripper provide real-time data to the robot's controller, which then commands the gripper's servo so that the right amount of force is generated.

Sensors placed in robotic grippers are also important when it is necessary to handle objects which have specific stability, rigidity, and orientation requirements. There is no reason to expect that one will always have objects of one type, and a truly versatile system should be able to handle a variety of shapes and sizes.

It is clear that as external sensors are more heavily utilized, the information provided by them will increase the computational burden placed on the robot's computer systems. This has already been demonstrated in Chapter 6, where vision systems and the computational considerations were discussed.

7.6 ROBOT PROGRAMMING

As discussed in Chapter 2, the most sophisticated robot control systems have a programming capability that allows for elemental decision making, a capability needed to coordinate a robot's actions with ancillary devices and processes (i.e., to interface with its environment). Branching is the ability of the software to transfer control during program execution to an instruction other than the next sequential command. At a specific point in a task cycle, the robot will be programmed to anticipate a branching signal—a special electrical signal sent to the controller by a designated internal or external sensor. If such a signal is received, the program will follow a predetermined path or function (branching). If no signal is received, the program will continue to follow the main path. Thus a robot interacting with a group of machine tools will perform a given sequence of operations, depending on which steps have been completed. For example, after a raw part is loaded onto a press, the program will look for a branching signal. If the signal is received, the program will branch to a pause, causing the robot to wait while an ancillary machine works on that part. After the machine has completed the prescribed work, an external completion signal is sent to the controller by a sensor located on that ancillary machine. Then the robot is directed to take the part out of the press and transfer it to another machine. Decision making can also be used to correct an operational problem. For example, a program may have a branch to a taught subprogram for releasing a jammed tool.

Robot languages provide flexibility to the user in defining the task to be performed. Not only do they permit the motion of the task to be defined but they also provide the user with the ability to imbue intelligence in the control program. In its simplest forms, this intelligence may check binary sensors and change a location, or make a simple decision based on sensory information to handle an exception. As the capability of the language increases, the intelligence of the algorithm controlling the robot in a specific application can also increase. Thus corrections based on sensory inputs (such as vision or tactile sensors) are possible along with communication with other computers and data bases.

Historically, the initial applications of robots were relatively simple and accordingly, their controllers did not require or provide sophisticated sequence control. Typically, the following sequence was all that was needed:

- Move to a specified location in space
- Control the state of a gripper
- Control the state of output lines
- Provide sequence control based on the state of input lines

As applications became more complex, and computer technology more advanced, techniques were developed to take advantage of the newer computer architectures.

In the following section, techniques for robot control sequencing will be presented from three appropriately more progressive perspectives (fixed instruction sequence control, robotic extensions to general purpose programming languages, and robot-specific programming languages). This is followed by a summary of robot programming languages and two examples illustrating these methods are presented. The section concludes with a discussion of how points in space are taught or "demonstrated" to a robot.

7.6.1 Robot Control Sequencing

Robot sequencing can be accomplished in a variety of ways. As discussed in Chapter 2, there are certain features of functionality required by a robot control system in order to facilitate both the training (programming of the sequence of events) and its use with ancillary equipment. To someone familiar with general-purpose programming languages, it is obvious how certain aspects of this functionality can be easily provided by a computer language. What may not be as obvious is that most of the important functions needed for manipulator control and simple interfacing can be implemented by dedicated sequencers. These sequence controllers accept commands (possibly given by the setting of switches) and record the robot's joint positions. The sequencing of the manipulator is achieved by "playing back" the desired states at a later time. In a certain sense, these sequencers also possess the power of programming languages but without all the explicit commands and data structures associated with a formal programming language.

To contrast various "programming" methods, all of which permit the user to define the sequence of operations of a manipulator, three distinct implementations will be discussed. Specifically, they are:

- Fixed Instruction Sequence Control
- Robotic Extensions of General-Purpose Programming Languages
- Robot-Specific Programming Languages

The first is a relatively simple method which makes use of a fixed event sequence in each instruction. The second is based on extensions of programming languages which add robot-specific functions (or subroutines) to the standard library, or in which robot-specific commands have been added to the structure of the language. The third is a language tailored specifically to the programming or training of robots.

7.6.1.1 Fixed instruction sequence control

In this mode of implementation, the sequence of the robot's operation is defined by means of a "teach pendant" which provides the ability to position the tool point of the manipulator in space by means of buttons or a joystick. Additional controls allow the trainer to define the state of the gripper (open or closed) and the state of each of the output lines (on or off) as well as time delays and simple branching based on the state of input lines. By saving joint position, and other state data, a sequence of events can then be defined.

To better understand the nature of a fixed instruction sequence controller, the implementation used on the Mark I controller from United States Robots will be examined. In general, each program step consists of a series of actions. These are:

- Check the status of input lines
- Check for a subroutine call
- Perform a robot motion
- Delay a specified time interval
- Set the state of the gripper (open or closed)
- Set the state of output lines

To understand how this relatively simple structure can provide sufficient program control, and for the sake of discussion, let us assume that the controller already has a number of programs stored in its memory. A specific program is first selected (by number) utilizing a series of thumbwheel switches. To begin the sequence of actions defined by the program, a "start" switch is depressed which causes the first instruction to be obtained from memory. First, a logical "AND" of a subset of the input lines is performed against a "mask" stored in memory. It should be understood that the program will wait indefinitely until the specified input line(s) are asserted. Next, if the step is a subroutine (another series of program steps), then it is executed and the following program step is obtained from memory (note that the motion and subsequent steps are not performed in this case). If no subroutine call was indicated, then the robot controller causes the manipulator to move to a point in space defined by a set of joint variables stored in memory. Once this location is reached, the remaining actions (for the current program step) are executed. These include waiting a specified delay time, opening or closing the gripper, and the final action, which is the setting of the state of the output lines to a value defined in the programming sequence. Following this, the next program instruction (step) is fetched from memory and decoded as defined previously. After all the steps of a particular program are executed, the sequence repeats from the first step. That is, the controller keeps executing the program indefinitely.

Due to the nature of the fixed sequence of actions for each program step, it may be necessary to program additional steps to properly sequence the manipulator. For example, it is necessary to provide a delay to ensure gripper activation prior to arm motion. This is due to the fact that it takes a finite time for a gripper to reach its final state after its activating mechanism receives its control signal. Therefore, the trainer might want to insert a delay (on the order of a few hundred milliseconds) prior to the execution of any other manipulator motion. Since the action sequences of a program step without a subroutine call are check inputs, perform motion, delay, set gripper state, and set output line states, one easily sees that it is possible for the next program step to cause a motion (if the input conditions are satisfied immediately) before the gripper's state has stabilized. To accomplish a delay prior to the motion of this subsequent step, it is necessary to program an additional step in which no motion occurs but which makes use of the delay in the sequence of actions.

While this type of programming may require substantial human activity, it is still able to produce the desired results (i.e., sequencing a manipulator through a set of motions). The key to both successful and efficient programming of this type of controller is knowing the sequence of actions and how to take advantage of them.

As the complexity of the tasks being performed by robots increased, the demands for more advanced motion control and decision capability also increased, thereby requiring more sophisticated programming methods. In some cases, the simple sequencing controls could be expanded by adding more functionality to the teach pendant by means of multiple levels and added control switches. Besides increasing the complexity of the teach pendant, this approach also increased the programming time and required skill level of the trainer.

An outgrowth of such complex sequence controllers is a "menu-driven" programming system that permits the training of the robot using a fixed set of functions. The menu system differs from the "fixed instruction" sequence control in that instructions specific to each function are generated. Unfortunately the use of a menu system can be quite awkward and requires a trainer well versed in the concepts of computer programming.

One major advantage of a menu system, however, is that it may be easily extended to accommodate new functions and even provide interfaces to external sensors such as vision. It should be apparent that this concept can also be extended to a robot-specific language by adding a terminal interface and the typical language functionality such as syntax checking of instructions prior to execution (or during compilation).

Although extensions of fixed instruction sequence control could certainly have provided additional capability, they lacked flexible program control and data structures. Consequently, another approach was needed. This approach is discussed in the next section.

7.6.1.2 Robotic extensions of general-purpose programming languages

Another step in the evolution of robot programming was the incorporation of a language. The use of a general purpose programming language with extensions provides the user with the control and data structures of the language. The robot-specific operations are handled by subroutines or functions. Clearly this implies that the training of a robot now requires a person well versed in the concepts of computer programming.

Various permutations of this concept are possible, including the use of subroutines as compared to extensions of languages. The extensions to the language include robot-specific commands (and possibly new data types) in addition to the existing set of commands (and data types) while leaving the general syntax and program flow intact.

An advantage of using an extension of a general-purpose programming language is that the designers can concentrate on the problem at hand, designing a robot instead of spending time designing a sequencer, providing editing capabilities, and so on. The actual implementation may make use of a compiled or interpreted language depending on the nature of the base language chosen to be extended and the objectives of the design team. One other advantage in extending a language is that more sophisticated cell control can be handled by the robot controller. In this case, it now has more power to perform nonrobot input/output and has the ability to perform certain man-machine interfaces, e.g., statistical and error reporting.

An example program for the United States Robots' MAKER 22 Scara robot is illustrated in Table 7.6.3. (This example is treated in detail in Section 7.6.3.) It is interesting to note that this is the form used to program most Scara robots from Japan.

This programming method (as compared to the fixed instruction technique) makes use of program control, specifically the FOR-NEXT loop and the STOP statements.   One should also observe that there are statements that do not cause robot motion and the sequence of events is chosen by the programmer or trainer. Thus it is seen that some of the constraints imposed by the fixed event instruction are removed.

As the available technology became more sophisticated and manufacturing requirements grew, the limited flexibility of the language extension approach became obvious.   This provided the impetus for the development of robot-specific languages.

7.6.1.3 Robot-specific programming languages

A major motivating factor that led to the development of robot-specific programming languages was the need to interface the robot's control system to external sensors in order to provide "real-time" changes to its programmed sequence based on sensory information.   Other requirements such as computing the locations for a palletizing operation based on the geometry of the pallet, or being able to train a task on one robot system and perform it on another (with minor manual adjustment of the points) also were an impetus.   Additionally, requirements for off-line programming, CAD/CAM interfacing, and more meaningful task descriptions led to various language developments.

Table 7.6.2 shows a complete terminal session of a Westinghouse/Unimation robot using VAL 1.   This example, discussed more fully in Section 7.6.3, shows an entire environment for the training of the robot.   As shown in the table, the program is retrieved from a mass storage device, then listed, and the fixed positions defined in the program are displayed.   Finally, the program is executed and output, indicating the current cycle, is displayed on the terminal.   As the listing indicates, this language clearly provides more capability for complex robot control than that of the fixed instruction sequencer or the extended language examples described previously.

Section 7.6.2 presents various commercial and research robot programming languages and a table that compares program control, robot specific mathematics, and input/output capability for each language.   Once again, it should be noted that regardless of the complexity of the programming language, the objective is to define a sequence of operations that are needed to obtain successful control of the robot.

7.6.2 Selected Summary of Robot Languages

Currently, a large number of robot languages are available, although no standards for these exist.   The more common languages include:

- AL
- AML
- RAIL

- RPL
- VAL

Brief descriptions of each of these are given below. This summary is adapted from a paper by Gruver et al. [9].

7.6.2.1 AL

AL was the second-generation robot programming language produced at the Stanford University Artificial Intelligence Laboratory, an early leader in robot research. Based on concurrent Pascal, it provided constructs for control of multiple arms in cooperative motion. Commercial arms were integrated into the AL system. This language has been copied by several research groups around the world. Implementation required a large mainframe computer, but a stand-alone portable version was marketed for industrial applications. It runs on a PDP 11/45 and is written almost entirely in OMSI Pascal [9]. In the AL system, programs are developed and compiled on a PDP-10. The resulting p-code is downloaded into a PDP-11/45, where it is executed at run time. High-level code is written in SAIL (Stanford Artificial Intelligence Language). The run-time system is written in PALX. The PDP 11/45 has a floating-point processor, no cache memory, a single terminal, and 128 kilobytes of RAM memory. Two PUMA 600's and two Stanford Scheinman arms were controlled at the same time by this language.

7.6.2.2 AML

A manufacturing language (AML) was designed by IBM to be a well-structured, semantically powerful interactive language that would be well adapted to robot programming. The central idea was to provide a powerful base language with simple subsets for use by programmers with a wide range of experience. An interpreter implements the base language and defines the primitive operations, such as the rules for manipulating vectors and other "aggregate" objects that are naturally required to describe robot behavior. A major design point of the language was that these rules should be as consistent as possible, with no special-case exceptions. Such a structure provides a ready growth path as programmers and applications grow more sophisticated. AML is being used to control the RS/1 assembly robot, a Cartesian arm having linear hydraulic motors and active force feedback from the end effector. The computer controller on the RS/1 assembly robot consists of an IBM series/1 minicomputer with a minimum of 192-kilobyte memory. Peripherals include disk and diskette drive, matrix printer, and keyboard/display terminals. A subset of AML was employed on the Model 7535 robot that was controlled by the IBM personal computer. However, the features of this version are not included here since the 7535 is no longer being marketed by IBM.

7.6.2.3 RAIL

RAIL was developed by Automatix, Inc. of Bilerica, Massachusetts as a high-level language for the control of both vision and manipulation. It is an interpreter, loosely based on Pascal. Many constructs have been incorporated into RAIL to support inspection and arc-welding systems, which are a major product of Automatix. The central processor of the RAIL system is a Motorola 68000. Peripherals include a terminal and a teach box. RAIL is being supplied with three different systems: vision only, no arm; a custom-designed Cartesian arm for assembly tasks; and a Hitachi process robot for arc welding.

7.6.2.4 RPL

RPL was developed at SRI International to facilitate development, testing, and debugging of control algorithms for modest automatic manufacturing systems that consist of a few manipulators, sensors, and pieces of auxiliary equipment. It was designed for use by people who are not skilled programmers, such as factory production engineers or line foremen. RPL may be viewed as LISP cast in a FORTRAN-like syntax.

The SRI Robot Programming System (RPS) consists of a compiler that translates RPL programs into interpretable code and an interpreter for that code. RPS is written mostly in Carnegie-Mellon's BLISS-11 and cross-compiles from a DEC PDP-10 to a PDP-11 or LSI-11. The programs written in this language run under RT-11 with floppy or hard disks. The RPL language is implemented as subroutine calls. The user sets up the subroutine library and documents it for people who must write RPL programs. Previously, SRI operated the Unimate 2000A and 2000B hydraulic arms and the SRI vision module with this language.

7.6.2.5 VAL

VAL is a robot programming language and control system originally designed for use with Unimation robots. Its stated purpose is to provide the ability to define robot tasks easily. The intended user of VAL will typically be the manufacturing engineer responsible for implementing the robot in a desired application.

Eight robot programming languages are compared in Table 7.6.1. Prior programming knowledge is helpful but not essential. VAL has the structure of BASIC, with many new command words added for robot programming. It also has its own operating system, called the VAL Monitor, which contains the user interface, editor, and file manager. The central monitor contains a DEC LSI-11/03, or more recently, the LSI-11/23. In a Puma 550 robot, each of the joints is controlled by a separate 6503 microprocessor. The monitor communicates with the user terminal, the floppy disk, the teach box, a discrete I/O module, and an optional vision system. VAL is implemented using the C language and the 6502 assembly language. It has been released for use with all PUMA robots and with

TABLE 7.6.1 LANGUAGE-COMPARISON TABLE

	AL	AML	HELP	JARS	MCL	RAIL	RPL	VAL
Language Modalities								
Textual	×	×	×	×	×	×	×	×
Menu		×						
Language Type								
Subroutines				×			×	
Extension					×			
New language	×	×	×			×		×
Geometric Data Types								
Frame (pose)	×			×	×	×		×
Joint angles		×		×				×
Vector	×	×		×	×			
Transformation	×			×	×			×
Rotation	×			×	×			
Path						×		
Control Modes								
Position	×	×	×		×	×	×	×
Guarded moves	a	a	a					
Bias force	×							
Stiffness/compliance	×			b				
Visual servoing	c			c		c	c	c
Conveyor tracking				×	×			
Motion Types								
Coordinated joint between two points	×	×	×	×		×	d	×
Straight line between two points	e			×	×	×	d	×
Splined through several points	×		×	×		×	d	×
Continuous path ("tape recorder" mode)								
Implicit geometry circles					×			
Implicit geometry patterns					×			
Signal Lines								
Binary input	0	64	f	0	242	6	32	32
Binary output	0	64	f	2	242	10	32	32
Analog input	64	0	f	0	242	0	32	0
Analog output	4	0	0	0	242	0	64	0
Display and Specification of Rotations								
Rotation matrix	g			h				
Angle about a vector	×			h				
Quaternions								
Euler angles	i	×		×		×		×
Roll-pitch-yaw	j		×		×			
Ability to Control Multiple Arms								
Multiple arms	×		×		×			
Control Structures								
Statement labels		×	×	×	×		×	×
If-then	×	×	×	×	×	×	×	×
If-then-else	×	×	×	×	×	×	×	
While-do	×	×	×	×	×	×	×	
Do-until	×	×		×		×	×	

540

TABLE 7.6.1 (Continued)

	AL	AML	HELP	JARS	MCL	RAIL	RPL	VAL
Case	×			×		×	×	
For	×	×		×		×	×	
Begin-end	×	×		[k]	[l]			
Cobegin-coend	×		[m]		[l]			
Procedure/function/ subroutine	×	×	×	×	×	×	×	×
Successful Sensor Interfaces								
Vision	×	[n]	×	×	×	×	×	×
Force	×	×	×					
Proximity								
Limit switch	×	×		×	×	×	×	×
Support Modules								
Text editor	[p]	×	[o]	[o]		×	×	×
File system	[p]	×	[o]	[o]		×	×	×
Hot editor		×						
Interpreter	×	×	×					
Compiler	×			×	×		×	
Simulator	×	[q]			×			
MACROs	×		×		×			
INCLUDE statement	×	×						
Command files				×			×	
Logging of sessions	×							
Error logging	×							
Help functions	×	×						
Tutorial dialogue		×						
Debugging Features								
Single stepping		×	×			×	×	
Breakpoints	×	×				×	×	
Trace		×	×		×		×	
Dump	×		×		×		×	

Source: Reprinted courtesy of the Society of Manufacturing Engineers. Copyright 1983 from the ROBOTS 7/13th ISIR Conference Proceedings.

[a]Using force-control or limit-switch action.

[b]Currently being implemented at Jet Propulsion Laboratory.

[c]Uses visual inputs to determine set points but does not specifically perform visual servoing.

[d]Relies on the VAL controller.

[e]Currently being implemented at Stanford University.

[f]Custom for each system.

[g]AL displays rotations as a rotation matrix.

[h]Normally, JARS does not display these forms; however, the user may write a routine to print them because JARS has the forms available internally.

[i]AL accepts directly the specification of an orientation by three Euler angles (or by an angle about a vector).

[j]AL orientations could also be specified by roll–pitch–yaw angles.

[k]Since it is a language based on subroutines added to Pascal, JARS has all the structures of Pascal.

[l]MCL can invoke tasks in parallel using INPAR.

[m]HELP permits the simultaneous activation of several tasks.

[n]Reported by the IBM T. J. Watson Research Center, Yorktown Heights, New York; not commercially available.

[o]JARS and HELP use the systems support features of the RT-11 operating system.

[p]AL uses the support features of the PDP-10 operating system.

[q]A simulator has been developed at the IBM T. J. Watson Research Center, Yorktown Heights, New York.

541

the Unimate 2000 and 4000 series.   The languages described above as well as three others, HELP, JARS, and MCL, are compared in Table 7.6.1 and have been adapted from Gruver et al. [9].

7.6.3 Sample Programs

The following examples illustrate the use of two different robot programming languages, VAL and one employed on a particular Scara-type manipulator.

EXAMPLE 7.6.1 VAL Example

Assume that it is desired to pick up identical objects from a known location and then stack the objects on top of each other to a maximum stacking height of four.   Figure 7.6.1 shows the application.

Let us consider this application and its implementation in the VAL programming language.   Table 7.6.2 is a listing of a session on the terminal, which includes loading and listing the program, viewing the value of the stored locations, and finally, executing the program.

The dot (.) in the leftmost column is the prompt, which tells the user that VAL is ready to accept a command.   The first command given to the

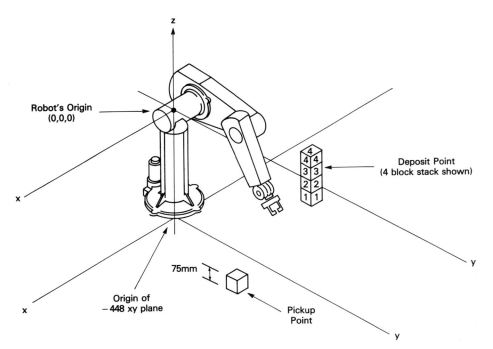

Figure 7.6.1.   Workspace for VAL programming example. The pickup and deposit points are on the *xy*-plane offset (in *z*) from the robot's origin by −448 mm.

TABLE 7.6.2. LIST OF A VAL TERMINAL SESSION

```
.LOAD STACK
     .PROGRAM STACK
     .LOCATIONS
OK
.LISTP STACK

.PROGRAM STACK

     1.              REMARK
     2.              REMARK    THIS PROGRAM PICKS UP PARTS FROM A FIXED
     3.              REMARK    LOCATION CALLED PICKUP, THEN DEPOSITS THEM AT A
     4.              REMARK    LOCATION CALLED B.   IT IS ASSUMED THAT 4 PARTS
     5.              REMARK    ARE TO BE STACKED ON TOP OF ONE ANOTHER.
     6.              REMARK
     7.              OPENI
     8.              SET B = DEPOSIT
     9.              SETI COUNT = 0.
    10.       10  APPROS PICKUP, 200.00
    11.              MOVES PICKUP
    12.              CLOSEI
    13.              DEPARTS 200.00
    14.              APPRO B, 200.00
    15.              MOVES B
    16.              OPENI
    17.              DEPARTS 200.00
    18.              SETI COUNT = COUNT + 1
    19.              TYPEI COUNT
    20.              REMARK COUNT INDICATES THE TOTAL NUMBER OF ITEMS STACKED
    21.              IF COUNT EQ 4 THEN 20
    22.              REMARK       MOVE THE LOCATION OF B UP BY 75.00 MM.
    23.              SHIFT B BY 0.00, 0.00, 75.00
    24.              GOTO 10
    25.       20  SPEED 50.00 ALWAYS
    26.              READY
    27.              TYPE *** END OF STACK PROGRAM ***
     .END
     .LISTL
```

	X/JT1	Y/JT2	Z/JT3	O/JT4	A/JT5	T
DEPOSIT	−445.03	130.59	−448.44	−87.654	88.890	−180.000
PICKUP	163.94	433.84	−448.38	178.006	88.896	−180.000

```
     EXEC STACK
     COUNT    =    1.
     COUNT    =    2.
     COUNT    =    3.
     COUNT    =    4.
*** END OF STACK PROGRAM ***
PROGRAM COMPLETED:   STOPPED AT STEP 28
```

robot controller, LOAD STACK, tells the system to recall the program and any location data from the disk. The system response is on the next three lines, indicating successful completion of this request. The following command to the controller is LISTP STACK, which tells VAL to list the program which is called STACK. This particular version also delimits the program listing by printing .PROGRAM STACK at the beginning and .END at the end. Two more commands that are used in the table are (1) LISTL, which commands the controller to print all the locations that the controller knows about (in this case there are two such locations, DEPOSIT and PICKUP), and (2) EXEC STACK, which tells the controller to execute the program called STACK, which is stored in its memory. Following the EXEC command is the output generated by the program STACK. This output is the value of the variable COUNT as the program is executed. Note that the value of COUNT is used to terminate execution of the program when the desired number of items have been stacked.

Examination of the program listing shows that each line has a number associated with it (i.e., 1 through 27). These numbers are used to identify a line so that the program may be edited. VAL has an editor that allows the user to create programs and store them in the controller. Once stored, a program may be modified by referring to its line numbers. The modifications include inserting, deleting, or modifying lines.

The operation of the robot based on the program steps will now be described.

- Lines 1 through 6 are comments.
- Line 7 tells the gripper to open immediately and then wait a small amount of time to ensure that the action took place.
- Line 8 equates the location of the variable B to a defined location called DEPOSIT. This step is necessary since the value of B will be modified each time a new item is stacked.
- Line 9 sets an integer variable called COUNT to zero. The variable COUNT is used to terminate the program when the proper number of items have been stacked (i.e., 4 items).
- Line 10 has a label (10) associated with it. It commands the robot to move from wherever it is along a straight line to a location 200 mm above the point called PICKUP. At the end of the motion, the approach vector of the gripper will be pointing downward. Recall that the approach vector is defined so that moving along it causes objects to go toward the inside of the gripper.
- Line 11 tells the robot to move its gripper in a straight line toward the position defined by PICKUP. In this example, the motion will be along the approach vector since the gripper is pointing downward. The po-

sition defined by PICKUP is such that when motion ends, the object will be inside the gripper's jaws.

- Line 12 commands the system to close the gripper and wait a sufficient amount of time for the action to occur. In some cases it may be necessary to add an additional delay if that provided by the command is insufficient.

- Line 13 tells the manipulator to move along its approach vector in the direction opposite from which it originally came to a point 200 mm above the pickup point.

- Line 14 tells the manipulator to move to within 200 mm of point B, aligning its approach vector downward.

- Line 15 commands the manipulator to move in a straight line until its tool point is coincident with location B.

- Line 16 tells the gripper to open so that the part can be deposited. This also includes some time delay for the action to occur. As stated previously, additional delay may be necessary to compensate for the actual valves and mechanics used to implement the gripper and to permit the manipulator to settle to the desired location.

- Line 17 tells the manipulator to move back along the approach vector so that it is 200 mm above location B.

- Lines 18 and 19 increment the variable COUNT and display its value.

- Line 20 is a comment.

- Line 21 is a test to see if COUNT is equal to 4. If so, go to the statement with label 20; otherwise, go to the next line.

- Line 22 is a comment.

- Line 23 modifies the location defined by B so that its z coordinate is increased by 75.0 mm.

- Line 24 forces the program to go to label 10.

- Line 25, which is labeled, tells the controller to reduce the speed of motions to 50%.

- Line 26 tells the controller to move the manipulator to its ready position, which is defined as all of the links in a straight line pointing upward.

- Line 27 tells the controller to print a message to the terminal.

From the description of the program, one can easily see the power implemented by the instructions. Commands exist to cause the manipulator to move in a straight line and to manipulate position data. (Note that the "S" in the statement indicates that straight-line motion is desired.) For example, the variable B, which represents a location (i.e., a set of six joint variables) is modified by a single statement in line 23. Similarly, the commands APPROS and DEPARTS are quite interesting because they actually define positions relative to a variable but do not make it necessary for the user to define the actual positions for each move that the robot has to make. This concept is quite important for robot training, since we have really defined only two positions, PICKUP and B. However, we can move to many positions relative to them. Using this approach, if it is necessary to modify either of the points (PICKUP or B), the changes made to them will automatically be reflected in the intermediate points (selectively by the robot's path planner), which are defined solely on these two positions.

The following example illustrates the programming language used by the MAKER 22 (a 4-axis SCARA robot, see Figure 1.3.13) robot from United States Robots. The reader should contrast the power of this language with that of Example 7.6.1.

EXAMPLE 7.6.2 Scara Programming Example

The MAKER 22 is programmed in a language similar to BASIC, with robot-specific extensions. For example, positions in space may be referenced by a single-variable name of the form Pxxx, where xxx is a three-digit number from 000 to 999. In order that position variables may be referenced by an index, it is possible to catenate the P with an integer variable such as A and refer to the point PA. Whatever the value (from 000 to 999) specified by the programmer, A will then reference the actual position variable. Certain operations may be performed on these position points, such as addition and subtraction. Additionally, provisions exist to multiply or divide a position by a scalar. Only two types of moves are provided in the language: MOV, which causes the manipulator to move in a joint-interpolated fashion; and CP, which causes the robot to move in a continuous-path fashion. Whenever a CP command is encountered, the controller will move the manipulator from its current location to the point which is the argument of the command while also looking ahead for the next CP command and its argument. The occurrence of the next such command tells the controller to continue moving toward this next specified position once it has come close to the location defined by the previous CP command. This process continues until the end of the program or a MOV command is encountered. It is clear that if one wanted the manipulator to follow a specific path, all that would be necessary is to define a sufficient number of points for the path and then write a program that uses CP moves to connect them.

The example that we explore illustrates the use of topics discussed in the previous paragraphs. It is desired to cause the MAKER 22 to move in a straight line. For our discussion, we will assume that two positions have

TABLE 7.6.3   MAKER 22 PROGRAMMING EXAMPLE

10: "STRAIGHT LINE"	Label with a comment
N = 10	Number of intermediate points plus 1
P100 = P1	Copy P1 to P100
P101 = P2 − P1	P101 is distance to be moved
P101 = P101 / N	Incremental distance
MOV P100	Set manipulator at first point
For L = 1 TO N	Beginning of loop
P100 = P100 + P101	Compute intermediate point
CP P100	Do CP move to point
NEXT L	End of loop
STOP	

been defined previously, P1 and P2*, and that we wish to have the manipulator move in a straight line starting from P1 and ending at P2.

Table 7.6.3 shows a listing of the program and comments defining the purpose of the instructions. The program in Table 7.6.3 takes the difference between the initial and terminal points of the line and divides by the number of intermediate points plus 1 to compute an incremental distance. It then instructs the manipulator to move to the first point, P100. After attaining this position, it computes intermediate points by adding P101 to P100 and then instructs the robot to move in a continuous-point fashion connecting the 10 points to form an approximation to a straight line. Note that the last point is P2.

It should be apparent that the robot programming language for the MAKER 22 does not contain as high a level of expression as indicated in the example using VAL. This is obvious if one recognizes that a straight line is achieved with *one instruction* using VAL whereas it requires the *entire* program in Table 7.6.3 to perform the identical maneuver with the Maker 22. However, the same functionality, that is, the ability to move in a straight line, is provided by both languages.

After reviewing these two examples and the discussion on robot programming languages, it is suggested that Section 2.4 on the functionality of a robot controller be reviewed in order to relate the desired design functionality to this material.

---

*P1 and P2 are Cartesian coordinate points in (x, y, z) space.

7.6.4 Demonstration of Points in Space

To program a servo-controlled robot, a skilled operator often breaks down the assigned task into a series of steps so that the manipulator/tool can be directed through these steps to complete the task (a program). This program is played back (and may be repeated several times, i.e., it can be used as a subroutine) until the task cycle is completed. The robot is then ready to repeat the cycle. The robot's actions may be coordinated with ancillary devices through special sensors and/or limit switches. These, in conjunction with the controller, send "start work" signals to, and receive "completion" signals from other robots or interfacing devices with which that robot is interacting.

A servo-controlled robot can be "taught" to follow a program which, once stored in memory, can be replayed, causing the controller to be instructed to send power to each joint's motor, which in turn, initiates motion. This teaching process may require that the operator "demonstrate" points in space by causing the end effector to move (using one of a number of possible methods) to a series of locations within the work cell.

The robot can also be taught its assembly tasks from a CAD/CAM data base. Here, the desired points in space are down loaded from such a data base, rather than being taught (on the robot) by an operator. This has the advantage of not occupying the robot for teaching of points and also permits the optimization of the path using simulation techniques. In addition, it is also likely that within the next few years artificial intelligence (AI) techniques will permit robot teaching to be more generalized. For example, AI will allow the robot to place filled bottles in a case or pallet, without having to be explicitly taught a predetermined pattern and/or having specific points actually demonstrated by an operator or down loaded from a CAD/CAM system. Before discussing this topic, however, we will consider more standard techniques of demonstrating points to a robot.

There are several methods currently in use. The method employed depends on the manufacturer's specifications, control system software, and the robot's computing/memory capabilities. Teaching typically involves one of the following methods: continuous path, via points, or programmed points. Each of these is now briefly discussed.

7.6.4.1 Continuous path (CP)

With the CP method, the operator releases all joint brakes and enables an automatic sampler. The manipulator is then manually moved through each of the positions required to perform the task. The controller "remembers" or stores the coordinates of all the joints for every position. In this manner, complex three-dimensional paths may easily be followed. Teaching may be done at a speed different from that speed needed for real-time operation (i.e., playback may be set at other speeds, allowing for different cycle times). This method requires minimal debugging, allows for continuous-path programming, and requires minimal knowledge of robotics. However, a thorough understanding of the assigned task is a prerequisite, and editing requires reprogramming from the error point. This method is typically used with robots employed in spray-painting and arc welding applications.

7.6.4.2 Via points (VP)

Teaching with the VP method does not require that the operator physically move the manipulator; rather, it is remotely controlled by either a computer terminal or, more commonly, a teach pendant—a device similar to a remote control box with the additional capability to record and play back stored commands. The teach pendant is plugged into the controlling computer during programming (the on-line method), and the operator then presses the appropriate buttons to position the arm, with small incremental motion for precise positioning. When the correct position is achieved, a switch is activated to inform the computer to read and store positions for all joints. This process is repeated for every spatial point desired to be "taught." Essentially, only the endpoints of the motions are demonstrated.

The VP method is often employed to program discrete points in space (through which the end effector is required to pass) and is most commonly used for point-to-point robots. The teach pendant is most commonly used for heavy-duty robots and in those lightweight robots that have sophisticated control systems.

There are more advanced systems that allow for the movements and endpoints to be recorded in an unspecified order. This enables new programs to be created by calling out the points in a sequence that differs from the original order of input, thus facilitating programming and editing. These systems also allow the programmer to define velocity and acceleration or deceleration between points. However, such advanced systems have an inherent danger; that is, the path resulting from a new sequence of movements may inadvertently bring the end effector in contact with nearby machinery. For this reason, manufacturers recommend that once the program is complete, the program should be played back at a very slow speed to minimize the possibility of damage to the robot or other equipment.

7.6.4.3 Programmed points (PP)

The PP method is also an on-line system. The robot operates via a prerecorded program (i.e., without manual intervention), with the program sequence having been set up externally. Applications of the PP method of using decision making include orienting (i.e., aligning workpieces in designated positions) for assembly operations and material-handling work using conveyers. In addition to the techniques used for programming a robot as described above, there is a new methodology emerging. This is discussed next.

7.6.5 Artificial Intelligence and Robot Programming

The discipline known as artificial intelligence (AI) is becoming more practical as new developments in computer hardware and software evolve. Higher memory density, faster processors, and new languages are bringing the tools of artificial intelligence to practice. There are "expert systems" development environments that execute on nominally priced personal computers, and these are already having an impact in many areas previously the exclusive domain of the human thought process. Experience is showing that in a complex equipment maintenance milieu, in certain classes of medical diagnosis, theorem proving, biochemical analysis, and a plethora of other fields, AI is contributing to productivity. The much touted nationalized Japanese fifth-generation computer project is directed toward creating AI techniques that will reduce software production to a blue-collar job. Whether or not the Japanese will succeed is yet to be determined, but even if the goal is not fully reached, there will be significant technological fallout from the effort.

In the programming of robotic systems, the use of AI techniques is certain to have an impact because of the availability of data base information that can be used to plan a robot's task efficiently. Although there is no integrated system available today, laboratory demonstration such as the assembling of simple structures from randomly presented and available parts is already accomplished. Moreover, a number of laboratory facilities are currently implementing AI/expert systems in a variety of mobile robots. Intended for use in the nuclear power industry and by the military, these devices are being employed as testbeds for practical results in the areas of autonomous navigation, collision avoidance, maintenance and repair, assembly, reconnaissance, and perimeter monitoring.

7.7 PATH PLANNING

Path planning is a critical aspect of robotic manipulator control. Two specific aspects of path planning are discussed here.

7.7.1 Coordinated Motion

Path planning or trajectory planning algorithms are concerned with the generation of the intermediate points along a manipulator's trajectory. These are the points (or positions) that must be fed to the control system so that the joints can be commanded to move to the correct locations necessary to position the end effector properly. In addition, it is often desired to start and stop all robotic axes at the same time. This behavior is referred to as coordinated motion and will modify the path-planning algorithm.

In a robot, the initial path position is inferred (from the current position) while the final path position is specified. Along with the final point, some rule defining the trajectory must be specified and may include the following options:

1. Joints of robot to start and stop at the same time as the end effector moves from the initial to the final point (not exceeding physical constraints or robot specifications). However, the actual path taken is not specified. This is called joint interpolated motion.

2. The "tool point" is to move along a straight line. This is sometimes referred to as world motion. Note that this implies that all axes start and stop at the same time.

3. The tool (or end effector) is to move along a straight line defined by extending the approach, normal, or orientation vectors associated with the tool point. This is called tool motion (see Chapter 8 and Section 7.6.4).

4. The end effector may be told to follow a straight line as in world motion, while the initial and final orientation of the gripper may be required to change.

5. The acceleration or velocity may be specified prior to the motion, or may be commanded to change during the motion based on some external input.

The mathematics to accomplish these types of motion is discussed in Chapter 8. It should be apparent that the speed at which the computations need to be made must be in "real time." This implies that they are completed as soon as (or a significant time before) the information is needed by the joint servos. Although the mathematics for accomplishing this is formulated in terms of matrices, the implementation may take advantage of certain properties or simplifications. That is, the actual implementation may involve little or no matrix multiplication. Additionally, and depending on the various types of implementations, it is possible that the computations may take too long and certain motions may be impossible or speed limited.

It is also possible to define a series of points that determine the trajectory of a manipulator. This can be accomplished in a number of ways, including:

1. A series of points are taught or demonstrated (defining a complicated curve). The manipulator is expected to pass through these points as closely as possible and perform some interpolation between them in order to faithfully reproduce the path. Of course, the more points taught, the better the curve is reproduced.

2. Three points defining a circle are demonstrated and the manipulator is expected to be able to compute any necessary intermediate points so that it can draw the circle.

3. An equation is defined (in Cartesian, cylindrical, or other space) that the robot is commanded to follow.

Various curve-fitting routines or series expansions can be used to implement these features. Once again, it is important to note that time is critical and an approximation may be needed. For instance, a reasonable trade-off may be made on the algorithms used and the actual control resolution and accuracy of the manipulator. It may only be necessary to define a joint angle to the nearest minute since the resolution of the encoding device may be similarly limited.

The ideas discussed above need to be implemented in whatever type of programming language is to be used for a manipulator. Some languages previously described provide for terse powerful commands. For example, the statement from VAL

MOVES POINT1

allows the programmer to command the robot to move from its current location to one defined as POINT1 in a straight line. (The motion will be joint interpolated if the "s" is deleted.) Other languages may provide the mathematical capability to compute intermediate points necessary to move in a straight line but without an explicit command.

7.7.2 Automatic Programming and World Modeling

The concept of automatic programming is associated with a robot "teaching itself." Essentially, the robot is assigned a task and must develop a plan to accomplish it autonomously. Various AI techniques are utilized to define the steps that the robot must take to accomplish the task. However, before any algorithms are used, it is first necessary to define a three-dimensional world model. This defines the environment in which the robot must perform its designated task. It includes models of the pieces or parts that the robot must manipulate as well as any physical constraints or obstructions in the workcell.

The complexity of the modeling process is staggering and compromises must be made. For example, if a printed circuit board with 100 holes were modeled, one would define the plane and the location of the center of the holes and their sizes. The model probably would not compensate for manufacturing errors such as misplaced holes or errors in roundness. Even based on this simple example, it can be seen that the model is rarely, if ever, the same as the real world. This is where the problem begins since the things assumed correct are usually the items that will affect the reliability of the process.

Automatic programming is currently a laboratory tool which is only recently being adapted to real-world applications. It also includes the concept of collision avoidance, in which the robot ensures that it does not hit anything in the process of doing its job. It is expected, however, that as processes become more complex and higher performance is desired, future generations of industrial robots will have this capability.

7.8 THE ROBOT'S COMPUTER SYSTEM

As discussed previously in this chapter, the role of the computer in a robot's controller can be quite varied. What we will attempt to do in this section is to discuss the requirements of a hypothetical robot controller from the computational point of view using as a model the generic architecture of a robotic controller, as shown in Fig 2.2.2, the capabilities discussed in Section 2.4, and the servo control loop discussed in Appendix C.

Starting from the designer's and implementer's points of view, a computer architecture similar to Figures 2.2.3 and 4.1.1 is chosen. The use of distributed microprocessors to implement the controller has many advantages for the development phase. Specifically:

- A suitable processor can be chosen for each application, and thereby cost and complexity can be kept to a minimum.
- The software for each processor can be designed, coded, and tested independently of the other processors.
- The system can be designed to be modular in nature, and the complexity of the controller can be reduced when less functionality is needed.
- If the functionality is distributed in the proper way, the task of troubleshooting the final system is reduced to checking only those modules responsible for functions that are not operational.
- If single-board computers with identical computational architecture are used to implement some or all of the processors (which in fact actually execute completely different software), we obtain commonality of hardware and the possibility of swapping cards in the field to facilitate troubleshooting.

Of course, we must also understand the disadvantages of employing distributed microprocessors. These include the following:

- The communications between the processors must be clearly defined.
- Provisions must be made so that testing of each processor can be done independently of the others. Thus, both hardware and software may be necessary so as to emulate signals and data from nonexistent pieces of the system.
- If all the processors must be debugged at the same time, multiple logic analyzers or other test equipment must be available.

For our hypothetical system, the following additional specification will also be included:

- The robot will be programmed using any commercially available language and a library of subroutines which perform functions associated with robot control.

This particular specification makes the system implementation quite simple since only subroutines (or functions) associated with robot control have to be developed and the remaining control structure of the language, such as looping, data structures, and syntax are already available. To simplify the design further, if we choose a commercial operating system that will run in the processor and support the language, we have already accounted for the "housekeeping features" detailed in Section 2.4 since the operating system should provide for file maintenance and a commercially available editor can be used to create or edit programs. At this point, the elegant simplicity of our robot controller should be apparent.

In addition to what has been described above, the remaining pieces that need to be included are the specialized interfaces to the electronics that control the physical hardware of the manipulator or interface ancillary devices (such as binary inputs and outputs).

The computer we have chosen certainly includes an interface to the outside world either in the form of a standard bus (such as VME or STD), serial ports such as RS-232C, or a custom interface. In any event, these interfaces become the medium of communication between the control program and the hardware-specific interfaces.

Figure 7.8.1 shows the proposed architecture of our robot controller. We have assumed that some type of relatively high-speed interface exists between our "central control unit" or "host" and each piece of specific hardware. Here the term "central control unit" or "host" is used to encompass the functionality of sequencer, memory, and computational unit defined in Chapter 2.

Note that in this system a separate computer is used to control each servo associated with the robot. This processor essentially executes code as defined in Table C.4.1 (without the profile generator). Therefore, the only information it needs is the set point data, which will come from the host over the "common bus" at a fixed rate. To synchronize the "joint processors," another message is sent to all of the processors simultaneously, telling them to execute their algorithms using the new set point data. A timing diagram depicting this data transfer is shown in Figure 7.8.2. As pointed out in Appendix C, it may not be possible to send set point data to the joints as fast as we would like (e.g., every 1 ms) because of the host's computational limit. In this case it is possible that data are sent every 16 ms and the joints themselves have a linear interpolator used to generate intermediate set point information.

Following the concept of data transfer between the host and joint processor just discussed, other processors on the "common bus" also need to receive commands and data from the host and send data back to it. The host may use the information directly or route it to one of the other processors of the system. In keeping with the communication scheme of Figure 7.8.2, these data are transmitted to and received from the processors at fixed times. Thus Figure 7.8.2 may be modified to include slots for communication to the other computers. While on the surface, this type of communication scheme may seem to contradict the need for extremely fast response, the update rates are generally much faster than the

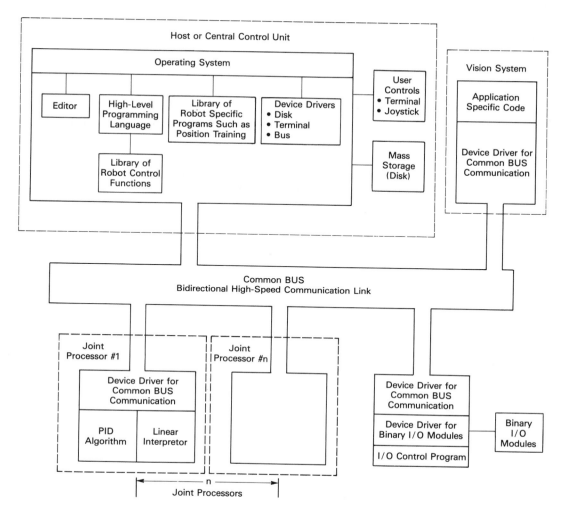

Figure 7.8.1. Controller architecture from computer perspective.

mechanical devices they are controlling and should not present a problem. Also as noted earlier, the servos are interpolating data and are performing at a faster rate (e.g., 1 msec updates) than the set point information is being sent from the host (e.g., 16 msec updates).

This scheme is also advantageous in coupling job-specific hardware devices into the system. For example, if we wished to add vision, it would be a self-contained system and would merely send or receive position information over the common bus for use by the host when necessary. Additionally, if we wanted a force-controlled gripper, hardware specific to this task could be added, which could perform such tasks as signal processing (for use with a force sensor) and servo control of the gripper's actuator. The host would merely send a signal to the

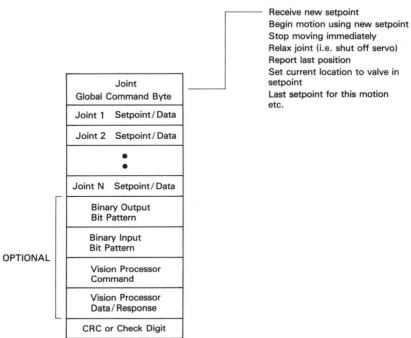

DATA INTERCHANGE STRUCTURE (DIS)

Joint Global Command Byte	— Receive new setpoint Begin motion using new setpoint Stop moving immediately Relax joint (i.e. shut off servo) Report last position Set current location to valve in setpoint Last setpoint for this motion etc.
Joint 1 Setpoint / Data	
Joint 2 Setpoint / Data	
• •	
Joint N Setpoint / Data	
Binary Output Bit Pattern	
Binary Input Bit Pattern	
Vision Processor Command	
Vision Processor Data / Response	
CRC or Check Digit	

OPTIONAL

TIMING OF DATA EXCHANGE BETWEEN HOST AND JOINT PROCESSORS

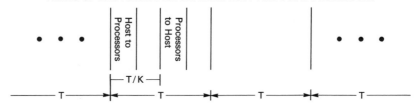

- Every T sec host writes DIS to common bus and all joint processors on bus receive it simultaneously
- Each processor computes CRC as a validity check
- Each processor interprets global command byte and takes its appropriate data
- Any information that the processors are to send to the host are written to the appropriate location (such as a response to the last command byte)
- T/K sec after T the host interprets the information written into the DIS
- The cycle repeats forever

Figure 7.8.2.   Data transfer between host and joint processors.

computer to close and exert a force of say 8 oz. Once the processor received the command, it would perform its task independently of the "central control unit" and upon completion of its task send back a signal indicating success or the reason for failure.

To facilitate the "teaching" of points, the system may include another program by which the trainer uses a joystick, keys, or other control to cause the manipulator to move either in a joint-by-joint fashion or along a straight line. In either case, set points are generated and fed to the servos, which in turn must report their positions back to the control program. When the trainer is satisfied with the position of the manipulator, the values of all the variables relating to the position of the manipulator are saved and given a name such as "this_point." While the name "this_point" may be associated with a specific position of the manipulator, it should be understood that this may be a complex data structure and will carry information not readily apparent to or needed by the trainer.

Once this type of system has been designed and debugged, it will be used by people who may be less skilled in the art than its designers. From the user's point of view, the functionality described in Section 2.4 is implemented primarily by high-level programming instructions. For example, using the programming language, the trainer may command the robot to go in a straight line from its current location to another simply by using a function such as

```
move_straight_to(next_point);
```

The argument of the function "next_point" may have been created by the "teaching program" discussed previously.

As pointed out earlier, the function "move_straight_to( )" was defined by the designers, and the details of its exact implementation are probably of little interest to the user, whose primary interest is to be able to use the function along with some data to cause the manipulator to perform a specific function. Of course, other functions must also exist in the library. These instructions provide the ability to perform the following actions:

- Wait a specific amount of time.
- Set the binary outputs to a particular state.
- Read the binary inputs (from external sensors) and the state of the gripper (i.e., opened or closed).

Besides the functions provided for specific robot control, the commercial operating system gives the user the ability to perform "housekeeping functions" and supports any other programs needed to support the robot programming environment.

In summary, this section has attempted to give the reader some insight as to how computational elements and software come together to form a robot controller.

Of paramount importance was the need to implement a certain level of functionality with standard software and hardware so that the ultimate end user could have the ability to define a robot task.

7.9 SUMMARY

In this chapter we have discussed many topics relevant to computer considerations for robotic systems. The picture presented here is a snapshot of numerous technological considerations that are changing rapidly, and thus the specific material in the chapter may be quickly outdated. The general topics treated here will not become outdated, however, and for this reason one must develop a general set of methods to evaluate new advances in robotic software, communications, cell controllers, and other robotic computer-related subjects. Although the specific robot languages or the specific interface protocol may change, the role that these technological components play will be more or less consistent.

The authors cannot emphasize enough the importance of evaluating the computer elements in a robotic system, and being aware of changes in technology that may alter the specific value of a specific technique. For this reason the authors expect that the reader will need to be aware of the general importance of these issues and can analyze them as technology changes.

7.10 PROBLEMS

7.1 Design the software architecture of a robot controller that utilizes a multitasking operating system to provide the following:

- Service a terminal
- Interpret terminal commands
- Perform computations
- Wait for the specific events
  —joints at set points
  —intrusion detection
- Execute the instructions of a robot program

Modify the design so that an editor may run and a program be entered while the robot controller is running another program and executing its other defined tasks. If you had to prioritize activities, which would be the most important; the least important?

7.2 Pick a specific processor (such as a 68000 or 8086) and investigate what commercially available bus architecture, operating systems, and programming languages are available. Based on robotic considerations as defined in Chapters 2 and 7, what combinations provide the most support for the desired functionality at the least cost? Which provide the most flexibility?

7.3 Based on the discussions in Chapter 8, determine the time it takes to multiply two 4 × 4 matrices using languages such as BASIC, FORTRAN, Pascal, and C; also determine the time factor if coded in Assembly language. Use a computer with support for more than one of these languages to simulate this. Consider fixed-point and floating-point numbers along with a general matrix multiplication routine and one that is specific to DH matrices.

7.4 Using the fixed instruction sequence control definition of the MAKER 100 Mark 1 controller (see Section 7.6.1.1), program the stacking application (which was accomplished in VAL (see Table 7.6.2) and described in Section 7.6.3). Assume that there are eight binary outputs and that straight-line motion is possible. Use mnemonics to define positions (remember that each position must be taught previously) and other operations such as gripper state and the states of output lines. Since no terminal device is available, use a subset of the binary outputs to indicate which object has been stacked, and a special one to indicate that the program has completed its task. Comment on the fixed instruction sequence versus the VAL implementation.

7.5 Obtain descriptions of at least two commercial programming languages from their manufacturers. Use these languages to reprogram the stacking application (see Table 7.6.2 and Section 7.6.3). Compare and comment on the similarity of the functionality of the different languages (for instance, straight line motion, operations on positions in space, terminal display, and so on). Are any of the languages better for performing this object-stacking application? Next consider the programming skill required for each language and comment on whether people with explicit computer programming skills could easily understand the construction of the language and effectively use it. For the languages you have chosen, are any applications better suited for implementation by one versus the others?

7.6 Using a personal computer, program a robot simulator using two-dimensional graphics and a fixed instruction sequence control scheme. For example, assume that we graphically show the x-y plane of a r-θ manipulator and the annulus of its workspace. Also, the display will be used to trace the trajectory of the tool tip of our simulated robot. Moreover, the state of each of the output lines will be shown along with any other pertinent information.

Define a fixed instruction sequence set (similar to that of the MAKER 100 described in Section 7.6.1.1). One possible suggestion is to input the robot program into a file which is read by your simulation program. As each instruction is read from the file, update the display to show the manipulator position, output line state, gripper state, and so forth. Note that it may be necessary to query for the input states prior to updating the display.

7.7 Find commercial examples of robotic extensions to computer languages. Comment on the similarity of the extensions in terms of functionality.

7.11 FURTHER READING

1. Artwick, Bruce A., *Microcomputer Interfacing*, Englewood Cliffs, N.J.: Prentice-Hall, Inc., 1980.

2. Bonner, Susan and Kang G. Shin, "A Comparative Study of Robot Languages," IEEE Computer Society, Computer, Vol. 15, No. 12, 12/82, pp. 82–96.

3. Booth, Taylor L., *Introduction to Computer Engineering, Hardware and Software Design*, Third Edition, New York: John Wiley and Sons, 1984.

4. Corti, Pierluigi, Gini, Giuseppina, Gini, Maria, and Marco Somalvico, "Problem Solving and Automatic Emergency Recovery: Towards the Design of Intelligent Robots," Cybernetica (Belgium), Vol. 23, No. 1, pp. 37–45, 1980.

5. Critchlow, Arthur J., *Introduction to Robotics*, New York: Macmillan Publishing Co, 1985.

6. Doty, Keith L., *Fundamental Principles of Microcomputer Architecture*, Portland: Matrix Publishers, Inc., 1979.

7. Gault, James W. and Russell L. Pimmell, "Microcomputer-based Digital Systems," New York: McGraw-Hill Book Company, 1982.

8. Goldman, Ron, "Recent Work with the AL System," Fifth International Joint Conference on Artificial Intelligence, MIT, August, 1977, Vol. 2, pp. 733–735.

9. Gruver, William A., Barry I. Soroka, John J. Craig, Timothy L. Turner, "Evaluation of Commercially Available Robot Programming Languages," Thirteenth International Symposium on Industrial Robots and Robots 7, Conference Proceedings, Future Directions, Vol. 2, pp. 12-58–12-68, April 17–21, 1983, Society of Manufacturing Engineering, Dearborn, Michigan.

10. Ish-Shalom, Jehuda, "The CS Language Concept: A New Approach to Robot Motion Design," The International Journal of Robotics Research, Vol. 4, No. 1, Spring, 1985, pp. 42–58.

11. Lam, Herman and John O'Malley, *Fundamentals of Computer Engineering Logic Design and Microprocessors*, New York: John Wiley and Sons, Inc., 1988.

12. Lawrence, Peter D. and Konrad Mauch, *Real-Time Microprocessor System Design: An Introduction*, New York: McGraw-Hill Book Company, 1987.

13. Lozano-Perez, Tomas and Patrick H. Winston, "LAMA: A Language for Automatic Mechanical Assembly," Fifth International Joint Conference on Artificial Intelligence, MIT, August 1977, Vol. 2, pp. 710–716.

14. Mano, M. Morris, *Computer System Architecture*, Second Edition, Englewood Cliffs, N.J.: Prentice-Hall, Inc., 1982.

15. Mujtaba, M. Shahid, "Current Status of the AL Manipulator Programming System," Tenth International Symposium on Industrial Robots, Fifth International Conference on Industrial Robot Technology, Proceedings, 1980, pp. 119–127.

16. Rajaraman, V. and T. Radhakrishnan, *An Introduction to Digital Computer Design*, Second Edition, Englewood Cliffs, N.J.: Prentice-Hall, Inc., 1978.

17. Ruoff, C. F., "PACS—An Advanced Multitasking Robot System," The Industrial Robot, June 1980, pp. 87–98.

18. Shimano, Bruce, "VAL: A Versatile Robot Programming and Control System," Proceedings of COMPSAC—The IEEE Computer Software and Applications Conference, Chicago, Nov. 1979, pp. 878–883.

19. Tocci, Ron, *Digital Systems Principles and Applications*, Fourth Edition, Englewood Cliffs, N.J.: Prentice-Hall, Inc., 1988.

8

Transformations and Kinematics

8.0 OBJECTIVES

The purpose of this chapter is to introduce the concepts of "homogeneous coordinates" and "coordinate transformations." When utilized in the framework of the kinematics of robots, these concepts provide a concise mathematical formulation to describe the position and orientation of a manipulator and open up ways for generating more complex motions than can be obtained by moving joints individually or in a "joint-interpolated" fashion. Recall from Chapter 2 that joint-interpolated motion refers to a motion in which all joints start and stop simultaneously, and whose trajectory (path traced by the tip of the arm) does not generally follow a well-defined Cartesian path such as a line or the arc of a circle. However, straight-line motion algorithms and task descriptions are readily described in terms of homogeneous coordinates and transformations without the need for considering the particular geometry of a robot until the joint positions are needed. In addition, positions defined in terms of homogeneous coordinates may be related to one another by simple matrix transformations. This concept provides the means to easily determine the location of objects whose exact position may be unknown but which have been defined with respect to some other known locations or objects.

In this chapter we present the following topics:

- Homogeneous coordinates
- Transformations

- Coordinate reference frames
- Description of points and objects in terms of transformation matrices
- Kinematic equations (the forward solution)
- Reverse (back) solution
- Transformation matrices applied to motion control
- The Jacobian
- Continuous path algorithms
- Trajectory control
- Controller architecture

8.1 MOTIVATION

In describing the position and orientation of a manipulator, one may choose to use various types of coordinate reference frames. The configuration of the robot may be such that the joint motions correspond naturally to the variables associated with Cartesian, cylindrical, or spherical reference frames (see Chapter 1). Other configurations may produce more intricate geometric relationships. In addition, the manipulator may provide more degrees of freedom than are associated with position in three-dimensional space. Specifically, once the end of the manipulator is located, its orientation may also be changed. Therefore, in describing the position of a robot and the gripper or tool it is carrying, we define both the position and orientation of the tool with respect to some reference frame.

Since the control strategy used in most robots is based on the ability to control the position of the joints (typically, most feedback devices are located in a manner so that they monitor joint position and velocity), the most natural reference frame for a robot is defined by its joints. With the control system monitoring and controlling joint positions and velocities, it is quite simple to move the robot about by changing the positions of each of its joints. A particular position of the manipulator or its tool may be defined in terms of a set of joint positions. Figure 8.1.1 shows a five-joint manipulator with a set of "rulers" attached to each of its joints. By selecting a point on each of the rulers as a reference point, the position of each joint may be defined. We define the term *joint space* as the space in which the position of the manipulator is described by a *joint vector* whose components are the positions of each joint. A point in a robot's work volume is in its joint space. Figure 8.1.2 shows the robot of Figure 8.1.1 tracing out a plane in its workspace.

The joint vector is defined as a column matrix containing the positions of the joints or a value associated with each joint variable at a given point in the workspace with respect to some calibration point. This point sets the reference for measuring

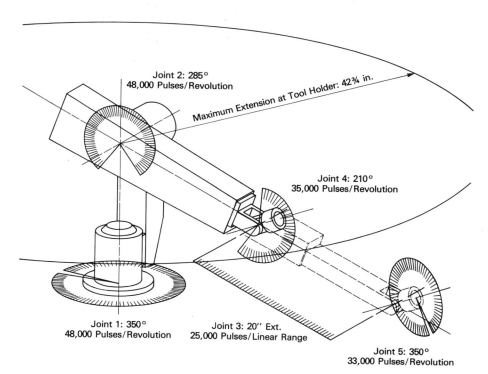

Figure 8.1.1.   Robot and joint measurements.

joint angles or joint displacements.   We define a joint vector **J** (not to be confused
with inertia) as follows:

$$
\mathbf{J} = \begin{bmatrix} \theta_1 \\ \cdot \\ \cdot \\ \cdot \\ \theta_n \\ r_1 \\ \cdot \\ \cdot \\ \cdot \\ r_m \end{bmatrix}
\qquad (8.1.1)
$$

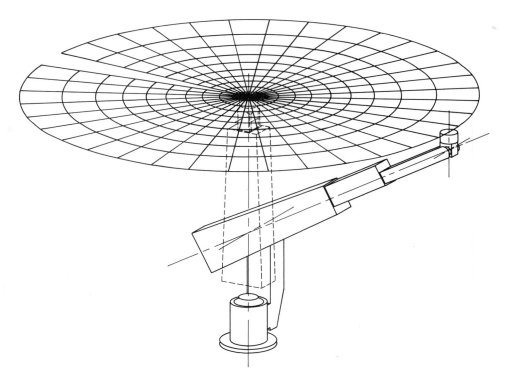

Figure 8.1.2.   Robot and plane in its workspace.

where θ_1 through θ_n define the angular positions of rotary joints and r_1 through r_m define the linear extensions of prismatic joints.   The dimension of the joint vector is dependent on the number of axes of the manipulator.

Using this type of position representation with position servo control of each joint, one may teach locations in space by manually moving each joint until the tip of the manipulator is at the desired location.   Once the location is reached and has stabilized, the value of the joint vector may be recorded for later reference. Typically, the values corresponding to the joint positions are stored in memory for use at a later time.   This is the approach used to teach most robots.

If the only way we have to position a manipulator in its workspace is by manipulating the joint positions, the limitations of positioning or teaching the manipulator become quite clear.   Consider the problem of attempting to trim up the position of the manipulator to a desired final endpoint located a few inches directly in front of the tool tip's current position.   In the case of a manipulator with six rotary joints (refer to the robot configurations in Chapter 1), it is clear that this simple motion may require movement of all six revolute joints and that the required changes in each depend on the exact location of the tool tip.   The difficulty of this particular motion will tend to change with different robot configurations.

A more complicated example to illustrate this problem further is to require the tip of the manipulator to move along a straight line defined only by its two endpoints, which would correspond to two joint vectors. Once again, depending on the configuration of the manipulator, it may be extremely difficult to generate a simple algorithm to compute the intermediate points working only in joint coordinates.

A solution to the positioning problem is to be able to represent positions and orientations of the tool point or tool tip (see Chapter 2) in a convenient coordinate system rather than being forced to work solely in the coordinate system defined by the joints of the manipulator. This lets us use the most convenient coordinate representation for the task at hand and simplifies the generation of curves and straight lines by the tool (or end effector).

The most natural coordinate system for human beings is the standard Cartesian system. We tend to think in straight lines, for example, "move up a few inches" or "move along a diagonal" to some point. Similarly, most machine tools do the same thing and reference coordinates to a "datum." This datum establishes an x-y or xyz frame to a workpiece and allows a machinist to drill holes in the correct locations. Although Cartesian coordinates (or distances as pointed out above) may be advantageous in certain circumstances, other coordinate systems, such as cyclindrical, spherical, or even the joint space of the robot itself, should not be dismissed.

To establish the location of the tool tip of the robot in the coordinate frame being used (consider Cartesian), the joint vector of the manipulator must be mapped into Cartesian space. This mapping provides an additional method of recording points, in that the location may now be stored as Cartesian coordinates as opposed to joint coordinates.

Regardless of the complexity of the desired coordinated motion (e.g., straight line in Cartesian space, or a circle defined in cylindrical space) to ultimately control the robot, the joints must receive the appropriate position values. This establishes the need for a method of mapping the particular reference frame in which the motion is defined into joint space. As we shall see later on in this chapter, the mapping from joint space to Cartesian space may not be unique.

Homogeneous transformations provide the mathematics for describing points with preferred orientations in three-dimensional space. Matrix operators applied to homogeneous coordinates allow points to be rotated about axes, translated in space, and referenced relative to other points and to other reference frames. We may also define relationships that map the coordinates of spherical or cylindrical frames into Cartesian, or vice versa.

Using the framework of homogeneous transformations, the serial links of a manipulator may be modeled as a set of reference frames whose relative positions and orientations are determined by the values of the joint variables. By relating successive frames to each other (i.e., starting at the base of the robot and ending at the gripper), the "forward solution" (joint vector to Cartesian representation) is obtained. This provides a means of mapping an n-dimensional joint vector to Cartesian space, giving both the position and orientation of the gripper with respect to a predefined Cartesian coordinate system. The "inverse" (or "back") solution maps Cartesian position and orientation to the n-dimensional joint vector. This solution is essential for implementing complex coordinated motions such as straight line or positioning the manipulator if the position is defined in any reference frame other than joint.

Coupling the matrix operations, homogeneous transformation representation of positions, and forward and reverse solutions along with some motion algorithms provides the entire framework for complicated robot control and motion trajectories. Each of these subjects is now considered in turn.

8.2 HOMOGENEOUS COORDINATES

Recall that a vector is a quantity that has both magnitude and direction. It is usually represented by an arrow of length equal to its magnitude and pointing in the appropriate direction. An alternative notation utilizes a linear combination of a set of basis vectors. For vectors in Cartesian coordinates, the basis is a set of unit vectors directed along the orthogonal x, y, and z axes used to represent this space. Thus a vector $\mathbf{v}$ can be written as

$$\mathbf{v} = x\hat{\mathbf{i}} + y\hat{\mathbf{j}} + z\hat{\mathbf{k}} \tag{8.2.1}$$

where the unit vectors $\{\hat{\mathbf{i}}, \hat{\mathbf{j}}, \hat{\mathbf{k}}\}$ have been weighted by the appropriate constants

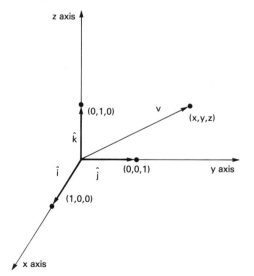

Figure 8.2.1. Vector and basis set in Cartesian reference frame.

(x, y, z) (see Figure 8.2.1). Our notation for vectors will be a boldface upper or lowercase letter (sometimes with a subscript). Unit vectors will be boldface lowercase letters with a circumflex (^) above them.

In the special case where the vector originates at the origin of the coordinate system and terminates on a particular point (x, y, z), we may also think of this as a representation of the point. Throughout the rest of our discussion this will usually be the case.

Another way of representing a three-dimensional vector is by using homogeneous coordinates. In this representation a fourth component which acts as a scaling factor is added. The physical coordinates [which were the basis coefficients in Eq. (8.2.1)] are obtained by dividing each component in the homogeneous representation by the scale factor. Since we would like this representation to generate the same vector as our previous method [i.e., Eq. (8.2.1)], the components are also multiplied by this scaling factor. The homogeneous coordinate representation for the vector defined by Eq. (8.2.1) is written as the column matrix

$$\mathbf{v} = \begin{bmatrix} x' \\ y' \\ z' \\ s \end{bmatrix} \tag{8.2.2}$$

where:

$$x' = sx$$

$$y' = sy$$

$$z' = sz$$

and s is the scaling factor.

If the value of the scaling factor s is set to 1, the components of the homogeneous and standard Cartesian representations are identical. The importance of the fourth element will become clear when we introduce matrix operators.

A numerical example illustrating the two types of vector representations is shown below.

EXAMPLE 8.2.1: VECTOR REPRESENTATIONS

Cartesian basis representation:

$$\mathbf{v} = 3\hat{\mathbf{i}} + 4\hat{\mathbf{j}} + 5\hat{\mathbf{k}} \tag{8.2.3}$$

Homogeneous representations:

$$\mathbf{v} = \begin{bmatrix} 3 \\ 4 \\ 5 \\ 1 \end{bmatrix} = \begin{bmatrix} 6 \\ 8 \\ 10 \\ 2 \end{bmatrix} = \begin{bmatrix} -30 \\ -40 \\ -50 \\ -10 \end{bmatrix} \tag{8.2.4}$$

This example points out that the homogeneous coordinate representation for a vector is not unique. In Eq. (8.2.4), three of the many possible column matrices which could represent **v** are shown. Each of these column matrices represents the same vector in three-dimensional space.

When the scaling factor s is set to 0, the vector in homogeneous coordinates represents a direction. This is easily seen if we imagine a limiting process where we let s approach zero. As s approaches zero, the vector components become infinite. Thus direction vectors pointing along the x, y, and z axes can be represented as follows:

$$\hat{\mathbf{i}} = \begin{bmatrix} 1 \\ 0 \\ 0 \\ 0 \end{bmatrix} \tag{8.2.5}$$

$$\hat{\mathbf{j}} = \begin{bmatrix} 0 \\ 1 \\ 0 \\ 0 \end{bmatrix} \tag{8.2.6}$$

$$\hat{\mathbf{k}} = \begin{bmatrix} 0 \\ 0 \\ 1 \\ 0 \end{bmatrix} \tag{8.2.7}$$

In the case of the representing direction, it is a good idea to ensure that the entries x', y', z' have been normalized. This prevents unnecessarily large numbers from being carried. Recall that the normalization for vectors represented by Eq. (8.2.1) is accomplished by dividing each component of the vector by the square root of the sum of the squares of the individual components (i.e., the vector norm). This ensures that the resulting magnitude of the vector (i.e., the square root of the sum of the squares is equal to 1).

The null vector representing a point at the origin of a reference frame is defined as

$$\begin{bmatrix} 0 \\ 0 \\ 0 \\ s \end{bmatrix} \tag{8.2.8}$$

where s, the scaling factor, can be any number other than zero. We will usually use scaling factors of 1 for our work. Thus the null vector will also use a scaling factor of 1. Physically, the null vector represents the origin of a coordinate reference frame. A vector is *undefined* if all the entries are zero.

8.2.1 Vector Operations

Many operations may be performed on vectors. These include the typical ones, such as addition, subtraction, and multiplication by a scalar. Also, two common vector operations are the dot product (**a** · **b**) and the vector cross product (**a** × **b**). The former generates a scalar result, while the latter results in a vector. These operations are important in determining the normal to a plane and whether points are in a plane. Various texts on vector analysis describe these tests and operations in detail [3], [11].

Recall that the homogeneous representation included a scaling factor, s. Thus, when adding or subtracting vectors defined in this manner, it is necessary to compute the coefficients of the basis set and then perform the operation since the scale factors of the two vectors may not be equal. The scale factor for the resulting vector can be set to any nonzero value as long as each of the coefficients of the result is adjusted properly. The final result is placed in the format of the homogeneous coordinate column vector.

The multiplication of a vector by a scalar results in each component of the vector being multiplied by the scalar. In the case of the homogeneous representation, however, it is important to note that if a scalar multiplies all four components, we have not changed the vector, as should be evident from the results of Example 8.2.1. Therefore, when performing multiplication by a scalar it is important that the scale factor s should not be multiplied by the scalar. Otherwise, all that is accomplished is to obtain an equivalent representation as shown in Example 8.2.1. An alternative way to perform scalar multiplication of a vector represented by homogeneous coordinates is to adjust the scale factor and leave the original components alone. For example, to multiply the vector of Eq. (8.2.4) by 2, all that is necessary is to divide s by 2.

The proper use of the scale factor for the dot and cross products is illustrated in the following examples.

EXAMPLE 8.2.2: DOT PRODUCT EQUATIONS

Obtain the Cartesian and homogeneous representations for the dot product of the two vectors **x** and **y**.

Cartesian basis representation

$$\mathbf{x} = a\hat{\mathbf{i}} + b\hat{\mathbf{j}} + c\hat{\mathbf{k}} \tag{8.2.9}$$

$$\mathbf{y} = d\hat{\mathbf{i}} + e\hat{\mathbf{j}} + f\hat{\mathbf{k}} \tag{8.2.10}$$

Using the definition of the vector dot product, we find that

$$\mathbf{x} \cdot \mathbf{y} = (ad + be + cf) \tag{8.2.11}$$

Homogeneous representation. The same vectors represented in homogeneous coordinates are given by

$$\mathbf{x} = \begin{bmatrix} a' \\ b' \\ c' \\ w \end{bmatrix} \qquad (8.2.12)$$

$$\mathbf{y} = \begin{bmatrix} d' \\ e' \\ f' \\ r \end{bmatrix} \qquad (8.2.13)$$

where:

$$a' = aw$$

$$b' = bw$$

$$c' = cw$$

$$d' = dr$$

$$e' = er$$

$$f' = fr$$

By using the appropriate coefficients (a, b, c, d, e, f) from Eqs. (8.2.12) and (8.2.13) in Eq. (8.2.11) we obtain Eq. (8.2.14) which defines the dot product in terms of all the column entries.

$$\mathbf{x} \cdot \mathbf{y} = \frac{a'd' + b'e' + c'f'}{wr} \qquad (8.2.14)$$

EXAMPLE 8.2.3: CROSS-PRODUCT FORMULAS

Obtain the Cartesian and homogeneous representations for the cross product of the two vectors $\mathbf{x}$ and $\mathbf{y}$ of Example 8.2.2.

Cartesian basis representation. The vector cross product is defined as

$$\mathbf{x} \times \mathbf{y} = (bf - ce)\hat{\mathbf{i}} + (cd - af)\hat{\mathbf{j}} \ (ae - bd)\hat{\mathbf{k}} \qquad (8.2.15)$$

Homogeneous representation. By substituting the coefficients $(a'$ through $f')$ from Eqs. (8.2.12) and (8.2.13) into the definition of the vector cross product [Eq. (8.2.15)], the following representation is obtained:

$$\mathbf{x} \times \mathbf{y} = \begin{bmatrix} b'f' - c'e' \\ c'd' - a'f' \\ a'e' - b'd' \\ uw \end{bmatrix} \qquad (8.2.16)$$

8.2.2 Matrix Operators: Transformations

In addition to performing the vector operations outlined previously, it is possible to define a class of matrix operators that can perform simultaneous vector operations resulting in the translation or rotation of a vector. The matrix H will be considered such an operator, which transforms a point or vector $\mathbf{x}$ into another point or vector $\mathbf{y}$.

$$\mathbf{y} = H\mathbf{x} \qquad (8.2.17)$$

Recall that the order of a matrix is defined as the number of rows r by the number of columns c. Thus the order of a matrix, having r rows and c columns is $(r \times c)$. A (4×1) column matrix for $\mathbf{y}$ is obtained by premultiplying a (4×1) column matrix, $\mathbf{x}$ by a (4×4) H matrix. This follows from the requirement that H and $\mathbf{x}$ must be conformable; that is, the number of columns in H is equal to the number of rows in $\mathbf{x}$.

8.2.2.1 Translational transformations

We define the translational transformation Trans(a, b, c) as the operator that moves the point defined by the original vector $\mathbf{x}$ to a new point $\mathbf{y}$ whose location is given by the vector addition of $\mathbf{x}$ and a translation vector defined by the components a, b, and c. This is essentially vector addition. An interesting interpretation can be obtained by considering moving a point from its current location for some distance along a diagonal. The distance is the vector magnitude, while the diagonal is defined by the direction of the vector. The operator transforms the coordinates of the original vector into the coordinates of a new vector. This matrix operator is defined as follows:

$$\text{Trans}\,(a, b, c) = \begin{bmatrix} 1 & 0 & 0 & a \\ 0 & 1 & 0 & b \\ 0 & 0 & 1 & c \\ 0 & 0 & 0 & 1 \end{bmatrix} \qquad (8.2.18)$$

The values a, b, and c represent the components of the vector which are to be added to those of the operand $\mathbf{x}$ of Eq. (8.2.17). From our prior discussion concerning matrix addition, this matrix will operate properly on any homogeneous representation of a vector, independent of the value of its scaling factor.

The following example illustrates both the vector addition concept and the motion along a vector for a prescribed distance.

EXAMPLE 8.2.4: TRANSLATION OF A POINT

Suppose that a point p lies at $(x, y) = (5, 5)$ in a two-dimensional reference frame. It is desired to move the point along the diagonal corresponding to an angle of 45° for a distance of 10 units. What are the coordinates of the final point p' (x_2, y_2)? Figure 8.2.2 illustrates the initial and final positions.

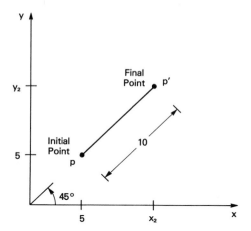

Figure 8.2.2. Translational transform example.

The original point may be represented as a vector in homogeneous coordinates. Its value is

$$\mathbf{p} = \begin{bmatrix} 5 \\ 5 \\ 0 \\ 1 \end{bmatrix} \tag{8.2.19}$$

Since it is desired to move the point along the 45° radial extension, the unit vector corresponding to this direction is

$$\mathbf{u} = \begin{bmatrix} 0.707 \\ 0.707 \\ 0 \\ 1 \end{bmatrix} \tag{8.2.20}$$

To get a motion of 10 units it is necessary to multiply the components of **u** by 10, *excluding the scaling factor*, as mentioned previously. The new vector **u** becomes

$$\mathbf{u} = \begin{bmatrix} 7.07 \\ 7.07 \\ 0 \\ 1 \end{bmatrix} \tag{8.2.21}$$

which can be considered a point 10 units from the origin along the radial line at 45°. Using Eqs. (8.2.17) and (8.2.18) we obtain the expression for the translated point $\mathbf{p}'$.

$$\mathbf{p}' = \begin{bmatrix} 1 & 0 & 0 & 7.07 \\ 0 & 1 & 0 & 7.07 \\ 0 & 0 & 1 & 0 \\ 0 & 0 & 0 & 1 \end{bmatrix} \begin{bmatrix} 5 \\ 5 \\ 0 \\ 1 \end{bmatrix} \tag{8.2.22}$$

where we have substituted the components of **u** for a, b, and c. Carrying out the multiplication in Eq. (8.2.22) yields

$$\mathbf{p}' = \begin{bmatrix} 12.07 \\ 12.07 \\ 0 \\ 1 \end{bmatrix} \qquad (8.2.23)$$

This simple example points out the two ways that one may view the translational transformation (i.e., as vector addition or as "motion along a line") and is extremely useful in indicating why the scaling factor is important. Consider the generation of one element of the **p**′ vector of Eq. (8.2.23), say the x component. In Eq. (8.2.22), the first row of the Trans (a, b, c) matrix selects the x component from vector **p** and then adds the x component of the translation vector times the scale factor of **p**. Finally, when the scale component for **p**′ is computed, it is the original scaling factor of the input vector **p**. Thus before adding the vectors, the components of the last column of Trans (a, b, c) are scaled by the proper constant so that like quantities are added. Finally, the actual value is generated and the original scaling factor is preserved.

8.2.2.2 Rotational transformations

In addition to moving points along vectors, it is possible to generate new positions in space by rotating them about an axis. Before investigating the rotational matrix operator, we will derive the rotation formula for a set of orthogonal axes rotated in a plane. Our derivation is based on a geometric interpretation as given in [4].

Consider a point P located at x_1, y_1 with respect to the x-y reference axis, which we shall call x-ref and y-ref (see Figure 8.2.3). In addition, assume initially that a second set of axes is lying on top of the reference axes. This set will be called x-rot and y-rot and the relationship of the point P with respect to these axes will remain fixed. Now assume that the plane containing the point P (located

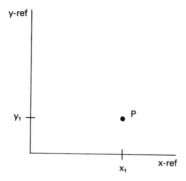

Reference axis

Figure 8.2.3.   Point in reference frame.

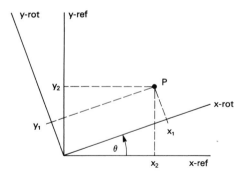

Figure 8.2.4. Point rotated with respect to reference frame.

in the frame defined by x-rot and y-rot) is rotated counterclockwise by an angle θ with respect to the reference axes. We would like to determine the location of the point P with respect to the reference axes (x-ref and y-ref as shown in Figure 8.2.4).

Figure 8.2.5 shows the geometric construction necessary to determine the coordinates. Line segment PA is the perpendicular from the point P to the x-ref axis. Line segment PB is the perpendicular from the point P to the x-rot axis. From the diagram we may write the following relationships:

$$x_2 = |OA| = |OE| - |AE|$$

$$y_2 = |AP| = |AC| + |CP|$$

$$|OE| = |OB| \cos \theta$$

$$= x_1 \cos \theta$$

$$|AC| = |EB| = |OB| \sin \theta$$

$$= x_1 \sin \theta$$

$$|AE| = |CB| = |BP| \sin \theta$$

$$= y_1 \sin \theta$$

$$|CP| = |BP| \cos \theta$$

$$= y_1 \cos \theta$$

Combining these equations yields the desired results:

$$x_2 = x_1 \cos \theta - y_1 \sin \theta$$
$$y_2 = x_1 \sin \theta + y_1 \cos \theta$$

(8.2.24)

The relationships above allows us to determine the location of points affixed to movable frames of reference in terms of a fixed reference frame. The derivation in the x-y plane is easily extended to either the y-z or z-x plane.

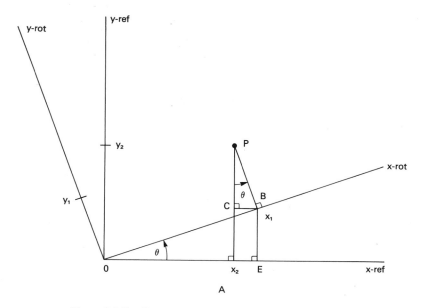

Figure 8.2.5. Geometric construction for rotation of axes.

Rotational transformations which operate on homogeneous coordinates and perform rotations about a given axis of the reference coordinate system can be shown to be:

1. Rotation θ degrees about the x axis:

$$\text{Rot}\,(x,\,\theta) = \begin{bmatrix} 1 & 0 & 0 & 0 \\ 0 & \cos\theta & -\sin\theta & 0 \\ 0 & \sin\theta & \cos\theta & 0 \\ 0 & 0 & 0 & 1 \end{bmatrix} \qquad (8.2.25)$$

2. Rotation θ degrees about the y axis:

$$\text{Rot}\,(y,\,\theta) = \begin{bmatrix} \cos\theta & 0 & \sin\theta & 0 \\ 0 & 1 & 0 & 0 \\ -\sin\theta & 0 & \cos\theta & 0 \\ 0 & 0 & 0 & 1 \end{bmatrix} \qquad (8.2.26)$$

3. Rotation θ degrees about the z axis:

$$\text{Rot}\,(z,\,\theta) = \begin{bmatrix} \cos\theta & -\sin\theta & 0 & 0 \\ \sin\theta & \cos\theta & 0 & 0 \\ 0 & 0 & 1 & 0 \\ 0 & 0 & 0 & 1 \end{bmatrix} \qquad (8.2.27)$$

TABLE 8.2.1 POSITIVE ANGLE
REFERENCES

Cross product	Rotation
$\mathbf{x} \times \mathbf{y} = \mathbf{z}$	Positive rotation about z
$\mathbf{y} \times \mathbf{z} = \mathbf{x}$	Positive rotation about x
$\mathbf{z} \times \mathbf{x} = \mathbf{y}$	Positive rotation about y

Examining the matrices in Eqs. (8.2.25) through (8.2.27) shows that the entries of the last column are all zero except for the scaling factor, which has been set to 1. This is reasonable since we do not wish to do any translation. The first three columns represent the directions of the x, y, and z axes of the rotated frame. Observe that the scaling factor is set to zero in each case. For Eqs. (8.2.25) through (8.2.27) the axis about which the rotation is performed has a unit entry in the appropriate column showing the direction of that particular axis. For example, in Eq. (8.2.27) the 1 in the third column shows that rotation is about the z axis. The columns corresponding to the direction of the axes forming the plane in which rotation takes place contain the transcendental terms needed to accomplish the rotation. For example, in Eq. (8.2.25) columns 2 and 3 indicate a movement of the y-z plane.

To visualize the concept of rotating a point about an axis, the reader should imagine a perpendicular drawn from the point to the axis about which it will be rotated. If one were to rotate the initial perpendicular, an angular displacement would result. The angle of rotation may be positive or negative and is defined by using the *right-hand rule*. Rotation is positive if the cross product defined by the initial and final vectors is in the same direction as the axis about which the rotation was performed. The cross products of the x, y, and z axes are shown in Table 8.2.1.

The following example shows the effects of operating on a point or vector with a rotational operator.

EXAMPLE 8.2.5: ROTATION OF A POINT ABOUT A REFERENCE FRAME

It is desired to rotate the point u represented by a column vector about the z axis and determine the coordinates in terms of the original reference frame. Figure 8.2.6 shows this operation. The vector $\mathbf{u}$ will be given as

$$\mathbf{u} = \begin{bmatrix} 1 \\ 2 \\ 3 \\ 1 \end{bmatrix} \tag{8.2.28}$$

One may also think of this problem as the rotation of a plane located

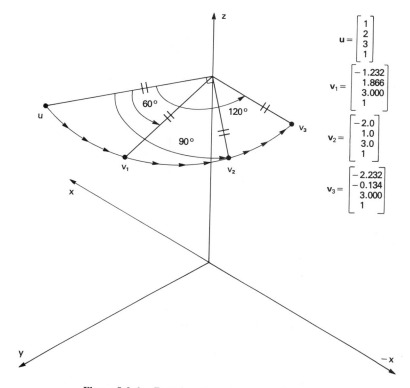

Figure 8.2.6. Rotation of a point about the z axis.

at a height of 3 units above the *x-y* reference plane which contains the point $(1, 2)$. The relationship of the point $(1, 2)$ remains the same to the reference frame of the plane in which it is located but is different from the base or original *x-y* reference frame located at height $z = 0$. Using Eq. (8.2.27) we will consider rotations of 60°, 90°, and 120°. The operators corresponding to these angles are given as

$$\text{Rot}\,(z,\,60) = \begin{bmatrix} 0.500 & -0.866 & 0 & 0 \\ 0.866 & 0.500 & 0 & 0 \\ 0 & 0 & 1 & 0 \\ 0 & 0 & 0 & 1 \end{bmatrix} \tag{8.2.29}$$

$$\text{Rot}(z,\,90) = \begin{bmatrix} 0 & -1 & 0 & 0 \\ 1 & 0 & 0 & 0 \\ 0 & 0 & 1 & 0 \\ 0 & 0 & 0 & 1 \end{bmatrix} \tag{8.2.30}$$

$$\text{Rot}(z, 120) = \begin{bmatrix} -0.500 & -0.866 & 0 & 0 \\ 0.866 & -0.500 & 0 & 0 \\ 0 & 0 & 1 & 0 \\ 0 & 0 & 0 & 1 \end{bmatrix} \qquad (8.2.31)$$

If each of the operators defined by Eqs. (8.2.29) through (8.2.31) is applied to the vector $\mathbf{u}$, using Eq. (8.2.17), one obtains

$$\mathbf{v_1} = \begin{bmatrix} -1.232 \\ 1.866 \\ 3.000 \\ 1 \end{bmatrix} \qquad (8.2.32)$$

$$\mathbf{v_2} = \begin{bmatrix} -2.0 \\ 1.0 \\ 3.0 \\ 1 \end{bmatrix} \qquad (8.2.33)$$

$$\mathbf{v_3} = \begin{bmatrix} -2.232 \\ -0.134 \\ 3.000 \\ 1 \end{bmatrix} \qquad (8.2.34)$$

where $\mathbf{v_1}$, $\mathbf{v_2}$, $\mathbf{v_3}$ represents a vector to the rotated point after 60°, 90°, and 120° of rotation, respectively.

Based on the three rotational transformations [Eqs. (8.2.25) through (8.2.27)] about the primary axis, we may begin to define compound rotational transformations. These will consist of rotations of a point or vector about an axis, followed by another rotation of the result. Since we have chosen a matrix and vector representation for these operations, it is possible to string matrix operators together to perform these multiple rotations without the need to compute intermediate results. The following example illustrates this.

EXAMPLE 8.2.6: MULTIPLE ROTATIONS

Considering the results of Example 8.2.5, suppose that it is desired to rotate the point $\mathbf{v_1}$ (obtained by rotating $\mathbf{u}$ 60° about the z axis) $-90°$ about the y axis, resulting in a new vector, $\mathbf{w}$. The operation becomes

$$\mathbf{w} = \text{Rot}(y, -90)\mathbf{v_1} \qquad (8.2.35a)$$

$$\mathbf{w} = \text{Rot}(y, -90)(\text{Rot}(z, 60)\mathbf{u}) \qquad (8.2.35b)$$

$$\mathbf{w} = (\text{Rot}(y, -90)\ \text{Rot}(z, 60))\mathbf{u} \qquad (8.2.35c)$$

Using the associative law as illustrated in Eqs. (8.3.35a), through (8.3.35c),

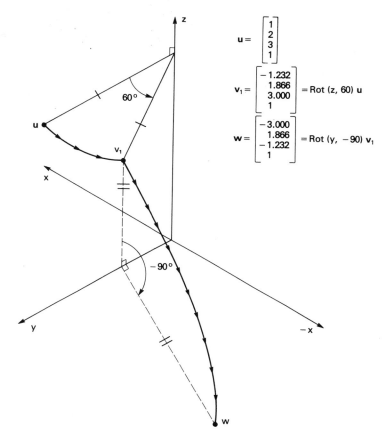

$$\mathbf{u} = \begin{bmatrix} 1 \\ 2 \\ 3 \\ 1 \end{bmatrix}$$

$$\mathbf{v_1} = \begin{bmatrix} -1.232 \\ 1.866 \\ 3.000 \\ 1 \end{bmatrix} = \text{Rot (z, 60) } \mathbf{u}$$

$$\mathbf{w} = \begin{bmatrix} -3.000 \\ 1.866 \\ -1.232 \\ 1 \end{bmatrix} = \text{Rot (y, } -90) \text{ } \mathbf{v_1}$$

Figure 8.2.7.   Rot (y, $-90°$) Rot (z, $60°$).

we may generate the transformation Rot $(y, -90)$ Rot $(z, 60)$. Thus

$$\text{Rot}(y, -90) \text{ Rot}(z, 60) = \begin{bmatrix} 0 & 0 & -1 & 0 \\ 0.866 & 0.500 & 0 & 0 \\ 0.500 & -0.866 & 0 & 0 \\ 0 & 0 & 0 & 1 \end{bmatrix} \quad (8.2.36)$$

Using this compound operator on vector **u** yields the new vector:

$$\mathbf{w} = \begin{bmatrix} -3.000 \\ 1.866 \\ -1.232 \\ 1.0 \end{bmatrix} \quad (8.2.37)$$

Figure 8.2.7 shows these rotation operations.

Since matrix operations are generally not commutative, the order in which the rotations are performed becomes quite important. In general, the product of elementary rotational transformations taken in different order will yield different results. Consider first a rotation of $-90°$ about the y axis followed by a rotation of $60°$ about the z axis. The product of the two elementary transformations is given by

$$\text{Rot}(z, 60) \, \text{Rot}(y, -90) = \begin{bmatrix} 0 & -0.866 & -0.500 & 0 \\ 0 & 0.500 & -0.866 & 0 \\ 1 & 0 & 0 & 0 \\ 0 & 0 & 0 & 1 \end{bmatrix} \quad (8.2.38)$$

The result of this transformation operating on **u** is shown in Figure 8.2.8. The new vector becomes

$$\mathbf{w}' = \begin{bmatrix} -3.232 \\ -1.598 \\ 1.0 \\ 1.0 \end{bmatrix} \quad (8.2.39)$$

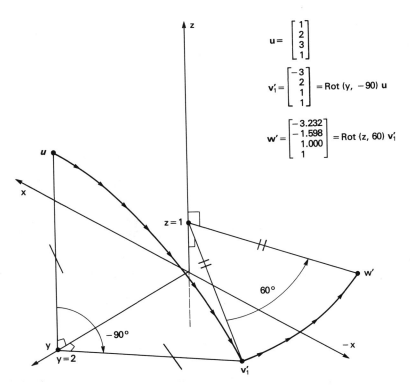

Figure 8.2.8. Rot (z, 60°) Rot (y, −90°).

Comparing the results of the operators defined by Eqs. (8.2.36) and (8.2.38), it is seen from Figures 8.2.7 and 8.2.8 or Eqs. (8.2.37) and (8.2.39) that they are not the same. The order in which the rotations are performed generates different results. This is expected since matrix multiplication is generally not commutative.

An interesting application of the rotational transformation is to provide additionally incremental angular motion. Suppose it was desired to rotate a point 60° about the z axis followed by another 60°. Application of the operator Rot(z, 60) will accomplish the first motion. Reapplication will provide the second motion. The reader can verify this applying the results of Example 8.2.5. The reader should also verify the general relationship for rotation about a given axis, that is,

$$\text{Rot(axis, } \theta_1 + \theta_2 + \cdots + \theta_n)$$

$$= \text{Rot(axis, } \theta_1) \text{ Rot(axis, } \theta_2) \cdots \text{Rot(axis, } \theta_n) \qquad (8.2.40)$$

In Eq. (8.2.40) the operators may be commutative if we are concerned only with the final result and the rotation is about only a single axis. It is important to understand that as soon as multiple rotations about different axes are performed, the results above do not generally hold.

8.3 COORDINATE REFERENCE FRAMES

The translational and rotational operators introduced in Section 8.2 are examples of homogeneous transformations. The most general form of the homogeneous transformation performs both the operations of translation and rotation when applied to a point. The power of this matrix operator lies in its ability to transform points from one frame of reference to another. In addition, the 4 × 4 transformation matrix [refer to Eqs. (8.2.17), (8.2.18), and (8.2.25) through (8.2.27)] may be interpreted as the representation of a frame in space with respect to another frame, or even as a point in space. These concepts, extremely important in describing the position of a robotic manipulator, permit the introduction of relative points and frames of reference. As a consequence, the amount of explicit information necessary to describe or teach a robot task is greatly reduced.

In Section 8.2 all the vectors (which we used to describe points) were located in an orthogonal reference frame. Elementary matrix operators [i.e., Rot ($\cdot$) and Trans ($\cdot$)] operating on these vectors moved them about in space with respect to the given reference frame. Let us now consider a space consisting of two orthogonal reference frames whose origins are displaced as shown in Figure 8.3.1. The origin of frame 2 is at coordinate (x, y, z) with respect to frame 1, as shown by the vector **p**. Note that the axes of each frame are aligned in different directions and for simplicity are either parallel or antiparallel to axes of the other frame. This assumption is for illustrative purposes only and the results obtained in the following derivation will apply to any other orientations.

What is sought is a way to reference points which are described in terms of

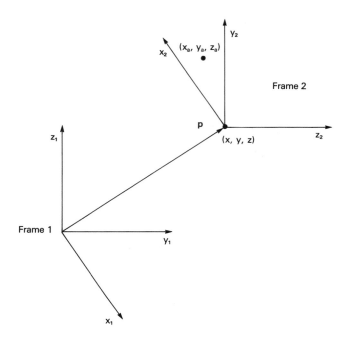

Figure 8.3.1. Three-dimensional space with two reference frames.

the coordinates of frame 2 in terms of the coordinates of frame 1, and vice versa. This gives us the ability to refer to a point using the coordinates of the most convenient reference frame. The application of these ideas will be illustrated in the sections that follow.

The derivation begins by assuming that two coordinate frames are initially aligned in such a way that their origins and axes are all coincident. These two frames will be located at frame 1 of Figure 8.3.1. One set of axes will remain stationary, while the second set will be moved through space. The motion of the second axis is obtained by assuming that a particular point, say the origin, is operated on by the rotational and translational operators introduced in Section 8.2.2. For these manipulations, besides rotating and translating the point about the reference axes, the coordinate frame associated with this point will remain affixed to it. For example, if the origin was rotated by 90° about the z axis, the reference frame would be rotated with respect to the fixed frame by 90°.

Table 8.3.1 shows two of three possible methods of using elementary homogeneous transformations to cause the second (or movable) axes to be aligned with the coordinate axes labeled frame 2. The motivation for doing this is that if we know a vector (or a point) referenced to the movable frame, the relationship between that frame and the vector remains fixed, regardless of the location of that frame in space. Consider some vector referenced with respect to frame 2, say

TABLE 8.3.1   OPERATIONS TO LINE UP TWO FRAMES

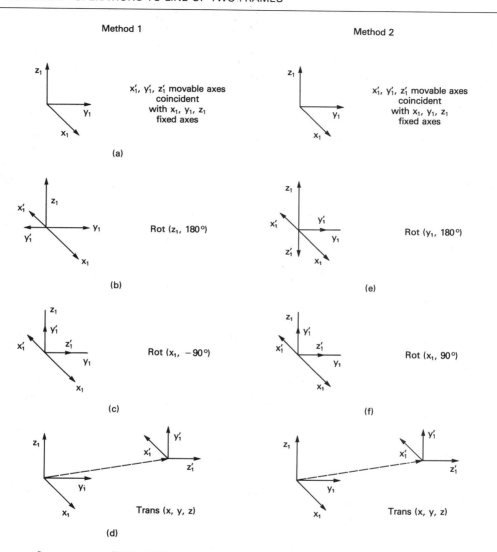

Method 1

x_1', y_1', z_1' movable axes coincident with x_1, y_1, z_1 fixed axes

(a)

Rot $(z_1, 180°)$

(b)

Rot $(x_1, -90°)$

(c)

Trans (x, y, z)

(d)

Method 2

x_1', y_1', z_1' movable axes coincident with x_1, y_1, z_1 fixed axes

Rot $(y_1, 180°)$

(e)

Rot $(x_1, 90°)$

(f)

Trans (x, y, z)

$[x_a, y_a, z_a, 1]^{tr}$.*   This vector always has the same coordinates in terms of the movable frame. Using the fixed frame as a point of reference we wish to know the coordinates of the vector (attached to the movable frame) in terms of this fixed frame. If the movable frame is made coincident with the fixed frame, the coordinates of the vector are the same in either frame of reference. Since the location

---

*tr indicates the matrix (vector) transpose.

and orientation of frame 2 with respect to 1 are known (see Figure 8.3.1) by performing operations on the point $[x_a, y_a, z_a, 1]^{tr}$ (described in terms of frame 1), it is possible to make that point coincident with the point having the same coordinates with respect to frame 2. Assuming that the movable reference frame stays with the point during these operations, the final location of the movable reference frame will be coincident with frame 2. The operations performed (in a particular order) constitute the transformation that maps the coordinates of points referenced to the coordinate axes of frame 2 into the coordinate system of frame 1.

Referring again to Table 8.3.1, we note that the rotation matrices are different for both cases. However, the product of the two rotation and one translation matrices of each method is identical.

The product of transformations Trans (x, y, z), Rot $(x, -90)$, Rot $(z, 180)$ or Trans (x, y, z), Rot $(x, 90)$, Rot $(y, 180)$ generates a single 4×4 compound transformation which if applied to any vector defined with respect to frame 2 will give the coordinates of the vector in terms of the axes of frame 1. Note that it is always possible to make two reference frames coincident by performing at most two rotations about different axes of one frame and a translation.

At this point we can define the homogeneous transformation matrix as the 4×4 matrix having the property of transforming a vector from one coordinate system (or frame of reference) to another by means of performing simultaneous rotation and translation. Later we investigate a special form of this matrix called the Denavit–Hartenberg (D-H matrix) transformation matrix, which is used to relate to one another frames attached to each of the links of a robotic manipulator.

Table 8.3.2 shows how the general form of the homogeneous transformation can be partitioned into submatrices. The entries consist of a 3×3 submatrix responsible for the rotation and a 3×1 submatrix responsible for the translation. In addition, two other partitions may be identified as the perspective transformation and scaling factor. Typically, we set the scaling factor to unity, as discussed in the translational transformation example (see Section 8.2.2.1). The perspective transformation is set to $(0, 0, 0)$ so that the columns of the rotation matrix taken along with the appropriate entry of the perspective transformation define direction vectors [such as Eqs. (8.2.5) through (8.2.7)]. In our applications the values of these two submatrices will always be defined as above. Additional details about the perspective transform can be found in [1].

TABLE 8.3.2 PARTITIONING OF
THE HOMOGENEOUS
TRANSFORMATION

$$
\left[
\begin{array}{c:c}
\text{Rotation matrix } (3 \times 3) & \text{Position vector } (3 \times 1) \\
\hdashline
\text{Perspective transform } (1 \times 3) & \text{Scaling factor } (1 \times 1)
\end{array}
\right]
$$

TABLE 8.3.3 VECTOR AND POSITION
NOMENCLATURE FOR HOMOGENEOUS
TRANSFORMATION MATRIX

$$
H = \begin{bmatrix} n_x & o_x & a_x & p_x \\ n_y & o_y & a_y & p_y \\ n_z & o_z & a_z & p_z \\ \hline 0 & 0 & 0 & 1 \end{bmatrix} \qquad (8.3.1)
$$

The real power of this matrix is seen by interpreting its columns. With the perspective transform and scaling factor set to the values mentioned above, the homogeneous transformation matrix may be interpreted as defining a relationship between two frames. Equation (8.3.1) shows the labeling of the individual vector components and positions. We may think of this matrix as containing the direction vectors for the x, y, and z axes of frame 2 written in terms of the directions of the x, y, and z axes of frame 1. For example, the elements of the first column (n_x, n_y, n_z) are the components of the unit vector defining the x axis of frame 2, written in terms of the three unit vectors defining the axes of frame 1. The second column defines the y axis of frame 2, while the third column defines the direction of the z axis. The fourth column is the location of the origin of frame 2 with respect to 1 and may be thought of as a vector (having coordinates p_x, p_y, p_z) drawn from the origin of frame 1 to the point defining the origin of 2.

In a later section we will use this matrix notation to define both the position and orientation of a robotic manipulator with respect to a reference frame. For now it is sufficient to say that a set of coordinate axes is attached to the end of the arm and the position of the manipulator is given by the relationship of these axes and the distance of their origin to the reference frame. When describing the position of a manipulator, conventional terminology refers to the z axis as the *approach vector*, the y axis as the *orientation vector*, and the x axis as the *normal vector*—thus the use of n, o, a, and p in Table 8.3.3. The subscript indicates the x, y, and z components of that vector. This terminology will become more familiar and appropriate in a subsequent section when the kinematics of the manipulator is discussed.

Based on the interpretation of the column vectors it is possible to write by *inspection* a transformation matrix that relates points referenced in one frame to that of another. The following example illustrates this concept.

EXAMPLE 8.3.1: FRAME-TO-FRAME TRANSFORMATIONS BY INSPECTION

Suppose it is desired to find the homogeneous transformation that relates points in frame 2 to frame 1 as shown in Figure 8.3.1.

Based on the interpretation of the homogeneous transformation matrix mentioned above, we proceed by filling in the entries of the matrix in Table

8.3.3. The first column corresponds to the direction vector for the x axis of frame 2. Since the x axis of frame 1 points in the opposite direction from the x axis of frame 2, the first column of the transformation matrix is

$$\begin{bmatrix} n_x \\ n_y \\ n_z \\ 0 \end{bmatrix} = \begin{bmatrix} -1 \\ 0 \\ 0 \\ 0 \end{bmatrix} \tag{8.3.2}$$

The second column corresponds to the y axis of frame 2. From Figure 8.3.1 we note that the z axis of 1 is parallel to the y axis of 2. Thus the second entry is

$$\begin{bmatrix} o_x \\ o_y \\ o_z \\ 0 \end{bmatrix} = \begin{bmatrix} 0 \\ 0 \\ 1 \\ 0 \end{bmatrix} \tag{8.3.3}$$

The third column, representing the direction of the z axis of frame 2, is set to indicate the direction of the y axis of 1. In this case they are parallel. Thus we obtain

$$\begin{bmatrix} a_x \\ a_y \\ a_z \\ 0 \end{bmatrix} = \begin{bmatrix} 0 \\ 1 \\ 0 \\ 0 \end{bmatrix} \tag{8.3.4}$$

Finally, the fourth column is the distance of the origin of frame 2 from frame 1, using 1's origin as the point of reference. Thus

$$\begin{bmatrix} p_x \\ p_y \\ p_z \\ 1 \end{bmatrix} = \begin{bmatrix} x \\ y \\ z \\ 1 \end{bmatrix} \tag{8.3.5}$$

Combining Eqs. (8.3.2) through (8.3.5) into the matrix defined by Eq. (8.3.1) yields the same transformation matrix shown in Table 8.3.1, namely:

$$\begin{bmatrix} -1 & 0 & 0 & x \\ 0 & 0 & 1 & y \\ 0 & 1 & 0 & z \\ 0 & 0 & 0 & 1 \end{bmatrix} \tag{8.3.6}$$

We will now introduce notation for a homogeneous transformation that relates the coordinates of two frames m and n. The symbol A_{nm} will represent matrices of the form of Eq. (8.3.1), where n is the reference frame to which all points will be referenced and m is the frame where the points or vectors are located. Thus

A_{12} relates points in frame 2 to frame 1. We may refer to Eq. (8.3.6) of Example 8.3.1 as

$$A_{12} = \begin{bmatrix} -1 & 0 & 0 & x \\ 0 & 0 & 1 & y \\ 0 & 1 & 0 & z \\ 0 & 0 & 0 & 1 \end{bmatrix} \tag{8.3.7}$$

Assuming that we have applied Eq. (8.3.7) to some arbitrary vector **u**, we would now have **u** referenced to frame 1 ($\mathbf{u}' = A_{12}\mathbf{u}$). To get **u**′ back to frame 2, the matrix operator A_{21} would be applied to **u**′ ($\mathbf{u} = A_{21}\mathbf{u}'$). The transformation A_{21} which maps points defined in frame 1 to the coordinates of frame 2 is the inverse of matrix A_{12}. From Figure 8.3.1 and the rules defining the entries of Eq. (8.3.1), transformation A_{21} is written as

$$A_{21} = \begin{bmatrix} -1 & 0 & 0 & x \\ 0 & 0 & 1 & -z \\ 0 & 1 & 0 & -y \\ 0 & 0 & 0 & 1 \end{bmatrix} \tag{8.3.8}$$

By comparing Eqs. (8.3.7) and (8.3.8), it seems that the rotation part of A_{12} and A_{21} is identical. It is important to understand that this is not generally true. In Section 8.4.1, we elaborate on the inverse of transformation matrices.

It is possible to use the mechanics of the homogeneous transformation to work with multiple reference frames in space. Essentially, all that is necessary is to expand on the previous concepts and relate frames to one another via the A_{nm} matrices. In addition, note that in relating one frame of reference to another, it is possible to introduce an intermediate one that is related to the others. This intermediate frame has no effect on the relationship between the original frames and will not appear in the final result. To illustrate this, consider three frames in space 1, 2, and 3 and the associated A_{nm} matrices relating them (A_{12}, A_{23}). A_{13} can be generated by the relationship

$$A_{13} = A_{12} \times A_{23} \tag{8.3.9}$$

This result or "chain rule" can be extended in general. Thus the final A matrix relating the first and last frames of a series of frames is given by

$$A_{1n} = A_{12} \times A_{23} \times A_{34} \times \cdots \times A_{(n-2)(n-1)} \times A_{(n-1)n} \tag{8.3.10}$$

The following example illustrates multiple frames and the use of Eq. (8.3.10).

EXAMPLE 8.3.2: MULTIPLE REFERENCE FRAMES IN SPACE

Figure 8.3.2 shows four reference frames located in three-dimensional space. Such a configuration may arise by attaching frames to the individual links of a three axis robot or by assigning frames to interrelated objects. It is desired

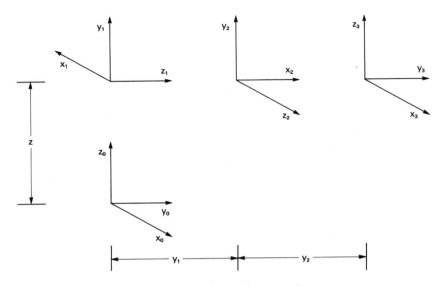

Figure 8.3.2. Four reference frames in three-space.

to find the transformation that relates points in each frame (e.g., 1, 2, or 3) to the reference frame, 0. The following transformations are obtained by inspection:

$$A_{01} = \begin{bmatrix} -1 & 0 & 0 & 0 \\ 0 & 0 & 1 & 0 \\ 0 & 1 & 0 & z \\ 0 & 0 & 0 & 1 \end{bmatrix} \qquad (8.3.11)$$

$$A_{12} = \begin{bmatrix} 0 & 0 & -1 & 0 \\ 0 & 1 & 0 & 0 \\ 1 & 0 & 0 & y_1 \\ 0 & 0 & 0 & 1 \end{bmatrix} \qquad (8.3.12)$$

$$A_{23} = \begin{bmatrix} 0 & 1 & 0 & y_2 \\ 0 & 0 & 1 & 0 \\ 1 & 0 & 0 & 0 \\ 0 & 0 & 0 & 1 \end{bmatrix} \qquad (8.3.13)$$

These matrices relate consecutively numbered frames to each other. It is also possible to generate the transformations A_{02}, A_{13}, or A_{03} either by inspection or by using the chain rule of Eq. (8.3.10). When done by inspection, one need only consider the two frames of interest, as any intermediate-numbered one would be absorbed as illustrated in the chain rule.

The transformations relating frames 1, 2, and 3 to the reference frame can be shown to be given by Eqs. (8.3.14) through (8.3.16):

$$A_{01} = \begin{bmatrix} -1 & 0 & 0 & 0 \\ 0 & 0 & 1 & 0 \\ 0 & 1 & 0 & z \\ 0 & 0 & 0 & 1 \end{bmatrix} \tag{8.3.14}$$

$$A_{02} = \begin{bmatrix} 0 & 0 & 1 & 0 \\ 1 & 0 & 0 & y_1 \\ 0 & 1 & 0 & z \\ 0 & 0 & 0 & 1 \end{bmatrix} \tag{8.3.15}$$

$$A_{03} = \begin{bmatrix} 1 & 0 & 0 & 0 \\ 0 & 1 & 0 & (y_1 + y_2) \\ 0 & 0 & 1 & z \\ 0 & 0 & 0 & 1 \end{bmatrix} \tag{8.3.16}$$

Example 8.3.2 illustrates the ability of the general transformation to reference points from one frame to another. In addition, it illustrates that we may relate points directly from frame to frame by ignoring intermediate ones and only using the spatial relationships between the two frames of interest. For illustrative purposes it is important to note that the axes between frames were related by multiples of 90°. However, if one can write the direction vectors of one frame by using the direction vectors of the preceding one, the inspection method also works extremely well.

8.4 SOME PROPERTIES OF TRANSFORMATION MATRICES

This section defines some additional properties of the transformation matrices.

8.4.1 Formula for the Inverse of the Transformation Matrix

In Section 8.3 we identified the inverse of a homogeneous transformation. Specifically, given a transformation A_{nm} relating frame m to n, the transformation A_{mn} relates points in frame n to frame m. A_{mn} is by definition the inverse of A_{nm}. Thus the inverse of a homogeneous transformation may be indicated by

$$(A_{mn})^{-1} = A_{nm} \tag{8.4.1}$$

Utilizing the notation of Eq. (8.3.1) for a homogeneous transformation consisting of column vectors, and standard matrix techniques, we can show that the inverse of the transformation A_{nm} is

$$(A_{nm})^{-1} = \begin{bmatrix} n_x & n_y & n_z & -(\mathbf{p} \cdot \mathbf{n}) \\ o_x & o_y & o_x & -(\mathbf{p} \cdot \mathbf{o}) \\ a_x & a_y & a_z & -(\mathbf{p} \cdot \mathbf{a}) \\ 0 & 0 & 0 & 1 \end{bmatrix} \tag{8.4.2}$$

where $\mathbf{p}$, $\mathbf{n}$, $\mathbf{o}$, and $\mathbf{a}$ are the column vectors of Eq. (8.3.1).

Since the product of any square matrix and its inverse equals the identity matrix, I, we have

$$A_{nm} \times A_{mn} = I \tag{8.4.3}$$

This relationship is easily verified by substituting Eqs. (8.3.1) and (8.4.2) into Eq. (8.4.3) and performing the multiplication. The known orthogonality relationships of the various vectors must also be used.

8.4.2 Objects and Transformations

An interesting application of homogeneous coordinates and transformations is in the field of graphics. The extension to robotics will become quite apparent throughout our discussion. The discussion in this section is a modification of that presented in [1].

Objects may be represented by a collection of vertex points. These points are defined with respect to a reference frame affixed to the object. The relationship of the vertex points to the frame remains fixed once it is established. As an example, consider the wedge shown in Figure 8.4.1. As shown in Figure 8.4.1, the points (or vectors) describing the object are related to the orthogonal frame (x', y', z') by some fixed relationship. If the object is moved, the relationship between (x', y', z') and the vertex points does not change (i.e., the object does

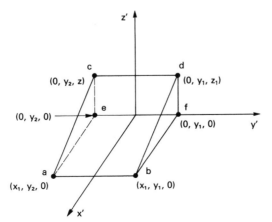

Figure 8.4.1. Wedge with reference frame affixed.

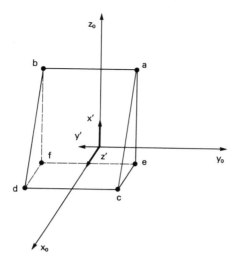

Figure 8.4.2. Wedge repositioned in space.

not move with respect to its internal reference frame).   Therefore, performing transformations on the reference frame of the object or to the points describing the vertices has the effect of moving the object about in space.

Let us consider two operations on the wedge of Figure 8.4.1.   The first is concerned with performing rotations necessary to stand the wedge on its wide end with the sloping part facing the reader.   This can be accomplished by first rotating the wedge $-90°$ about the y axis, followed by a rotation of $180°$ about the z axis. The results of these operations are shown in Figure 8.4.2.   Note that the origin of the reference frame has not moved.

The next step that we wish to perform is to move the wedge onto a table. Suppose that a translational transformation called "*TABLE*" relates points in the coordinate system attached to the wedge to a coordinate system located on the top of the table.   Figure 8.4.3 shows the locations of the two reference frames.   From this figure the transformation *TABLE* is defined as

$$TABLE = \begin{bmatrix} 1 & 0 & 0 & \Delta x \\ 0 & 1 & 0 & \Delta y \\ 0 & 0 & 1 & \Delta z \\ 0 & 0 & 0 & 1 \end{bmatrix} \qquad (8.4.4)$$

By premultiplying the *repositioned* vertex points (shown in Figure 8.4.2) by the transformation *TABLE* these vertex points are moved such that they will trace the outline of the wedge on the reference frame attached to the table (see Figure 8.4.3). Obviously, the entire operation could be accomplished by multiplying the rotation and translation operators together in the appropriate order.

In this discussion we introduced some of the concepts needed to describe

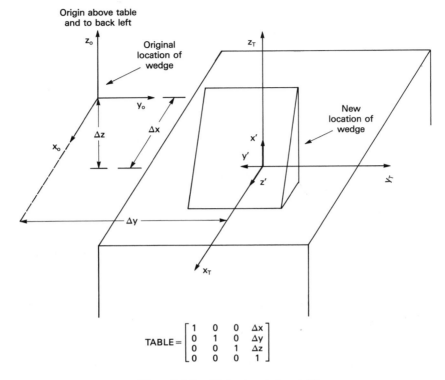

$$\text{TABLE} = \begin{bmatrix} 1 & 0 & 0 & \Delta x \\ 0 & 1 & 0 & \Delta y \\ 0 & 0 & 1 & \Delta z \\ 0 & 0 & 0 & 1 \end{bmatrix}$$

Figure 8.4.3. Wedge erected on table.

tasks in a robot-independent environment. This type of description is sometimes called task description.

8.4.3 Effects of Pre- and Postmultiplication of Transformations

We shall now describe the effects of premultiplying and postmultiplying points in a given frame by homogeneous transformations. Let us begin with an arbitrary frame located in space, for example, frame 1. The transformation relating points in this frame to our base or reference frame (normally referred to as frame 0) is A_{01}. In addition, we have shown that if we know the transformation (relating a second frame, 2, to frame 1) A_{12}, then postmultiplying A_{01} by A_{12} will relate points in the second frame to the base reference.

If we now premultiply the transformation A_{01} by another transformation, the effect is "moving the base reference frame" (i.e., we will now have the points in the coordinate system of frame 1 in terms of a new reference). Assuming that we wished to re-reference frame 1 to a different reference frame (e.g., frame $0'$), we would first generate the transformation $A_{0'0}$ relating the new reference frame to the original one, frame 0. Premultiplying A_{01} by $A_{0'0}$ would then generate a 4×4 transformation $A_{0'1}$ which would relate points in frame 1 to frame 0'. As stated previously, we can now completely discard all references to frame 0. This particular relationship is extremely useful in describing robot tasks independent of the actual working position and is investigated in Section 8.5.

The importance of this last concept is illustrated in the following discussion. Tasks to be performed by a robot can be described with respect to some fixed repeatable reference frame which we will call the base frame. If it is desired to perform the same task somewhere else in space, all that is needed is to find the transformation relating the new coordinate system to the coordinate system of the base frame. Premultiplying all the positions describing the point of the original task by this transformation permits the task to be performed with respect to the new coordinate system. Note that in theory, no reteaching or trimming of points is necessary since the original points have been transformed to the new frame. Once a series of points have been taught or computed with respect to some coordinate system, they may be transformed into a new coordinate system and their interrelationship will be preserved.

Two special cases of the homogeneous transformation, simple rotations and translations, have been introduced in Sections 8.2.2.1 and 8.2.2.2. We will now discuss the effects of pre- and postmultiplying transformations relating frames by these operators. Recall the operations performed in Table 8.3.1 to align a movable coordinate system initially coincident with a fixed frame to a secondary frame displaced and oriented differently from the fixed frame. Operations of Rot (z_1, 180), Rot (x_1, −90), and Trans (x_a, y_a, z_a) were performed and always referenced to axes on the fixed reference frame. We may consider the fixed frame to be the *identity matrix which represents a transformation* that keeps points referenced to the same frame. This is premultiplied by Rot (z, 180), causing the points to be rotated about the z axis. The resulting points are then rotated −90° about the x axis by premultiplying by Rot (x, −90). Finally, the points are translated by x_a, y_a, z_a with respect to the reference frame by premultiplying by Trans (x_a, y_a, z_a). This example indicates that premultiplication of a transform representing the relationship between frames by a transformation representing a translation or rotation causes that translation or rotation to be made with respect to the base or reference coordinate frame.

In the case of postmultiplication, the rotation or translation is made with respect to the axes of the second frame. In other words, postmultiplying the transformation A_{nm} by a rotation operator will perform the operation about the coordinate frame m, while premultiplying will perform the operation about the coordinate frame n.

It is important to visualize that chaining these operations changes the reference frames. That is, once a transformation is postmultiplied by a rotation or translation operator, we should think of that frame being transformed into another frame. Subsequent operations are now performed on this new frame, whose axes may bear no relationship to the directions of the axes of the original frame.

8.5 HOMOGENEOUS TRANSFORMATIONS AND THE MANIPULATOR

Before describing the mechanics of obtaining the forward and inverse solutions of a robotic manipulator, we will utilize the properties of homogeneous coordinates and homogeneous transformations to aid in describing the position of the manipulator and to define concisely the mathematics required to define the positions of the manipulator in situations where they must be computed. This approach was chosen since these concepts are independent of the robot configuration and serve to illustrate the power of describing the manipulator with homogeneous transformations. In addition, they provide some insight into what type of computations and data manipulations must be performed in a robotic control system in order to define terminal and intermediate positions required for motion control.

8.5.1 The Position of the Manipulator in Space

The only concept lacking in our background of homogeneous coordinates and robotic manipulators is that of representing a *point* by a transformation. We will use the term point to describe the position of the manipulator with respect to a reference frame. This point is the information that must be saved by the robot's controller for later use. By recalling these points and extracting the correct data for the servos, the manipulator can be ordered to go to various positions in space.

The position of the manipulator will be defined as the position and orientation of its end effector or gripper with respect to a reference frame usually attached to the base of the manipulator. Homogeneous transformations will be used to define these data concisely.

The term *base* or *base reference* will be used to define the frame to which the gripper is referenced. Sometimes it is convenient also to refer to this as the *world frame*. In our discussions, the terms *base reference* and *world reference* are usually interchangeable, and any exceptions will be noted.

Recall that in Section 8.3 we defined the homogeneous transformation as a 4 × 4 matrix that could be used to relate points located in different coordinate systems. Its columns corresponded to the direction of the *x, y,* and *z* axes of one frame, written in terms of the *x, y,* and *z* direction vectors of another. In addition, the last column was the position of the origin of the first set of orthogonal axes with respect to the second. This concept may be utilized to describe the position of a robot. By attaching a reference frame to the end effector, a homogeneous transformation that relates this frame to a frame attached to the base of the robot arm may then be defined. This matrix, called the *hand matrix*, describes the position and orientation of the end effector with respect to the base reference of the robot. Figure 8.5.1 shows a schematic of a parallel-jaw type of gripper with a coordinate frame attached.

The *x, y,* and *z* directions of the gripper are called the *normal, orientation,*

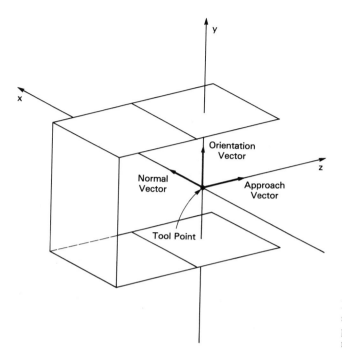

Figure 8.5.1. Schematic of gripper showing a reference frame with the tool point, approach, orientation, and normal vector directions.

and *approach vectors*, respectively—thus the reasoning for the symbols used in Eq. (8.3.1). The origin of the gripper's frame is called the *tool point* and is usually positioned at some location in the jaws of the gripper. The location of the tool point defines the point on the manipulator that will be positioned at the coordinates defined by the fourth column of the transformation matrix. In addition, the axes of the coordinate frame will be aligned with the base frame as defined by the first three columns. For example, if the tool point is defined to be in the center of the gripper as shown in Figure 8.5.1, and Eq. (8.3.1) defines the gripper's position and orientation, the center of the gripper would be located at (p_x, p_y, p_z). If the tool point was defined on the center of the backplate of the gripper, this is the point that would be positioned at (p_x, p_y, p_z). Such a scheme is referred to as *tool point control*. When controlling the position of a robot, for example, as in curve-following or straight-line motion, it is the tool point that is made to follow the desired path. By utilizing tool point control, we may align other frame origins with the physical location defined by this point. Additionally, the tool point can be moved mathematically to compensate for picking up objects of various heights or to adjust for tool wear.

The symbol T is used to designate the hand matrix. This matrix is the product of multiple transformations which define frames relating the positions of the various parts of the manipulator to one another and to the base [see Eq. (8.3.10)]. By this definition, T may also be thought of as A_{0n} for a system with $(n + 1)$ frames. Thus we may write

$$HAND = T = A_{0n} \qquad (8.5.1)$$

For the sake of simplicity in the following sections, we will assume that our manipulator is ideal, that is, one that can be positioned (and oriented) anywhere within its workspace, so that, in effect, the robot workspace is unbounded. In addition, we will assume that given any hand matrix representing its position, the robot will be able to achieve this desired configuration. For example, positioning its gripper in the center of one of its members, or moving through itself, will be permissible. Obviously, this is not always the case for real-world manipulators. The configuration, degrees of freedom, and mounting locations can vary greatly and therefore can limit the positions and orientations that the robot may achieve.

It is important to note that a position and orientation defined by a homogeneous transformation may be unattainable by a given manipulator. The ones always guaranteed to be achievable are those computed from the joint angles defining a known position and orientation of the manipulator. The results of matrix operations on the hand matrix may define positions and orientations that the manipulator cannot attain. Therefore, in utilizing the framework of transformations in robotic control and task description, one must be aware of the physical limits of the manipulator and its workspace.

8.5.2 Moving the Base of the Manipulator via Transformations

In Section 8.4.5 the concept of premultiplication of a transformation matrix that represented a coordinate system by another homogeneous transformation was discussed. The ability to re-reference the coordinate system in which a task was originally demonstrated or computed provides a means of minimizing the number of points that have to be retaught or recomputed in order to describe the same task in another location. The following discussion illustrates this concept.

Consider the top view of a workspace as shown in Figure 8.5.2. The base reference of the manipulator is at the origin of the x-y axes shown in the center of the Figure 8.5.2. The z axis (not shown) is directed out from the plane of the paper, with its direction being defined by the "right-hand rule." Let us assume that the manipulator's task consists of drawing the letter A in each of the six sectors shown. Additionally, the A is to be positioned on the various line segments, such as ac, as shown in the figure.

For this example the tool consists of a spring-loaded pen, which is colinear with the robot's approach vector. The tool point will be defined so that if the orientation and normal vectors are placed in a plane, there will be sufficient pressure for the pen to write. Note that with this definition, the pen will always be oriented so that it is perpendicular to the work surface. Also, it is assumed that sector 0 will be the area in which the robotic task has initially been taught or where the

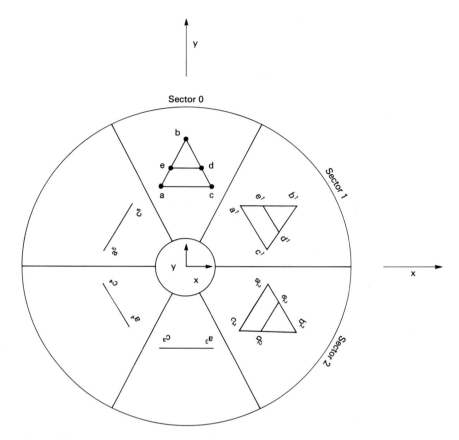

Figure 8.5.2.   Task using a change in manipulator's reference frame.

locations of points *a, b, c, d,* and *e* defining the character are known. We would like an easy way to perform the task in the remaining five sectors without having to teach additional points (e.g., a^1, b^1, c^1, and d^1). By using coordinate transformations to find the points a^1, b^1, c^1, and d^1 the task performed in sector 1 will be identical to that demonstrated in sector 0 since the relationship of the points will be preserved. It should be obvious that if the points were taught in each sector there might be discrepancies. Using transformation techniques to demonstrate the points only one time and then transferring them to another location in space makes the relationship among the various points consistent (i.e., the distance from point *a* to point *b* and so on).

To get some idea of the complexity of the number of positions necessary to perform the task, consider what must be done to draw a single letter. The points and operations are defined in Table 8.5.1.

To repeat the same task in sector 1, all that is required is to transform points

TABLE 8.5.1 OPERATIONS REQUIRED TO DRAW THE LETTER A IN SECTOR 1 OF FIGURE 8.5.2

1. Initially, manipulator is positioned at some "reset" or "home" location.
2. On receipt of a start signal, the robot moves the pen to a position located directly above point a, so that the pen is normal to the plane of the work surface but not touching it.
3. Robot moves its tool point down so that it is in the plane of the work surface. The pen is now perpendicular to the work surface and touching point a.
4. Robot moves pen in a straight line to b (still touching the work surface).
5. Pen moves in a straight line to c.
6. Pen is lifted off surface.
7. Pen moves to a position above point d.
8. Pen moves down to touch surface at d.
9. Pen moves in a straight line to e.
10. Pen is lifted off surface
11. Robot moves to the "reset" position (as in step 1).
12. Robot waits for another command.

(a, b, c, d, e) to $(a^1, b^1, c^1, d^1, e^1)$. This is easily accomplished by *premultiplying* all of the taught points (as outlined in Table 8.5.1) by the homogeneous transformation, which rotates points 60° in the clockwise direction about the z axis of the base. The operator Rot $(z, -60)$ [see Eq. (8.2.27)] performs this operation. Observe that the rotation is in the negative direction because the cross product of the two vectors, each drawn from the origin to a and a^1, respectively, defines a new vector in the direction opposite to the z axis of the base.

Once the operation in sector 1 is complete, the same transformation, Rot $(z, -60)$ is applied to the points a^1 through d^1, transforming them into points a^2 through d^2. Note that we could have gone directly from the positions defined in sector 0 to the corresponding ones in sector 2 by rotating 120° clockwise. We may think of the operation just described as equivalent to physically rotating the base of the manipulator by 60° clockwise before performing the task originally defined in sector 0. However, physical motion was not necessary.

It is interesting to note that besides rotating about the z axis of the base, we could have used transformations to compensate for position offsets. This situation could arise if the base of the robot was not coincident with the coordinate frame of the work surface. In the case where the z axis of the manipulator's base is not perpendicular to the work surface, we could have compensated for this tilt by the application of an appropriate rotation matrix.

8.5.3 Moving the Tool Position and Orientation

In Section 8.4.3 we noted that the postmultiplication of a homogeneous transformation representing a frame resulted in rotations and/or offsets with respect to the coordinate system of that frame. This concept provides a means of defining new tool positions and orientations based on the current value of the hand matrix.

Assume that we have defined the position and orientation of a manipulator at some point in space by a suitable hand matrix. Further, suppose that it is desired to know the value of some of the intermediate positions that result if the tool is moved in a straight line along its approach vector. From the results of previous sections we know that postmultiplying the hand matrix by Trans$(0, 0, dz)$ will move the origin of the tool point a distance dz from its current location. Repeated postmultiplications of the current point by Trans$(0, 0, dz)$ will produce points evenly spaced (i.e., equal increments of dz) along the direction of the tool's approach vector. From the resulting matrices, the intermediate positions along the straight-line trajectory can be determined.

Application of the operators Trans $(dx\ 0, 0)$ and Trans $(0, dy, 0)$ can be used to generate points along the straight-line paths parallel to the normal and orientation vectors of the tool reference frame. In the same fashion, one can define transformations such as Rot $(z, 180)$, which will give the position of the manipulator that corresponds to a rotation of the gripper 180° about its approach vector. Transformations such as Rot $(y, \Delta\theta)$ can be used to find the position of the gripper for incremental rotations about the orientation vector. The operations just described are associated with tool motions.

It is often important for the gripper to move so that the tool point remains parallel to the robot's reference axes (i.e., the world axes) while maintaining the orientation of the gripper during the motion. World motions are defined as motions in which the tool point moves in a direction parallel to a defined world x, y, or z axis and maintains the orientation of its tool frame to this world frame. Points for these motions are generated by modifying the appropriate variable in the position vector, that is, the fourth column of the hand matrix. For example, to move along a path parallel to the x axis of the world reference frame, the entry p_x of Eq. (8.3.1) would be the only variable modified.

EXAMPLE 8.5.1: OPERATIONS ON THE POSITION OF A MANIPULATOR

It is desired to compute the value of the hand matrix, which defines the position of the manipulator displaced dz units along its approach vector, followed by a rotation of 180° about the same vector. This could correspond to the endpoint of a motion in which a robot moves a container forward (dz units) and then dumps out its contents by rotating the container 180°.

The base of the manipulator and the position of its gripper are given by the same frames as shown in Figure 8.3.1. Frame 1 is the robot's base or world frame, while frame 2 shows the position of its gripper. The transformation relating the gripper position and orientation to the base is

$$T = \begin{bmatrix} -1 & 0 & 0 & x \\ 0 & 0 & 1 & y \\ 0 & 1 & 0 & z \\ 0 & 0 & 0 & 1 \end{bmatrix} \qquad (8.5.2)$$

Postmultiplication of Eq. (8.5.2) by Trans $(0, 0, dz)$ results in the matrix

$$T = \begin{bmatrix} -1 & 0 & 0 & x \\ 0 & 0 & 1 & (y + dz) \\ 0 & 1 & 0 & z \\ 0 & 0 & 0 & 1 \end{bmatrix} \tag{8.5.3}$$

From the geometry of Figure 8.3.1, we note that motion along the approach vector will result in an increase of the variable y since the tool z axis is parallel to the y axis of the base or world frame. This is also clearly seen in Eq. (8.5.3). Postmultiplication of Eq. (8.5.3) by Rot $(z, 180)$ yields the desired result:

$$T = \begin{bmatrix} 1 & 0 & 0 & x \\ 0 & 0 & 1 & (y + dz) \\ 0 & -1 & 0 & z \\ 0 & 0 & 0 & 1 \end{bmatrix} \tag{8.5.4}$$

Example 8.5.1 demonstrated in a simple manner the power of postmultiplication. By using this technique it is possible to generate positions for motion or other points related to known positions but not explicitly taught. To illustrate this, consider the discussion of Section 8.5.2. Recall that to perform the desired task (i.e., drawing the letter "A") we also needed a point directly above the point a. If a had been explicitly taught, a point directly above it could have been computed using the techniques just discussed.

Transformations can be used to compensate for differences between the location (and orientation) of the original tool (used to teach or demonstrate the program points) and the actual tool used for performing the task. Typically, robot programming languages provide for features called *tool transforms*. Such transforms are a convenient method of compensating for wear or for using different grippers or tools. Tool transforms make it possible to teach a task with one gripper and then perform the same task with a different one which may have longer jaws or be offset from its original mounting position on the robot. Postmultiplication of all the points of a given task by an appropriate transformation will provide the necessary compensation for the worn tool or new gripper.

Another interesting application of postmultiplying the hand matrix by a transformation is to vary the relationship between an object and the tool point of the robot. Consider the task of stacking 1 in.-per-side blocks on a tabletop. A block is placed in the gripper and the robot is positioned so that the block is at the desired location at the bottom of the stack. The tool point will be some location above the table, since the block is assumed to protrude from the jaws. The hand matrix, T, corresponding to the bottom of the stack is recorded. To cause the manipulator to place the next block directly on top of the first, all that is required is to postmultiply T by the matrix Trans $(0, 0, -1)$. This causes the tool point to stop 1 inch above the point originally taught so that the robot can deposit the second block correctly.

8.5.4 Relative Points and Reference Frames

8.5.4.1 Reference frames

Palletizing is a common application of robots whereby objects are placed into a two-dimensional array. A common example is placing soda bottles into a partitioned cardboard carton similar to the discussion in Section 2.3. *Depalletizing* is the inverse process, concerned with unloading the regular array. Figure 8.5.3 shows the schematic drawing of a pallet consisting of 12 locations.

The pallet itself can be considered to have its own reference frame, as is shown in this figure. One corner is chosen as the origin with two perpendicular edges defining the x and y axes. The compartments or holes define the locations where objects are to be placed. The center of each hole may be defined with respect to the coordinate system affixed to the pallet. It is assumed that these

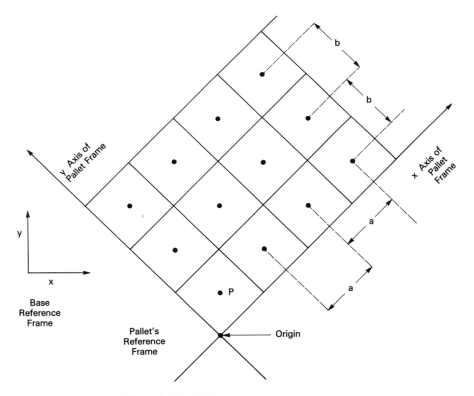

Figure 8.5.3. Pallet with frame and locations.

holes are evenly spaced in the x and y dimensions, and that their locations are fixed with respect to the coordinate system of the pallet. With the location of the first hole known, in terms of the pallet's coordinate system, all other locations can be determined.

Now consider placing the pallet into the workspace of a robot. Even if we know the spacing between the hole centers and the absolute location of the first hole in the coordinates of the pallet's reference frame, we still do not have enough information to guide the robot to the pallet locations. Although it is known how to move the gripper, either with respect to the robot's base or with respect to its tool axes, we cannot relate the robotic workspace to the coordinate frame of the pallet. We need the transformation that relates the base or world frame of the manipulator to the coordinate system of the pallet. Given this relationship, the gripper can be commanded to go to the appropriate locations.

Three possible methods of determining the transformation relating the two reference frames are outlined in the following discussion. The first approach simply consists of "teaching" the manipulator the first pallet point, P, of Figure 8.5.3. This information allows us to compute the location of the pallet's frame with respect to that of the robot. Using the chain rule we obtain

$$T = A_{\text{base,pallet\_frame}} A_{\text{pallet\_frame, first\_pallet\_point}} \qquad (8.5.5)$$

In Eq. (8.5.5), T, the hand matrix, relates the base reference of the robot to the tool frame, which has been positioned on the first pallet point and oriented in some known manner to the pallet's reference frame. The second A matrix, relating the pallet frame to the first pallet point, can be found from the geometry of the pallet. The remaining A matrix is the desired result. Solving Eq. (8.5.5) for the relationship between the manipulator's base frame and the pallet's frame yields

$$A_{\text{base,pallet\_frame}} = T A_{\text{first\_pallet\_point,pallet\_frame}} \qquad (8.5.6)$$

Although Eq. (8.5.6) yields the correct results, (i.e., is mathematically correct), physically, the alignment of the gripper with the first pallet point so that both position and orientation (to the pallet frame) are correct is a difficult task. The reason for this ranges from interferences between the gripper and pallet, to coarseness of the robot's control resolution, to operator indifference, among others. It should be obvious that any small angular error may introduce gross errors as we try to compute the pallet bin's locations farther from the pallet's coordinate axes.

A similar approach is to align the frame attached to the gripper so that it is coincident with the frame of the pallet. This essentially makes the transformation defining the hand matrix (with respect to the reference frame of the robot) the same as the transformation relating the base of the robot to the frame of the pallet. Accomplishing this alignment may be difficult if the frame on the pallet and that of the gripper interfere mechanically. Additionally, both the frame of the gripper and the frame of the pallet may not be well defined, and thus lining up the gripper would be only an approximation. If the two coordinate frames are not perfectly aligned, the transformation matrix may describe a plane for the pallet which could be tilted in as many as three axes.

A more reliable method does not rely on the frames of the pallet and tool. Instead, the robot is "shown" three points on the pallet, thus defining a plane. This will be referred to as a *three-point alignment method*. By showing the origin, and points on the *x* and *y* axes as far from the origin as possible, we can define vectors for the *x* and *y* axes of the pallet frame with respect to the base reference frame of the robot and thereby calculate the transformation relating the frames. Three points are necessary in that they also give information on the tilt of the frame if it is not flat on the work surface. The cross product of the *x* and *y* axes will define the normal to the pallet and the *z* axis of the frame.

When teaching these points to the robot, we have the ability to define both their position and orientation. As mentioned before, since it may be quite difficult to show points with exact position and orientation, we utilize only position information. This is accomplished by moving the tool tip (or a known point on the gripper) to the points of interest. Note that the tool transform may be used to modify the offset if the tool point is not used to show the points. Then the last column of the hand matrix, T [Eq. (8.3.1)] contains the position of the point using the world coordinates. From the three points and the vector techniques of Section 8.2.1, the desired transformation is computed.

Once the frame transformation is defined, the only remaining concern is to calculate the positions of the gripper that correspond to the holes in the carton. Clearly, both the location of the tool point and its orientation need to be defined. It is quite obvious that if the gripper had an object in its jaws, we would want the approach vector to point downward. This is necessary so that when the gripper is opened, the part falls into the partition. Additionally, let us assume that we want the parts placed in the box so that the orientation vector was parallel to the *y* axis of the pallet.

To determine the hand matrix corresponding to each of the compartments in the box, one can make use of the translational and rotational transformations of Section 8.2.2. For each compartment of the box, one can define a transformation $A_{\text{pallet\_frame,hole}}$. This transformation has, as its first three columns, the identity matrix while the last column is the location of a particular hole with respect to the pallet's reference frame. This is essentially Trans(a,b,c), where a and b are the spacing between holes and c is a value that makes the point some distance above the bottom of the box. Finally, if one postmultiplies this matrix by Rot(y,180), the resultant transformation will have the tool point located in the center of the hole with its orientation vector parallel to the *y* axis of the pallet and its approach vector antiparallel to the *z* axis of the pallet (i.e. pointing downward).

The computation of the hand matrix based on the discussion above is as follows:

$$T = A_{\text{base,pallet\_frame}}\ A_{\text{pallet\_frame,hole}}\ \text{Rot}(y,180) \tag{8.5.7}$$

Equation (8.5.7) generates a hand matrix which is the product of the A matrix relating the base frame to the pallet, and the A matrix relating the origin of the pallet to a particular hole. The Rot matrix orients the gripper so that it points downward while keeping the orientation vector of the gripper parallel to the y axis of the pallet.

Although the preceding discussion may seem complicated, the benefits are quite useful. After the operations on the first pallet have been completed, it can be removed. Subsequent pallets do not have to be placed at exactly the same location in the workspace, since their location with respect to the base of the robot may be recomputed once the three-point alignment described above is performed. This approach eliminates the need for complex or expensive fixturing for parts presentation.

8.5.4.2 Relative points

The preceding discussion illustrates a problem and one possible solution in the application of using transformation techniques. Recall that we assumed that the locations were known in terms of the frame of the pallet. In the case where the task to be performed can be defined in a reference frame, but the locations must be "taught" (possibly due to irregularity), the use of *relative points* becomes attractive.

Relative points are positions of the manipulator defined as offsets in both position and orientation from some frame. These points are saved (i.e., stored in the controller's memory) as the homogeneous transformations which relate the point to a particular frame. Thus if the frame moves in space relative to, for example, the world reference frame, we can always find the coordinates of the point in question with respect to this reference frame. Equation (8.5.8) shows the relationship that would be used to accomplish such an operation:

$$T = A_{\text{world,frame}} \, A_{\text{frame,point}} \qquad (8.5.8)$$

As long as $A_{\text{world,frame}}$ is defined, the hand matrix T can be computed by post-multiplying the first transformation by $A_{\text{frame,point}}$, which relates the position and orientation of the relative point to its reference coordinate system. As seen by Eq. (8.5.8), relative points must always be referred to a frame.

Utilizing this concept in the previous palletizing example, the locations of the pallet holes of Figure 8.5.3 would be defined with respect to the frame of the pallet. That is, the relative point [form of Eq. (8.3.1)] is written in terms of the axes of the pallet frame. The pallet frame is defined with respect to the world frame of the manipulator by using the three-point alignment method. The hand matrix, T, used to position the gripper, is computed as in Eq. (8.5.8). Alternatively, if the actual relationship of each point with respect to the pallet's frame is unknown but the transformation relating the pallet frame to the world frame can be defined, it is possible to compute the relative points. This is accomplished by moving the robot's gripper to the points of interest and then solving Eq. (8.5.8) for the transformation, $A_{\text{frame,point}}$. Note that the data saved are not referenced to the base or world frame but are referenced to the relative frame. Thus to define the position of the gripper, we need two pieces of information: (1) the transformation relating the world frame to the reference frame, and (2) the transformation relating the reference frame to the relative point. If these relative points are then applied to another orientation of the frame, the end effector will move to the correct location.

An extension of this concept is to chain relative points by defining relationships between objects or locations in terms of frames. This provides a dynamic relationship that will adjust to the environment provided that the correct reference frames are applied. In a real-world robot, relative points applied to frames other than those to which they were referenced may yield unattainable solutions.

The following example illustrates the use of relative points and the three-point alignment method.

EXAMPLE 8.5.2: FRAMES AND RELATIVE POINTS

Figure 8.5.4 shows a two-dimensional workspace consisting of a base (or world) frame, a four-cell pallet A with cell center points (p_1, p_2, p_3, p_4) and a second pallet, B, which is identical to A but located at another position in the workspace. The second pallet contains the points (p_1', p_2', p_3', p_4').

It is desired to perform a task consisting of having the gripper transfer each object from pallet A to some processing station, such as a grinding operation, and then to move the finished part to the corresponding location on pallet B. We will assume that pallet A is fixed with respect to the world

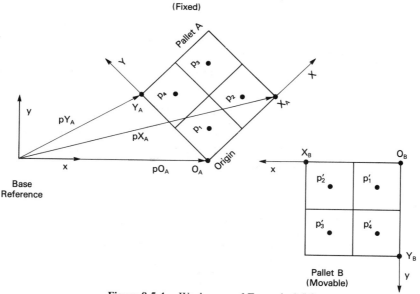

Figure 8.5.4. Workspace of Example 8.5.2.

axes, while pallet B will be removed and replaced with an identical unit after the task has been performed. The points on the second pallet are not to be explicitly "taught" since the location of pallet B can change each time the task is performed.

To accomplish the foregoing task, the procedure to be used will be to find a frame, frame A, to which points p_1 through p_4 can be referenced. Once this is done, relative transformations, which we will call LOC_1 through LOC_4, can be determined by the use of Eq. (8.5.8). Using the three-point alignment technique, a transformation relating the frame of pallet B to that of the world is defined and can be used with the transformations LOC_1 through LOC_4 to find points p_1' through p_4' located in pallet B. The steps are outlined below:

1. Manually move the manipulator's gripper to points p_1 through p_4 and record the corresponding hand matrices. This is the tool point's location and orientation with respect to the base reference frame. For each point, T of Eq. (8.5.8) is now defined. Let T_n designate the hand matrix for the demonstrated point n ($n = 1, 2, 3, 4$).

2. Manually move the manipulator to points O_A, X_A, and Y_A on the first pallet and record the hand matrices corresponding to these points. This provides the data for the three-point alignment algorithm.

3. Compute the 4×4 transformation matrix representing frame A with respect to the world reference frame using the information in Table 8.5.2. This transformation will be designated by $A_{w,f}$. The notation pX_A means to use the vector $\mathbf{p}$ $(p_x, p_y, p_z, 1)^{tr}$ of the transformation matrix for the computation (i.e., the fourth column of the hand matrix obtained in step 2 for the point X_A). In addition, the function "mag" means to compute the magnitude of the vector. These steps define the four columns of the transformation matrix relating the world frame to the frame of pallet A.

4. Using the relationship given by Eq. (8.5.8), the values of LOC_n are computed. Thus

$$LOC_n = T_n A_{w,f}^{-1} = T_n A_{f,w} \qquad (8.5.9)$$

We now have the value of the (relative) points that relate the positions of each of the cells of the pallet to the pallet's coordinate frame.

5. Using the three-point alignment method, establish the transformation for

TABLE 8.5.2 STEPS FOR FINDING $A_{\omega,f}$

1. x-direction vector: $(pX_A - pO_A)/\text{mag}\,(pX_A - pO_A)$
2. y-direction vector: $(pY_A - pO_A)/\text{mag}\,(pY_A - pO_A)$
3. z-direction vector: $\mathbf{x} \times \mathbf{y}$
4. Position vector: pO_A

pallet B with respect to the world frame. This will be designated $B_{w,f}$. The procedure is similar to step 3, except that X_B, Y_B, and O_B are used.

6. Compute points on pallet B using the relationship

$$T_{n'} = B_{wf} LOC_n \qquad (8.5.10)$$

After this procedure has been followed, all the information necessary to position the manipulator at the points is available, and the task may be performed. Once pallet B has been used, it may be removed and another pallet moved into the workspace. Remember that the new location of the empty pallet can be anywhere and its orientation is also not critical. The entire task can be repeated by performing the three-point alignment to re-define the transformation $B_{w,f}$, and computing the appropriate values of $T_{n'}$. Note that it is unnecessary to recompute LOC_1 through LOC_4, since once established they remain constant. The transformation $A_{w,f}$ also remains unchanged unless pallet A is moved.

This example has shown how reference frames and relative points may be utilized to aid in defining a robotic task. Their ability to compensate automatically for changes in the location of objects within the workspace allows tasks to be done without the need for complex or costly fixturing to ensure the exact repeatable placement of an object, such as pallet B of the example.

8.5.5 Vision and Reference Frames

Most vision systems are two-dimensional in nature and look at a portion of the workspace of the manipulator. These systems usually give position information which defines the location of an object. This positional information is with respect to the reference frame established for the field of view. To use vision in a robotic system it is necessary to calibrate the camera's frame of reference to that of the robot so that the robot may move to the correct locations. The calibration or alignment is necessary since it may be difficult to guarantee the exact placement of the vision system's frame of reference with respect to that of the base reference of the robot.

Various procedures may be used to perform this calibration or alignment. Whatever method is selected, it should be simple, repeatable, and perhaps require the use of no special fixtures. Of course, the nature of the problem will dictate other criteria or modify those mentioned.

From the previous discussions and examples in this section, it should be evident that the procedure will generate a transformation relating the frame of the vision system to the base reference of the robot. Thus using the properties of transformations (see Section 8.3), points in the frame of the vision system may be related to the world frame of the manipulator. For this discussion the camera will be assumed to remain fixed in relationship to the world frame of the manipulator.

One possible method of calibrating the vision system's frame of reference to that of the manipulator is to have the manipulator position an object (that the vision system has been trained to recognize) in the field of view of the camera. The corresponding hand matrix will once again be called T. T may also be thought of as the transformation

$$T = A_{\text{base,object}} \tag{8.5.11}$$

Equation (8.5.11) relates the center of the top surface of the object (assuming that the tool point is at the center of the back wall of the gripper and the object is seated in the gripper) to the base reference of the robot. The robot is then instructed to release the object and move out of the vision system's field of view. Next, the vision system is instructed to "find the object." It reports back a transformation defined as follows:

$$A_{\text{vision\_frame,object}} \tag{8.5.12}$$

where the coordinates given by the fourth column of Eq. (8.5.12) define the position of the object in the vision system's field of view. It is extremely important to note that the position reported by the vision system (such as the centroid of the object) must be the same point defined by the hand matrix of the robot; otherwise, undesirable offsets may occur. One simple method to ensure they are the same is to require that the object the robot places in the camera's field of view has a marking (such as a solid circle) on its top surface. When the object is placed in the robot's gripper, the center of the circle is aligned with the tool point either manually or by some other method such as self-centering jaws.

We should note that, typically, vision systems only report x,y coordinates with respect to their frame of reference. Therefore, the pz entry of the fourth column can be set to zero and the first three columns of the transformation of Eq. (8.5.12) may be defined as the unit vectors in the x, y, and z directions (which are consistent with the coordinate system used by the vision system to define position). Vision systems typically use coordinate frames with the z axis going into the plane of the paper (see Chapter 6).

For a two-dimensional imaging system, no z information exists if the vision system is looking at a part of the x-y plane of the base reference system. Note that it is possible to look at the y-z or z-y planes, and if this is the case, the algorithm must be adjusted accordingly.

It is desired to extract a frame transformation from Eqs. (8.5.11) and (8.5.12) which will be a constant as long as the base of the manipulator and the vision system do not move in the workspace. The transformation that relates the base reference of the robot to the reference of the vision system is defined by the following relationship:

$$T = A_{\text{base,object}} = A_{\text{base,vision\_frame}} A_{\text{vision\_frame,object}}$$
$$= A_{\text{b,vf}} A_{\text{vf,o}} \tag{8.5.13}$$

Using matrix techniques, the required frame transformation $A_{b,vf}$ may be obtained. This particular transformation is a constant and changes only if the mounting of the vision system in relation to the manipulator changes. If $A_{b,vf}$ is applied to subsequent information from the vision system (of the form $A_{vf,o}$), Eq. (8.5.13) may be used to compute the value of T, required to position the manipulator correctly.

8.5.6 Summary of What Transformations May Represent

Table 8.5.3 summarizes various interpretations for the homogeneous transformation matrix as discussed so far. It is important to remember that regardless of the interpretation, the form of the matrix, as given by Eq. (8.3.1), is the same. Thus it is impossible to determine what the matrix represents by just looking at its entries. Its actual use must be considered. Otherwise, one could confuse a relative point with an operator.

TABLE 8.5.3 INTERPRETATIONS OF HOMOGENEOUS TRANSFORMATION MATRIX

1. Rotation or translation of a vector with respect to a fixed reference frame (see Sections 8.2.2.1 and 8.2.2.2)
2. Transformation of points from one frame of reference into another (see Section 8.3)
3. Representation of a reference frame in space (see Section 8.3)
4. Representation of the position of a manipulator in space (see Section 8.5)
5. Operator that can change the position and orientation of a manipulator (see Sections 8.5.2 and 8.5.3)
6. Representation of a point relative to a reference frame (see Section 8.5.4)

8.6 THE FORWARD SOLUTION

The forward or direct solution of a robotic manipulator maps the value of the joint vector [Eq. (8.1.1)] to the transformation matrix relating the gripper's frame to the robot's world reference frame. The values of the joint positions are the variables in the forward solution, while certain other parameters, such as the distance between the axes of rotation, remain constant. Thus the forward solution will produce a matrix of the form of Eq. (8.3.1), where the entries will be functions of the joint variables and constants. The forward solution is essential in defining the kinematic relationship of the manipulator.

Kinematics is defined as the study of motions ignoring such concepts as forces, torques, and inertias. The types of motions discussed in the examples of Section 8.5.3 deal only with the position of the manipulator and do not take into account the load it is carrying or the fact that the inertia, as seen by the joints, changes as the robot's position varies. Consequently, these examples were kinematic in na-

ture. Similarly, relating the frame of the tool to the base of the robot is also a
kinematic problem, since none of the dynamic quantities affecting the tool's motion
are taken into consideration.

The following sections define the terminology and definitions that are used
in the Denavit–Hartenberg (D-H) representation of kinematic chains. This method
uses matrices to describe the relationship among reference frames attached to
various points on the manipulator. Once these matrices are defined, the chain
rule [Eq. (8.3.10)] is used to find the relation between the base and tool frames.

8.6.1 Background Information and Terminology

A robotic manipulator consists of a sequence of rigid bodies called links which are
connected to each other by joints. Together, the links and joints form a *kinematic
chain*. The types of joints used in most manipulators are revolute (rotary) or
prismatic (linear) and are driven by some type of actuator (see Chapter 4). Motion
of a joint causes the link attached to move with respect to the link containing the
joint's actuator. The position of the joint may be obtained from a sensing device,
such as an optical encoder (see Chapter 5), which monitors the position of the
joint either directly or indirectly (e.g., by monitoring the drive motor's shaft position). Figure 8.6.1 shows the links, joints, and actuators of a kinematic chain
composed entirely of rotary joints.

A serial link manipulator is one in which the links and joints are arranged in
an alternating fashion, with the additional constraint that closed loops are not
formed. Figure 8.6.1a depicts a serial link kinematic chain, while Figure 8.6.2
shows an example of a closed kinematic chain. In the latter figure, note that the
linear motion of the prismatic joint causes link 1 to rotate about joint 1. For a
serial chain, each link is connected to at most two others, and the joints and links
are arranged to prevent the formation of closed loops. This is the form of most
robotic manipulators and the one that will be considered in the following sections.
In the case of manipulators composed of mechanical structures that form closed
loops (such as the parallelogram configuration shown in Chapter 1), it is possible
to use adaptations of the techniques discussed in this text to model these manipulators as a serial chain.

Each pair of joints and links constitutes a degree of freedom. Thus in a four-
degree-of-freedom manipulator we must be able to identify four links and four
joints. The joints and links are numbered sequentially starting from the base,
which is considered link 0 and is not counted in defining the degrees of freedom.
The joint connecting the base to the first link in the serial chain is joint 1, while
the link that follows the first joint is link 1. Using this numbering scheme, we
note that no joint exists at the end of the final link. Figure 8.6.3 shows the
numbering of the joints and links of the kinematic chain of Figure 8.6.1a.

The significance of links, from a kinematic perspective, is that they maintain
a fixed relationship between the manipulator's joints present at the end of the link.
This relationship is used in assigning reference frames to the manipulator.

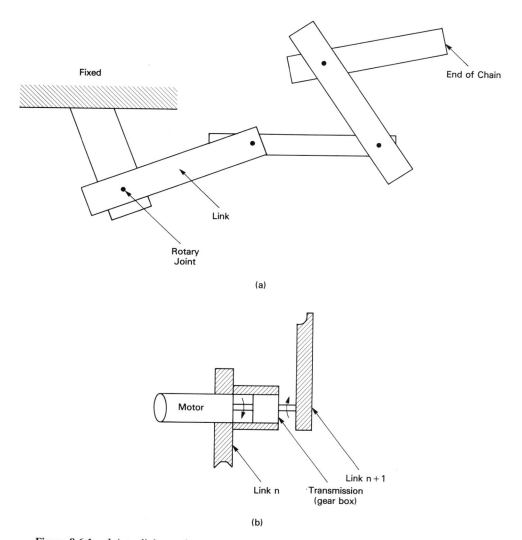

Figure 8.6.1. Joints, links, and actuators of a serial kinematic chain: (a) joints and links;
(b) concept arrangement of actuators and links.

Four parameters can be associated with every link of a manipulator. The
first two, a_i and α_i, define the structure of the link, while the second two, d_i and
θ_i, determine the position of the neighboring link. Each of these four parameters
is defined with respect to the two joint axes attached to a particular link. For a
rotary joint, the axis is defined along the axis of rotation, with its positive direction
being assigned in an arbitrary fashion. For a prismatic joint, the axis is defined

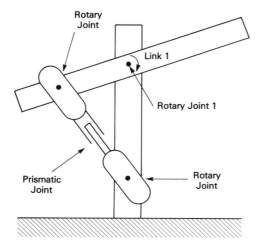

Figure 8.6.2. Closed kinematic chain.

to be coincident with the direction of linear motion. In this case the positive direction is taken in the direction of increasing length.

Figure 8.6.4 shows a_i, the common normal distance between the joint axes at each end of link i and is directed from the axis of joint i to joint $(i + 1)$. α_i, the angle between the two joint axes, is measured in a plane perpendicular to a_i. It may be easier to visualize this angle by drawing a line parallel to joint axis i so that it intersects joint axis $(i + 1)$ and the common normal. This is shown as the dashed line in Figure 8.6.4. The angle is measured from axis i to axis $(i + 1)$ using a right-hand rule about a_i. The parameters a_i and α_i, called the *length* and *twist* of the link, respectively, define the structure of the link.

Figure 8.6.5 shows two links and three joints of a serial chain. Every joint axis has two normals to it, one for each of the links it connects. Thus the axis of

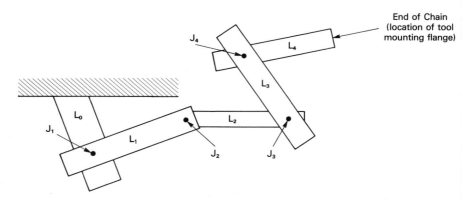

Figure 8.6.3. Numbering of joints and links.

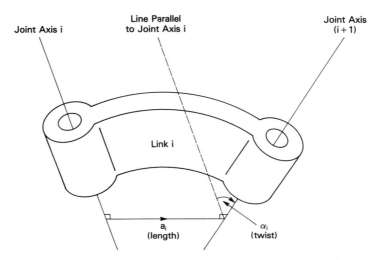

Figure 8.6.4. Length and twist of a link.

joint i has both a_i and a_{i-1} perpendicular to it.   The relative position of the two links is given by d_i, which is the distance between the two normals (a_i and a_{i-1}) measured along the ith joint axis. θ_i is the angle between these two normals measured in a plane perpendicular to the axis.   The parameters d_i and θ_i are called the *distance* and *angle* between the links, respectively, and are used to define these relationships.

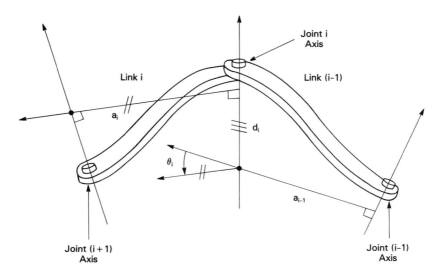

Figure 8.6.5. Distance and angle between links.

TABLE 8.6.1 RULES FOR ESTABLISHING LINK COORDINATE FRAMES

1. The z_{i-1} axis lies along the axis of motion of the ith joint.
2. The x_i axis is normal to the z_{i-1} axis directed toward the z_i axis.
3. The y_i axis is defined by the cross product of z_i and x_i, so that the three axes form a right-handed system.

8.6.2 Establishing Link Coordinate Frames

Based on our knowledge of homogeneous transformations, if reference frames are established on each link and then related to one another, the tool point or the frame at the outermost link can be described in terms of the coordinate system established at the base of the manipulator. Denavit and Hartenberg [13] developed this relationship in terms of the four link parameters just introduced in Section 8.6.1 and the joint axes.

To understand how this is done, it is first necessary to define each set of reference axes. The coordinate frames are established based on the rules shown in Table 8.6.1. Figure 8.6.6 shows the coordinate system along with links and joint axes.

These rules are all that is necessary to establish coordinate frames on each link at its joint axis [5]. There is some implicit information generated by them, however, which will now be discussed.

1. Since the x_i axis is, by definition, perpendicular to the z_i axis (rule 3), rule 2

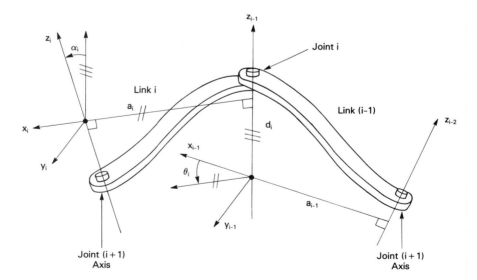

Figure 8.6.6. Link coordinate frames and joint parameters.

implies that the x_i axis is along the common normal, a_i, of the z_{i-1} and z_i axes with its direction from the z_{i-1} to the z_i axes.

2. The *origin* of the ith coordinate frame (i.e., the frame attached to link i) is located at the intersection of the common normal of joint axis i and joint axis $(i + 1)$ and the axis of joint $(i + 1)$.

3. The location of the origin of frame 0 may be chosen anywhere as long as the z axis lies along the axis of motion of the first joint.

4. The last coordinate frame (attached to the tool or hand) can be placed anywhere in the hand as long as the x axis is normal to the z axis of the joint linking the last frame to the manipulator.

Although the rules of Table 8.6.1 define the assignment of coordinate frames, some of the most common joint configurations that occur in robotic manipulators do not fit into this model. The following clarifies some common exceptions.

1. *Consecutive joint axes intersect.* This configuration is probably the rule more than the exception. If two joints are arranged so that their joint axes intersect when extended, there can be no common normal connecting them. Thus the parameter a_i is zero. In this case, the x_i axis may be chosen along either of two opposite directions normal to both the z_{i-1} and z_i axes. The origin of the ith frame is then selected as the point of intersection of the two joint axes.

2. *Consecutive joint axes are collinear.* Any direction normal to the z_{i-1} axis is a valid choice for the x_i axis. The origin of the ith frame may be placed anywhere along the common joint axes.

A careful examination of the rules presented above reveals that the *assignment of reference frames is not necessarily unique.* Typically, manipulators are designed so that some joint axes do intersect. The particular geometry of the manipulator and type of inverse solution sometimes dictate the best choice if an axis may be selected in more than one way.

8.6.3 The Denavit–Hartenberg Matrix

Once we have established the reference frames for the links at the joint axes, we can determine the four parameters, $(a_i, \alpha_i, d_i, \text{ and } \theta_i)$, describing each joint–link pair. These parameters are then used to define terms in a transformation matrix relating consecutive frames of the serial manipulator. This matrix will be called the Denavit–Hartenberg matrix to distinguish it from other forms of the transformation matrix. The D-H matrix is solely dependent on the four link parameters [8].

In the following discussion, a reference to an axis assumes reference to its positive direction. Also it is assumed that one mentally extends axes to find their

TABLE 8.6.2 LINK PARAMETERS FROM LINK COORDINATE FRAMES

1. θ_i is the angle from the x_{i-1} to the x_i axis measured about the z_{i-1} axis. This is defined using a right-hand rule since both x_{i-1} and x_i are perpendicular to z_{i-1}. The direction of rotation is positive if the cross product of x_{i-1} and x_i defines the z_{i-1} axis. θ_i is the joint variable if the joint i is revolute. In the case of a prismatic joint it is a constant that may or may not equal zero.

2. The distance between links, d_i, is the distance from the x_i to the x_{i+1} axis measured along the z_{i-1} axis. If the joint is prismatic, d_i is the joint variable. In the case of a revolute joint, it is a constant that is not necessarily equal to zero.

3. The common normal distance, a_i, is the shortest distance between the z_{i-1} and z_i axes. It is measured as the distance along the direction of x_i from the intersection of z_{i-1} and x_i to the origin of the ith coordinate frame. For intersecting joint axes the value of a_i is zero. It has no meaning for prismatic joints and is set to zero in this case.

4. The offset angle, α_i, is measured from the z_{i-1} to the z_i axes about the x_i axis, again using a right-hand rule. For most commercial manipulators the offset angles are multiples of $90°$.

points of intersection when necessary. With respect to Figure 8.6.6, the link parameters can be defined in terms of the link coordinate frames (see Table 8.6.2).

With coordinate frames established at each link and the link parameters determined, the relationship between successive frames $(i - 1)$ and i can be determined by a series of translations and rotations similar to those defined in Section 8.3 to derive the transformation matrix relating the coordinates of two frames in space. Our derivation and nomenclature are similar to [1], [6], and [8]. Referring to Figure 8.6.6, the operations in Table 8.6.3 will move the coordinate frame at joint $(i - 1)$ to the frame at joint i.

The product of the elementary transformations defined in Table 8.6.3 when taken in order is the Denavit–Hartenberg matrix (D-H matrix). Equation (8.6.1) shows the product of these operators.

$$A_{(i-1),i} = \text{Rot}(x_i, \alpha_i) \, \text{Trans}(a_i, 0, 0) \, \text{Trans}(0, 0, d_i) \, \text{Rot}(z_{i-1}, \theta_i) \qquad (8.6.1)$$

TABLE 8.6.3 TRANSFORMATION MATRICES TO RELATE SUCCESSIVE LINK COORDINATE FRAMES

1. $\text{Rot}(z_{i-1}, \theta_i)$. Rotate by an angle θ_i about z_{i-1}. This sets the x_i and x_{i-1} axes parallel to each other although they are not necessarily in the same plane.

2. $\text{Trans}(0, 0, d_i)$. Translate the origin of frame $(i - 1)$ a distance d_i. This sets the x_i and x_{i-1} axes coincident.

3. $\text{Trans}(a_i, 0, 0)$. Translate the origin of frame $(i - 1)$ a distance a_i. This sets the origins of frame $(i - 1)$ and i coincident.

4. $\text{Rot}(x, \alpha_i)$. Rotate an angle α_i about the x_i axis. This sets the z_i and z_{i-1} axes (and, therefore, the y axis) coincident.

Evaluation of Eq. (8.6.1) using the link parameters as variables yields Eq. (8.6.2).

$$A_{(i-1),i} = \begin{bmatrix} \cos\theta_i & -\cos\alpha_i\sin\theta_i & \sin\alpha_i\sin\theta_i & a_i\cos\theta_i \\ \sin\theta_i & \cos\alpha_i\cos\theta_i & -\sin\alpha_i\cos\theta_i & a_i\sin\theta_i \\ 0 & \sin\alpha_i & \cos\alpha_i & d_i \\ 0 & 0 & 0 & 1 \end{bmatrix} \qquad (8.6.2)$$

Of the four parameters in the D-H matrix, a_i and α_i are constants while either d_i or θ_i is the joint variable. Revolute joints have variables in terms of θ_i while terms that are functions of a_i, α_i, and d_i are constants. In the case of a prismatic joint, a_i is zero, θ_i and α_i are constants, and d_i is the joint variable. Thus for the linear joint, all the entries of the D-H matrix, excluding d_i, are zero or constants.

The transformation matrix of Eq. (8.6.2) relates points defined in frame i to frame $(i - 1)$. By defining the coordinate frames and determining the entries for the D-H matrices, we may determine the position of the tool or gripper (last link and frame) with respect to the base (0 link and joint 1) as a *function of the joint variables*. Recalling the previous definition of the hand matrix in Eq. (8.5.1), this matrix may now be defined by the successive frames established on each link of the manipulator. For an n-degree-of-freedom manipulator, the expression for the hand matrix is given by

$$T = A_{0n} = A_{01}A_{12}A_{23}A_{34} \cdots A_{(n-1)n} \qquad (8.6.3)$$

Each of the A matrices in Eq. (8.6.3) will be of the form of Eq. (8.6.2). A particular A matrix will contain functions of the appropriate joint variable and some constants. The T matrix will be a function of *all* the joint variables. In later discussions it will be necessary to distinguish a T matrix that is a function of the joint variables from one which contains known entries. We will call the T matrix containing variables the *symbolic T matrix* and use the term *numeric T matrix* for the matrix that results when a symbolic T matrix is evaluated for a set of joint variables, or is defined in terms of known position and orientation vectors.

Using the results above we are now in a position to generate the forward solution. Table 8.6.4 summarizes the steps needed to do this.

8.6.4 Comments on Forming the Forward Solution

In applying the steps in Table 8.6.4 to determine the symbolic T matrix, it would be advantageous to be able to check the intermediate results before doing the

TABLE 8.6.4   STEPS TO OBTAIN MATRIX DEFINING FORWARD SOLUTION

1. Establish coordinate systems for each link as outlined in Section 8.6.2.
2. Determine the parameters a_i, α_i, d_i, and θ_i for each link as defined in Table 8.6.2.
3. Evaluate the individual A matrices relating each successive link coordinate system as given by Eq. (8.6.2).
4. Evaluate the symbolic T matrix as given by Eq. (8.6.3).

algebraic manipulations required to evaluate Eq. (8.6.3). To accomplish this, we will draw on the results of Section 8.3 to evaluate the transformation matrix that relates two coordinate systems.

Before assigning coordinate frames and determining the link parameters leading to the A and T matrices, it is suggested that one position the manipulator in a "convenient configuration." This configuration should be such that one can easily determine "numeric" entries of the hand matrix with respect to the manipulator's world axes. Remember that the world z axis will be along the axis of motion of the first joint.

Possible examples of the robot's initial configuration include aligning the tool vectors (approach, normal, and orientation) to be parallel or antiparallel to the world axes, and positioning the tool point at a measurable or known location with respect to the world origin. The preferred initial positioning is to have the tool point at a position along either the x or y world axis at some height z. This type of alignment of the manipulator allows one to determine the hand matrix by inspection.

Once an initial orientation is chosen, an arrow diagram similar to those in Examples 8.6.1 and 8.6.2 is made. The steps of Table 8.6.4 are followed and the A matrices may be determined. Before computing the symbolic T matrix of Eq. (8.6.3), it is suggested that the numeric values of the A matrices be computed and then multiplied together in the appropriate order. The numeric T matrix (evaluated from the A matrices) should be compared to the hand matrix obtained by inspection. If no errors exist in either procedure, the two matrices should be equal.

Additional tests should be made by moving a joint or some combination of the joint's known increments that cause the tool tip to move to other "convenient locations." A table of hand matrices versus joint variables may be obtained for these joint vectors. The numeric values of the A matrices for each of these joint vectors and the corresponding numeric T matrix is generated and compared to the hand matrix. One should verify the solution for combinations of joint variables that cause the rotary joints to span 360°. This ensures that the signs of the sine and cosine functions are exercised and that the signs and direction of rotation defined when the axes were assigned are consistent.

Once we are satisfied with the results of the preceding exercise, the symbolic T matrix is computed. The multiplication is easily checked by evaluating at a known joint vector and comparing the result with a known corresponding hand matrix.

When multiplying the symbolic A matrices together to obtain Eq. (8.6.3), it is sometimes convenient to group the matrices in a particular order which generates intermediate results which may be used in one form of the inverse solution, as discussed in Section 8.7.2. For a six-degree-of-freedom manipulator, the preferred intermediate matrices are defined in Table 8.6.5. In this table, the superscript preceding the T indicates that it transforms from frame 6 (assumed) to the number

TABLE 8.6.5 PREFERRED
INTERMEDIATE A MATRIX PRODUCTS
FOR FORMING THE SYMBOLIC T
MATRIX

$$^5T = A_{56}$$
$$^4T = A_{45} \quad ^5T = A_{45}A_{56}$$
$$^3T = A_{34} \quad ^4T = A_{34}A_{45}A_{56}$$
$$^2T = A_{23} \quad ^3T = A_{23}A_{34}A_{45}A_{56}$$
$$^1T = A_{12} \quad ^2T = A_{12}A_{23}A_{34}A_{45}A_{56}$$
$$T = A_{01} \quad ^1T = A_{01}A_{12}A_{23}A_{34}A_{45}A_{56}$$

designated by the superscript. For instance, the third entry of the table is the transformation relating the coordinate system of link 6 to that attached to link 3.

8.6.5 Examples of the Forward Solution

The following examples show two common robotic manipulators, possible coordinate axis assignments, and the corresponding A matrices.

EXAMPLE 8.6.1: THE STANFORD ARM

Figure 8.6.7 shows the Stanford arm and a corresponding arrow diagram [5]. This is a spherical coordinate, six-jointed manipulator consisting of five revolute and one prismatic joints. The first three joints are used to move the tool point to its desired position, while the last three adjust the orientation of the end effector. A parallel-jaw gripper is attached to the tool plate and the origin of the coordinate frame attached to the gripper is offset along the tool z direction a distance d_6, as shown in the figure. The last three joints, which orient the tool plate, and hence the end effector, can be thought to have frames whose origins are coincident (this concept is developed in Section 8.7).

Note that the structure of the manipulator is such that offsets exist between the links. Specifically, the axis along the rotation of the first joint does not intersect the z axis corresponding to the linear motion of the prismatic joint (joint 3). This offset manifests itself in a nonzero d_2 term between frames 1 and 2. It should be noted that these offsets increase the computational complexity in both the forward and inverse solutions. However, they may be unavoidable because of certain design constraints incurred when developing a real-world manipulator.

Table 8.6.6 shows the link parameters for the manipulator, while Table 8.6.7 shows the symbolic A matrices of the form of Eq. (8.6.2) evaluated for the known values of the link parameters. The joint variables are θ_1, θ_2, d_3, θ_4, θ_5, and θ_6. Note that these appear as variables in the appropriate A

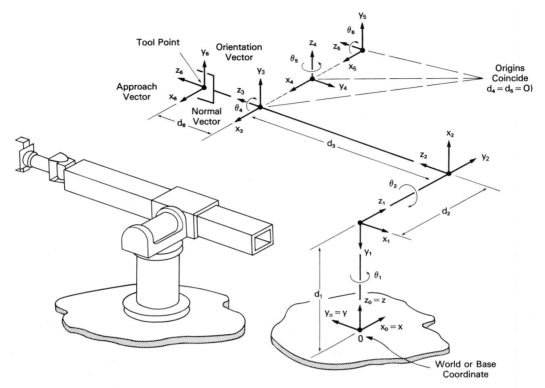

Figure 8.6.7.   Coordinate frames for the Stanford arm. (Reproduced with the permission of SRI International, Menlo Park, CA.)

matrix.   It is important to understand that the parameters d_1, d_2, and d_6 are known lengths of the kinematic configuration and hence not joint variables.

The configuration of this manipulator is such that "consecutive" joint axes intersect or are collinear.   Recall from Section 8.6.2 that this implies that x_1 could have been chosen in the opposite direction with equal validity.

TABLE 8.6.6 LINK PARAMETERS FOR THE STANFORD ARM

Joint	θ_i	α_i	a_i	d_i
1	-90	-90	0	d_1
2	-90	90	0	d_2
3	NA	0	0	d_3
4	0	-90	0	0
5	0	90	0	0
6	0	0	0	d_6

NA, not applicable.

TABLE 8.6.7 SYMBOLIC A
MATRICES FOR THE STANFORD
MANIPULATOR OF FIGURE 8.6.7

$$A_{01} = \begin{bmatrix} \cos\theta_i & 0 & -\sin\theta_1 & 0 \\ \sin\theta_1 & 0 & \cos\theta_1 & 0 \\ 0 & -1 & 0 & d_1 \\ 0 & 0 & 0 & 1 \end{bmatrix}$$

$$A_{12} = \begin{bmatrix} \cos\theta_2 & 0 & \sin\theta_2 & 0 \\ \sin\theta_2 & 0 & -\cos\theta_2 & 0 \\ 0 & 1 & 0 & d_2 \\ 0 & 0 & 0 & 1 \end{bmatrix}$$

$$A_{23} = \begin{bmatrix} 0 & 1 & 0 & 0 \\ -1 & 0 & 0 & 0 \\ 0 & 0 & 1 & d_3 \\ 0 & 0 & 0 & 1 \end{bmatrix}$$

$$A_{34} = \begin{bmatrix} \cos\theta_4 & 0 & -\sin\theta_4 & 0 \\ \sin\theta_4 & 0 & \cos\theta_4 & 0 \\ 0 & -1 & 0 & 0 \\ 0 & 0 & 0 & 1 \end{bmatrix}$$

$$A_{45} = \begin{bmatrix} \cos\theta_5 & 0 & \sin\theta_5 & 0 \\ \sin\theta_5 & 0 & -\cos\theta_5 & 0 \\ 0 & 1 & 0 & 0 \\ 0 & 0 & 0 & 1 \end{bmatrix}$$

$$A_{56} = \begin{bmatrix} \cos\theta_6 & -\sin\theta_6 & 0 & 0 \\ \sin\theta_6 & \cos\theta_6 & 0 & 0 \\ 0 & 0 & 1 & d_6 \\ 0 & 0 & 0 & 1 \end{bmatrix}$$

Reproduced with the permission of SRI
International, Menlo Park, CA.

Referring to Table 8.6.6, the change of direction for x_1 would mean that y_1's direction must be adjusted to form a right-handed system. In addition, the values of θ_2 and α_1 both need to be redefined as $+90°$ instead of $-90°$, as is shown.

EXAMPLE 8.6.2: THE SIX-JOINT PUMA

The Puma robot shown in Figure 8.6.8 is a jointed spherical manipulator having six revolute joints. The values of the link parameters are shown in Table 8.6.8. Table 8.6.9 shows the symbolic A matrices derived from these

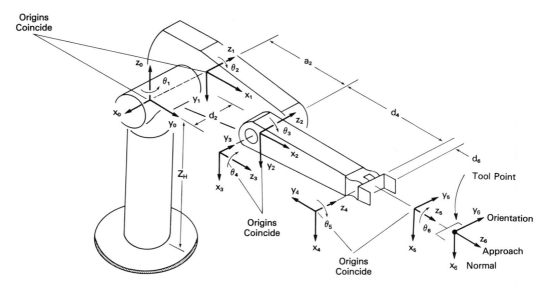

Figure 8.6.8. Coordinate frames for the six-joint PUMA. (Reprinted courtesy of the Society of Manufacturing Engineers. Copyright © 1983, from the ROBOTS 7/13th ISIR Conference Proceedings.)

link parameters. The A matrices and directions of axes are as presented in [14].

The reader should note that the base or world reference frame (i.e., the one fixed to link 0) was chosen above the mounting flange. An equally valid choice is any other location along the axis of rotation of the first joint. The d_1 term is zero since the z_0 and x_1 intersect at a common origin. The value of a_2 is nonzero since the axes of joints 2 and 3 do not intersect. Once

TABLE 8.6.8 LINK PARAMETERS
FOR THE SIX-JOINT PUMA

Joint	θ_i	α_i	a_i	d_i
1	90	-90	0	0
2	0	0	a_2	d_2
3	90	90	0	0
4	0	-90	0	d_4
5	0	90	0	0
6	0	0	0	d_6

TABLE 8.6.9 SYMBOLIC A MATRICES FOR THE SIX-JOINT PUMA OF FIGURE 8.6.8

$$A_{01} = \begin{bmatrix} \cos\theta_1 & 0 & -\sin\theta_1 & 0 \\ \sin\theta_1 & 0 & \cos\theta_1 & 0 \\ 0 & -1 & 0 & 0 \\ 0 & 0 & 0 & 1 \end{bmatrix}$$

$$A_{12} = \begin{bmatrix} \cos\theta_2 & -\sin\theta_2 & 0 & a_2\cos\theta_2 \\ \sin\theta_2 & \cos\theta_2 & 0 & a_2\sin\theta_2 \\ 0 & 0 & 1 & d_2 \\ 0 & 0 & 0 & 1 \end{bmatrix}$$

$$A_{23} = \begin{bmatrix} \cos\theta_3 & 0 & \sin\theta_3 & 0 \\ \sin\theta_3 & 0 & -\cos\theta_3 & 0 \\ 0 & 1 & 0 & 0 \\ 0 & 0 & 0 & 1 \end{bmatrix}$$

$$A_{34} = \begin{bmatrix} \cos\theta_4 & 0 & -\sin\theta_4 & 0 \\ \sin\theta_4 & 0 & \cos\theta_4 & 0 \\ 0 & -1 & 0 & d_4 \\ 0 & 0 & 0 & 1 \end{bmatrix}$$

$$A_{45} = \begin{bmatrix} \cos\theta_5 & 0 & \sin\theta_5 & 0 \\ \sin\theta_5 & 0 & -\cos\theta_5 & 0 \\ 0 & 1 & 0 & 0 \\ 0 & 0 & 0 & 1 \end{bmatrix}$$

$$A_{56} = \begin{bmatrix} \cos\theta_6 & -\sin\theta_6 & 0 & 0 \\ \sin\theta_6 & \cos\theta_6 & 0 & 0 \\ 0 & 0 & 1 & d_6 \\ 0 & 0 & 0 & 1 \end{bmatrix}$$

again observe that other locations of frames will generate equally valid solutions.

8.7 THE INVERSE OR BACK SOLUTION

The *back*, *reverse*, or *inverse* solution maps the position of the manipulator, represented by a hand matrix [see Eqs. (8.3.1) and (8.5.1)] into the corresponding joint vector [Eq. (8.1.1)]. Recall that the joint vector contains the values of the rotary joints' angles and the lengths of the prismatic joints. Since the location and orientation of the robot's tool point are defined by the angles (or lengths) of the joints, in order to command the gripper to achieve some position defined by a hand matrix a corresponding set of joint variables must be known.

Sections 8.5.2 and 8.5.3 illustrated that we can make modifications to the position of the manipulator using matrix operators. For example, if the hand matrix were premultiplied by some transformation (for the purpose of rotating the base of the manipulator), or if a tool transform postmultiplied the hand matrix (corresponding to the linear motion of the tool along one of its axes), the joint variables needed to position the manipulator so as to correspond to the new tool point and orientation need to be determined. The inverse solution will compute the appropriate joint variables needed to position the arm as defined by the new hand matrix.

The back solution algorithm is generally more difficult than the forward solution and may require insight into the geometry of the manipulator. Additionally, the equations generated are nonlinear and may not possess obvious solutions. Typically, the more complex the kinematic chain, the more complex the reverse solution. Some of the "problems" with the reverse solution for a particular numeric hand matrix are as follows:

1. No solution may be possible, due to limitations inherent in the manipulator (i.e., the rotation of a particular joint is limited).
2. Multiple solutions are possible since the manipulator can be positioned at the same "point" by more than one joint vector.
3. The complexity of the solution may make real-time computation difficult.

We now discuss these difficulties.

8.7.1 Problems of Obtaining an Inverse Solution

In the real world, the differences between the configurations of manipulators may limit the possible orientations that the tool tip may achieve. Therefore, positions explicitly demonstrated on one type of manipulator may not be achievable by a manipulator with a different kinematic configuration. In addition, the degrees of freedom greatly limit the "dexterity" of a manipulator. To illustrate this, consider locking θ_4 of the Stanford manipulator (Figure 8.6.7). This change will reduce the number of degrees of freedom from six to five and limits the possible orientations that the coordinate axes attached to the tool point may take. Although five degrees of freedom may be sufficient for certain tasks, the dexterity of the arm is now limited (i.e., it cannot achieve the same number of positions and orientations in space that the six-degree-of-freedom configuration can).

Physical limitations dictated by the actual implementation of the kinematic chain may also reduce the number of positions and orientations possible. For example, consider attempting to have the approach vector of the Stanford arm pointing toward the origin of the base reference frame and parallel to the prismatic joint.   Since the ranges of motion of the various joints are determined by the particular implementation of the mechanics, it is impossible in a configuration such as the Stanford arm to have joint 5 with a range of 360°.   These types of positioning problems will be true of all real manipulators.   Regardless of the configuration, the physical implementation will dictate the maximum ranges of motions of the joints and therefore limit the possible positions and orientations of the manipulator.

Another real-world constraint is based on the resolution, repeatability, and accuracy of the joints.   Chapter 3 deals with these definitions in detail.   However, since in most typical control implementations where the working range of the joints is divided into discrete positions, the coarseness or resolution of the control system may not allow the manipulator to be positioned at exactly the desired position as would be determined by calculation.   The magnitude of this positioning error is defined by the absolute accuracy of a given manipulator.

Based on the previous discussions, it should be apparent that transformations applied to the matrix representing the position of a manipulator may result in new hand matrices that define positions and orientations impossible to attain.   In designing the motion and position control software for a robot, it may be required to perform the reverse solution for each of the points the arm will be commanded to go to, or even along the entire set of points defining a desired trajectory of some particular motion (prior to moving the manipulator) to determine if the joint values are within their acceptable ranges.   Heuristic approaches may be needed to ensure valid solutions or to force some predefined constraint in the event that an impossible solution occurs.

In general, the reverse solution will always generate valid joint vectors if the hand matrices were computed from a valid joint vector: that is, the manipulator was physically moved to some position in space, its joint variables were obtained, and then the forward solution was applied.   However, there is no guarantee that matrix operations performed on a hand matrix defined by a valid joint vector will produce a hand matrix that has a back solution with a valid joint vector.

The problem of multiple solutions for a particular "point" stems from trigonometric solutions that can yield two possible sets of angles describing a particular "point."   Referring again to the Stanford arm in Figure 8.6.7, the same point (tool position and orientation) shown could be achieved by rotating joints 4 and 6 each 180° in opposite directions.

Another illustration of multiple joint vectors that result in the same position and orientation of the tool would be to move joint 2 an angular distance of 180° and then swing the arm via joint 1 so that the same tool point is achieved.   Any other joints could then be modified to attain the correct position and orientation. In fact, this last manipulation introduces the idea of "right" and "left"-handed ways of approaching a point.   The configuration of Figure 8.6.7 could be considered "right-handed" since joint 2 corresponds to a "shoulder."   Rotating joint 2, 180° followed by a joint 1 rotation of 180° would correspond to a "left"-handed configuration.   Other ideas, such as having the "elbow" of a manipulator (such as joint 3 of Figure 8.6.8) "up" or "down," can also be conceived.   These preferred configurations can be used to resolve the joint vector ambiguity heuristically when multiple solutions are possible and force the manipulator to stay in a particular configuration to ensure smooth motion.

The complexity and speed of computation requires a trade-off study when the reverse solution is being implemented. Items such as update rate and bandwidth requirements come into consideration (refer to Chapter 4). Compromises such as reducing the frequency content of the velocity profile at which the manipulator moves, or performing some type of interpolation in lieu of computing the reverse solution as often, may have to be considered and compared against cost and performance considerations in the actual implementation of the control system. Another possible compromise is to keep the sum of certain joint angles constant during a motion. Thus knowing the constant and assuming that one of the angles is found by a reverse solution algorithm, the other angle may be found by simple subtraction.

8.7.2 Techniques for Obtaining the Inverse Solution

There are numerous techniques and approaches to finding the inverse solution. The common attribute is that their difficulty increases with the complexity of the kinematic chain. The approaches may be applied individually or collectively to obtain the information required to specify the joint variables in terms of the entries of a hand matrix. Ultimately, we want an algorithm that works for any numeric hand matrix described in terms of a set of variables as given by Eq. (8.3.1).

For this discussion we assume a six-degree-of-freedom manipulator configured so that the last three joints primarily control the wrist, and therefore the orientation of the gripper, while the first three joints are used to position the link to which the wrist is attached. Recall the first three joints and links are called the *major linkages*, while the last three joints are called the *minor linkages*. A description of four techniques for finding the back solution follows.

8.7.2.1 Direct approach

This method obtains the equations from the symbolic hand matrix. To use the method, one equates a T matrix containing *variables* representing the known elements of the hand matrix to the symbolic T matrix [Eq. (8.6.3)], as shown in Eq. (8.7.1).

$$\begin{bmatrix} nx & o_x & a_x & p_x \\ ny & o_y & a_y & p_y \\ nz & o_z & a_z & p_z \\ 0 & 0 & 0 & 1 \end{bmatrix} = \begin{bmatrix} \text{functions} \\ \text{of the joint} \\ \underline{\text{variables}} \\ 0 \quad 0 \quad 0 \quad 1 \end{bmatrix} \qquad (8.7.1)$$

It is possible, by equating corresponding elements in the two matrices, to solve for the joint angles or lengths (θ_1 through θ_n) as a function of the variables (n_x, n_y, n_z, o_x, o_y, o_z, a_x, a_y, a_z) representing the entries in the hand matrix.

Although this is the most obvious approach, a careful examination of a typical *T* matrix will show that this approach can generate a set of up to 12 equations which may be highly nonlinear. However, this method should not be dismissed for simple cases or for obtaining some of the joint variables.

8.7.2.2 Geometric approach

This technique determines the joint variables by making use of the geometric relationship between the various links and joints of a manipulator. Depending on the points of reference chosen, the solutions for the joint variables may generate results that have to be re-referenced (usually, by adding an offset angle or distance) so that they correspond to the joint variables which are used to compute the hand matrix.

To simplify the problem, the kinematic chain of the manipulator is separated into two parts. This is possible for most industrial robots. The first part (consisting of the first three joints or major linkages) positions a point on the manipulator (and ultimately its tool point), while the second part (consisting of the last three joints or minor linkages) determines the tool's orientation. It is important to mention that there may be interaction between the major and minor linkages when positioning the tool point. In general, the first three joints will grossly position the tool point, while the last three joints will move its position (in a finer manner) and orient the tool coordinate frame.

Figure 8.7.1 shows the vector construction used to separate the kinematic chain into the two parts. Vector **P** terminates at the tool point, vector **R** points to the origin of the frame attached to the last link of the major linkage, and vector **S** (coincident with the approach vector) is directed from the tip of vector **R** to the tip of vector **P**.

As an aid in visualizing why the vector construction of Figure 8.7.1 is possible, the following model may prove helpful. If the last three joints are all rotary and their joint axes intersect at a common point, we may take advantage of reducing the last three joints to a compound joint. It can be shown [5], [6] that the last three joints and links, which normally provide a roll–sweep–roll action in positioning the last frame, may be combined and modeled as a ball-and-socket joint attached to a link of length d_6, which terminates in the gripper. Figure 8.7.2 [5] shows a six-joint manipulator and a representation which uses a ball-and-socket joint in place of the minor linkage. A ball-and-socket joint has three degrees of freedom, representing the joints of the wrist it is used to model. The frames attached to the joints forming the wrist are considered to be coincident and located in the center of the rotating ball.

Referring again to Figure 8.7.1, it is an easy task to evaluate the coordinates of the three vectors (**P**, **R**, and **S**). The coordinates of vector **P** are known from the hand matrix, since the coordinates of the tool point are the fourth column of the hand matrix. Note that **S** is directed along the approach vector, whose direction is completely defined by the third column of the hand matrix and whose magnitude (or length) is defined by the distance from the origin of link 3 to the tool point (a

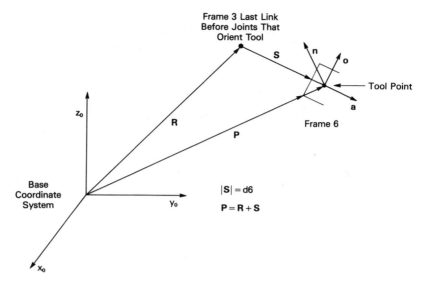

Figure 8.7.1. Geometric approach for the inverse solution using vector relationships.

known parameter of the kinematic chain). Knowledge of both **P** and **S** (obtained directly from the hand matrix) permit calculation of **R** by subtraction. This vector represents the location from the base reference frame to the last link before the joints and links that orient the tool.

Once vector **R** is determined, the values of the first three joint variables may be found using known parameters of the manipulator and geometric techniques. Reference [7] tabulates all the geometric solutions for permutations of rotary and prismatic joints in various configurations for the major linkages. The same approach can be used to solve for the last three joint variables with respect to the frame at the end of link 3 (see Figure 8.7.1).

Once a geometric solution for the arm has been obtained, it should be tested against the forward solution to ensure that the joint variables are properly referenced.

Reference [15] provides a detailed geometric approach to finding the inverse solution of a PUMA robot. It is noteworthy to mention that this approach makes use of arm configuration data [i.e., the preferred orientation of the arm (left or right), the elbow (above or below the arm), and the wrist (up or down)] to resolve multiple solutions.

8.7.2.3 Geometric approach with coordinate transformation

A modification of the geometric approach involves separating the kinematic chain as described previously and then using the A matrices of the forward solution.

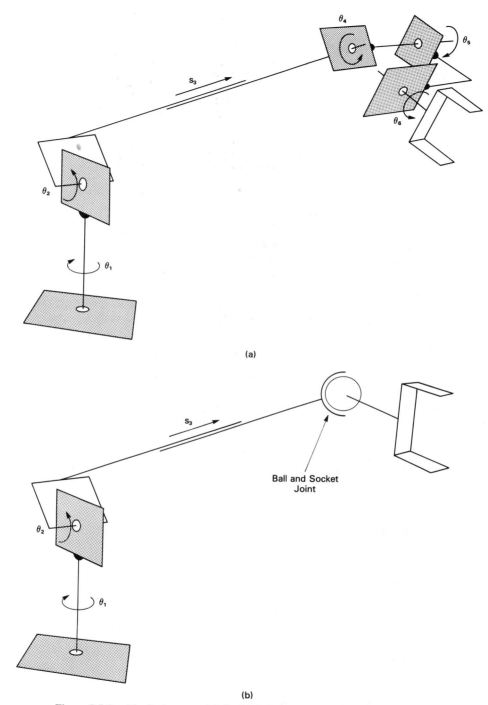

(a)

(b)

Figure 8.7.2. Manipulator modeled with ball and socket joint: (a) six-joint manipulator R-R-P-R-R-R; (b) equivalent four-joint manipulator R-R-P ball. (Reproduced with the permission of SRI International, Menlo Park, CA.)

Vector **R** defines numeric values for the last column of the D-H matrix A_{03}; the three resulting equations may be solved for the first three joint variables. These variables may then be substituted into the first three columns of A_{03} to define the position and orientation of the frame attached to link 3. After a numerical value for A_{03} is obtained, using the chain rule and the value of the hand matrix, the value of A_{36} can be found using the relationship given in Eq. (8.7.2).

$$T_{\text{numeric}} = A_{06_{\text{numeric}}} = A_{03_{\text{numeric}}} A_{36_{\text{numeric}}} \qquad (8.7.2)$$

By equating the numeric value of A_{36} to its symbolic representation, we may obtain equations for the last three joint variables.

An alternative approach for determining the joint variables of the minor linkages is to use vector techniques. The vectors (normal, orientation, and approach) of the hand matrix and A_{03} numeric matrix may be used to determine the angles of the joints corresponding to the minor linkages. The angles between axes corresponding to the coordinate system attached to the end of the major linkages and the approach, normal, and orientation vectors associated with the gripper (or link 6) are used to determine the variables for the last three joints.

8.7.2.4 Manipulations of symbolic *T* and *A* matrices

This method [1] makes use of the symbolic T matrices, as shown in Table 8.6.5. It is the most general technique in that *no geometric insight* is needed and it uses the forward solution as the basis of all back-solution algorithms. The procedure for using this approach is defined as follows:

1. Equate a T matrix containing variables for all of its entries to the symbolic T matrix [Eq. (8.6.3)], which is the product of the A matrices relating successive frames to one another, i.e.,

$$T_{\text{variable}} = T_{\text{symbolic}} = A_{06} \qquad (8.7.3)$$

2. Look for equations (which are obtained by equating elements of the two matrices) that are functions of only one joint variable. Use any information present to evaluate expressions or make substitutions for variables. Typical expressions obtained may contain transcendental functions of only one joint variable in terms of variables associated with hand matrix entries.

3. To reduce the complexity of the terms on the right-hand side of Eq. (8.7.3), premultiply the *variable T* matrix by the inverse of the appropriate A matrices and equate to the corresponding $^n T$ matrix. For example, premultiplying by the inverse of A_{01} gives

$$A_{10}T_{\text{variable}} = {}^1T = A_{12}A_{23}A_{34}A_{45}A_{56} \qquad (8.7.4)$$

At this point the left side of Eq. (8.7.4) contains the variables corresponding to the *variable T* matrix and functions (usually transcendental) of the first joint variable. If the first joint variable was previously defined, the left-hand side of Eq. (8.7.4) is known. The right-hand side contains equations which are functions of the remaining joint variables (θ_2, θ_3, θ_4, θ_5, and θ_6). By equating corresponding terms, it may be possible to find equations that have constants on the right side or may be expressed in terms of known quantities.

Once a solution is obtained for a joint variable, it may now be used to solve for the others. For example, assuming that we found the first joint variable, we now look for terms on the right-hand side of Eq. (8.7.4) that are functions of another joint variable. Typically, we are interested only in those expressions which are functions of a single joint variable and possibly some constants. Since the left-hand side is fully defined, an expression for the joint variable on the right can be expressed in terms of those quantities. Once all the information in this first equation is exhausted, the equation's complexity is again reduced, as described in the step below.

4. The "unknown variable elimination" process is carried on by premultiplying both sides of Eq. (8.7.4) by the inverse of A_{12}, yielding

$$A_{21}A_{10}T_{\text{variable}} = {}^2T = A_{23}A_{34}A_{45}A_{56} \qquad (8.7.5)$$

Since previous manipulations determined the value of joint variable θ_1, the left-hand side is now only a function of joint variable θ_2, while the right-hand side of Eq. (8.7.5) contains the remaining unknown joint variables. The process of equating corresponding entries is repeated to try to eliminate the second joint variable.

5. The process is continued as far as necessary using information from each step to simplify as many entries as possible until all unknown joint variables are found.

It should be noted that the equations may not simplify as easily as one would hope, and the method above may generate intermediate equations such as Eq. (8.7.5), which yield no simplifying information. If this occurs, it is necessary to premultiply by the next corresponding inverse of the A matrix until simpler expressions are generated.

The techniques discussed are not foolproof. However, they do provide a starting point for determining the back solution. A combination of the geometric approach for the major linkages and the matrix technique (discussed last) seems to provide a good balance. It should be noted that once a back solution algorithm is developed, one would like to optimize it (with respect to minimum computational time or effort) if it is to be used in a real-time control application.

8.7.3 Some Comments on Solution Methods

Typically, the angles encountered in kinematic chains are not limited to one or two quadrants. When solving for an unknown joint variable it is necessary to

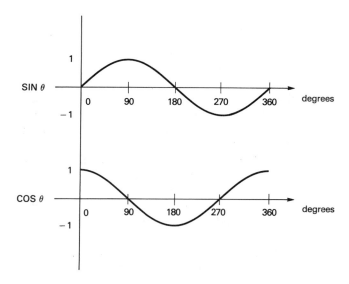

Figure 8.7.3. Sine and cosine curves.

define the *quadrant* of the angle correctly with respect to the positive x axis. Examination of the sine and cosine curves shown in Figure 8.7.3 reveals that over the interval from 0 to 360°, there is not a unique mapping of the independent variable to the dependent variable. Therefore, using the *arcsine* or *arccosine* function alone will not correctly define the angle. Similar problems also exist with the tangent function. The arctangent of (y/x) and $(-y/-x)$ produces the same angle since both ratios are positive. Similarly, the ratios $(-y/x)$ and $(y/-x)$ both generate the same angle.

One way to overcome this ambiguity is to use information from both the sine and cosine since the arccosine uniquely maps the range (1 to -1) into (0 to 180°) and the sine is always greater than or equal to zero over this range. The following algorithm will uniquely define the angle:

Algorithm that determines the correct quadrant of an angle given the sine and cosine of the angle

Given the values of $\sin \theta$ and $\cos \theta$

$$\theta_{\text{candidate}} = \text{arccosine}[\cos \theta]$$

IF $\sin \theta \leq 0$

$$\theta = 360 - \theta_{\text{candidate}}$$

ELSE

$$\theta = \theta_{\text{candidate}}.$$

The function ATAN2 usually available in most math packages performs this function, returning the values of the angle in the range $(-180 \le \theta \le 180)$. This function takes into account quadrant information defined by the signs of its arguments. ATAN2 is a double-argument function defined as

$$\theta = \text{ATAN2}\,(y, x) \tag{8.7.6}$$

$$\theta = \text{ATAN2}\,(\sin\theta, \cos\theta) \tag{8.7.7}$$

When obtaining expressions for an unknown joint variable it is usually easy to obtain expressions for both the sine and cosine of the unknown angle. These expressions can then be used in the ATAN2 functions to define the angle uniquely.

Another important point to remember is that care must be taken in defining solutions that contain division by transcendental terms that can be equal to zero or negative numbers under radicals.

8.7.4 Forward and Inverse Solutions of a Three-Axis Manipulator

The following example will illustrate the concepts just described without the need to resort to cumbersome mathematics. For solutions of some common and more complex manipulators, the reader is referred to [1], [2], [5], [7], [8], and [15].

EXAMPLE 8.7.1: FORWARD AND INVERSE SOLUTION OF A SIMPLE MANIPULATOR

Figure 8.7.4 shows the side view of a simple three-axis manipulator. The configuration is essentially that of a cylindrical coordinate robot. The addition of the third axis ensures that the orientation of the part at deposit may

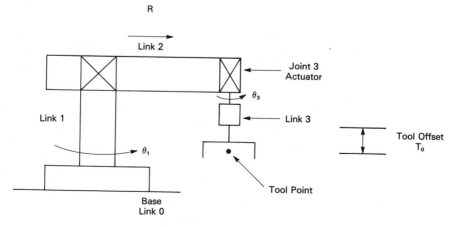

Figure 8.7.4. Three-axis manipulator.

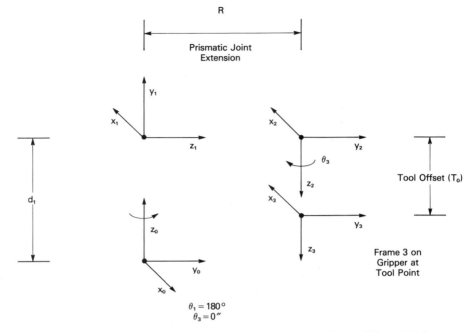

Figure 8.7.5. Coordinate frames for the three-axis manipulator of Figure 8.7.4.

be made the same as it was at pickup. With this configuration and the addition of an axis for vertical motion, the robot could pick up an object in its workspace and place it at another location with the same orientation that it originally had with respect to the base reference frame (see Chapter 9 for a detailed application). Note that this is being done in planes parallel to the x-y plane of the base reference axes.

Figure 8.7.5 shows the assignment of coordinate frames to each link. It is important to take note of the position of the angles and their direction of positive rotation. These are defined when the frames are attached to each link by the rules given in Section 8.6.2. For the prismatic joint of the manipulator shown in Figure 8.7.4, it is physically impossible for the link parameter, d, to be equal to zero. Referring to the figure, a value of $d = 0$ corresponds to the approach vector of the gripper being coincident with z_0. In reality, d takes on a minimum value corresponding to the fully retracted case and a maximum value corresponding to the fully extended case. The reader should recall that during calibration (see Chapter 5), the internal registers representing the value of the joint variables are set to appropriate values corresponding to some predefined and reproducible physical reference. In this particular case, the maximum or minimum value of d would be a logical choice for the prismatic axis. Table 8.7.1 defines the link parameters, while

TABLE 8.7.1 LINK
PARAMETERS FOR THREE-AXIS
MANIPULATOR

Link	θ_i	d_i	a_i	α_i
1	θ_1	d_1	0	90
2	0	R	0	90
3	θ_3	T_0	0	0

Table 8.7.2 defines the A matrices.   Multiplication of the A matrices generates the T matrices shown in Table 8.7.3.

At this point we will investigate the relationship of the joint variables of the forward solution to the base reference frame.

Figure 8.7.6 shows an overhead view of the manipulator base super-imposed on the x and y axes.   For the case shown, the joint variables are: $\theta_1 = 180°$ and $\theta_3 = 0°$.   The R axis and the gripper's y axis are parallel to the y axis of the base reference frame.   The corresponding T matrix evaluated at these joint variables is given by Eq. (8.7.9).

$$T = \begin{bmatrix} -1 & 0 & 0 & 0 \\ 0 & 1 & 0 & R \\ 0 & 0 & -1 & (d_1 - T_0) \\ 0 & 0 & 0 & 1 \end{bmatrix} \qquad (8.7.9)$$

Note that if the zero reference for the joint variables is not convenient, it can be set to any arbitrary reference as long as the value used in the forward

TABLE 8.7.2 *A* MATRICES FOR THREE-AXIS
MANIPULATOR

$$A_{01} = \begin{bmatrix} \cos\theta_1 & 0 & \sin\theta_1 & 0 \\ \sin\theta_1 & 0 & -\cos\theta_1 & 0 \\ 0 & 1 & 0 & d_1 \\ 0 & 0 & 0 & 1 \end{bmatrix}$$

$$A_{12} = \begin{bmatrix} 1 & 0 & 0 & 0 \\ 0 & 0 & -1 & 0 \\ 0 & 1 & 0 & R \\ 0 & 0 & 0 & 1 \end{bmatrix}$$

$$A_{23} = \begin{bmatrix} \cos\theta_3 & -\sin\theta_3 & 0 & 0 \\ \sin\theta_3 & \cos\theta_3 & 0 & 0 \\ 0 & 0 & 1 & T_0 \\ 0 & 0 & 0 & 1 \end{bmatrix}$$

TABLE 8.7.3 *T* MATRICES

$$
{}^{2}T = A_{23} = \begin{bmatrix} \cos\theta_3 & -\sin\theta_3 & 0 & 0 \\ \sin\theta_3 & \cos\theta_3 & 0 & 0 \\ 0 & 0 & 1 & T_0 \\ 0 & 0 & 0 & 1 \end{bmatrix}
$$

$$
{}^{1}T = A_{12}A_{23} = \begin{bmatrix} \cos\theta_3 & -\sin\theta_3 & 0 & 0 \\ 0 & 0 & -1 & -T_0 \\ \sin\theta_3 & \cos\theta_3 & 0 & R \\ 0 & 0 & 0 & 1 \end{bmatrix}
$$

$$
T = \begin{bmatrix} \cos\theta_1\cos\theta_3 + \sin\theta_1\sin\theta_3 & \sin\theta_1\cos\theta_3 - \cos\theta_1\sin\theta_3 & 0 & R\sin\theta_1 \\ \sin\theta_1\cos\theta_3 - \cos\theta_1\sin\theta_3 & -\sin\theta_1\sin\theta_3 - \cos\theta_1\cos\theta_3 & 0 & -R\cos\theta_1 \\ 0 & 0 & -1 & (d_1 - T_0) \\ 0 & 0 & 0 & 1 \end{bmatrix}
$$

$$(8.7.8)$$

solution is corrected to be consistent with the coordinate frames and link parameters chosen.

The back solution for this example will be obtained in two ways. The first corresponds to Section 8.7.2.1, the direct approach of equating a numeric *T* matrix to the symbolic *T* matrix [Eq. (8.7.8)]. The necessary equations for the back solution algorithm are obtained by manipulation of terms containing joint variables until they are simplified. The second approach utilizes the method of Section 8.7.2.4, manipulations of symbolic *T* and *A* matrices, to find the joint variables.

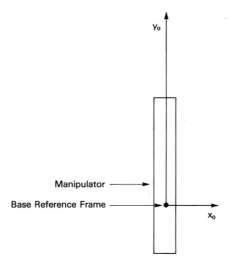

Figure 8.7.6. Base coordinate frame and manipulator of Figure 8.7.4.

Method 1: Back solution by direct approach.   Before attempting the back solution, the problem is simplified by reducing the general expression of the numeric T matrix so that its entries contain variables corresponding only to terms affected by the joint variables.   Since certain entries of the symbolic T matrix are constant, the numeric T matrix will be represented as

$$T_{\text{numeric}} = \begin{bmatrix} n_x & o_x & 0 & p_x \\ n_y & o_y & 0 & p_y \\ 0 & 0 & -1 & (d_1 - T_0) \\ 0 & 0 & 0 & 1 \end{bmatrix} \qquad (8.7.10)$$

Equating corresponding elements in the matrices given in Eqs. (8.7.8) and (8.7.10), we obtain six equations in three unknowns (R, θ_1, and θ_3). Since the normal and orientation vectors are by definition orthogonal to one another, we could eliminate two of the equations.   That is, given n_x and n_y we know o_x and o_y, and vice versa. This leaves us with four equations in three unknowns which are nonlinear functions of the joint variables.   From previous discussions recall that a solution of one of the equations in terms of only sines or cosines does not yield sufficient information to determine the angle uniquely.   Therefore, it is necessary to obtain an expression for both the sine and cosine of a particular rotary joint variable to determine its angle correctly.   The procedure used follows.

The equations corresponding to the p_x and p_y entries of Eqs. (8.7.10) and (8.7.8) are

$$p_x = R \sin \theta_1 \qquad (8.7.11)$$

$$p_y = -R \cos \theta_1 \qquad (8.7.12)$$

Noting that p_x is proportional to $\sin \theta_1$ by the factor R and that p_y is proportional to $\cos \theta_1$ by the factor $-R$, dividing both sides of each equation by the appropriate factor yields the sine or cosine of joint variable θ_1. θ_1 is determined by the double-argument arctangent function with p_x/R and $-p_y/R$ as its argument.   Since the factor R is removed when the tangent is defined, it may be dropped and the expression for θ_1 becomes

$$\theta_1 = \text{ATAN2} (p_x, -p_y) \qquad (8.7.13)$$

If it had been possible for joint variable R to take on the value zero, this approach would not have been applicable and an alternative would have been necessary. However, as discussed previously, this is physically impossible for this manipulator.

The value of R is readily obtained by squaring the expressions for p_x and p_y and taking the square root of the sum.

$$R = \sqrt{p_x^2 + p_y^2} \qquad (8.7.14)$$

Once we have θ_1 and R or expressions for them in terms of the values assigned to the numeric T matrix, the solution for θ_3 proceeds as follows.   Selecting the symbolic expressions corresponding to n_x and n_y, we obtain the following pair of equations:

$$n_x = \cos\theta_1 \cos\theta_3 + \sin\theta_1 \sin\theta_3 \tag{8.7.15}$$

$$n_y = \sin\theta_1 \cos\theta_3 - \cos\theta_1 \sin\theta_3 \tag{8.7.16}$$

Substituting the expressions for $\cos\theta_1$ and $\sin\theta_1$ yields the following two equations in two unknowns ($\sin\theta_3$ and $\cos\theta_3$) :

$$Rn_x = -p_y \cos\theta_3 + p_x \sin\theta_3 \tag{8.7.17}$$

$$Rn_y = p_x \cos\theta_3 + p_y \sin\theta_3 \tag{8.7.18}$$

Solving for the unknown transcendental functions of θ_3 gives

$$\cos\theta_3 = \frac{-p_y n_x}{R} + \frac{p_x n_y}{R} \tag{8.7.19}$$

$$\sin\theta_3 = \frac{p_x n_x}{R} + \frac{p_y n_y}{R} \tag{8.7.20}$$

The value of θ_3 is obtained by using the double-argument arctangent function.

$$\theta_3 = \text{ATAN2}\,(\sin\theta_3, \quad \cos\theta_3)$$
$$\theta_3 = \text{ATAN2}\,((p_x n_x + p_y n_y), \quad (p_x n_y - p_y n_x)) \tag{8.7.21}$$

Note that in Eq. (8.7.21) the variable R was dropped from the expression.

Thus given any numeric T matrix of the form of Eq. (8.7.10) and any known constants of the system, the joint variables θ_1, R, and θ_3 can be found by the algorithm consisting of Eqs. (8.7.13), (8.7.14), and (8.7.21).

Method 2: Solution by matrix manipulation method. To illustrate this approach, let us once again equate the symbolic T matrix [Eq. (8.7.8)] to the numeric T matrix. This yields the following matrix equation:

$$
\begin{bmatrix}
n_x & o_x & 0 & p_x \\
n_y & o_y & 0 & p_y \\
0 & 0 & -1 & (d_1 - T_0) \\
0 & 0 & 0 & 1
\end{bmatrix}
$$
$$
=
\begin{bmatrix}
\cos\theta_1\cos\theta_3 + \sin\theta_1\sin\theta_3 & \sin\theta_1\cos\theta_3 - \cos\theta_1\sin\theta_3 & 0 & R\sin\theta_1 \\
\sin\theta_1\cos\theta_3 - \cos\theta_1\sin\theta_3 & -\sin\theta_1\sin\theta_3 - \cos\theta_1\cos\theta_3 & 0 & -R\cos\theta_1 \\
0 & 0 & -1 & (d_1 - T_0) \\
0 & 0 & 0 & 1
\end{bmatrix}
\tag{8.7.22}
$$

We wish to reduce the complexity of this equation by essentially removing all references to a particular joint variable from the right side by moving it to the left side. Premultiplying both sides of Eq. (8.7.22) by A_{01}^{-1} (or A_{10}) will yield the new equation:

$$\begin{bmatrix} \cos \theta_1\, n_x + \sin \theta_1\, n_y & \cos \theta_1\, o_x + \sin \theta_1\, o_y & 0 & \cos \theta_1\, p_x + \sin \theta_1\, p_y \\ 0 & 0 & -1 & -T_0 \\ \sin \theta_1\, n_x - \cos \theta_1 n_y & \sin \theta_1\, o_x - \cos \theta_1\, o_y & 0 & \sin \theta_1\, p_x - \cos \theta_1\, p_y \\ 0 & 0 & 0 & 1 \end{bmatrix}$$

$$= {}^1T = \begin{bmatrix} \cos \theta_3 & -\sin \theta_3 & 0 & 0 \\ 0 & 0 & -1 & -T_0 \\ \sin \theta_3 & \cos \theta_3 & 0 & R \\ 0 & 0 & 0 & 1 \end{bmatrix}$$

$$(8.7.23)$$

The equation defined by entry (1, 4) is chosen since the expression generated has a zero on the right side and the left-hand side is in terms of only one joint variable, θ_1. This equation can be manipulated to obtain

$$\tan \theta_1 = \frac{\sin \theta_1}{\cos \theta_1} = \frac{-p_x}{p_y} \qquad (8.7.24)$$

Equation (8.7.24) and the quadrant information defined by the values of p_x and p_y in the numeric T matrix are sufficient to define completely variable θ_1 (this is the same information as yielded by ATAN2). Another possible approach for the solution of θ_1 is to use both the entries (1, 4) and (3, 4). Since R is nonzero, one may generate the matrix equation

$$\begin{bmatrix} \cos \theta_1 \\ \sin \theta_1 \end{bmatrix} = \frac{\begin{bmatrix} p_x & -p_y \\ p_y & p_x \end{bmatrix} \begin{bmatrix} 0 \\ R \end{bmatrix}}{(p_x^2 + p_y^2)} \qquad (8.7.25)$$

Equation (8.7.25) yields solutions for both the sine and cosine of θ_1 in terms of p_x, p_y, and R. As in the generation of Eq. (8.7.13), the same argument may be used to remove R from the equations and obtain θ_1.

With θ_1 defined, entry (3, 4) of Eq. (8.7.23) may be used to solve for R, yielding the expression

$$R = \sin \theta_1\, p_x - \cos \theta_1\, p_y \qquad (8.7.26)$$

The final joint variable, θ_3, is obtained by using entries (1, 1) and (3, 1) or (1, 2) and (3, 2) to define the sine and cosine of this angle.

The methods of Section 8.7.2.4 becomes quite helpful in the case of more complex symbolic T matrices. Its use in this simple example was intended to illustrate the procedure.

8.8 MOTION GENERATION

Once a manipulator has been described by both the forward and inverse solutions, the problem of defining its position in space in terms of the joint variables has been solved.   In Section 8.5 we discussed how a desired position of the manipulator in terms of its hand matrix could be computed by using transformations.   What was not discussed was how the motion from its current position to a desired endpoint was achieved.   This section shows how the intermediate points between the initial and final position of a manipulator may be computed.

8.8.1 Position Trajectories and Velocity Profiles

As discussed in Chapter 2, certain control over the motion of a manipulator is desired.   For instance, we might like to specify the maximum velocity and maximum acceleration for a particular displacement, or the maximum time for a movement. One method for describing this is using a velocity profile which shows the velocity of a particular motion as a function of time.   In one of its simpler forms, the velocity profile is trapezoidal in nature.   A trapezoidal profile shows a distinct acceleration, constant velocity, and deceleration period.   Various relationships can be developed for this waveform so that given three parameters (such as the total displacement, a maximum allowable acceleration, and a maximum allowable velocity, the remaining parameters describing the profile can be computed.   Figure C.2.2 in Appendix C shows a set of waveforms describing position, velocity, and acceleration as a function of time, while Eqs. (C.2.1) through (C.2.3) define the position as a function of the maximum desired velocity and the three time intervals used to describe the motion.   Based on these waveforms, given the desired displacement (area under the velocity curve), the desired maximum velocity, and the three time intervals for a motion, the position can be computed as a function of time.   Other polynomials may also be used to describe the velocity and position waveforms; however, for our discussion we will use the trapezoidal velocity profile due to its simplicity.

   To control the velocity of a single joint, all that is necessary is to feed the position versus time data as the setpoint of a position servo loop, and (if the servo is designed properly) the joint should "track" the waveform.   Note in this case that position information was explicitly sent to the joint but the velocity profile was implicitly sent (since velocity is the first derivative of position).

   This concept can be extended to joint interpolated motion (i.e., all joints start and stop at the same time) by simply forcing the total time for each motion to be the same while adjusting the maximum velocity for each joint based on a fixed motion time and the desired displacement of each joint.   The total time for the motion is obtained by looking at each motion individually.   Since each joint could conceivably have a different displacement, then based on a desired maximum velocity and acceleration, each joint should require a different time to achieve its motion.   By forcing all the joints to use the longest time, those axes whose motion times would have been shorter will require reduced velocities (as opposed to moving individually) to make the total area under their velocity profile the required displacement.

8.8.2 Displacement Profiles Run through an Inverse Solution

When using tool point control to position a manipulator, it is necessary to first generate a trajectory (position versus time based on a velocity profile description) in Cartesian space, and then use the inverse solution to obtain the joint variables as a function of time. The joint variable displacements as a function of time are the inputs to the servos. Assuming that all the servos can track these profiles, the joints will move with the correct trajectories so that the tool point moves as described in Cartesian coordinates. It should be apparent that if a trapezoidal velocity profile is specified for a motion, the profiles of the individual joints may look nothing like it since the inverse transform is nonlinear. This is also true for the other waveforms, i.e., position, acceleration, and jerk. To illustrate this, consider Figure 8.8.1. This shows a straight line drawn from the coordinate (x_o, y_o) to (x_f, y_f) in terms of a Cartesian reference frame.

Clearly, to accomplish this straight line motion with a Cartesian manipulator, all that is necessary is to produce a displacement profile for the x axis that goes from x_o to x_f and a displacement profile for the y axis that goes from y_o to y_f. As long as both profiles start and stop at the same time, the tip will trace out the line drawn in the figure.

Now let us assume that the Cartesian coordinate manipulator is replaced with a cylindrical one (θ-R manipulator). The task is still the same—draw the diagonal

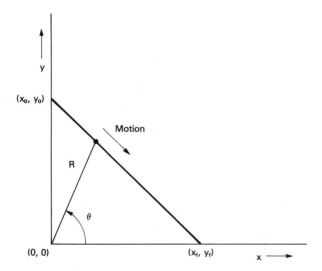

Figure 8.8.1. Diagonal line in Cartesian frame shown with θ-R manipulator.

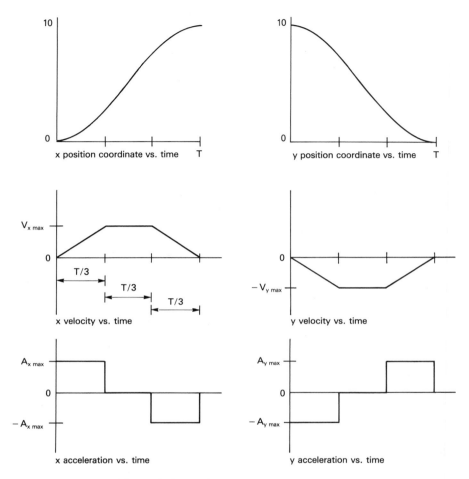

Figure 8.8.2. Position, velocity, and acceleration in Cartesian coordinates for motion from (0, 10) to (10, 0).

line from (x_o, y_o) to (x_f, y_f)—but in this case we need to generate θ versus time and R versus time.   To accomplish this we will use the relationships:

$$R(t) = \sqrt{x^2(t) + y^2(t)} \qquad\qquad (8.8.1)$$

$$\theta(t) = \text{ATAN2}\,[y(t), x(t)] \qquad\qquad (8.8.2)$$

Figure 8.8.2 shows $x(t)$ and $y(t)$ for a motion starting at (0,10) and ending at (10,0) with the corresponding velocity and acceleration waveforms while Figure 8.8.3 shows $R(t)$ and $\theta(t)$ with their corresponding velocity and acceleration waveforms. In the case of a six-axis manipulator, there will be six sets of position, velocity, and acceleration waveforms (corresponding to each joint).

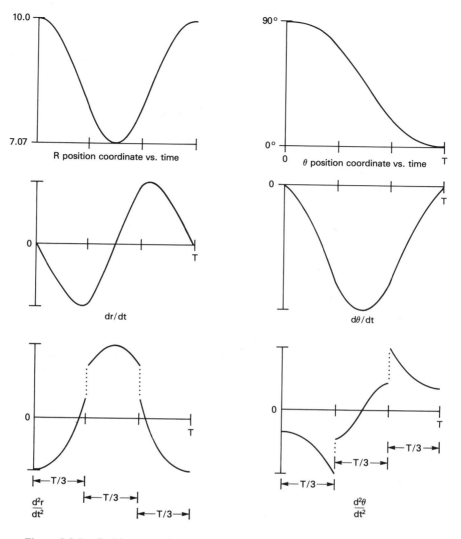

Figure 8.8.3. Position, velocity, and acceleration in cylindrical coordinates for motion from (0, 10) to (10, 0).

8.8.3 Cartesian Motions of a Manipulator

To cause the manipulator to follow a trajectory in Cartesian space, one must first obtain a set of displacement curves for the desired motion (starting and stopping at the same time) similar to the discussion in Section 8.8.2. Then depending on the type of motion, the appropriate entries of the hand matrix are modified, or the hand matrix is operated on by a transformation for each discrete-time increment describing the trajectory. An inverse solution is performed, and the joint variables fed to the servos.   This operation is repeated until all the displacement curves are satisfied.

The following describes the types of motion that can be achieved in Cartesian coordinates.

1. Motions parallel to world axes

To move the tool tip parallel to either of the three axes of the base reference frame, all that is necessary is to modify the appropriate position entry of the fourth column and perform the back solution.   Note that the orientation of the gripper will remain the same throughout the motion.   The update to the entry is the incremental change in position between the next and current updates, that is the difference between $x[(n+1)T]$ and $x(nT)$.

2. Motions parallel to tool axes

To cause the manipulator to move along a path parallel to one of the tool axes, all that is necessary is to postmultiply the hand matrix by the appropriate translation matrix Trans(x,0,0), Trans(0,y,0), or Trans(0,0,z) and perform the back solution.   Once again the orientation of the gripper will remain the same throughout the motion, and the value of x, y, or z in the translation matrix corresponds to the incremental change between updates.

3. Rotation about tool axes

In this case, postmultiplying the hand matrix by a rotation matrix will cause the tool point to remain essentially fixed in space and the gripper to be rotated about the specified axes.   In this situation the orientation of the gripper changes. A displacement curve for the angle versus time needs to be generated and the incremental changes used as the argument for the Rot operator.

4. Simultaneous motion of tool point and tool axes

It may be desirable to cause simultaneous motion of the tool point and the orientation of the tool axes.   This is the case if the endpoint of a motion needs an orientation of the tool axes different from the current orientation defined by the hand matrix.   To accomplish this the appropriate profiles are generated for moving the tool point, and a profile is generated that occurs for the same amount of time as the Cartesian motion but generates an angle corresponding to the desired rotation of a tool axis.   After updating the Cartesian position, a rotation matrix postmultiplies the hand matrix before the inverse solution is found and the servos are updated.   Once again, the entry in the rotation matrix should be the incremental change in rotation since the last update.

These particular examples show control of the tool tip. This is typically done since the end effector or gripper is attached to the end of the arm. It is also possible to control points other than the tool point (such as the origin and orientation of any frame used to describe the manipulator). Of course more complex motions are also possible, for instance motion along (1) an arbitrary vector, (2) an arbitrary curve, and (3) a straight line with gripper orientation changes.

8.8.4 A Continuous Path Algorithm

In certain applications it is desired to teach a series of points that the manipulator is to move through without stopping. This may correspond to an application in which a bead of glue or gasketing material is placed on the edge of a cabinet. There are many possible ways to accomplish this, including utilizing curve fitting algorithms in order to obtain a mathematical function of the curve. An alternative method presented in this section is simple yet effective and requires no complex curve fitting or time-consuming computations. Thus it is well suited for *real-time control* as implemented in robotic manipulators.

To begin our discussion, assume that we have three points in Cartesian space as shown in Figure 8.8.4. The starting point is (1,1), the intermediate point is (10,5) and the ending point is (5,10). Our objective is to trace a trajectory that comes as close as possible to the intermediate point without stopping the manipulator.

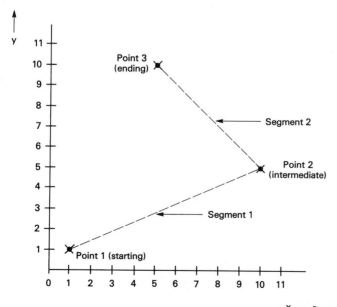

Figure 8.8.4. Model for continuous path algorithm.

Consider now what would be done in Cartesian space if we wanted two distinct motions—that is a motion from point 1 to point 2 and another from point 2 to point 3. For the first segment, the total displacements for both x and y would be computed, in this case a distance of 9 and 4 respectively. Second, a velocity profile would be chosen that meets some criteria with the stipulation that both axes start and stop at the same time. Finally the displacement functions $x(t)$ and $y(t)$ would be generated. If one were to plot $y(t)$ versus $x(t)$ the result would be a straight line from point 1 to point 2. Now let us assume that the functions are discretized such that we obtain $x(nT)$ and $y(nT)$. In this case the time increment between each point plotted would be the same while the spacial distance would change as a function of the velocity of the point. During slow motions, the points would be closer together while at higher speeds the points would be spaced further apart. The same concept holds for the second segment except that the displacements in x and y become -5 and 5. Note that the time for motion of segment 1 is not necessarily the same time as segment 2. This may be due to a maximum velocity limit or an arbitrary choice.

One method of preventing the motion from stopping as we go from segment

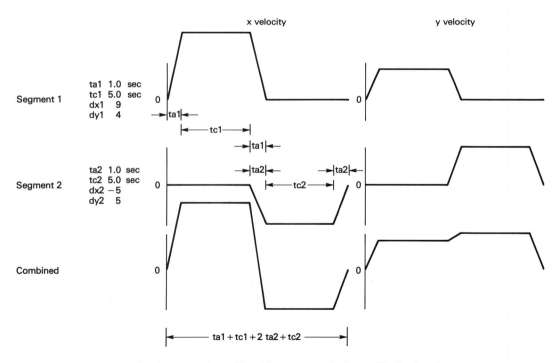

Figure 8.8.5. Continuous path profile with segment velocity profiles having the same acceleration, constant velocity, and deceleration times.

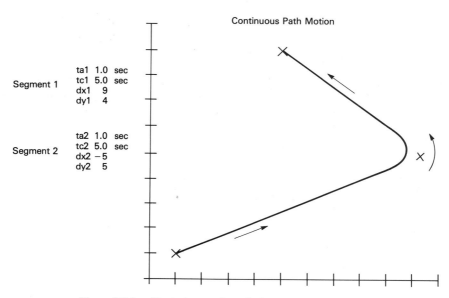

Continuous Path Motion

Segment 1   ta1 1.0 sec
             tc1 5.0 sec
             dx1   9
             dy1   4

Segment 2   ta2 1.0 sec
             tc2 5.0 sec
             dx2 − 5
             dy2   5

Figure 8.8.6.   *X-y* trajectory for velocity profiles of Figure 8.8.5.

1 to segment 2 is to add the two velocity profiles together. This is accomplished by starting the motion for segment 1. Therefore we obtain $x_1(nT)$ and $y_1(nT)$ and feed these positions as setpoints to the position servo of each axis in order to accomplish the motion. As soon as we reach the time when deceleration starts, we begin the motion of the second segment $x_2(nT)$ and $y_2(nT)$. During the time of deceleration for the first segment, the setpoints fed to the servos also include the motion for the second segment. Therefore as segment 1 is decelerating, segment 2 is accelerating, and the net effect is a continuation of motion. The following figures show the velocity profiles for the independent axis motion of segment 1 and segment 2 and the combined velocity profile for continuous motion from point 1 to point 2 and then to point 3. Also shown are *x-y* plane trajectories. As one can see from examining Figures 8.8.5 and 8.8.7, the combined velocity profiles may be obtained by adding the slopes of the individual velocity profiles for each segment.

At this point it should be quite obvious how this concept can be extended to a series of N-line segments by simply starting the $(n+1)$th segment's motion at the deceleration point of the nth segment.

The performance of this approach with respect to its ability to pass closely to the intermediate point is dependent on the velocity of each segment and the areas under the deceleration and acceleration curves of segment 1 and segment 2 respectively. In general the slower the motion, the nearer the trajectory comes to the intermediate point. Clearly, one can compensate for increases in speed by

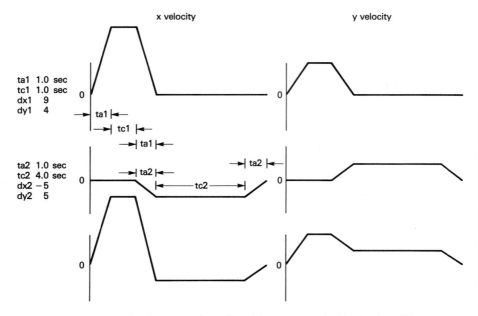

x velocity   y velocity

ta1 1.0 sec
tc1 1.0 sec
dx1 9
dy1 4

ta2 1.0 sec
tc2 4.0 sec
dx2 −5
dy2 5

Figure 8.8.7. Continuous path profile with segment velocities having different constant velocity times.

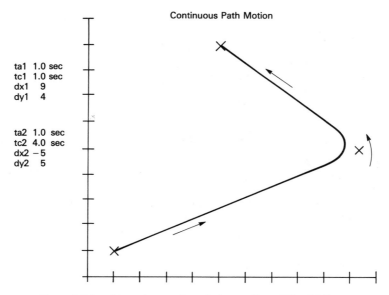

Continuous Path Motion

ta1 1.0 sec
tc1 1.0 sec
dx1 9
dy1 4

ta2 1.0 sec
tc2 4.0 sec
dx2 −5
dy2 5

Figure 8.8.8. *X-y* trajectory for velocity profiles of Fig. 8.8.7.

initially teaching a series of points and running the robot using this continuous path algorithm. If the position trajectory misses the points, they can be moved (inward or outward) until an acceptable performance is achieved.

8.9 THE JACOBIAN

We have previously described the forward and inverse solutions for a serial link manipulator whereby both the location and orientation of the tool point are defined in terms of joint variables or conversely, given the location and orientation of the tool point the corresponding joint variables may be computed. This particular approach when coupled with encoders and position servo loops can provide the means to control the tool point of a manipulator in space. Section 8.8 discussed how one could achieve motions with respect to both the tool point and base coordinates while Section 8.10 will discuss the architecture of such a control system.

Another method for controlling the joints of a manipulator in a coordinated fashion is *Jacobian control*. This control method is based on relating the rates of the variables in one coordinate system to those in a secondary coordinate system. For example, a convenient coordinate system for a manipulator is that defined by its joint vector, Eq. (8.1.1), while a convenient coordinate system for a human observing the robot could be the coordinate frame associated with the hand or gripper as shown in Figure 8.5.1.

The Jacobian relates the velocities of the joints to those of the tool point. Essentially, it allows the computation of a differential change in the tool coordinate frame due to a differential change in the position of the manipulator's joints. As one would expect, by making use of the inverse Jacobian, the differential changes in the manipulator's joints needed to accomplish a differential change in the tool's coordinate frame can also be computed.

If x_j is a coordinate of the tool point (location given by the x, y, z coordinates referenced to the manipulator's base and rotations about the x, y, or z axes associated with the hand or tool), then the generalized function relating each coordinate of the tool point to the joint variables, θ_n, can be written as:

$$x_j = f_j(\theta_1, \ldots \theta_n) \tag{8.9.1}$$

where the maximum value of j corresponds to the number of Cartesian coordinates (typically 6), and n corresponds to the number of joints of the particular manipulator.

Taking the total differential of Eq. (8.9.1) yields:

$$\dot{x}_j = \sum_{i=1}^{i=n} \frac{\delta[f_j(\theta_1, \ldots \theta_n)]}{\delta\theta_i} \dot{\theta}_i \tag{8.9.2}$$

$$\dot{x}_j = \sum_{i=1}^{i=n} J_{ji} \dot{\theta}_i \tag{8.9.3}$$

Assuming that both j and n equal 6, the coefficients Jji can be arranged in matrix form so that Eq. (8.9.3) can be written as:

$$\begin{bmatrix} \dot{x}_1 \\ \dot{x}_2 \\ \dot{x}_3 \\ \dot{x}_4 \\ \dot{x}_5 \\ \dot{x}_6 \end{bmatrix} = \begin{bmatrix} dx \\ dy \\ dz \\ \delta x \\ \delta y \\ \delta z \end{bmatrix} = \begin{bmatrix} \dfrac{\delta f_1}{\delta \theta_1} & \dfrac{\delta f_1}{\delta \theta_2} & \dfrac{\delta f_1}{\delta \theta_3} & \dfrac{\delta f_1}{\delta \theta_4} & \dfrac{\delta f_1}{\delta \theta_5} & \dfrac{\delta f_1}{\delta \theta_6} \\ \dfrac{\delta f_2}{\delta \theta_1} & \dfrac{\delta f_2}{\delta \theta_2} & \dfrac{\delta f_2}{\delta \theta_3} & \dfrac{\delta f_2}{\delta \theta_4} & \dfrac{\delta f_2}{\delta \theta_5} & \dfrac{\delta f_2}{\delta \theta_6} \\ \dfrac{\delta f_3}{\delta \theta_1} & \dfrac{\delta f_3}{\delta \theta_2} & \dfrac{\delta f_3}{\delta \theta_3} & \dfrac{\delta f_3}{\delta \theta_4} & \dfrac{\delta f_3}{\delta \theta_5} & \dfrac{\delta f_3}{\delta \theta_6} \\ \dfrac{\delta f_4}{\delta \theta_1} & \dfrac{\delta f_4}{\delta \theta_2} & \dfrac{\delta f_4}{\delta \theta_3} & \dfrac{\delta f_4}{\delta \theta_4} & \dfrac{\delta f_4}{\delta \theta_5} & \dfrac{\delta f_4}{\delta \theta_6} \\ \dfrac{\delta f_5}{\delta \theta_1} & \dfrac{\delta f_5}{\delta \theta_2} & \dfrac{\delta f_5}{\delta \theta_3} & \dfrac{\delta f_5}{\delta \theta_4} & \dfrac{\delta f_5}{\delta \theta_5} & \dfrac{\delta f_5}{\delta \theta_6} \\ \dfrac{\delta f_6}{\delta \theta_1} & \dfrac{\delta f_6}{\delta \theta_2} & \dfrac{\delta f_6}{\delta \theta_3} & \dfrac{\delta f_6}{\delta \theta_4} & \dfrac{\delta f_6}{\delta \theta_5} & \dfrac{\delta f_6}{\delta \theta_6} \end{bmatrix} \begin{bmatrix} \dot{\theta}_1 \\ \dot{\theta}_2 \\ \dot{\theta}_3 \\ \dot{\theta}_4 \\ \dot{\theta}_5 \\ \dot{\theta}_6 \end{bmatrix} \qquad (8.9.4)$$

The 6×6 coefficient matrix relating the hand velocities to the joint velocities is called the Jacobian. Note that depending on the degrees of freedom and the number of joints, the Jacobian matrix may not be square. That is, it may be an $n \times m$ matrix where n is the number of joints of the manipulator and m is the dimensionality of the Cartesian coordinates of the hand under consideration.

A careful inspection will show that the Jacobian is a time-varying quantity since it must be evaluated for instantaneous joint velocities and the actual position of the joints. It is clearly dependent on the kinematic structure of the arm and the coordinate system used to express the hand or tool coordinate frame.

If the matrix is invertible, then it is possible to calculate joint velocities given the velocities associated with the displacement and orientation of the tool point. Recall that only square matrices that are nonsingular may be inverted, and therefore it is quite important to determine whether the Jacobian or its inverse can be singular. In the case of nonsquare Jacobians, pseudo inverses may be used to obtain inverse relationships.

Singularities are points in the workspace where the velocity of a joint must become extremely large in order to maintain the proper relationship between itself and the other axes in order to perform the desired trajectory. Some possible solutions to the problem of manipulator singularities are: (1) staying away from the singularity points; (2) adding heuristic constraints when in the vicinity of the singularity to produce a constrained velocity; or (3) adding an additional, redundant degree of freedom.

By examining the Jacobian or its inverse, singular points in the workspace may be identified. The presence of entries that can make the determinant zero indicate the nonexistence of a corresponding inverse and can be used to define positions or areas in the workspace where the use of a Jacobian and its inverse are illegal. We should also examine conditions that have a tendency to make the determinants approach zero.   For these cases while the condition that forces the determinant to be zero may not be physically attainable by the manipulator, velocities of some joints may reach large values which may be beyond the capabilities of the control system.   Of course, heuristics may be substituted for the Jacobian and inverse Jacobian relationships near these areas.   Example 8.9.1 shows a condition where the determinant of a Jacobian approaches zero.

Logically, one may surmise that if the hand position is expressed in Cartesian coordinates then the Jacobian will be nonsingular since by design, a small motion of each joint will not cause large changes in the hand position coordinate.   Keep in mind that these incremental motions of the joints do not correspond to Cartesian trajectories of the tool tip. For the case of the inverse Jacobian, it should be evident that singularities may exist due to the nonlinear nature of the relationships.

EXAMPLE 8.9.1: THE JACOBIAN OF AN R-θ-Z MANIPULATOR

For an R-θ-Z manipulator similar to Figure 8.7.4 but with vertical motion in place of rotary motion at the last joint), the equations relating the robot's joint space to Cartesian space may be written as:

$$x = r \cos \theta$$
$$y = r \sin \theta \qquad (8.9.5)$$
$$z = z$$

The set of relationships given by Eq. (8.9.5) corresponds to Eq. (8.9.1).

Taking the total derivative yields a relationship similar to Eq. (8.9.2).

$$dx = dr \cos \theta - r \sin \theta \, d\theta$$
$$dy = dr \sin \theta + r \cos \theta \, d\theta \qquad (8.9.6)$$
$$dz = dz$$

Writing Eq. (8.9.6) in matrix form yields:

$$\begin{bmatrix} dx \\ dy \\ dz \end{bmatrix} = \begin{bmatrix} \cos \theta & -r \sin \theta & 0 \\ \sin \theta & r \cos \theta & 0 \\ 0 & 0 & 1 \end{bmatrix} \begin{bmatrix} dr \\ d\theta \\ dz \end{bmatrix} \qquad (8.9.7)$$

The inverse relationship is given by:

$$\begin{bmatrix} dr \\ d\theta \\ dz \end{bmatrix} = \begin{bmatrix} \cos \theta & \sin \theta & 0 \\ -\dfrac{\sin \theta}{r} & \dfrac{\cos \theta}{r} & 0 \\ 0 & 0 & 1 \end{bmatrix} \begin{bmatrix} dx \\ dy \\ dz \end{bmatrix} \qquad (8.9.8)$$

It is possible to rewrite the inverse Jacobian in Eq. (8.9.8) solely as functions of x, y, and z to facilitate computations and if we are working in Cartesian coordinates it makes more sense to use these variables. The result becomes:

$$\begin{bmatrix} dr \\ d\theta \\ dz \end{bmatrix} = \begin{bmatrix} x/r & y/r & 0 \\ -y/r^2 & x/r^2 & 0 \\ 0 & 0 & 1 \end{bmatrix} \begin{bmatrix} dx \\ dy \\ dz \end{bmatrix} \tag{8.9.9}$$

Investigating the determinant shows that J becomes singular if r equals zero. While mathematically possible, in a physical robot it would be quite difficult for r to become zero and, for all intents, taking the physical situation into account this Jacobian is nonsingular. However, near positions where r approaches zero, large velocities are required on the θ axis. For example, assume that we wish to trace a trajectory in an x-y plane along the line $x = a$ with a constant velocity in the y direction, V_y and of course a velocity in the x direction that is zero. Using the inverse Jacobian [Eq. (8.9.9)] we obtain the following relationships:

$$\dot{r} = y/r \, V_y$$

$$\dot{\theta} = x/r^2 \, V_y \tag{8.9.10}$$

$$\dot{z} = 0$$

Since x is a constant, one sees that $\dot{\theta}$ will be a maximum when r is at its minimum value or when $y = 0$. Thus the maximum value of $\dot{\theta}$ is found to be V_y/x which in our example is V_y/a. Now as the line moves closer to the y axis the magnitude of $x = a$ grows smaller and the velocity needed on the θ axis increases to a large value.

As a matter of completeness, the determinant of J^{-1} is given by:

$$\det J^{-1} = 1/r \tag{8.9.11}$$

which substantiates the previous results—that motion passing through the point corresponding to $r = 0$ requires an infinite velocity on the θ axis.

8.9.1 The Jacobian in Terms of D-H Matrices

To relate the position of the tool point of a manipulator (as defined in Cartesian coordinates) to the joint angles of the manipulator, one may make use of the forward solution and obtain a "hand" or "T" matrix. Recall, that while the T matrix provides a wealth of information, the orientation information, describing the three vectors comprising the coordinate frame attached to the tool plate, is given in terms of the basis set attached to link zero of the manipulator (or some fixed coordinate frame). As shown in Eq. (8.9.4), the Jacobian is defined with respect to incremental changes in position (dx, dy, dz) and rotations about the tool frame $(\delta x, \delta y, \delta z)$. Therefore we need a method of relating incremental changes in these variables to the hand matrix.

Reference [1] presents a method that relates differential changes in joint variables to the differential changes in the tool point's position and orientation. The following discussion uses the concepts of the approach in [1].

To begin our discussion, consider a hand matrix T and the result of a differential change in its entries, dT. The resultant hand matrix $(T + dT)$ may be computed using matrix operators as shown below.

$$T + dT = T \text{ Trans}(dx,dy,dz) \text{ Rot}(x,\delta x) \text{ Rot}(y,\delta y) \text{ Rot}(z,\delta z) \qquad (8.9.12)$$

Equation (8.9.12) shows translation and rotation operators postmultiplying a T matrix and therefore performing their operations about its established coordinate frame. In the case of differential rotations, if second order terms (such as $\delta x \delta y$) are neglected and all angles are considered small, then the following approximations may be used:

$$\sin \delta x \approx \delta x \qquad (8.9.13)$$

$$\cos \delta x \approx 1 \qquad (8.9.14)$$

When Eqs. (8.9.13) and (8.9.14) are used and second order terms are set to zero, rotation operators become commutative (and therefore the order of matrix multiplication becomes immaterial).

For small rotational changes, the product of the three rotation operators causing rotation about the x, y, and z axes (or normal, orientation, and approach vector directions) becomes:

$$\begin{bmatrix} 1 & -\delta z & \delta y & 0 \\ \delta z & 1 & -\delta x & 0 \\ -\delta y & \delta x & 1 & 0 \\ 0 & 0 & 0 & 1 \end{bmatrix} \qquad (8.9.15)$$

where δx, δy, and δz correspond to the differential angular change around the normal, orientation, and approach vectors associated with the hand coordinate system. Of course, these are expressed in radians.

If Eq. (8.9.15) is premultiplied by the differential Trans operator, we obtain:

$$\begin{bmatrix} 1 & -\delta z & \delta y & dx \\ \delta z & 1 & -\delta x & dy \\ -\delta y & \delta x & 1 & dz \\ 0 & 0 & 0 & 1 \end{bmatrix} \qquad (8.9.16)$$

The dx, dy, and dz entries correspond to incremental positional changes in the tool point with respect to the previous location of the tool point.

Utilizing Eq. (8.9.16) to solve Eq. (8.9.12) for dT results in:

$$dT = T \text{ (Trans Rot Rot Rot } - I) = T \Delta \tag{8.9.17}$$

where Δ is given by:

$$\Delta = \begin{bmatrix} 0 & -\delta z & \delta y & dx \\ \delta z & 0 & -\delta x & dy \\ -\delta y & \delta x & 0 & dz \\ 0 & 0 & 0 & 0 \end{bmatrix} \tag{8.9.17a}$$

At this point, one should recognize that we have the ability to compute the differential change in the T matrix, dT, if we have knowledge of differential changes in the position and orientation of the tool point. All that is needed is a way of relating differential changes in joint positions to differential changes in the tool position so that Eq. (8.9.17) may be computed.

Recall that by definition (see Section 8.6.3), the hand matrix is the product of A matrices that relate successive frames of a serial link manipulator together [Eq. (8.6.3)]. Furthermore, each A matrix has a single joint variable associated with it, and follows the rule that if it is rotary, the rotation is about the preceding frame's z axis, while if the joint is prismatic, the translation is along the preceding frame's z axis. Using this information, we may write the following equation for changes in the T matrix as a result of an incremental change of a single joint variable. To aid in visualizing what is happening, our derivation will concentrate on a kinematic chain with six joints, with the incremental changes occurring in joint 3. Therefore, we may write:

$$dT = T \Delta_3 \, dq_3 = A_{01} A_{12} \, {}^2\Delta_3 A_{34} A_{45} A_{56} \, dq_3 \tag{8.9.18}$$

Equation (8.9.18) says that dT, the change in the hand matrix, can be represented by postmultiplying the hand matrix by some unknown Δ_3 as given by Eq. (8.9.17a), or inserting a differential rotation (or translation) matrix, ${}^2\Delta_3$, in the appropriate location of Eq. (8.6.3). Remember, that while the matrix ${}^2\Delta_3$ is of the form of Eq. (8.9.16) the only valid entries are δz or dz.

Thus the change in the hand matrix resulting from a differential change in joint 3 is

$$dT/dq_3 = T \Delta_3 \tag{8.9.19}$$

where

$$\Delta_3 = T^{-1} [A_{01} A_{12} \, {}^2\Delta_3 A_{34} A_{45} A_{56}] \tag{8.9.20}$$

Since the T matrix is the product of the A matrices, Eq. (8.9.20) may be written as:

$$\Delta_3 = A_{65} A_{54} A_{43} A_{32} A_{21} A_{10} A_{01} A_{12} \, {}^2\Delta_3 A_{34} A_{45} A_{56} \tag{8.9.21}$$

which simplifies to

$$\Delta_3 = A_{65} \, A_{54} \, A_{43} \, {}^2\Delta_3 \, A_{34} \, A_{45} \, A_{56} \qquad (8.9.22)$$

Using the notation of Table 8.7.5, Eq. (8.9.22) may be written as:

$$\Delta_3 = {}^6T^3 \, {}^2\Delta_3 \, {}^3T^6 \qquad (8.9.23)$$

The result we have attained allows us to compute the differential change matrix which would postmultiply the T matrix to produce dT. Another way of looking at this result is that we have found the contribution to each possible nonzero entry of Δ_3 due to a differential change in joint 3. Since the changes which result from ${}^2\Delta_3$ are restricted to rotations or extensions about the z axis, we can relate these changes which correspond to a single joint variable to differential changes in position and rotation which occur in Δ_3.

In this way we have also found the third column of Eq. (8.9.4). The next step is to obtain the actual entries for this column. Our solution will take both translations and rotations into account. However, as noted previously, since only a translation or rotation may occur, terms corresponding to the nonmoving direction will be set to zero.

Starting with Eq. (8.9.23), we replace Δ_3 with the general incremental change matrix of Eq. (8.9.17a) and ${}^2\Delta_3$ with Eq. (8.9.17a) having set the terms $\{\delta x, \delta y, dx, dy\}$ equal to zero. Finally, we use the symbolic representation of a general transformation matrix as given in Table 8.3.3 for ${}^3T^6$ and the formula for the inverse of a transformation matrix in terms of the symbolic entries as given by Eq. (8.4.2) for ${}^6T^3$.

After performing the appropriate matrix multiplications and reducing terms such as:

$$-o_y \, a_x + o_x \, a_y = n_z \qquad (8.9.24)$$

by noting cross product relationships, we are left with the following entries in Δ_3 due to either a rotational or translational change in joint 3.

$$\begin{bmatrix} 0 & -a_z\delta_j & o_z\delta_j & d_j \, n_z + (n_x \, p_y + n_y \, p_y)\delta_j \\ a_z\delta_j & 0 & -n_x\delta_j & d_j \, o_z + (o_x \, p_y + o_y \, p_x)\delta_j \\ -o_z\delta_j & n_z\delta_j & 0 & d_j \, a_z + (a_x \, p_y + a_y \, p_x)\delta_j \\ 0 & 0 & 0 & 0 \end{bmatrix} \qquad (8.9.25)$$

The term d_j corresponds to the incremental change in length while δ_j is the rotary change. Since only one change is possible, either δ_j or d_j must be set to zero. The entries of Eq. (8.9.25) correspond to those of Eq. (8.9.17), i.e., $\{\delta x, \delta y, \delta z, dx, dy, dz\}$ for a change due to joint 3.

It should be apparent that equations similar to Eq. (8.9.18) can be written for each joint of the manipulator. Additionally, a relationship such as Eq. (8.9.25) can be generated in terms of the entries of equations similar to Eq. (8.9.23).

Specifically the columns of the Jacobian of Eq. (8.9.4) are determined as follows:

$$\text{Col } 1 = f(^0T) = f(T)$$

$$\text{Col } 2 = f(^1T)$$

$$\text{Col } 3 = f(^2T)$$

$$\text{Col } 4 = f(^3T)$$ (8.9.26)

$$\text{Col } 5 = f(^4T)$$

$$\text{Col } 6 = f(^5T)$$

The method of using the Jacobian then consists of forming Eq. (8.9.4) as discussed previously, evaluating the actual Jacobian for the current known joint positions, $\{\theta_1 \ldots \theta_6\}$ premultiplying the vector of differential joint changes, and obtaining the vector $\{dx, dy, dz, \delta x, \delta y, \delta z\}$. The results of the multiplication can then be placed in Eq. (8.9.17) and used to find dT.

8.9.2 Jacobians in Force Control

The transpose of the Jacobian relates the hand frame forces or torques to those at each of the joints.

$$[\Gamma] = J^{\text{tr}} [F]$$ (8.9.27)

where the column vector, Γ has entries corresponding to torques of forces depending on the type of joint and the column vector, F has entries of force or torque relating to the degrees of freedom chosen. For a six-degree-of-freedom situation, it consists of three forces corresponding to the x, y, and z directions and three torques (or moments) corresponding to rotations about the x, y, and z frames of the hand. Equation (8.9.27) may be derived using the method of virtual force [1].

If the transpose of the Jacobian is invertible, then it is possible to solve for the forces and torques applied by the hand with respect to the forces or torques applied by the joints.

$$[F] = (J^{\text{tr}})^{-1} [\Gamma]$$ (8.9.28)

By measuring the torque or force applied by each actuator and using some type of filtering technique to reduce noise (as is inherent in the measurement of the current flowing in a dc motor due to such factors as brushes or coupled noise), Eq. (8.9.28) can be used to determine the forces applied by the hand.

EXAMPLE 8.9.2: STATIC FORCE OF AN R-θ-Z MANIPULATOR

Example 8.9.1 presented the Jacobian and its inverse for a manipulator similar to Figure 8.8.4. Using Eq. (8.9.27) we obtain the following relationship:

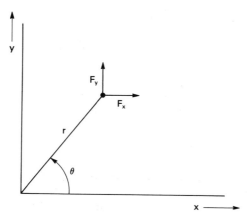

Figure 8.9.1. R-θ manipulator applying forces F_y and F_x.

$$\begin{bmatrix} F_r \\ T_\theta \\ F_z \end{bmatrix} = \begin{bmatrix} \cos\theta & \sin\theta & 0 \\ -r\sin\theta & r\cos\theta & 0 \\ 0 & 0 & 1 \end{bmatrix} \begin{bmatrix} F_x \\ F_y \\ F_z \end{bmatrix} \qquad (8.9.29)$$

Figure 8.9.1 shows a schematic in a fixed z plane. If one drew two force vectors as shown, then the torque on the θ axis and force on the r axis to generate these forces are given by Eq. (8.9.29).

For example, if one wished to generate a force of F_x as shown with F_y equal to zero, the torque to be supplied by the θ axis and force to be supplied by the r axis are given as:

$$F_r = \cos\theta\, F_x$$

$$T_\theta = -r\sin\theta\, F_x$$

As a numerical example, assume that $\theta = 45°$, and $r = 10.0$ in. The force that must be exerted in the outward direction of the r axis (by its actuator) is $0.707\, F_x$ while the torque that must be generated by the θ axis's actuator is $-7.07\, F_x$ (which is in the clockwise direction).

8.10 CONTROLLER ARCHITECTURE

In implementing motion control for a robotic manipulator various approaches may be chosen. We will discuss three concepts which result in defining setpoints to the joint servos.

- Joint position control (JPC)
- Resolved motion position control (RMPC)
- Resolved motion rate control (RMRC)

8.10.1 Joint Position Control

The architecture of a joint position control scheme is shown in Figure 8.10.1. This particular architecture is suitable for:

- Individual joint control
- Multiple joint motions stopping at various times
- Joint interpolated motion (point-to-point)
- Continuous path control of the joints

Note that each joint of the manipulator is controlled by a position servo loop. The joint trajectory control algorithm accepts a new joint setpoint, J_N, and trajectory parameters (such as velocity and acceleration), and based on system parameters and constraints computes a position profile for each joint of the manipulator.

The total displacement that each joint moves is its new setpoint minus its current position. It should be obvious that to move a single joint, the new setpoint

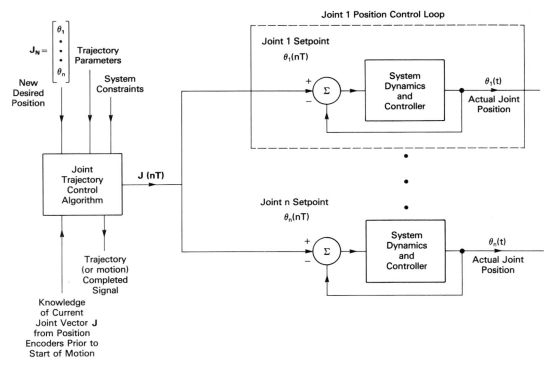

Figure 8.10.1. Joint position control.

J_N should contain the current position of all the joints except for the desired final position of the joint which is to move.

To implement the continuous path algorithm of Section 8.8.4, a command must be given to the trajectory control algorithm indicating that continuous path motion is desired along with a series of joint vectors. This is typically accomplished by having the master processor send the joint vectors along with a command string to the trajectory control algorithm. The command string specifies the trajectory parameters, the new joint vector, and whether the motion is point-to-point or continuous path.

It is interesting that this architecture is essentially open loop if one considers the actual joints' position and the joint trajectory control algorithm. Typically, when the joint trajectory algorithm has sent its last setpoint for a motion, it checks to see that the position error of each servo loop is within some specified tolerance. Once all of the joints are within their defined tolerance (a position error) the trajectory control algorithm signals the master that the trajectory is completed.

8.10.2 Resolved Motion Position Control

The architecture for resolved motion position control is shown in Figure 8.10.2. This topology is necessary to implement Cartesian control of the manipulator as discussed in Section 8.8.3. A comparison of Figure 8.10.2 with Figure 8.10.1 shows that the control of the joints is essentially the same. That is, a joint vector is sent to the position servos at every update. The primary difference between the two figures is the inclusion of the inverse solution. In Figure 8.10.2, the trajectory control algorithm works in Cartesian coordinates and then sends its position data to the back solution algorithm before updating the joint servos.

The input to the Cartesian trajectory control algorithm is a hand matrix defining the final position and orientation of the tool point after the motion. Depending on the type of motion (see Section 8.8.3) the appropriate intermediate hand matrices are generated and then passed to the inverse solution to get the joint vector at each update.

8.10.3 Resolved Motion Rate Control

The architecture of a controller using resolved motion rate control is shown in Figure 8.10.3. We have omitted the trajectory control algorithm in order to concentrate on the actual control signals sent to the joints and controller. However, note that the input, $T(nT)$, is essentially the output of a scheme similar to that of Figure 8.10.2.

The major difference between this control strategy versus joint position control and resolved motion position control is that the servos controlling the joints are velocity as opposed to position servos. The input to the system is a hand matrix, $T(nT)$, which may come from a trajectory control algorithm and is updated

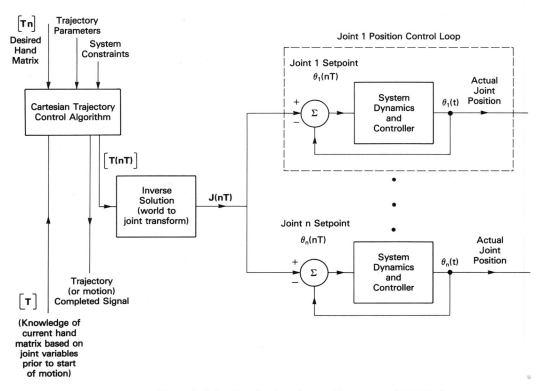

Figure 8.10.2. Resolved motion position control (RMPC).

every T seconds. This is immediately converted to a column vector with six entries corresponding to the position of the tool point and the angles of the three tool axes. The input to the inverse Jacobian is the difference between this setpoint vector and the current position of the tool as computed in the feedback loop. Recall that the inverse Jacobian relates Cartesian rates to joint rates, and therefore the output of the inverse Jacobian is a velocity command, essentially the derivative of the joint vector. The diagonal gain matrix K applies the appropriate gain to each velocity which is then fed into the velocity servos that control each joint. Encoders on each joint axis determine the position, which is then converted to a hand matrix and the column vector with the position of the tool point and its angles.

As opposed to the previous two schemes, this topology is essentially a closed loop control. It is important to note that the actual position of the manipulator is compared with its setpoint to generate the velocity control signal. One advantage in using this scheme is that the velocity commands will remain fixed between updates and therefore cause the joints to continue to move between updates. In the case of a position loop, once the setpoint has been reached, the joint will stop. Thus velocity loops generally produce a smoother motion for slow update rates. With

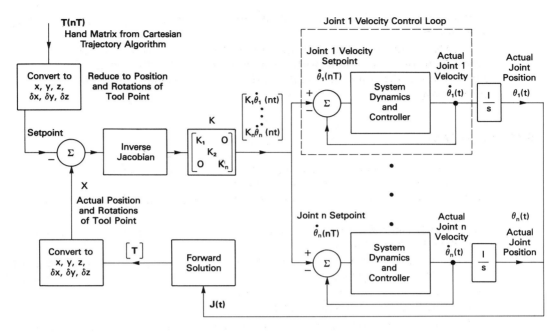

Figure 8.10.3.   Resolved motion rate control (RMRC).

today's extremely fast computational elements this advantage may not be as great, since the position setpoints can be generated at a rate that is faster than the arm dynamics. This can produce error signals which will not go to zero between updates to the position servos, and therefore the servos continue to drive the arm between updates due to a following error.

8.11 SUMMARY

In this chapter we have presented the concept of using homogeneous coordinates to represent the position of a manipulator in space. By using various matrix operators it was shown that the position of the manipulator's tool tip could be referenced to various frames and moved with respect to a base frame.

Rules were presented for establishing coordinate frames on the links of a kinematic chain, and a method of defining the relationship between successive links by matrices was described.

The forward solution (relating joint angles to Cartesian coordinates) and the inverse solution (relating Cartesian coordinates to joint angles) were defined.

The Jacobian that relates the rates between the variables of different coordinate systems was introduced. Additionally a method of computing the Jacobian in terms of the A matrices of the manipulator was presented. The use of the manipulator's Jacobian to relate joint torques or forces to forces at the tool tip was shown.

Algorithms to define the trajectory of the manipulator in Cartesian and joint space were discussed along with a simple continuous path algorithm. The representation of the controller architectures needed to achieve these algorithms was also shown.

8.12 PROBLEMS

8.1 Show how to add two vectors represented in homogeneous coordinates with different scale factors. What is the scale factor of the result?

8.2 Verify that to multiply the vector of Eq. (8.2.4) by 2, all that is necessary is to divide s by 2.

8.3 Redo Example 8.2.4 using a vector with a nonunity scale factor s. Is s preserved? What happens if the scale factor of Trans (a, b, c) is not 1?

8.4 Derive rotation in the y-z plane and the z-x plane using the geometric approach presented in Section 8.3.

8.5 Derive the rotation formula using vector techniques.

8.6 Verify Eqs. (8.2.25) through (8.2.27).

8.7 Using graphical techniques, prove that it is always possible to make two reference frames coincident by performing at most two rotations and one translation. Discuss the implications of the order of the operations.

8.8 Investigate the effects of multiplying matrices with nonzero perspective and nonunity scaling terms on points.

8.9 Verify that A_{12} and A_{21} are mathematical inverses. Use the argument of transforming from frame 1 to frame 2 and then from frame 2 to frame 1.

8.10 Verify Eqs. (8.3.11) through (8.3.16).

8.11 Prove Eq. (8.4.2).

8.12 Prove Eq. (8.4.3).

8.13 Perform the same operation as discussed in Section 8.4.2 to erect a wedge on a table, but use a different set of transformations.

8.14 Premultiply and postmultiply Eq. (8.3.7) by Rot $(z, 90)$. Plot the relative axes in space. Examine how the transformation affects points.

8.15 Referring to the task described in Section 8.5.2:
   a. Find the transformation matrix that will draw the A at a point in sector 2 where line segment ac is at a radial distance of $\frac{1}{2}$ the original (i.e., closer to the origin).
   b. Find the transformation that will draw the A upside down.

8.16 Describe the effects of premultiplying Eq. (8.5.2) by the operators Trans $(0, 0, dz)$ and Rot $(z, 180)$. How does the result differ from Example 8.5.1?

8.17 For the task defined in Table 8.5.1, determine the transformation needed to transform point a to some point above it. Discuss the sign of the variables in the matrix. What happens if the original sign is negated?

8.18 Devise and mathematically test a task that uses postmultiplications to pick up objects of various heights.   Consider the PUMA programming example given in Chapter 7 as a model for this task.

8.19   Establish numeric values for the positions of the pallets and pickup points of Example 8.5.4.   Plot on graph paper, and verify the procedure.

8.20   Referring to Section 8.5.5, demonstrate that $A_{\text{base,vision-frame}}$ is a constant.   Make a sketch on graph paper and do a numeric example.

8.21   For the Stanford arm shown in Figure 8.6.7, change the direction of x_1, redraw the schematic diagram, and determine the link parameters and A matrices.   What does this change do to the T matrix?

8.22   For the six-jointed Puma, select another valid set of coordinate systems, and determine the link parameters and symbolic A and T matrices.

8.23   Consider the forward solution of the Stanford arm.   Assume that angles may be resolved only to 1° and linear motion to increments of 0.01 in.   Compute the error for each element of the hand matrix for an arbitrary joint vector.

8.24   Illustrate the problem of multiple solutions using the Stanford arm of Figure 8.6.7. Specifically illustrate the same position as shown in the figure with joints 4 and 6 rotated 180° in opposite directions.   What other combinations achieve this result.   Consider the sum of the angles of joints 4 and 6.

8.25   Given a two-joint-and-link kinematic chain with coordinate frames attached at the extreme ends (ground and link 2), using vector techniques (dot- and cross-product relationships), determine the value of the two joint variables.   Assume both joints are rotary and the link lengths are d_1 and d_2. See Figure 8.6.3.

8.26   Write a computer program that can perform joint interpolated motion for the manipulator of Example 8.7.1.   The inputs are to be the desired final position given in terms of a legal hand matrix along with a maximum motion time.   Assume a maximum acceleration and velocity for each joint.   Output should be in the form of velocity profiles for each joint.

8.27   Extend the program of Problem 8.26 to include the capability of continuous path motion.   In this case the inputs should be a set of legal hand matrices (defining the desired positions) along with a maximum time for passing through each segment of the motion.   Once again outputs should be in the form of velocity profiles.

8.28   Obtain the Jacobian for the manipulator of Example 8.7.1 in terms of the method shown in Section 8.9.1.

8.29   Block out a control strategy using joint position control, (JPC), and resolved motion position control, (RMPC), for the manipulator of Example 8.7.1.

8.30   Define all the relationships needed along with the control structure to implement resolved motion rate control, (RMRC), for the manipulator of Example 8.7.1.

8.13 REFERENCES AND FURTHER READING

1. Paul, Richard C., *Robot Manipulators: Mathematics, Programming and Control*, Cambridge, Mass, The MIT Press, 1981.
2. Brady, M., Hollerbach, J., Johnson, T., Lozano-Perez, T., Mason, M., eds., *Robot Motion Planning and Control*, Cambridge, Mass., The MIT Press, 1982.

3. Anton, Howard, *Elementary Linear Algebra*, New York, John Wiley & Sons Inc., 1973.

4. Crouch, R., Herr, A., Sasin, D., *Calculus with Analytic Geometry*, Boston, Mass., Prindle, Weber Schmidt Inc., 1971.

5. Rosen, C., Nitzan, D., Agin, G., Andeen, G., Berger, J., Eckerle, J., Gleason, G., Hill, J., Kremers, J., Meyer, B., Park, W., Sword, A., *Exploratory Research in Advanced Automation: Second Report covering 10/1/73 through 6/30/74*, Prepared for National Science Foundation by Stanford Research Institute, Menlo Park, Calif.

6. Piper, D. L., *"The Kinematics of Manipulators under Computer Control,"* Ph.D., Dissertation, Stanford University Computer Science Department, Stanford, California, (October 1968).

7. Milenkovic, V., Huang, B., *Kinematics of Major Robot Linkages*, Conference Proceedings of the 13th International Symposium on Industrial Robots and Robots 7, Vol. 2, 1983.

8. Paul, R. L., *"Modeling Trajectory Calculations and Servoing of a Computer Controlled Arm,"* Ph.D., Dissertation, ARPA Order No. 457, SAIL Memo AIM-177, Computer Science Department Report CS-311, Stanford University, Stanford, California (November 1972).

9. Whitney, D. E., *"Resolved Rate Control of Manipulators and Human Prostheses,"* IEEE Transactions on Man-Machine Systems, Vol. MMS-10, (June 1969).

10. Paul, R. L., *"Manipulator Path Control,"* IEEE Proceedings of 1975 International Conference on Cybernetics and Society, September 23-25, 1975.

11. Davis, H., Snider, A., *Introduction to Vector Analysis*, 3rd edition, Boston, Mass., Allyn and Bacon Inc., 1975.

12. Pugh, A., *Robotic Technology*, Peter Peregrinus Ltd., London, U.K., 1983.

13. Hartenberg, R. S., and Denavit, J., *Kinematic Synthesis of Linkages*, McGraw-Hill, New York, 1964.

14. Lee, C. S. G., Ziegler, M., *A Geometric Approach in Solving the Inverse Kinematics of PUMA Robots*, Conference Proceedings of the 13th International Symposium on Industrial Robots and Robots 7, Vol. 2, 1983.

15. Lee, C. S. G., *A Geometric Approach in Solving the Inverse Kinematics of PUMA Robots*, IEEE Transactions on Aerospace and Electronic Systems, Vol. AES–20, no. 6, pp. 696–706, (November 1984).

9

Design Example

9.0 OBJECTIVES

The objective of this chapter is to bring many of the concepts presented in previous chapters to bear in solving a practical problem. The reader will be challenged to utilize this information in solutions that will permit the authors' integrated approach to be more fully appreciated.

9.1 MOTIVATION

The motivation for this chapter is to provide a complete example whereby the reader has the opportunity to consider many aspects of the application of the technology of robotics and associated fields. The example that has been selected provides a simple yet rich framework that permits the utilization of many of the principles presented in previous chapters. The reader is encouraged to design, in as complete a way as possible, practical solutions to the problems presented. The text suggests possible solutions to the application problems, but the reader is challenged to improve on these, either by creating simpler ones or by creating more general and more sophisticated solutions.

9.2 INTRODUCTION

To illustrate the various topics and subject matter throughout this book, a specific example has been selected that will allow a variety of considerations to be made.

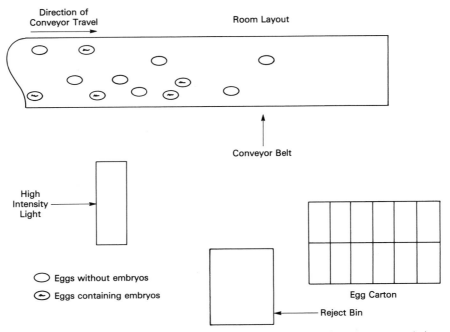

Figure 9.2.1. General room layout of an egg packing system. (Not drawn to scale.)

This example, involving an egg packing system, is described below and illustrated in Figure 9.2.1. A detailed solution to the problem of eliminating "bad eggs" is presented in this chapter.

Egg Packing Problem

The flow of activities required of such a system is:

1. Pick up an egg from the conveyor belt.
2. Hold the egg up to a bright light and determine if the egg is clear (i.e., has no embryo growth). This process is known as "candling."
3. Place the egg into a reject bin or a receptacle in an egg carton, depending on whether or not it contains an embryo.

Although the problem statement itself is straightforward enough, there are clearly many subproblems associated with the actual implementation of its solution. Figure 9.2.2 shows one example of a general schematic layout for an egg packer/checker. Before implementation of this system can be achieved, however, a number of additional assumptions and/or specifications are required. These include assumptions as to the conveyor speed, egg weight, carton size, robot geometry, type of sensors, gripper geometry, and so on. In this respect it is suggested that the reader consider Problems 9.1 through 9.4 given at the end of the chapter before proceeding to the next section.

In the following sections, a possible solution will be presented. In particular, the remainder of the chapter will consider the following:

- Robot design and process specifications
- Motor control sequences
- Motor and drive mechanism selection for all axes
- Encoder selection
- Computer architecture and control structure
- Vision system considerations, including camera resolution, camera selection, use of photo-optical interrupters, colorizer/density measurements, and calibration

Figure 9.2.2 illustrates in more detail the general layout for the egg packer/ checker. Some assumptions that the text solution presumes are the following:

- A photointerrupter will be used to stop the conveyor when an egg interrupts the light beam.

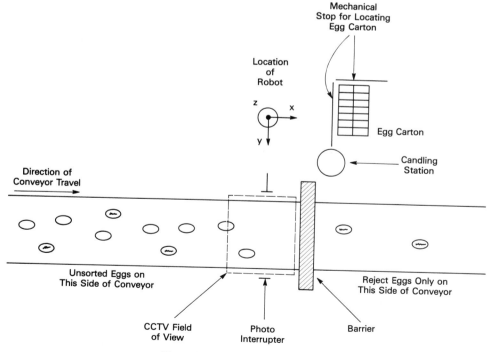

Figure 9.2.2. General layout for egg packer/checker.

- A CCTV will be used to locate the egg in the field of view.
- Candling of the egg will be accomplished by transilluminating the egg with a bright light and then using a vision sensor to read the egg's density pattern and/or color.
- Egg placement will either be in the egg carton (if the egg passes candling) or on the back portion of the conveyor (if the egg fails candling), which is designated as the reject area.

9.3 ROBOT DESIGN AND PROCESS SPECIFICATIONS

Although there are many solutions to the problem described above, in what follows we will suggest one of them so that an appreciation may be gained of the many factors that influence the final design. Obviously, others are possible and the reader is encouraged to investigate one or more of these (see Problems 9.1 through 9.7).

In what follows we have assumed that the robot has four axes: r, θ, z, and ϕ. Moreover, the control and mechanical specifications of the process to accomplish the candling task (and how they influence the robot itself) are presented below. The actual motion sequence is then stated and a discussion of the mechanical description of the robot given. Finally, the actual design of the robot actuators is included to conclude this section.

9.3.1 System Specifications

1. Cycle time ≤ 3.0 s maximum.
2. Three distinct moves per cycle:
   a. Move to conveyor and acquire an egg
   b. Move to the candling station
   c. Place the egg in the carton or into the reject area
3. Three pauses during a cycle:
   a. 0.20 s required to close gripper (grasp the egg) or open gripper
   b. 0.05 s to perform the candling operation
   c. 0.20 s required to place egg in carton by opening gripper
4. Weight of egg ≤ 3 oz.
5. Weight of gripper ≤ 13 oz.
6. Positional resolution: at least 0.05 in.
7. All axes should move simultaneously and finish moving at the same time (i.e., coordinated motion is required).

9.3.2 Mechanical Description of the Robot

Although there are many manipulator designs that will meet the specifications, for purposes of illustration we will work with a specific one. The reader is encouraged to modify the assumptions and repeat the calculations.

At first glance, a robot that appears to meet the specifications given in Section 9.3.1 should have one revolute and two prismatic joints. In fact, one additional rotary axis is needed to permit orientation of the egg, which is lost when the arm moves. Thus although a θ-r-z manipulator can pick up the egg, a ϕ joint whose axis of rotation is parallel to the primary θ axis is also needed. With reference to the general layout in Figure 9.2.2, specifications for these axes will be assumed to be:

a. r: 18-to-24 in. reach
b. θ: $\pm 90°$ waist rotation
c. ϕ: 360° wrist roll
d. z: 2 in. vertical travel

What is not clear is how to configure these axes.

From the standpoint of load, the major axis of the robot will be θ since it carries the entire robot's mechanism (including the other joints) in addition to the gripper and payload (egg). As a consequence and not surprisingly, it will be seen shortly that this joint will require the largest servomotor to drive it. This actuator will be coupled to a harmonic drive and the entire assembly mounted in the fixed base so that it will be unnecessary to move its weight. The r- or reach-axis motion will be obtained using a servomotor mounted in the shoulder assembly. It will be coupled to an Acme thread lead screw 7 in. in length. The z axis will be located at the end of the r linkage so that its tip velocity is perpendicular to the r axis motion. Its travel will be obtained using a bang-bang actuator (e.g., a pneumatic cylinder) since only two distinct vertical positions are needed, one for travel and one for both pickup and deposition of an egg. The wrist roll joint (ϕ) is located at the opposite end of the z cylinder and will require a small servomotor coupled to a simple gear train. Each of the three motors will have an optical incremental encoder mounted on its end bell so as to permit the shaft positions to be monitored.

It will be assumed that the overall weight of the portion of the robot that the θ axis must move is about 6 lb (including the load). The weight breakdown is shown in Table 9.3.1. A sketch of this manipulator and the special gripper is shown in Figure 9.3.1.

TABLE 9.3.1 MANIPULATOR AXIS LOAD
BREAKDOWN

---

1. Payload (egg) $\leq$ 3 oz
2. Gripper (including open/close mechanism) $\leq$ 13 oz
3. ϕ gearbox $\leq$ 18 oz
4. ϕ axis motor $\leq$ 7 oz
5. z axis actuator $\leq$ 7 oz
6. r axis structure (24 in. long) $\leq$ 25 oz
7. r axis lead screw $\leq$ 16 oz
8. r axis motor $\leq$ 7 oz

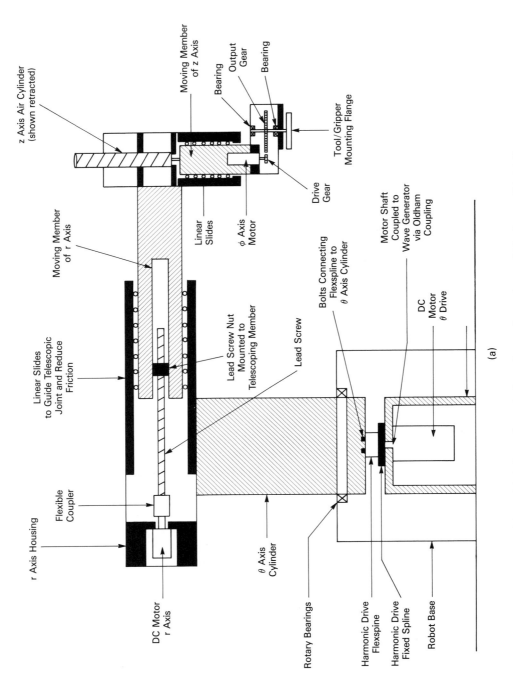

Figure 9.3.1. (a) A four-axis manipulator that can be used to pack/check eggs; (b) details of a gripper that handles eggs and is used on this robot.

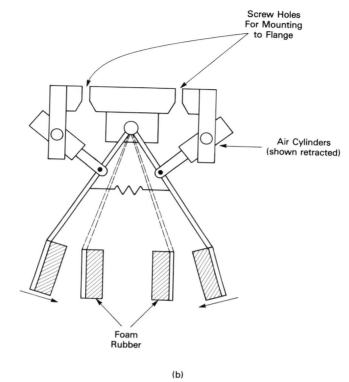

Screw Holes
For Mounting
to Flange

Air Cylinders
(shown retracted)

Foam
Rubber

(b)

Figure 9.3.1 *(continued)*.

9.3.3 Motion Sequence

The assumed sequence of worst-case moves that permits a single egg to be candled
and packed is now described as a series of "primitives" that might be used by a
robot controller (see Table 9.3.2).   Figure 9.2.2 may be used to visualize these
actions.   In what follows, we assume that the robot's gripper is initially located
over the egg carton.

The motion sequence described above can be summarized in a time-and-
motion state diagram (see Figure 9.3.2) that indicates when each of the axes is to
move and also when the candling operation and gripper motion is to take place.
The figure resulting from this describes the states of the individual quantities or
variables.   It is observed that there are two or three states associated with each
of the actuation axes.   The uncertain states (4, 5, 6, 13, 14, 15) are indicated by
dashed lines.   It is important to note that even though one axis could finish its
motion in a shorter period of time than is indicated (e.g., r or ϕ), the requirement

TABLE 9.3.2   MOTION SEQUENCE SUMMARY

Action	Time
1. Move θ axis clockwise to egg pickup location (worst-case distance = 150°)	
2. Extend r axis to egg pickup location (worst-case distance = 3 in.)	Worst-case time for steps 1, 2, and 3 is 750 ms
3. Move φ axis to orient gripper so that egg is picked up on long axis (worst-case distance = 360°)	
4. Move z axis down to egg pickup height	200 ms
5. Close gripper and acquire egg	200 ms
6. Move z axis up to travel height	200 ms
7. Rotate θ axis counterclockwise to candler location (worst-case distance is 75°)	
8. Retract r axis to candler location (worst-case distance is 3 in.)	Worst-case time for steps 7 and 8 is 500 ms
9. Candle egg	50 ms for vision "snapshot"
10. Rotate θ axis counterclockwise to egg drop-off or clockwise to reject point (worst-case distance for either move is 75°)	
11. Extend r axis to egg drop-off (or reject) point (worst-case distance = 3 in.)	
12. Rotate φ axis back to egg drop-off (or reject) point (worst-case distance is 360°)	Total time for steps 10 through 12 is 500 ms
13. Move z axis down to egg drop-off height	200 ms
14. Open gripper to deposit candled egg	200 ms
15. Move z axis up to travel height to get ready for next cycle	200 ms

of coordinated motion means that the axis that takes the longest time "paces" the other axes.

Additionally, it may be interesting to note that the actual motion times may be less than indicated, when worst-case motions are not required. In general, the cycle time for completion of packing an egg carton will be less than the simple summation of the worst-case times. The frequency of occurrence of eggs with embyros will also affect the cycle time for egg packing.

9.3.4 Motor and Drive Mechanism Selection

The design of the z, θ, r, and ϕ axes will be considered separately and in that order. It will be assumed in what follows that the z axis motion is obtained by using an air (pneumatic) cylinder or solenoid to move the gripper up or down by a small amount. Actually, the purpose of this axis is permit the grasped egg to

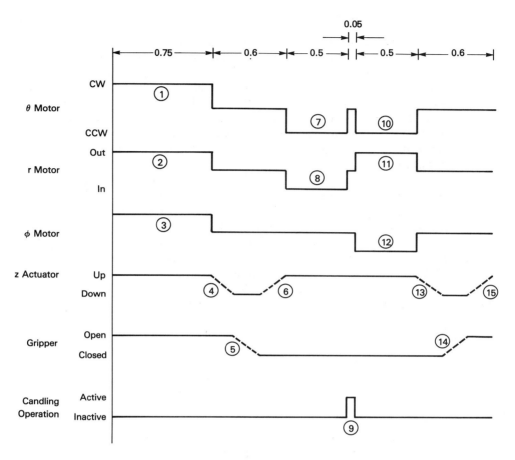

Figure 9.3.2.   Time and motion state diagram (shown for a "good egg"). *Note:* The numbers in circles correspond to the actions defined in Table 9.3.2.

clear the tabletop. This procedure has the advantage of eliminating a servomotor at the tool end of the manipulator, thereby reducing weight and cost.

θ is chosen as the first servo-controlled axis because its actuator must move the largest load and, as a consequence, may create the greatest problems in meeting the timing specifications.   In what follows, the motor selection procedure detailed in Appendix B will be utilized.   It is important to understand that if there did not exist a complete specification of all of the component weights, it would be necessary to start the calculations at the end effector and work backward so as to attempt to meet the motion/distance requirements while minimizing the size (and hence the cost) of the θ axis actuator.   However, in the present case, this is not necessary since the maximum weights are given for the components.

9.3.4.1 axis

From Section 9.3.3 it is apparent that three separate angular motions are required per cycle. The first necessitates a rotation of up to 150° clockwise (from a location on the egg carton or 75° from the "bad egg drop-off point," i.e., the reject site) in order to position the fingers of the gripper about the egg. At this time, the z axis moves down and the gripper is closed (this action takes about 0.4 s), the z axis moves up to the travel height, and the manipulator is rotated approximately 75° (counter-clockwise) in order to place the egg within the candling station. After a pause of 0.05 s, the arm is again rotated counterclockwise (by as much as an additional 75°) so that the egg can either be placed in a carton or deposited into the reject bin and hence onto the bad egg conveyor. Drop-off requires a 0.2 s pause. These motions are summarized in Figure 9.3.3.

It is observed from this figure that a trapezoidal (or triangular) velocity profile has been assumed (for simplicity) and that the acceleration and deceleration times are both 0.25 s. Although these times were selected to permit the specifications in Section 9.3.1 to be met, the time and motion state diagram shown in Figure 9.3.2 was also employed to determine the appropriate starting and stopping times for each of the three motions required during a single cycle.

Utilizing the motor selection procedure in Appendix B, it is found that a

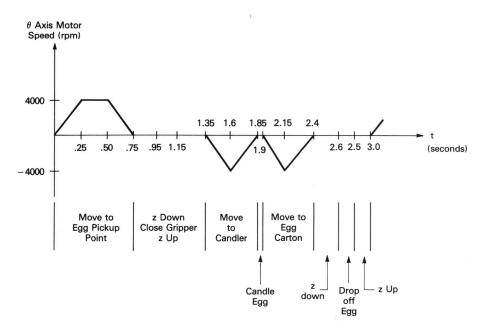

Figure 9.3.3. Assumed theta axis velocity profile for one complete motion cycle (worst case).

TABLE 9.3.3   SPECIFICATIONS FOR ELECTROCRAFT
E-532

Property	Symbol	Value
Weight	W	3 lb
Peak torque	T_{PEAK}	355 oz-in.
Continuous stall torque	T_{rms}	50 oz-in.
Armature inertia	J_M	0.063 oz-in.-s^2
Torque constant	K_T	14.81 oz-in./A
Voltage constant	K_E	10.95 V/1000 rpm
Armature/brush resistance	R_a	2.04 Ω
Armature inductance	L_a	5.65 mH
Friction torque	T_f	3 oz-in.
Thermal resistance	R_{th}	4.18°C/W

motor that will handle the task is an Electrocraft E-532. (This is demonstrated below.) Its specs are given in Table 9.3.3. It is important to note that the choice of this motor is not unique and that a number of others will perform adequately.

The selection of the speed-reduction ratio N of the harmonic drive is extremely important. A large value of N will reduce the reflected inertia (as the inverse square of N). This is clearly an advantage since it means that we can utilize a smaller motor to accelerate and decelerate the load. However, as the ratio increases, more actuator turns will be required in order to reach the (same) desired location. For the given timing constraints, this would imply that a very large maximum speed would be needed. Also, the power supply voltage would have to be larger.

Once again, after a number of tries, it was determined that a reduction ratio of 80:1 would be a good compromise. In addition, it was assumed that a reasonable dynamic efficiency η for the harmonic drive was 70%. (For the harmonic drive, the range of efficiency is usually 40 to 80% and depends on the input speed.)*

With these considerations as preliminaries, the following calculations can be perfomed and the results used as inputs to the motor selection procedure (together with the velocity profile of Figure 9.3.3 and the information about the harmonic drive):

1. Reflected Static Torque Load.   As there is no torque introduced by gravity for this axis, and the bearing and harmonic drive frictions are relatively small, we will assume that $T_L = 0$.

2. Reflected (Effective) Inertia.   We will assume that the 3 lb shoulder (r

---

*When the efficiency η of a mechanical coupling device is less than 100%, more torque is required in order to accelerate an inertial load. The reduced efficiency causes the motor to "see" a larger inertia. This is referred to as the "effective" reflected inertia.

axis) assembly weight acts at the midpoint of the fully extended r axis (i.e., at r_1 = 12 in.).   Also, the combined weight (3 lb) of the ϕ and z axes together with the gripper and load is assumed to be located at the endpoint of the reach (i.e., r_2 = 24 in.).   Thus we are employing a point-mass approximation.   Then the effective reflected load inertia is

$$J_L = \frac{(W_1/g)r_1^2 + (W_2/g)r_2^2}{\eta N^2}$$

$$= \frac{[(3 \times 16)/(32.2 \times 12)](12^2 + 24^2)}{0.7 \times 80^2}$$

$$= 0.01996 \text{ oz-in.-s}^2$$

3.   Total Inertia.   For the E-532 motor, J_M = 0.0063 oz-in.-s^2.   Thus the total reflected inertia is found to be

$$J_T = J_M + J_L$$

$$= 0.0063 + 0.01996$$

$$= 0.02626 \text{ oz-in.-s}^2$$

Note that we have neglected the inertia of the harmonic drive's wave generator and that of the robot's trunk in these calculations (see Problem 9.8).

4.   Angular Acceleration (and Deceleration).   Since the maximum speed for the E-532 motor is 6000 rpm, a conservative peak speed in this application would be 4000 rpm.   Thus the peak angular velocity is

$$\omega_{\text{pk}} = (\text{rpm}/60 \text{ s/min}) \times (2\pi \text{ rad/revolution})$$

$$= \frac{4000}{60} \times 2\pi$$

$$= 418.88 \text{ rad/s}$$

Then the angular acceleration (and deceleration) is

$$\alpha = \frac{\omega_{\text{pk}}}{t_a}$$

$$= \frac{418.88}{0.25}$$

$$= 1675.52 \text{ rad/s}^2$$

5.   Distances Traveled.   The final position can be found from the area under the velocity curve in Figure 9.3.3.   As the profile consists of either triangles or a trapezoid, this is a straightforward calculation.   Thus

a. Move to egg pickup from worst-case position on carton $=$

$$\frac{4000 \times 0.5/60}{80} \times 360° = 150°$$

b. Move to candling station $=$

$$\frac{4000 \times 0.25/60}{80} \times 360° = 75°$$

c. Worst-case move from candler to egg carton $=$

$$\frac{4000 \times 0.25/60}{80} \times 360° = 75°$$

These angular distances permit the cycle time and distance specifications to be met.   It is important to note, however, that if larger rotations or a shorter cycle time is desired, it may be likely that a different motor would be needed due to the larger maximum speed that almost certainly would be required.

6. Acceleration and Deceleration Torques.   During the acceleration and deceleration portions of each phase of the motion, the motor must move the total inertia and overcome any joint and/or motor friction.   For the constant-velocity portion of the profile, the motor must only overcome the friction term.   Finally, since there is no gravity term in this configuration, the motor torque is zero during the pause times.   As an example, consider the move to the candling station from the egg pickup point.   It is assumed that the joint friction as well as all the viscous frictions are zero.   Thus the only friction term is that due to the motor itself (i.e., $T_f = 3$ oz-in.).

$$T_{\text{accel}} = J_T \alpha + T_f$$
$$= 0.02626 \times 1675.52 + 3$$
$$= 47 \text{ oz-in.}$$
$$T_{\text{decel}} = J_T \alpha + T_f$$
$$= -41 \text{ oz-in.}$$
$$T_{\text{cv}} = T_f$$
$$= 3 \text{ oz-in.}$$

Since none of these values exceeds the peak torque T_{PEAK} (i.e., 355 oz-in.), the motor will not be demagnetized.   The overall torque profile is shown in Figure 9.3.4.   It is clear that T_{accel} and T_{decel} for the other two moves during any cycle are identical to those calculated above.

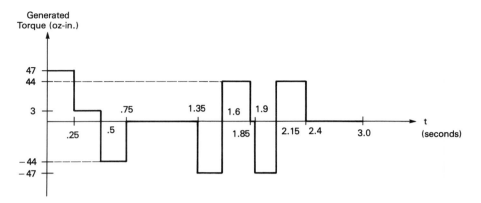

Figure 9.3.4. Overall torque versus time profile for the theta axis (only one complete cycle is shown).

7. RMS Torque. To prevent overheating, the developed rms torque must be less than the continuous stall torque of the motor (50 oz-in.). For a trapezoidal (or triangular) velocity profile (such as that shown in Figure 9.3.3), the calculation is quite straightforward because of the piecewise constant nature of the torque profile (see Figure 9.3.4). All that must be done in this case is to sum the squares of the individual torque segments each multiplied by the corresponding time duration (i.e., find the area under the torque squared curve).* This sum is then divided by the total cycle time and the square root taken. Thus

$$T_{\text{rms}} = \sqrt{\frac{3 \times 0.25 \times T_{\text{accel}}^2 + 3 \times 0.25 \times T_{\text{decel}}^2 + 0.25 \times T_{\text{cv}}^2}{3.0}}$$

$$= 32.2 \text{ oz-in.}$$

As this is below the continuous stall torque capability of the E-532, the motor will not overheat. In this expression the factor of three results from the fact that the acceleration (and deceleration) times are the same for all three portions of the motion profile. Also, it is noted that we have divided by 3.0 s, the overall cycle time. This assumes that there is no delay between candling operations. If there is, however, the cycle time must be increased accordingly and the resultant rms torque will decrease.

8. Size of Power Supply. The dc supply that is required to power the servo amplifier for this axis must be large enough to handle the back EMF generated by

---

*For nontrapezoidal profiles, the rms must be found by squaring the torque curve and then performing an actual integration.

the motor. For the E-532 running at a peak speed of 4000 rpm, the minimum voltage is

$$V_{min} = \frac{4000}{1000} \times K_E$$

$$= 4 \times 10.95$$

$$= 43.8 \text{ V}$$

In addition, the supply must be able to overcome the resistive drops in the armature and wiring. At the peak torque of 47 oz-in., a current of $47/14.91 = 3.15$ A is required. Thus there will be a drop of at least $3.15 \times 2.04 = 6.4$ V. Consequently, a supply of greater than $43.8 + 6.4 = 50.2$ V is necessary to meet the specifications. If this proves to be impractically large, a different motor winding could be utilized that had fewer turns and hence a smaller torque and back EMF constant. This would, of course, necessitate a larger peak current (not a problem here). It should be understood that 50.2 V is the value that must be supplied to the motor terminals. The actual supply voltage will be somewhat larger, due to ohmic losses in the cables and the printed circuit boards, the circuits themselves, and other distributed power-consuming elements.

9.3.4.2 *r* axis

As mentioned previously, the reach-axis motion will be achieved by having a servomotor drive an Acme-type lead screw. We will assume that on average, each of the three linear moves will be 3 in. It will be seen, however, that longer moves are quite easily accommodated even with the small motor selected. The velocity profile for this axis is illustrated in Figure 9.3.5. It is observed that the

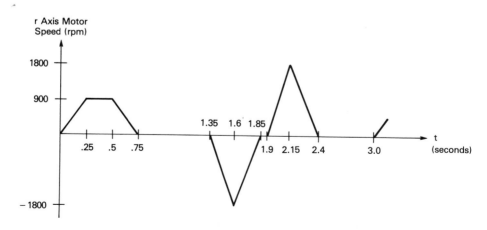

Figure 9.3.5. *r* axis velocity profile (one complete cycle).

TABLE 9.3.4 SPECIFICATIONS FOR PITMAN 9432
MOTOR

Property	Symbol	Value
Weight	W	7 oz
Peak torque	T_{PEAK}	13.8 oz-in.
Continuous stall torque	T_{rms}	3.16 oz-in.
Armature inertia	J_M	0.00059 oz-in.-s^2
Torque constant	K_T	2.2 oz-in./A
Voltage constant	K_E	1.93 V/1000 rpm
Armature/brush resistance	R_a	1.84 Ω
Armature inductance	L_a	0.1 mH
Friction torque	T_f	0.4 oz-in.
Thermal resistance	R_{th}	22.7°C/W

two shorter portions of this profile require higher peak angular velocities since the joint is required to move the same distance in a shorter time.

One motor that will drive the r joint adequately is a Pitman 9432 (winding 1) and has the specs given in Table 9.3.4. In addition, its diameter is 1.6 in.

It is found that a lead screw having a pitch P of 2.5 (i.e., 2.5 turns per inch) will produce the desired results. A 50% dynamic efficiency is assumed for this element (normally, the efficiency range for Acme lead screws is 25 to 85%).

Calculations similar to those of Section 9.3.4.1 are summarized below.

1. *Reflected static torque.*

$$T_L = 0 \text{ (no gravity in the } r \text{ direction)}$$

2. *Reflected inertia.* The actuator only has to move the combined weight of the gripper, load, and φ axis (i.e., 3 lb). Thus

$$J_L = \frac{W/g}{\eta(2\pi P)^2}$$

$$= \frac{3 \times 16/(32.2 \times 12)}{0.5(2\pi \times 2.5)^2}$$

$$= 0.001007 \text{ oz-in.-s}^2$$

3. *Total inertia.*

$$J_T = 0.00059 + 0.00159$$

$$= 0.00218 \text{ oz-in.-s}^2$$

Note that this calculation neglects the inertias due to the encoder disk, the lead screw (which could be fairly large), and any mechanical coupler.

From Figure 9.3.5 it is seen that a peak angular velocity of 900 rpm is needed during the move from the carton to the pickup point on the conveyor, whereas the other two moves require top speeds of 1800 rpm. Utilizing this information, the

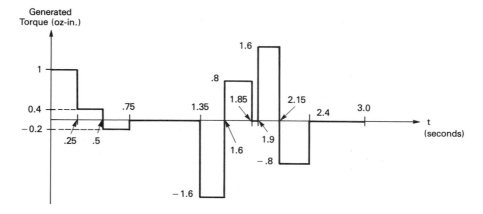

Figure 9.3.6.   Torque versus time curve for the *r* axis (one complete cycle).

torque versus time curve shown in Figure 9.3.6 results.   Once again it is quite simple to calculate the equivalent rms torque from such a diagram.   It can be shown that the motor must generate T_{rms} = 0.80 oz-in. which is about one-quarter of what it is capable of producing (i.e., 3.16 oz-in.).   Clearly, even a small motor such as the Pittman 9432 will have little trouble moving the load and φ-axis assembly.   In fact, even if longer motions are required, the motor will be adequate (see Problems 9.9 through 9.11).

Concerning the size of the power supply, it is clear that since K_E for the 9432 is much less than that of the E-532 and also since the peak velocities for this axis are also much lower, the voltage requirement of the *r* axis will be insignificant compared to that of θ.   Thus any supply large enough to handle the θ axis demands will be more than adequate for the *r* axis (see Problem 9.12).

9.3.4.3 φ axis

The φ or roll axis is needed in order to orient the gripper so that the egg is picked up on its long axis.   Although angular position is unimportant during the candling operation, it is necessary to orient the egg so that the largest end is placed on the edge of the appropriate location in the carton.   Consequently, we will assume that two 180° roll motions are required to achieve this and that the velocity profile for the φ axis is as indicated in Figure 9.3.7.

Using a trial-and-error approach, it is determined that a 10:1 (reduction) simple gear train having a minimum efficiency of about 50% and driven by another Pittman 9432 motor will be more than adequate for the assumed task.   To verify this, it is necessary to calculate the total reflected inertia J_L.   This axis is depicted in Figure 9.3.8.   From this figure it is observed that

$$J_L = J_{gear_1} + \frac{J_{gear_2} + J_{gripper} + J_{egg}}{\eta N^2}$$

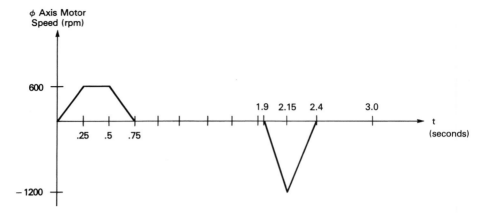

Figure 9.3.7. φ axis velocity profile (one cycle).

For simplicity, we will assume that both the gear train (with gear2 and gear1 having radii of 1.0 and 0.1 in., respectively, and corresponding weights of about 1.8 and 0.018 oz) are solid thin disks so that their inertias are given by the expression $0.5mr^2$. In addition, the gripper weighs 13 oz and consists of two solid thin disks (each of radius 1.5 in.) separated by a maximum distance of 2 in. and we may use the results developed in Chapter 3 to find the inertia (recall that the parallel axis theorem must be employed in this case). Finally, the 3 oz egg will be assumed to be a spherical solid (radius = 1 in.) so that its inertia is found from the formula $2/5 \times mr^2$. Therefore, we obtain

$$J_{\text{gear1}} = 0.5 \times 0.018 \times \frac{0.1^2}{32.2 \times 12}$$

$$= 0.000000233 \text{ oz-in.-s}^2$$

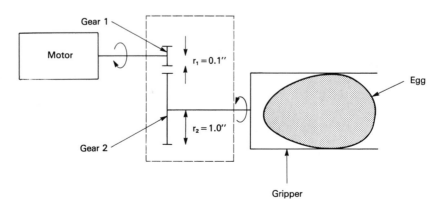

Figure 9.3.8. Details of the φ axis. The gripper (with an egg) is also shown.

$$J_{\text{gear2}} = 0.5 \times 1.8 \times \frac{1^2}{32.2 \times 12}$$

$$= 0.00233 \text{ oz-in.-s}^2$$

$$J_{\text{gripper}} = 2 \times \left(\frac{mr^2}{4} + md^2\right)$$

$$= 2\left(\frac{6.5}{32.2 \times 12}\right)\left(\frac{1.5^2}{4} + 1^2\right)$$

$$= 0.05257 \text{ oz-in.-s}^2$$

$$J_{\text{egg}} = \frac{2}{5} \times \frac{1^2}{32.2 \times 12}$$

$$= 0.003104 \text{ oz-in.-s}^2$$

and

$$J_L = 0.000000233 + \frac{0.00233 + 0.05257 + 0.003104}{0.5 \times 10^2}$$

$$= 0.001160 \text{ oz-in.-s}^2$$

Finally,

$$J_T = J_M + J_L$$

$$= 0.00059 + 0.001166$$

$$= 0.00175 \text{ oz-in.-s}^2$$

Utilizing these results, it can be shown that the torque versus time curve is that shown in Figure 9.3.9. From this it is found that T_{rms} is about 0.48 oz-in.,

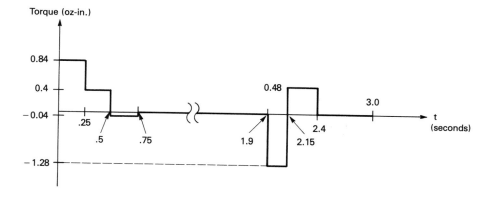

Figure 9.3.9.   Torque versus time curve for the ϕ axis (one cycle).

which is only about one-sixth of the maximum allowable continuous torque for the 9432 motor. In addition, the θ axis power supply can again be shown to be more than adequate for this axis (see Problem 9.12 and 9.13).

9.4 ENCODER SELECTION

To monitor position, the three servo-controlled axes of the robot (i.e., θ, r, and ϕ) will utilize optical encoders mounted on the motor end bells. The resolution of these encoders must be selected to meet the overall position resolution specification of 0.05 in.

For the θ axis, it is noted that a 100-line (per revolution) optical encoder will produce an output resolution of about 18.85 mils/count ($[2\pi/(100 \times 80)] \times 24$) in

TABLE 9.4.1 SUMMARY OF ROBOT DESIGN FOR EGG PACKING/CHECKING APPLICATION

Parameter	θ	r	ϕ	z
Axis reflected inertia (oz-in.-s²)	0.02626	0.00218	0.00175	NA
Peak acceleration (rad/s²)	1676	754	503	NA
Peak velocity (rpm)	4000	1800	1200	NA
Peak torque (oz-in.)	47	1.6	1.28	NA
RMS torque (oz-in.)	32.2	0.80	0.48	NA
Encoder resolution (lines/rev)	100	100	100	NA
Gear reduction or L.S. pitch	80:1	2.5:1	10:1	NA
Axis resolution (mils/count)	18.85	4.0	9.4	NA
Maximum excursion of motion	$\pm 90°$	18–24 in.	$\pm 180°$	2 in.
P.S. requirements, (V)	50.2	P	P	NA
P.S. requirements, (A)	3.15	P	P	NA

P, homework problem; NA, not applicable (z axis is pneumatic bang-bang, nonservo).

the worst case (i.e., with the *r* fully extended to 24 in.). Note that the factor of 80 results from the speed reduction provided by the harmonic drive. As will be seen, this is more than adequate for meeting the resolution spec.

Concerning the resolution of the *r* axis, a 100-line optical incremental encoder will yield a value of 4.0 mils/count [i.e, $1/(2.5 \times 100)$]. Thus the reach axis contributes considerably less to the overall position resolution of the load (the egg).

Finally, if a 100-line optical incremental encoder is used and one includes the effect of the 10:1 gear reduction, the resolution of the ϕ joint will be about 9.4 mils ($[2\pi/(100 \times 10)] \times 1.5$ in.). Thus the overall resolution of the robot (i.e., 32.25 mils) is well under the 50-mil spec.

A summary of the robot design is given in Table 9.4.1.

9.5 COMPUTER ARCHITECTURE AND CONTROL STRUCTURE

As discussed, the robot will have three servo-controlled axes and one bang-bang, nonservo axis. For this purpose, the multiprocessor structure described in Chapter 4 can be utilized for the servoed joints. Thus three separate joint (slave) processors along with a master (and perhaps a math coprocessor) will be required. It is also certainly possible that a single microcomputer could be used to handle the tasks of both the master and slaves (and perhaps even the math processor). The reason for this is that the geometry of the robot is fairly simple and extremely rapid and smooth motions are not critical in this application. In addition, the reader is encouraged to review the material presented in Section 7.8 where a model of a robot's computer system is discussed. Figures 7.8.1 and 7.8.2 are especially important in this respect.

Concerning the control structure, a PD control would be required at the very least. It is also possible that even though there is no disturbance due to gravity, friction will produce undesirable position errors. An integrator might then be required to compensate for this effect (i.e., a PID controller would be required for each joint). As discussed in Chapter 4, this device would be realized digitally in the software associated with the individual axes.

A small power amplifier would be needed to drive the *r* and θ axes. However, the θ joint would require a good-sized one capable of outputting at least 51 V at 3.5 A (i.e., almost 180 W). This is certainly not impossible, although the cost might be higher than would be warranted for such a robotic device.

Even though the *z* axis is pneumatically actuated (i.e., nonservo bang-bang), binary sensors should be incorporated to indicate when the axis is up and when it is down. These sensors would ensure that the axis is up before the manipulator begins to move and down before the gripper is opened or closed. In addition, the gripper itself could incorporate sensors to inform the controller when an egg has been successfully acquired or dropped off.

9.6 VISION SYSTEM CONSIDERATIONS FOR THE EGG PACKER

Recall that vision is to be used for two different purposes. The first is to locate the egg (on the conveyor) and the second is to test for the presence of an embryo. Note that two separate vision systems may be required to accomplish these tasks.

There are many considerations that are involved in selecting vision or photo-optical sensors for the egg packer/rejection system. These include:

- Vision system resolution
- Camera selection
- Photo-optical interrupter to stop the conveyor
- CCTV for egg location and orientation
- Colorizer/density measurement system
- Vision system calibration

Each of these will be treated in separate paragraphs for clarity, and where appropriate, pseudocode will be used to explain the authors' version of a solution. The reader is invited to be creative and establish personal solutions to the problems that will be posed.

9.6.1 Resolution

The area of the conveyor that is imaged by the CCTV is approximately 6 in. × 6 in. The pickup accuracy of the end of the arm tooling (i.e., at the robot's gripper) is about ±50 mils (actually, ±32.25 mils). From this information, the pixel density required can be computed as follows:

$$\text{pixel density} = \text{resolution}$$

$$= \frac{\text{linear extent}}{\text{pixel density}}$$

$$= \frac{6000 \text{ mils}}{50 \text{ mils/pixel}}$$

$$= 120 \text{ pixels}$$

This calculation shows that a video digitizer of 120 × 120 pixels will be sufficient to locate the egg to within ±50 mils. It is left as an exercise for the reader to determine the angular accuracy to which the egg's orientation can be found.

9.6.2 Camera Selection

The selection of a camera for this application depends on several requirements. These include:

- Resolution
- Linearity
- Gray-scale
- Acquisition speed
- Synchronization

Each is now discussed briefly.

9.6.2.1 Resolution

As has just been calculated, a resolution of 120×120 pixels is required to achieve the desired accuracy. Referring to the vision chapter (Chapter 6), it will be noted that standard RS-170 cameras have 525 lines per frame. If the camera is noninterlaced, there are 262.5 unique lines for imaging. These standard devices all have sufficient resolution for this application. Thus, practically any commercially available tube or solid-state RS-170 camera may be used based on the resolution requirement.

In addition to RS-170 cameras, there are several manufacturers of specialty non-RS-170 solid-state cameras that have resolutions of 128×128 that would be sufficient for this use. These units would be adequate for this application but are more expensive, since the market for them is not as large as for RS-170 standard CCTV cameras. One would use such a solid-state device if other factors, such as acquisition speed or synchronization with external events, are important requirements. These nonstandard cameras are able to acquire images faster or slower than RS-170 cameras, and may be required if ambient-light conditions are poor, if higher-speed response is needed, or if reduction of motion blur artifacts is important.

With regard to resolution, virtually any commercially available two-dimensional imaging camera is satisfactory. Others, such as line array devices, are inappropriate here because of additional expensive accessories that would be required, such as specialized lighting systems or mechanical transports for scanning the image orthogonally to the sensor's line-scan direction.

9.6.2.2 Linearity

No scanning device of any nature has perfect linearity. In the application being considered, the linearity of concern has to do with the accuracy of the scanning performed to acquire photo-optical information from the sensor. Solid-state sensors are manufactured using photolithographic processes and will generally be spatially correct to within 5 μm (millionths of a meter). The actual size of the photosensor is approximately 8.8 mm. Thus the solid-state sensor can hold a position placement of approximately 1 part in 1800 (0.05% accuracy) over the target area. This precision is generally swamped by lens and electronics inaccuracies. Tube-type cameras (e.g., vidicons) have more nearly continuous photosensitive material, but the scanning electronics introduces from 1 to 5% distortion in pixel placement accuracy, with the higher distortions generally being on the periphery of the field of view.

In this application, the resolution requirement of 50 mils over a 6000 mil field of view results in a 0.833% specification on resolution, so a solid-state camera is the preferred choice, since a tube-type camera of moderate cost will perform marginally.

9.6.2.3 Gray-scale

The gray-scale requirements of this application may be analyzed relative to the two separate vision requirements, first locating and then candling the egg. These will be treated as separate constraints.

Location of the egg can be accomplished by using an appropriate illumination system to create a binary image. Either toplighting or backlighting will permit this to be accomplished easily. Thus the camera will need to resolve only two gray levels. This requirement is easily met by any commercial CCTV camera. Solid-state sensor devices generally have 6- to 7.5-bit gray-scale resolving capabilities, whereas tube-type cameras generally have 4 to 5 bits of gray-scale resolving ability. It should be noted that television standards require only 10 gray levels (3 to 4 bits) for acceptable television imaging for viewing by human beings.

The candling application, however, requires detection of faint shadows of the embryo, its coloration, and its shading to be detectable. Since this requires differentiation of subtle amounts of a few percent changes in transmittance, the sensor, must be able to distinguish many gray shades. Since solid-state sensors will be able to provide between 0.5 and 1.5% gray-scale resolution and tube-type sensors will provide between 3.125 and 6.25% gray-scale resolution, the former camera is the logical choice. Another benefit of the solid-state sensor is that it typically has a higher response in the red end of the visible spectrum than the tube types. Since the candling operation to some extent will depend on blood specks around the embryo, the solid-state camera has yet one more advantage over the tube type.

9.6.2.4 Acquisition speed and synchronization

For this application, there are no special requirements in regard to these conditions, so either a solid-state or tube-type camera will suffice. It should be noted, however, that when both are operating at 60 Hz in a noninterlaced mode, they will require between 16.67 and 33.33 ms to acquire an image. The explanation of this is left as an exercise for the reader (see Problem 9.15).

Although not relevant here and well beyond the scope of this presentation, in some circumstances it is important to be able to synchronize the image acquisition to external events. In such cases, special electronically or mechanically shuttered cameras or stroboscopic illumination techniques are frequently used to solve these problems.

9.6.3 Photo-Optical Interrupter

The conveyor is used to bring eggs into the field of view of the CCTV. A simple photointerrupter will be used to stop the belt at the appropriate placement of the egg approximately halfway into the field of view of the CCTV. Pseudocode for the conveyor controller that may be translated into a computer program is given in Table 9.6.1.

9.6.4 CCTV for Egg Location and Orientation

The CCTV will be interfaced to a video digitizer whose output will then be processed by an appropriate software program. The pseudocode presented in Table 9.6.2 may be used to construct such a program.

9.6.5 Colorizer/Density Measurement System

The candling system that is needed must be able to detect changes in the density or color of the egg when it is transilluminated by a yellow/orange light (i.e., a "candle"). Blood formed in the embryo will absorb relatively more of the candle's light spectrum and the egg will appear relatively darker. A possible configuration for such a system might contain a bright light source in the end of arm tooling.

TABLE 9.6.1   CONVEYOR PSEUDOCODE

```
If "MOT REQ" = TRUE              ; "MOT REQ" = internal flag
    Then MOVE BELT
        Until VISION 0 = TRUE    ; VISION 0 = processed
    Set "MOT REQ" = FALSE        ; result of photo-optical
                                 ; interrupter
```

Note: MOT REQ = TRUE if cell controller says OK, i.e., field of
                CCTV is clear AND another EGG is needed
      MOT REQ = FALSE when an egg is in the CCTV field of view.

TABLE 9.6.2 PSEUDOCODE FOR OBJECT ID, ORIENTATION, AND PICKUP

```
Compute    A = area of egg
           P = perimeter of egg
           X = major axis of egg
           Y = minor axis of egg
    [ x,y ] = centroid of egg

;IDENTIFICATION SECTION
If A       NOT within size limits   THEN OBJECT = FALSE
If ·P      NOT within size limits   THEN OBJECT = FALSE
If P*P/A NOT within size limits    THEN OBJECT = FALSE
      Else OBJECT = TRUE
;ORIENTATION SECTION
Compute Angle of Major Axis   ; Arctan (Y/X)
;PICK-UP SECTION
Compute Centroid              ; [ x,y ]
END
```

The light will be activated as the arm brings the egg over the candling test point. A measurement of the brightness of the light passing through the egg will be used to accept or reject the egg.

It is left as an exercise for the reader to propose a specific candling system and a specific sensor to be used as well as the determination of the accept/reject logic. Among the things that should be considered in selecting such a system are:

- Resolution of the sensor
- Type of sensor (point, line, array)
- Normal density determination (i.e., statistics)

9.6.6 Vision System Calibration

The CCTV vision system must be calibrated and referenced to the robot coordinate system. Obviously, this should be done only after the robot has performed its own calibration routine. Vision calibration can be accomplished by moving the robot arm into the field of view of the CCTV and locating it at two separate positions. A test pattern stenciled onto the end of the arm can be used to pinpoint it by the vision system. For example, a high-contrast " + " on the end of the manipulator would be a convenient fiducial mark for the vision system to locate. The robot would be programmed to move the " + " mark to a specific portion of the field of view (e.g., the upper left corner). Then the vision system would be programmed to locate the " + " mark. The robot would then be commanded to move to another portion of the field of view (e.g., the lower left corner). The vision system would then locate the mark again.

Once the vision system has determined the two arm placements, a simple coordinate transformation may be used to relate the two frames (i.e., vision frame and robot frame), thus allowing for the vision coordinates to be translated into the robot system coordinates for egg pickup and subsequent placement. In particular, we will have

$$R = TV + O$$

where R and V are, respectively, the robot and vision frames, T is the linear matrix transformation between them and O is the system offset. This procedure relates the vision frame to the robot frame, in pixels.

The relationship of the pixels to actual dimensions must also be known. This is best done by placing a test object of known size (e.g., a disk of known diameter) in the field of view, and then having the vision system measure the length of the object at several places, several times, and then computing the pixel-to-distance calibration (see Problem 9.16).

9.7 SUMMARY

This chapter brings together many of the techniques and technologies that have been discussed throughout this book. Obviously, one moderately simple example cannot illustrate all the material presented. The solutions given are suggestions and the reader is challenged to create different (better?) solutions to the design and implementation of an egg packer. For example, one may choose to sort the eggs by size or color (e.g., brown versus white eggs). One may also choose to inspect the eggs for hairline cracks. The reader is also encouraged to construct a working model of the egg candler/packer robot, since in this way a true appreciation of the problems associated with robotics and its associated technologies can be developed and appreciated.

9.8 PROBLEMS

9.1 Complete a specification for the egg packing task described in Section 9.2. Make assumptions about all pertinent system considerations, such as speed of the conveyor belt, weight of an egg, size of cartons, and so on. Do not make trivial assumptions such as that the speed of the conveyor belt is constant, without adding tolerances: for example, "the speed of the conveyor belt is constant to within ± 1 cm/s." The specification should be complete enough so that a co-worker would be able to draw a flowchart of the process or construct a procedure for accomplishing the task. Pay attention to material-handling assumptions on parts entering and leaving the workcell.

9.2 a. Draw a structured procedure for completing the processes in Problem 9.1. The procedure should be able to be given to a fellow worker who should be able to complete the task from this procedure. You must use Figure 9.2.2 and assign reasonable physical dimensions to the room layout.

    b. Design at least two separate procedures, one utilizing a vision system capable of locating the egg on the conveyor belt, and one utilizing no vision system for location but incorporating a mechanical workpiece that can orient the eggs for pickup.

9.3 (Optional) Construct a wooden model (to size) of the robot mechanism and the mechanical gripper and sizing mechanism. Build the model so that the r, θ, ϕ, and z axes and the gripper may be manually moved into the proper position for parts handling. Work on a tabletop that simulates a conveyor, and use cardboard construction material to simulate vision sensors and other details. Place an incandescent bulb at a strategic place so that a simulated illumination source is available. Using this mock-up, follow your procedure developed previously, and walk through each process step. Based on this simulation, prepare a list of the weak points of your procedure. Correct or modify your procedure based on this walkthrough, and then allow a co-worker to run through the process without your prompting. How did your modified procedure hold up under this scrutiny? What are your conclusions?

9.4 For the room layout that you have constructed, specify the location required for a vision system and a robot having cylindrical coordinate geometry. This r, θ, ϕ, and z robot will also need end-of-the-arm tooling, so specify an appropriate workholder for gripping the egg, candling it, and placing it in the carton.

9.5 Assuming that it is desired to move the egg in an uncoordinated motion path from any point in space $P_1 = (x_1, y_1, z_1)$ to any other point $P_2 = (x_2, y_2, z_2)$, describe the control algorithm for accomplishing this action. Assume the simplest possible control algorithm that you can think of. Using reasonable motor constants, such as the ones suggested in the text, and assuming that dc servomotors are used to drive the r, θ, and ϕ axes, describe the profile of each motor axis output. What is the Cartesian space path that will be traversed?

9.6 Assuming that coordinated motion is to be accomplished in a straight line from P_1 to P_2 in Cartesian space, derive the r, θ, and ϕ drive signals for this requirement. Make sure to consider the work volume of the robot so that unreachable zones are not traversed.

9.7 Detail the interface signals and the information required among the following subportions of the problem:
    a. Interface between conveyor belt and robot controller.
    b. Interface between robot controller and vision system.
    c. Interface between egg carton output stage and robot controller.

9.8 Repeat the θ axis calculation and motor determination if the inertia of the harmonic drive wave generator is 0.01 oz-in.-s².

9.9 Show that T_{rms} is approximately $= 0.80$ oz-in. for the r axis of the robot designed in the text.

9.10 Assuming that all three distances for the r axis are the same during any cycle, determine the largest distance that can be achieved utilizing the Pitman 9432 motor. Assume a cycle time of 3 s and that all other times are unchanged (in Table 9.3.2).

9.11 Repeat the r axis calculations if an additional inertia of 0.002 oz-in.-s² due to the encoder disk and coupler is assumed.

9.12 Find the peak and rms torques for the ϕ axis.

9.13 Determine the maximum power supply required by the r and ϕ axes and demonstrate that in both cases it is much smaller than that needed to drive the θ axis.

9.14 Detail a structured algorithm for the vision system so that the standard processes described in Chapter 6 can be used to accomplish the job. Be careful to describe the video servoing necessary to locate the egg on the conveyor belt, and also consider the problems of differentiating between a brown and a white egg, and whether or not it has an embryo in it.

9.15 Show that the acquisition time from a random external event, for an image from an RS-170 CCTV camera operating in a noninterlaced fashion, is between 16.67 and 33.33 ms. What are the average and standard deviations of the acquisition time?

9.16 Why is a disk of known size a good test object for calibration of the pixel-to-distance ratio for a camera and image acquisition system? Describe a set of algorithms in a pseudocode form to perform the calibration of a vision system, especially considering the possibility of having different calibrations for the two axes.

9.17 Based on the resolution of the encoders chosen for the r and θ axes and the assumption that position is maintained within ± 1 pulse of the set point for an arbitrary location in space, compute the worst-case error. How would you reduce this error by a factor of 2? Be sure to include the effects of the ϕ axis in your analysis.

9.18 Consider the effect of the inertia of the lead screw used in the r axis. Assume that the steel screw is an ideal cylinder that is 7 in. long and weighs 16 oz. Find the radius and inertia of the lead screw. Use the result to determine the peak and rms torques that must be developed by the Pitman motor. Is this actuator still adequate?

9.19 Suppose that the time to perform steps 7 and 8 and also steps 10 through 12 in the motion sequence Table (9.3.2) is reduced from 500 ms to 374 ms. Recalculate the peak and rms torques that must be generated by the θ, ϕ, and r motors and determine whether they are "adequate." If so, is it still possible to reduce the cycle time still further?

9.20 Using the procedure illustrated in Example 8.7.1 as well as some of the relationships derived therein, obtain the forward and inverse solutions for the manipulator shown in Figure 9.3.1.

9.21 Using Figure 3.10.2 as a model for the r–θ axes of the manipulator of Figure 9.3.1 and the velocity profiles shown in Figures 9.3.3 and 9.3.5:

a. Compute the required torque (θ axis) and force (r axis) as a function of time using Eqs. (3.10.10) and (3.10.11).

b. Based on the results of part a determine if the motors selected are satisfactory in terms of peak and rms torque ratings.

Appendix A

Specifications of Commercial Robots

The following table lists some of the commercially available robots and describes their specifications as supplied by their manufacturers. The manipulator industry is in a rapid state of flux and the reader is cautioned that this table should be considered a rough guide. In using this table the reader should keep in mind that:

1. The performance characteristics quoted in the table may not actually be achievable.
2. Not all manipulators listed in the table may be available.
3. There is a wide variety of capability between manufacturers in regard to both current and potential future customer service.
4. Some of the manipulators have undergone little, if any field testing and development. Some may be shipped without adequate (at least 50 hours) life testing in the factory.
5. Some manufacturers may not carry a large inventory of spare parts. Others may be out of business in the near future.
6. This list is not complete and additional models (from these and other companies) are available.
7. Entries with an asterisk (*) are no longer being made.

Also, "opt." in the table means the feature or axis is optional.

ROBOT SPECIFICATION GUIDE*

Manufacturer, model	Telephone	Typical price (Thous. $)	Load carrying capacity (lbf)	Repeatability (in.)	Maximum tip speed, no load (in./s)	Coordinate system				Maximum movement						Type of drive			Control	
						Cartesian	Cylindrical	Spherical	General	Manipulator reach	Manipulator elevation	Manipulator rotation or translation	End effector pitch (deg)	End effector yaw (deg)	End effector roll (deg)	Electric	Hydraulic	Pneumatic	Nonservo	Servo (C, C*, P)
Acco Industries, Jolly 80*	203/371-5439	$120	220	0.012				✓		43"	70"	80"				✓				P
Accumatic Machinery, Nu-Man R1/R2	419/535-7997	$75	75/35	0.010	24				✓	72"	136"	144"	180	180	280		✓		✓	
Acrobe Positioning Systems, AG-4	312/273-4302	$30	10	0.001	45°/sec				✓	41"	70"	300"				✓				✓
Adaptive Intelligence Corp., AARM Series 1000		$75	18	0.001	40				✓							✓				
Advanced Robotics, Cyro 750	614/929-1065	$135	35	0.008	10	✓				30"	30"	80"	130		720	✓				C, P
Advanced Robotics, Cyro 820	614/929-1065	$85	22	0.008	30					32"	35"	240°	200		380	✓				C, P
Advanced Robotics, Cyro 2000	614/929-1065	$140	35	0.016	5	✓				78.7"	78.7"	78.7°	130		720	✓				C, P
Air Technical Industries, Electro-arm	216/951-5191	$40	150	0.060	120			✓				360°	280°	280°	280°	✓				✓
Air Technical Industries, Electrobot	216/951-5191	$75	200	0.060	120			✓			280°	360°	280°	280°	280°	✓				✓
Air Technical Industries, Robo-Master	216/951-5191	$56	300	0.060	120			✓		280°	78°	280°	280°	280°	280°	✓				✓
Air Technical Industries, Robo-Ton	216/951-5191	$40	2000	0.060	120		✓			60"	100°	360°				✓				✓
A.K.R., AKR 3000		$90	33	0.080	78		✓			±42°		±47°30'	±135	±135	±135		✓			C
Amatrol, Inc., Centari 830-HR2-R7	812/288-8285	$33	20	0.010	50			✓									✓			✓
Amatrol, Inc., Hercules 830-HR1-C5		$11	20	0.005	12		✓										✓			✓
The American Robot Corp., Merlin	919/748-8761	$62	50	0.001	60			✓								✓				✓
Anorad, Anorobot	516/231-1990	$50	50	0.0007	40	✓										✓				
Armax Robotics, Armax Robot*	313/478-9330	$50	150	0.050	50	✓				24"	24"	40°				✓				C,C,P
ASEA, 1Rb-6/2	313/528-3630	$75	13	0.008	75			✓		25.25"	31.5"	340°	180		360		✓			C, P
ASEA, 1Rb-60/2	313/528-3630	$102	132	0.016	100			✓		39"	45.5"	330°	195	300	360	✓				C, P

Control: Continuous path, controlled path, point-to-point (C, C*, P)

*Reproduced and modified with permission of H. R. Arum and Techno, Inc., New Hyde Park, NY.

Model	Phone	Price ($1000)		Repeat.	Speed			H	I	J	K	L	M		Type	
ASEA, MHU Senior	313/528-3630	$30	33	0.002	43		✓	43"	20"	360°						
ASEA, MHU Junior	313/528-3630	$20	11	0.002	39		✓	48"	64"	200°						
ASEA, MHU Minor	313/528-3630	$10	2	0.002	12		✓	7.8"		180°						
Automatix, AID 800	617/273-4340	$85	22	0.008	39	✓		49"	62"	370°	180		150	✓		✓
Automatix, AID 900	617/273-4340	$95	65	0.008	80			80"	110"	320°	350	220	350	✓		✓
Automatix, AID 600	617/273-4340	$80	17.6	0.003	39	✓		20.9"	15.7"	51.2°	216	—	360	✓		✓
Automation, Automation II*	313/471-0554	$75	100	0.008			✓	29"	37"	150°				✓		
Bendix, AA 620*	313/352-7700	$90	45	0.002	255		✓	60"	102"	210°×2	220	220	±360	✓	P	
Bendix, ML 670*	313/352-7700	$95	150	0.005	100			98"	165"	300°	103	103	20×360	✓	P	
Bendix, MA 510*	313/352-7700	$59	22	0.008	79			50"	68"	240°	180		360	✓	P	
Bendix, MA 503*	313/352-7700	$49	7	0.004		✓		40"	51"	240°	180		360	✓	P	
Bendix, SL330*	313/352-7700	$45	66	0.012			✓	48"	30"	70°			6"	✓	P	
Binks Manufacturing, 88-800	312/671-3000	$45	18	0.125				48"	84"	135°					C, P	✓
Cincinnati, Milacron, T3-726	513/932-4400	$65	14	0.006	40	✓		41"	67"	285°				✓		
Cincinnati Milacron, T3-746	513/932-4400	$90	70	0.010	25	✓		97"	130"	270°				✓		
Cincinnati Milacron, T3-566	513/932-4400	$80	100	0.050	50		✓	97"	154"	240°	180	180	270	✓	P	✓
Cincinnati Milacron, T3-586	513/932-4400	$85	225	0.050	35		✓	102"	154"	240°	180	180	270	✓	P	✓
Comet Welding Systems, Limat 2000	312/956-0126	$175		0.004	18	✓		79.6"	78.4"	59.3°	255		390			
Control Automation, Assembly Robot*	609/799-6026	$82	10	0.001		✓		20"	20"	56"				✓		
Copperweld Robotics, CR-5	313/585-5972		2.5	0.003	Variable		✓	20"	2"	200°						
Copperweld Robotics, CR-10	313/585-5972		8	0.003	Variable		✓	0"	2"	200°	—	—	270	✓		✓
Copperweld Robotics, CR-50	313/585-5972		25	0.010	Variable		✓	12"	5"	200°	—	—	270	✓	P	✓
Copperweld Robotics, CR-100	313/585-5972		11	0.002	Variable		✓	18"	12"	24°				✓	P	✓
Cybotech, G80	317/298-5136		175	0.008	40		✓	24"	39.4"	78.8°	335	210	334		C	
Cybotech, H8	317/298-5136	$90	17	0.004			✓	59"	20"	320°	210		345	✓	C	
Cybotech, H80	317/298-5136		175	0.008	1 rad/s		✓	20"	63"	±135°	335	210	344		C	
Cybotech, P15	317/298-5136		7	0.2	78		✓	135°	+55° to −35°	±115°					C	
Cybotech, V15 Electric Robot	317/298-5890		15	0.004	68		✓	66"	±40°		±105	±175	±175	✓	C	
Cybotech, V80	317/298-5136		175	0.008	1 rad/s		✓	78.8"	71"	±135°	335	335	±180	✓	P	
Cyclomatic Industries, Ironman I	714/292-7440		20	0.010	5		✓	36"	118.2"	96"	210	210	344	✓	C	

697

ROBOT SPECIFICATION GUIDE (continued)

Manufacturer, model	Telephone	Typical price (Thous. $)	Load carrying capacity (lbf)	Repeatability (in.)	Maximum tip speed, no load (in./s)	Cartesian	Cylindrical	Spherical	General	Manipulator reach	Manipulator elevation	Manipulator rotation or translation	End effector pitch (deg)	End effector yaw (deg)	End effector roll (deg)	Electric	Hydraulic	Pneumatic	Nonservo	Servo (C, C*, P)
DeVilbiss, TR-3500S	419/470-2169	$ 95	12—50		7		✓			124"	81"	93°	176	176	210		✓			C, P
DeVilbiss, TR-3500W	419/470-2169	$ 80	50	0.160	36				✓	124"	80.5"	93°	176	176	210		✓			C, P
Esched Robotec Ltd., Scorbot ER-111		$ 4	2.2	0.020	9.8			✓								✓				✓
Feedback, Inc., Armdraulic EHA 1052		$ 7.9	4	0.200				✓									✓	✓		✓
Gallsher Enterprises, Gemini Concept*	919/725-8494	$ 15	5	0.001	30	✓				24"	12"	270°	220		220	✓				P
Gametics Kelate Model 524*		$ 50	50	0.030	36		✓			26"	38"	40'	330	210	330	✓				C, P
GCA, XR100 Extended Reach	312/369-2110		2,240	0.020	23				✓	20'	84"	300°	120	300	180	✓				C, P
GCA, B1440	312/369-2110		110	0.020	39				✓	39.4"	27.6"	300°		300		✓				C, P
GCA, P300H	312/369-2110		11	0.004	66				✓	23"	3.9"	270°	210		300	✓				
GCA, P800	312/369-2110		66	0.020	39				✓	54"	91.8"	270°	210		270	✓				
GCA, P300V	312/369-2110		11	0.004	45.3				✓	28"	41.4"	51.2°	98	36	360	✓				P
General Electric, A12 Allegro*	203/382-2876		14	0.001	240	✓				12.2"	10.4"					✓				
General Electric, AW-7*	203/382-2876		1,323	0.040		✓				47"		360°					✓			
General Electric, MH33*	203/382-2876		33	0.100				✓		82"						✓				
General Electric, GP132*	203/382-2876		132	0.290					✓	46"						✓				
General Electric, GP66*	203/382-2876		66	0.039	39.4				✓	49.5"	63"	300°	±25		185	✓				
General Electric, P-5*	203/382-2876	$ 65	22	0.008					✓	76"	122"	150°	250	250	250	✓				C
General Electric, S-6*	203/382-2876	$ 75	6.6	0.190	69				✓						250	✓				C
General Numeric, A-0*	312/640-1595	$ 32	22	0.002			✓			11.8"	11.8"	300°				✓				C

Product	Phone	Price ($1000)	Cap.	Repeat.	Speed	Dim 1	Dim 2	Sweep	Axis 1	Axis 2	Opt.
General Numeric, Series M, Model 0*	312/640-1595	$ 20	22	0.020		5.9"	5.9"	180°	5.9"		P
General Numeric, Series M, Model 1*	312/640-1595	$ 26	44	0.039		72.6"	57.1"	210°			
General Numeric, Series M, Model 3*	312/640-1595	$ 70	110	0.039		47.2"	47.2"	300°	190	300	
GMF Robotics Corp., A-0	313/641-4122	$ 24.7	22	0.002							√
GMF Robotics Corp., M-0	313/641-4122	$ 21.5	44	0.020		11.8"		300°	190°	360°	√
GMF Robotics Corp., M-3	313/641-4122	$ 48	264	0.039							√
GMF Robotics Corp., S-108L	313/641-4122	$ 41	17	0.008		47"			190°	300°	√
GMF Robotics Corp., S-360L	313/641-4122	$ 76.2	132	0.020							√
GMF Robotics Corp., S-380L	313/641-4122	$ 82.7	176	0.020							√
Graco Robotics, OM5000*	313/261-3270	$125									
Hail Automation, Ramp Spraying	303/371-5868	$ 50	30	0.080	67	95"	83"	90°			
Hirata, AR300 Arm Base	317/846-8859	$ 20	4.5–15	0.002	98						
Hirata, Hi CNC	317/846-8859	$ 9	110	0.002		21.6"	4"	270°			
Hitachi, Assembly Robot 25	201/825-8000		4.4	0.002	59	7.9"		7.9°			
Hitachi America, "Mr. Aros" JP/SP	201/825-8000		11	0.040	236	32"/40"	20"/44"	60°	±100	−50 to +50	C
Hitachi America, Process Robot	201/825-8000		22.2	0.008	39.4	38"	46"	300°	+85/−95	±185	C
Hitachi America, Spray Painting Robot 6V/6H	201/825-8000		11	0.080	68.9	123"/52"	52"/123"	150"/110°	240	250	C
Hodges Robotics, H35	517/323-7427	$ 55	35	0.005	180	48"	72"	360°	180	250	C
IBM, 7535	800/327-0166	$ 28.5	13.2	0.002	57	25.6"	3"	200°	180	180	C, P
IBM, RS1	800/327-0166	$100	5	0.008	5	58"	17"	18°	180	270	P
Ikegai America, FX-RBT	312/397-3970	$ 39	11/22	0.020	120°/s	24"	7"	120°	180	270	
Industrial Automates, Industrial Automate	414/327-5656	$ 14	10	0.015	30						
Intarm, 100-S*	513/294-0834		35	0.015	100						P
Intarm, 200*	513/294-0834		50	0.040	250	14"	18.37"	320°	300		
International Robomation, IRI-M50 E	714/438-4424		50	0.040	3°/s						
I.S.I. Manufacturing, Modular Robotics*	313/294-9500	$ 75	150	0.005	90	80"	80"	540°	240	720	P

ROBOT SPECIFICATION GUIDE (continued)

Manufacturer, model	Telephone	Typical price (Thous. $)	Load carrying capacity (lbf)	Repeatability (in.)	Maximum tip speed, no load (in./s)	Coordinate system				Maximum movement						Type of drive			Control	
						Cartesian	Cylindrical	Spherical	General	Manipulator reach	Manipulator elevation	Manipulator rotation or translation	End effector pitch (deg)	End effector yaw (deg)	End effector roll (deg)	Electric	Hydraulic	Pneumatic	Nonservo	Servo
C. Itoh & Co., Taiyo Toffky-2300*	313/352-6570	$ 67	22	0.008		✓				33.9"	7.9"	29.5"				✓				
C. Itoh & Co., Nachi Uniman 8600 AK*	313/352-6570	$110	75	0.040					✓	75.5"	84.4"	270°				✓				
KUKA Weld Systems & Robot Corp., IR 200	313/478-7850	$125	132	0.040	134				✓	19.7"		8.2"				✓				
KUKA Weld Systems & Weld Corp., IR 662	313/478-7850	$ 90	220	0.048	146				✓			360°				✓				
Lloyd Tool & Mfg. Corp., Joblot 10*	313/742-1820	$ 90	155	0.010			✓			110"	69.2"	300°				✓				
Lamson, Robopal	315/432-5467	$ 40	100	0.125	Variable															
Lynch Machinery, E-Z Handler*	317/643-6671	$ 70	300 to 1000	Variable	20												✓			
Mack Corporation, B-A-S-E Robot, Mod 56	602/526-1120	$ 5.1	5	0.010	10	✓				6"	8"	10"	180, 90	180, 90	180, 90		✓	✓	P	
Manca, Fibró Manca*	201/767-7227	$180		0.003			✓				49°	360°				✓			✓	
Microbot, Alpha	415/968-8911	$ 10.5	1.5	0.020	20					18"	26"	330°	±90		±180	✓			P	
Microbot, MiniMover 5	415/968-8911	$ 2.2	1	0.030	7					17.5"	25.6"	180°	±90		±180	✓			P	
Mitsubishi Electric America, Melfa RH-211	312/298-1535	$ 66	22	0.002	63	✓										✓				✓
Mitsubishi Electric America, Melfa RV-242	312/298-1535	$ 86	132	0.040	79	✓										✓				✓
Mitsubishi Electric America, Melfa RW-252	312/298-1535	$144	22	0.008	39	✓										✓				✓
Mobot, Mobot*	619/275-4300	$ 45	600/1000	0.030	36	✓				N/A	15'	15' ×100"	360	360	360	✓		✓	P	
Mobot, Mobot Columnar*	619/275-4300	$ 45	1000	0.030	18		✓			3'	20'	0	360	270	0	✓		✓	P	P

Control: Continuous path, controlled path, point-to-point (C, C*, P)

Product	Phone	$	(payload)	(repeat.)	(speed)	3'	20'	100'	0	270	0	270	O	Prog.
Mobot, Mobot Vertical Bridge*	619/275-4300	$ 60	1000	0.030	18									P
Nordson, Nordson Robot System*	216/988-9411	$ 90	4	0.200	5.5	39"	85"	75°	240	240	240	240	240	C
Pentel of America, PUHA 1*	800/323-6653	$ 25	6.6	0.002		9.7"	23.4"	200°						
Pentel of America, PUHA 2*	800/323-6653	$ 30	13.2	0.002		24.2"	35.1"	195°						
Pick-O-Matic Systems FR-100*	313/939-9320	$ 35	70	0.0016		54"	54"	360°			40		360	P
Pick-O-Matic Systems FR200-4*	313/939-9320	$ 55	150			62.5"	20"	340°						
Planet, Armax*		$ 60	150		50									C, P
Positech, Probot	712/845-4548	$ 90	1250	0.020	30	69.5"	131"	270°	180	270			360	P
Positech, Custom Probots	712/845-4548	$ 90	1250	0.060	30	36"	24"	200°	—	—			280	C, P
Prab Robots, E	616/349-8761	$ 50	100	0.030	36	varies	varies	varies	180	270			360+	C
Prab Robots, FA	616/349-8761	$ 72	250	0.050	36	60/48"	60/48"	240°	opt.	opt.	opt.	opt.	opt.	P
Prab Robots, FB	616/349-8761	$ 80	600	0.050	36	60/48"	60/48"	270°	opt.	opt.	opt.	opt.	opt.	P
Prab Robots, FC	616/349-8761	$125	2000	0.080	36	60/48"	60/48"	270°	opt.	opt.	opt.	opt.	opt.	P
Prab Robots, 4200	616/349-8761	$ 32	125	0.008	40	42"	20"	250°	opt.	opt.			90	P
Prab Robots, 5800	616/349-8761	$ 34	100	0.008	Variable	58"	20"	250°	opt.	opt.			90	P
Precision Robots, PRI-1000	617/862-1124	$ 30	2	0.002	10									
Reeves Robotics, Reebot Mar IV	206/392-1447	$ 60	4	0.008	40									
Reis Machines, RR625	312/741-9500	$ 65	55	0.008	Variable	134"	47"	360°						
Rimrock, Rimrock*	614/471-5926	$ 50	40	0.010	35	36"		90°					90	C, P
Rob-Con. Pacer Double Arm*	313/591-0300	$125	60	0.030		81"	79"	206°					90	P
Rob-Con. Pacer II*	313/591-0300	$ 41	50	0.010		52"	40"	270°				270	270	
Rob-Con. Pacer V*	313/591-0300	$ 80	1600	0.100		110"	103"	270°				270	270	
Robogate Systems, Polar 6000*	313/368-4280	$100	132	0.040	40	54"	−23° to +27°	±105°	240	360			360	
Robotic Sciences Intl., GS-1*	714/979-6831		10	0.005	2									C, P
Robotics, Prog.-A-Spencer	518/899-4211	$100	10	0.005	40									
Robotics, Model I Robotiks*	215/674-2800		0.3	0.001	30									
Sandhu Machine Design, Rhino XR-2	217/352-8485	$ 3	5	0.050	1.7	22"	32"	360°						
Schrader-Bellows, Motion Mate	216/375-5202	$ 10	5	0.005	Variable	24.5"	9.84"	180°						P
Seiko Instruments, FMS Series	213/530-8777	$ 25	10	0.0005		12"	6"	210°					90	P
Seiko Instruments, PN-100	213/530-8777	$ 5.7	3.3	0.0004	15.6	7.9"	2"	—					180	P

Manufacturer, model	Telephone	Typical price (Thous. $)	Load carrying capacity (lbf)	Repeatability (in.)	Maximum tip speed, no load (in./s)	Coordinate system				Maximum movement						Type of drive			Control: Continuous path, controlled path, point-to-point (C, C', P)	
						Cartesian	Cylindrical	Spherical	General	Manipulator reach	Manipulator elevation	Manipulator rotation or translation	End effector pitch (deg)	End effector yaw (deg)	End effector roll (deg)	Electric	Hydraulic	Pneumatic	Nonservo	Servo
Seiko Instruments, PN-200	213/530-8777	$ 5.5	1.65	0.0004	180°/s		✓				0.39″	90°			180			✓	P	
Seiko Instruments, PN-400/400L	213/530-8777	$9.6–10.8	8.8/6.6	0.001/0.002	22/23		✓			15.7″/27.6″	3.5″	180°			180			✓	P	
Seiko Instruments, PH-700	213/530-8777	$ 9.6	2.2	0.001	12		✓			7.9″	1.6″	120°			180			✓	P	
Sigma Sales, Sigma Max*	714/974-0166	$ 4	1	0.025	30			✓			90°	180°	90	180	180	✓				
Sterling Detroit, Robotarm Model U	313/366-3500	$ 25	700	0.005			✓			120″	72″						✓			
Systems Control, Smart-Arms*	617/263-1767	$ 4.5	4.4	0.020	90°			✓		36″	72″	60°				✓				
Thermwood, Cartesian 5	812/937-4476	$ 50		0.010					✓	36″		60°					✓			C
Thermwood, Series Six	812/937-4476	$ 50	18	0.125	40			✓		46″	80″	135°	180	180	270		✓			C
Thermwood, Series Seven	812/937-4476	$ 37	25	0.060				✓		39″	76″	280°	180		300		✓			
Thermwood, Series 3	812/937-4476	$ 35	50	0.060	45°			✓		60″	100″	280°	210		360		✓			P
Tokico America, Spray-painting Robot	213/328-7484	$100	11	0.080	78				✓	81″	89″	100°	210	360			✓			C, P
Unimation, Unimate Apprentice	203/744-1800	$ 38	10	0.060					✓	64″	35″	64°	165	360		✓				
Unimation, Unimate Puma 260	203/744-1800	$ 47	2.5	0.002	39			✓		16″	29″	315°	240	535	575	✓				C, P
Unimation, Unimate Puma 500/600	203/744-1800	$ 47	5.5	0.004	22			✓		36″	62.5″	320°	200	532	300	✓				P
Unimation, Puma 760	203/744-1800	$ 65	25	0.008	40				✓			320°				✓				
Unimation, Unimate 1000	203/744-1800	$ 35	50	0.050	110°/s			✓		42″	56″	208°	90	90			✓			P
Unimation, Unimate 2000	203/744-1800	$ 60	300	0.050	110°/s			✓		42″	56″	208°	220	200	360		✓			P
Unimation, Unimate 4000	203/744-1800	$ 70	450	0.080	65°/s			✓		52″	51″	200°	226	320	370		✓			P
Unimation, 6000 Series	203/744-1800	$120	150	0.005					✓							✓	✓			
Unimation, 9000 Series	203/744-1800	$115	450	0.040	180°/s				✓							✓	✓			

702

Company / Model	Phone	Price													
United States Robots, Maker 110	215/825-8550	$ 45	10	0.0032	40			20"	20"	350°	210	—	350	✓	C. P
United Technologies, NIKO 150	313/593-9600	$ 58	33	0.006		✓		53.3"	93"	320°				✓	
Westinghouse, Series 1000 & 2000*	412/255-3329	$ 30	11	0.001		✓				360°				✓	
Westinghouse, Series 4000*	412/255-3329	$ 65	22	0.008	3.9	✓				370°				✓	
Westinghouse, Series 5000*	412/255-3329	$115	22	0.004	20	✓								✓	
Westinghouse, Series 7000*	412/255-3329	$115	20	0.016	9.6	✓								✓	
Wexford Robotics*	306/522-7429	$ 20	50	0.02	39.4									✓	
Yaskawa Electric, Motoman L3	312/564-0770	$ 42	6.6	0.004	70	✓	✓	40"	51"	240°	180		360	✓	C. P
Yaskawa Electric, Motoman L3C	312/564-0770	$ 50	6.6	0.004	70	✓		36.2"	47.8"	240°				✓	✓
Yaskawa Electric, Motoman L-10WC	312/564-0770	$ 74	22	0.008	55		✓	50"	68"	240°	180		360	✓	C. P
Yaskawa Electric, Motoman L60	312/564-0770	$ 68	132	0.012	40		✓	38"	30"	40°	—	—	—	✓	P

ROBOT SPECIFICATION GUIDE (continued)

	Memory devices				Programming method				Memory capacity	Material handling	Machine load/unload						Tool manipulation				Assembly	Inspection	Education
	Solid-state electronics	Magnetic tape or disk	Air logic	Mechanical step sequencer	Keyboard	Pendant	Walk through	Mechanical setup			Die casting	Forging	Plastic molding	Machine tools	Investment casting	General	Spray painting	Welding	Machining	Other			
Acco Industries, Jolly 80*						✓			512 steps	✓													
Accumatic Machinery, Nu-Man R1/R2	✓	✓				✓			4000 + steps/points		✓	✓	✓		✓	✓		✓			✓		✓
Acrobe Positioning Systems, AG-4	✓				✓					✓													
Adaptive Intelligence Corp., AARM Series 1000																				✓	✓		
Advanced Robotics, Cyro 750	✓				✓	✓	✓		64k bytes									✓					
Advanced Robotics, Cyro 820	✓				✓	✓			1000 points									✓				✓	
Advanced Robotics, Cyro 2000	✓				✓	✓	✓		64k bytes									✓					
Air Technical Industries, Electro-arm	✓				✓	✓			7k words		✓	✓		✓	✓			✓		✓		✓	
Air Technical Industries, Electrobot	✓				✓	✓			7k words		✓	✓		✓	✓			✓		✓		✓	
Air Technical Industries, Robo-Master	✓				✓	✓			7k words		✓	✓		✓	✓			✓		✓		✓	
Air Technical Industries, Robo-Ton	✓				✓	✓			7k words		✓	✓		✓	✓			✓		✓		✓	
A.K.R., AKR 3000	✓	✓			✓	✓	✓		32,000 +								✓						✓
Amatrol, Inc., Centari 830-HR2-R7	✓					✓			500 Prog lines														✓
Amatrol, Inc., Hercules 830-HR1-C5	✓				✓	✓			504 Prog lines														
The American Robot Corp., Merlin	✓	✓			✓	✓			500 steps/points	✓			✓	✓				✓			✓	✓	
Anorad, Anorobot	✓	✓			✓		✓		12,000 + char.	✓						✓							
Armax Robotics, Armax Robot*	✓	✓			✓	✓			1450 steps/points	✓	✓	✓	✓	✓	✓		✓	✓		✓	✓		
ASEA, IRb-6/2	✓	✓			✓	✓			500 points	✓	✓	✓	✓	✓				✓	✓			✓	
ASEA, IRb-60/2	✓	✓			✓	✓			500 points	✓	✓	✓	✓	✓				✓	✓			✓	
ASEA, MHU Senior					✓	✓			127 programs	✓						✓							
ASEA, MHU Junior					✓	✓			127 programs	✓													

Model	Memory Capacity
ASEA, MHU Minor	192 steps
Automatix, AID 800	1000 points
Automatix, AID 900	1000 steps/points
Automatix, AID 600	1000 points
Automation, Automation II*	
Bendix, AA 620*	up to 4800
Bendix, ML 670*	up to 4800
Bendix, MA 510*	2,200
Bendix, MA 503*	2,200
Bendix, SL 330*	2,200
Binks Manufacturing, 88-800	8 programs
Cincinnati Milacron, T3-726	3000 points
Cincinnati Milacron, T3-746	3000 points
Cincinnati Milacron, T3-566	450 points
Cincinnati Milacron, T3-586	450 points
Comet Welding Systems, Limat 2000	1024 steps/points
Control Automation, Assembly Robot*	16k bytes
Copperweld Robotics, CR-5	256 steps/points
Copperweld Robotics, CR-10	250 steps/points
Copperweld Robotics, CR-50	250 steps/points
Copperweld Robotics, CR-100	256 steps/points
Cybotech, C80	1000 points
Cybotech, H8	1500 points
Cybotech, H80	1500 points
Cybotech, P15	1500 points
Cybotech, V15 Electric Robot	1500 points
Cybotech, V80	1500 points
Cyclomatic Industries, Ironman I	128k bytes
DeVilbiss, TR-3500S	128 minutes
DeVilbiss, TR-3500W	128 minutes

	Memory devices				Programming method				Memory capacity	Potential application													
										Material handling	Machine load unload					General	Spray painting	Tool manipulation			Assembly	Inspection	Education
	Solid-state electronics	Magnetic tape or disk	Air logic	Mechanical step sequencer	Keyboard	Pendant	Walk through	Mechanical setup			Die casting	Forging	Plastic molding	Machine tools	Investment casting			Welding	Machining	Other			
Esched Robotec Ltd., Scorbot ER-111	✓				✓				variable														✓
Feedback, Inc., Armdraulic EHA 1052	✓					✓			512														✓
Gallsher Enterprises, Gemini Concept*					✓																	✓	
Gametics Kelate Model 524*	✓	✓				✓			150 steps	✓											✓		
GCA, XR100 Extended Reach					✓	✓			1500 points	✓						✓		✓	✓				
GCA, B1440					✓	✓			999 steps	✓						✓		✓					
GCA, P300H						✓	✓		511 steps	✓								✓			✓		
GCA, P800					✓	✓			999 steps	✓											✓		
GCA, P300V						✓	✓		511 steps	✓											✓		
General Electric, A12 Allegro*	✓	✓			✓	✓			464 steps							✓		✓	✓		✓	✓	
General Electric, AW-7*	✓	✓				✓			435 steps									✓					
General Electric, MH33*						✓			290 steps	✓						✓		✓	✓				
General Electric, GP132*						✓				✓						✓		✓					
General Electric, GP66*						✓				✓								✓	✓		✓		
General Electric, P-5*	✓	✓				✓			448 steps					✓									
General Electric, S-6*	✓	✓				✓			464 points								✓						
General Numeric, A-0*	✓				✓	✓			up to 6000 points	✓											✓		
General Numeric, Series M, Model 0*						✓			300 points	✓			✓	✓									
General Numeric, Series M, Model 1*						✓			300 points	✓			✓	✓							✓		
General Numeric, Series M, Model 3*	✓				✓	✓			up to 6000 points	✓				✓									

Model	Capacity
GMF Robotics Corp., A-0	12,500
GMF Robotics Corp., M-0	6000
GMF Robotics Corp., M-3	12,500
GMF Robotics Corp., S-108L	12,200
GMF Robotics Corp., S-360L	12,060
GMF Robotics Corp., S-380L	12,060
Graco Robotics, OM5000*	128 programs
Hail Automation, Ramp Spraying	
Hirata, AR300 Arm Base	1000 points
Hirata, Hi CNC	1000 points
Hitachi, Assembly Robot 25	
Hitachi America, "Mr. Aros" JP/SP	464 steps
Hitachi America, Process Robot	2000 steps
Hitachi America, Spray Painting Robot 6V/6H	1000 steps
Hodges Robotics, H35	8000 points
IBM, 7535	5 programs
IBM, RS1	
Ikegai America, FX-RBT	16 positions
Industrial Automates, Industrial Automate	1000 steps
Intarm, 100-S*	1024 steps/points
Intarm, 200*	1024 steps/points
International Robomation, IRI-M50	999 points
I.S.I. Manufacturing, Modular Robotics*	49 steps/points
C Itoh & Co. Taiyo Toffky-2300*	254 steps
C Itoh & Co. Nachi Uniman 8600 AK*	1000 points
KUKA Weld Systems & Robot Corp. IR 200	1000 points

Robot	Solid-state electronics	Magnetic tape or disk	Air logic	Mechanical step sequencer	Keyboard	Pendant	Walk through	Mechanical setup	Memory capacity	Material handling	Die casting	Forging	Plastic molding	Machine tools	Investment casting	General	Spray painting	Welding	Machining	Other	Assembly	Inspection	Education
KUKA Weld Systems & Weld Corp. IR 662	✓				✓				1000 points									✓					
Lloyd Tool & Mfg. Corp. Joblot 10*	✓	✓			✓				1800 instructions	✓						✓							
Lamson, Robopal	✓								over 200														
Lynch Machinery, E-Z Handler*	✓	✓		✓	✓	✓			48k bytes	✓					✓								
Mack Corporation, B-A-S-E Robot, Mod 56	✓			✓		✓		✓					✓	✓	✓								
Manca, Fibro Manca*										✓													✓
Microbot, Alpha	✓				✓	✓			227 points														✓
Microbot, MiniMover 5	✓					✓			53 points														✓
Mitsubishi Electric America, Melfa RH-211	✓								1000					✓							✓		
Mitsubishi Electric America, Melfa RV-242	✓					✓			1000		✓			✓	✓			✓		✓			
Mitsubishi Electric America, Melfa RW-252	✓					✓			6000									✓					
Mobot, Mobot*	✓	✓			✓			✓	unlimited	✓	✓	✓	✓	✓	✓	✓	✓					✓	
Mobot, Mobot Columnar*	✓	✓			✓	✓		✓	unlimited	✓	✓	✓	✓	✓	✓	✓					✓	✓	
Mobot, Mobot Vertical Bridge*	✓	✓			✓	✓		✓	unlimited	✓		✓	✓	✓	✓	✓					✓	✓	
Nordson, Nordson Robot System*	✓	✓					✓										✓			✓			
Pentel of America, PUHA 1*					✓	✓			150 steps					✓					✓		✓		
Pentel of America, PUHA 2*					✓	✓			150 steps					✓							✓		
Pick-O-Matic Systems FR-100*	✓				✓	✓			250 points	✓						✓							
Pick-O-Matic Systems FR-200-4*					✓	✓			250 points	✓						✓							
Planet, Armax*	✓						✓		1,700 points	✓		✓		✓				✓	✓			✓	
Positech, Probot	✓	✓				✓			1000 words +	✓	✓	✓		✓								✓	

Manufacturer / Model	Capacity
Positech, Custom Robots	
Prab Robots, E	1000 +
Prab Robots, FA	4000 points
Prab Robots, FB	4000 points
Prab Robots, FC	4000 points
Prab Robots, 4200	24 points
Prab Robots, 5800	24 points
Precision Robots, PRI-1000	500 steps/points
Reeves Robotics, Reebot Mar IV	1000 steps/points
Reis Machines, RR625	200,000 steps/points
Rimrock, Rimrock*	
Rob-Con, Pacer Double Arm*	
Rob-Con, Pacer II*	
Rob-Con, Pacer V*	1k byte
Robogate Systems, Polar 6000*	1000 steps
Robotic Sciences Intl., GS-1*	256 steps/points
Robotics, Prog.-A-Spencer	
Robotics, Model I Robotiks*	1000 steps/points
Sandhu Machine Design, Rhino XR-1	variable
Schrader-Bellows, Motion Mate	300 steps
Seiko Instruments, FMS Series	
Seiko Instruments, PN-100	varies
Seiko Instruments, PN-200	varies
Seiko Instruments, PN-400/400L	varies
Seiko Instruments, PN-700	varies
Sigma Sales, Sigma Max*	varies
Sterling Detroit, Robotarm Model U	1000 steps
Systems Control, Smart-Arms*	200 steps/points
Thermwood, Cartesian 5	
Thermwood, Series Six	3000 points

ROBOT SPECIFICATION GUIDE (continued)

Robot	Memory devices: Solid-state electronics	Magnetic tape or disk	Air logic	Mechanical step sequencer	Programming method: Keyboard	Pendant	Walk through	Mechanical setup	Memory capacity	Material handling	Machine load unload: Die casting	Forging	Plastic molding	Machine tools	Investment casting	General	Tool manipulation: Spray painting	Welding	Machining	Other	Assembly	Inspection	Education
Thermwood, Series Seven	✓	✓					✓		5 minutes	✓	✓		✓	✓	✓								
Thermwood, Series 3	✓	✓							1000 points	✓	✓		✓	✓	✓								
Tokico America, Spray-painting Robot						✓	✓		2800 points								✓						
Unimation, Unimate Apprentice						✓			27 ft of weld									✓					
Unimation, Unimate Puma 260	✓	✓			✓	✓			500 points									✓	✓	✓	✓	✓	✓
Unimation, Unimate Puma 500/600	✓	✓			✓	✓			500 points	✓				✓				✓	✓	✓	✓	✓	✓
Unimation, Puma 760		✓			✓	✓			varies									✓		✓	✓	✓	
Unimation, Unimate 1000		✓				✓			Up to 256 points	✓	✓	✓	✓	✓	✓			✓					
Unimation, Unimate 2000		✓				✓			2048 points	✓	✓	✓	✓	✓	✓			✓					
Unimation, Unimate 4000		✓				✓			2048 points	✓	✓	✓	✓	✓	✓			✓					
Unimation, 6000 Series	✓	✓			✓	✓			varies	✓	✓	✓	✓	✓	✓			✓					
Unimation, 9000 Series	✓	✓			✓	✓																	
United States Robots, Maker 110	✓	✓				✓			350 steps	✓										✓	✓	✓	
United Technologies, NIKO 150					✓				400 points							✓							
Westinghouse, Series 1000 & 2000*									1000 points	✓						✓		✓					
Westinghouse, Series 4000*	✓	✓			✓				variable				✓					✓			✓	✓	
Westinghouse, Series 5000*	✓	✓			✓				800 points						✓						✓	✓	
Westinghouse, Series 7000*					✓	✓			1000 steps/points	✓						✓		✓					
Wexford Robotics*					✓				1000 steps/points					✓							✓	✓	
Yaskawa Electric, Motoman L3	✓	✓			✓	✓			2200 points				✓	✓				✓		✓	✓	✓	✓
Yaskawa Electric, Motoman L3C					✓	✓			1000 points				✓	✓				✓			✓	✓	
Yasakawa Electric, Motoman L-1WC	✓	✓			✓	✓			2200 points				✓	✓				✓		✓	✓	✓	✓
Yasakawa Electric, Motoman L60	✓	✓			✓	✓			1000 points									✓					

REFERENCES AND FURTHER READINGS

1. Carrigan, B., *Robots, A Bibliography with Abstracts, Report for 1964–Jan. '80*, National Technical Information Center, PB80–803836, 1980.

2. Chin, F., *Automation and Robotics, A Selected Bibliography of Books*, Vance Bibliographies, Public Administration Series, Bibliography #P–969, May 1982.

3. Engelberger, J. F., *Robotics in Practice: Management and Application of Industrial Robots*, AMACOM, New York City, 1980.

4. Flora, P. C., Ed., *International Robotics Industry Directory*, Fourth Edition, Conroe, Texas, Technical Data Base Corp., 1984.

5. Gomersall, A. and Farmer, P., *Robotics Bibliography*, 1970–81, IFS Publications, U.K., 1981.

6. *Industrial Robots*, magazine published by International Fluidics Service, Ltd., U.K.

7. McGraw-Hill Research/Standard and Poor's Corporation and Hibbert Company, *Industrial Robots*, 54 pp. Trenton, N.J.

8. Peake, C. A. and Campbell, R. A., *Let's Discuss Robotics*, Gardner Publications, Cincinnati, OH, 1981.

9. *Proceedings of the 12th International Symposium on Industrial Robots*, Paris, France, 1982 (see also Proceedings of first eleven symposia).

10. *Robot Industry Directory*, La Canada, CA, 1983.

11. *Robotics Age*, magazine published in Peterborough, NH, Vol. 4, 1982.

12. Roth, B., "Robots," *Applied Mechanics Reviews*, Vol. 31, 1978, pp. 1511–19.

13. Safford, E. L., Jr., *The Complete Handbook of Robotics*, TAB Books, Inc., Blue Ridge Summit, PA, 1978.

14. Simons, G. L., *Robots in Industry*, NCC Publications, Manchester, 1980.

15. Tanner, W. R., editor, *Industrial Robots, Volume 1: Fundamentals, Volume 2: Applications*, 2nd Edition, Society of Manufacturing Engineers, 1981.

16. *Stock Drive Products Book 757*, Vol. 2, Techno, Inc., 1983.

Appendix B

Motor Selection in the Design of a Robotic Joint

B.0 MOTIVATION

After the mechanical structure of a robotic manipulator is determined, a suitable scheme for producing motion must be developed. This will usually include the specification of mechanical couplers and/or drives (e.g., gear trains, harmonic drives, etc.) as well as the actuators themselves. A discussion of the former is found in Chapter 3. In this appendix we are concerned primarily with developing an organized method for selecting the servomotors that drive the individual joints of a robot. What is to be presented here is by no means the only way to handle this problem but instead represents a conservative approach which ensures that the motor selected will not exceed physical limitations of any of its parameters. As such, this appendix is intended to act as a guide (hopefully useful) to people "just getting started." Most certainly, readers are encouraged to expand on this material and develop their own techniques.

B.1 INTRODUCTION

In the following procedure all load quantities, such as moment of inertia J_L, viscous damping B_L, and noninertial, nonviscous torque T_L (representing either a load or a disturbance), together with coupling quantities such as moment of inertia and damping, are assumed to be referred to the *motor shaft*. See Chapter 3 for a discussion of how such calculations are performed. In addition, it is assumed that the designer has compiled a list of motors (and their parameters) arranged in *ascending* order of unit cost.   If a particular motor has more than one winding available from a manufacturer, these windings will be arranged in ascending order of *torque constant*, K_T.*   A sample of such a list is given in Table B.3.1 at the end of Section B.3.   Furthermore, it is assumed that the motor parameters included are those normally specified by manufacturers at room temperatures (e.g., 25°C) so that no *active cooling* such as fans is considered.

The calculation procedure to be described is divided into four sections:

A. Peak and maximum continuous torque calculation
B. Peak current calculation
C. Armature temperature calculation
D. Power supply peak voltage requirement

If, in any of these sections, a particular condition is not met, the designer must go back to the list of motors and change some or all of the motor parameters by either selecting another motor on the list or another winding for the *same motor*.

A flowchart of the entire procedure from which a computer program could be written is given in Section B.4.   An example that makes use of this process is presented in Section B.5.   The final section includes a list of preferred units used in the procedure and factors for converting other units that are often encountered to the preferred ones.

Before beginning the process of motor selection in a particular problem, the following quantities must be specified by the designer:

$$\omega_{pk} = \text{peak motor shaft angular velocity (rad/s)}$$
$$t_a = \text{acceleration time (s)}$$
$$t_{cv} = \text{constant velocity time (s)}$$
$$t_s = \text{deceleration time (s)}$$
$$t_d = \text{dwell time (s)}$$
$$J_L = \text{load moment of inertia (oz-in.-s}^2\text{)}$$
$$B_L = \text{load viscous damping (oz-in.-s/rad)}$$
$$T_L = \text{applied load torque (noninertial, nonviscous, and assumed constant) (oz-in.)}$$
$$\theta_{amb} = \text{ambient temperature (°C)}$$
$$I_{PS_{max}} = \text{maximum power supply current (A)}$$
$$V_{PS_{max}} = \text{maximum power supply voltage (V)}$$

In addition, the load speed versus time profile is assumed to be *trapezoidal*, as shown in Figure B.1.1 and is also referenced to the motor shaft.†   The reader

---

*It is important to note that one could also use peak or rms torques as a method for ordering the motors.

†It is assumed here that the robot joint (the load) and the motor have the same velocity profile. That is, the coupling transmission permits them to move *in phase* with one another.

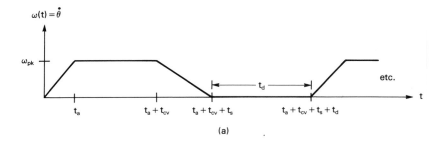

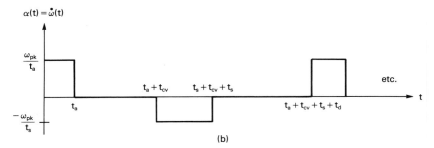

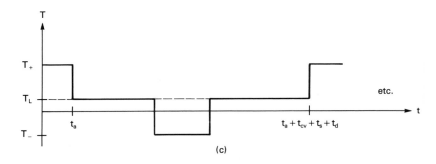

Figure B.1.1.: (a) motor shaft angular velocity profile (trapezoidal); (b) corresponding angular acceleration; (c) torque profile assuming a constant load torque. Here

$$T_+ = J_L \frac{\omega_{pk}}{t_a} + T_L \text{ and } T_- = -J_L \frac{\omega_{pk}}{t_s} + T_L.$$

should understand that this assumption is made for *computational convenience* since the associated acceleration is then a *piecewise constant* function of time. Thus calculation of the rms torque is easily done in closed form. See, for example, steps 2 and 3 in Section B.2.1. When, as is often the case, it is important to have a smoother velocity profile during the acceleration and deceleration phases, the rms torque must then be calculated by *integrating* the square of the torque versus time curve over the appropriate time interval(s), dividing by the time period (consisting of the total motion time plus any resting or "dwell" time t_d), and taking the square root of the result.

B.2 STEP-BY-STEP PROCEDURE FOR SELECTING A SERVOMOTOR

B.2.1 Peak and Maximum Continuous Torque Calculation

1. *Assume:* the motor parameters are all zero (i.e., $J_M = T_f = B = 0$). Therefore,

$$J_{\text{total}} = J_M + J_L = J_L$$
$$B_{\text{total}} = B + B_L = B_L$$
$$T_{\text{total}} = T_f + T_L = T_L$$

Compute:

$$T_{\text{accel}} = J_L\,\alpha_{\text{accel}} + T_L + B_L\,\omega_{\text{pk}}$$

$$= \frac{J_L\omega_{\text{pk}}}{t_a} + T_L + B_L\omega_{\text{pk}}$$

$$T_{cv} = T_L + B_L\omega_{\text{pk}}$$

$$T_{\text{decel}} = J_L\,\alpha_{\text{decel}} - T_L$$

$$T_{\text{dwell}} = T_L$$

$$T_{\text{rms}} = \sqrt{\frac{t_a T_{\text{accel}}^2 + t_{cv}T_{cv}^2 + t_s T_{\text{decel}}^2 + t_d T_{\text{dwell}}^2}{t_a + t_{cv} + t_s + t_d}}$$

2. Enter the LIST OF MOTORS (at winding number 1) (see Table B.3.1). Select the *least expensive* motor that satisfies all of the following conditions:

$$\text{(a)} \quad T_{\text{PEAK}} > T_{\text{accel}}$$

$$\text{(b)} \quad T_{\text{PEAK}} > T_{\text{decel}} \tag{B.2.1}$$

$$\text{(c)} \quad (T_{\text{cont}})_{\text{max}} > T_{\text{rms}}$$

3. Get J_M, T_f, and B from the list of motors for the motor selected in step 2. Recalculate T_{accel}, T_{cv}, T_{decel}, and T_{rms} as follows:

$$T_{accel} = (J_M + J_L) \frac{\omega_{pk}}{t_a} + T_f + (B + B_L) \omega_{pk} + T_L$$

$$T_{cv} = T_f + (B + B_L) \omega_{pk} + T_L$$

$$T_{decel} = (J_M + J_L) \frac{\omega_{pk}}{t_s} - T_f - T_L$$

$$T_{rms} = \sqrt{\frac{t_a T^2_{accel} + t_{cv} T^2_{cv} + t_s T^2_{decel} + t_d T_{dwell}}{t_a + t_{cv} + t_s + t_d}}$$

4. Check inequalities (B.2.1) in step 2. If the new values of J_M, T_f, and B do not produce a motor change, this motor satisfies the peak and continuous torque requirements. Then proceed to step 5. If inequalities (B.2.1) are not satisfied, move to the *next* motor on the list of motors and repeat steps 3 and 4 until there is no motor change required [i.e., inequalities (B.2.1) are satisfied]. Then proceed to step 5.

B.2.2 Peak Current Calculation

5. For the motor selected in step 4, obtain the K_T (the torque constant) and I_{demag} (the demagnetization current) for the appropriate winding number and compute the peak current as follows*:

$$I_1 = \frac{T_{accel}}{K_T}$$

$$I_2 = \left| \frac{T_{cv}}{K_T} \right|$$

$$I_3 = \left| \frac{T_{decel}}{K_T} \right|$$

$$I_{PEAK} = \text{max value } [I_1, I_2, I_3]$$

(B.2.2)

---

*Servomotors that employ *rare earth* magnets (e.g., samarium cobalt) usually have extremely large demagnetization currents which are virtually impossible to exceed during acceleration or deceleration of a motor shaft. In fact, for such motors, I_{demag} is normally not even specified by the manufacturer.

6. Is $I_{PEAK} < I_{demag}$?* If *yes*, go to step 7. If *no*, move to the *next motor* on the list of motors and repeat steps 3 through 6 [checking inequalities (B.2.1) in step 2 each time]. When inequalities (B.2.2) are satisfied, go to step 7.

7. Is $I_{PEAK} < I_{PS_{max}}$? If *yes*, go to step 8. If *no*, stay with the motor selected in step 6 until $I_{PEAK} < I_{PS_{max}}$. Now go to step 8. If all the windings are tried and I_{PEAK} is still greater than $I_{PS_{max}}$, move to the *next motor* on the list of motors (using winding 1) and repeat steps 3 through 7 [checking inequalities (B.2.1) and (B.2.2) each time] until $I_{PEAK} < I_{PS_{max}}$. Then go to step 8.

B.2.3 Armature Temperature Calculation

8. *Compute:*†

$$I_{rms} = \frac{T_{rms}}{K_T}$$

$$\theta_{rise} = \frac{R_a I_{rms}^2 R_{th}}{1 - R_a I_{rms}^2 R_{th}\,\psi} = \text{armature temperature rise above ambient}$$

$$\theta_{arm} = \theta_{rise} + \theta_{amb}$$

where ψ = resistance temperature coefficient = 0.00393 (copper wire)
                                = 0.00415 (aluminum wire)
R_a = armature winding resistance plus terminal resistance (at 25°C)
R_{th} = armature to ambient thermal resistance

9. Is $\theta_{arm} < \theta_{arm_{max}}$? If *yes*, go to step 10. If *no*, move to the *next winding* for the motor presently being used and repeat steps 5 through 9 until $\theta_{arm} < \theta_{arm_{max}}$. Now go to step 10. If all windings for this motor are tried and θ_{arm} is still greater than $\theta_{arm_{max}}$, move to the *next motor* (using the first winding) and repeat steps 3 through 9 until $\theta_{arm} < \theta_{arm_{max}}$. Then go to step 10.

---

*For rare earth motors, some arbitrarily large value of I_{demag} can be used in the list of motors so that step 6 can still be performed. Here, however, the "answer" will always be *yes*.

†Here it is assumed that the motor is mounted on a "suitable" heat sink. Often, this is specified (by the manufacturer) to be a block of metal 10 in. × 10 in. × 1/4 in. or equivalent. If this is not so, the motor temperature will be higher or lower than calculated.

B.2.4 Power Supply Peak Voltage Requirement

10. *Compute:*

$$R_h = R_a[1 + \psi(\theta_{\mathrm{arm}} - 25)]$$

$$= \text{``hot'' value of the armature resistance}$$

$$N_o = 9.593 \times 10^{-3}\, \omega_{\mathrm{pk}}$$

$$= \text{motor shaft speed (1000 rpm)}$$

$$V_{\mathrm{PEAK}} = \text{max. value} \begin{cases} V_1 = R_h I_1 + K_E N_o \\ V_2 = R_h I_2 + K_E N_o \\ V_3 = R_h I_3 + K_E N_o \end{cases}$$

11. Is $V_{\mathrm{PEAK}} < V_{\mathrm{PS_{max}}}$? If *yes*, go to step 12. If *no*, move to the *next motor* on the list of motors and repeat steps 3 through 11 [checking inequalities (B.2.1) and (B.2.2) each time]. Repeat until $V_{\mathrm{PEAK}} < V_{\mathrm{PS_{max}}}$. Then go to step 12.

12. Record the motor number, manufacturer's name, winding number, cost per unit, all other parameters, θ_{arm}, I_{PEAK}, V_{PEAK}, T_{PEAK}, and T_{rms}.

13. Stop.

B.3 MOTOR SELECTION PROCEDURE: FINAL CONSIDERATIONS

The reader should note that torque constant K_T is *temperature sensitive* and should be derated as follows:

$$K_T' = K_T [1 - \beta(\theta_{\mathrm{magnet}} - 25°\mathrm{C})]$$

where β = torque constant derating factor = 0.002 (ceramic magnets) = 0.005 (alnico magnets)
    θ_{magnet} = maximum motor temperature = θ_{arm}

This change in K_T will especially affect the demagnetization current check (step 6) but will become important only when θ_{arm} is near the maximum allowable value (often 155°C). Under these conditions, it may be wise to select the *next motor* on the list of motors. In fact, for safety sake, this would be good practice whenever θ_{arm}, I_{PEAK}, V_{PEAK}, T_{PEAK}, and/or T_{rms} are near the maximum values for the motor selected using the procedure. Note that this could automatically be built into the procedure by derating $V_{\mathrm{PS_{max}}}$ and $I_{\mathrm{PS_{max}}}$ and increasing θ_{amb} (and some or all of the load parameters by 10%, for example).

It is worthy of note that in steps 10 and 11, where the power supply peak voltage requirement is determined, it is assumed that there are *no voltage drops in the cables* that connect the supply to the motor armature terminals. In practice, however, the nonzero resistance of the wire produces a small to moderate (i.e.,

TABLE B.3.1 SAMPLE LIST OF MOTORS

Motor number	J_M (oz-in.-s²)	B (oz-in.-s/rad)	T_f (oz-in.)	T_{PEAK} (oz-in.)	$(T_{CONT})_{max}$ (oz-in.)	K_T (oz-in./A)	I_{demag} (A)	R_a^b (Ω)	R_{th} (°C/W)	ψ	θ_{armmax} (°C)	K_E (V/1000 rpm)	Winding number	N_{Wmax}^c	Manuf.[d]	Cost ($/unit)
540	0.0038	9.55 × 10⁻³	3	240	29	10.02	24	1.64	5.0	0.00393	155	7.41	A = 1	4	1	40
	0.0038	9.55 × 10⁻³	3	240	29	12.63	19	2.39	5.0	0.00393	155	9.34	B = 2	4	1	40
	0.0038	9.55 × 10⁻³	3	240	29	15.91	15	3.55	5.0	0.00393	155	11.76	C = 3	4	1	40
	0.0038	9.55 × 10⁻³	3	240	29	20.04	12	5.39	5.0	0.00393	155	14.82	D = 4	4	1	40
541	0.005	1.43 × 10⁻³	3	282	38	11.77	24	1.44	4.6	0.00393	155	8.70	A = 1	4	1	45
	0.005	1.43 × 10⁻³	3	282	38	14.83	19	2.31	4.6	0.00393	155	10.96	B = 2	4	1	45
	0.005	1.43 × 10⁻³	3	282	38	18.64	15	3.66	4.6	0.00393	155	13.81	C = 3	4	1	45
	0.005	1.43 × 10⁻³	3	282	38	23.54	12	5.81	4.6	0.00393	155	17.40	D = 4	4	1	45
2115	0.0044	2.86 × 10⁻³	3	150	30	6.12	24.5	.469	5.0	0.00393	155	4.5	A = 1	8	2	50
	0.0044	2.86 × 10⁻³	3	150	30	7.48	20.1	.718	5.0	0.00393	155	5.5	B = 2	8	2	50
	0.0044	2.86 × 10⁻³	3	150	30	9.52	16.2	1.14	5.0	0.00393	155	7.0	C = 3	8	2	50
	0.0044	2.86 × 10⁻³	3	150	30	11.90	12.6	1.84	5.0	0.00393	155	8.8	D = 4	8	2	50
	0.0044	2.86 × 10⁻³	3	150	30	15.00	10.0	2.87	5.0	0.00393	155	11.1	E = 5	8	2	50
	0.0044	2.86 × 10⁻³	3	150	30	18.70	8.0	4.52	5.0	0.00393	155	13.8	F = 6	8	2	50
	0.0044	2.86 × 10⁻³	3	150	30	23.00	6.5	7.01	5.0	0.00393	155	17.0	G = 7	8	2	50
	0.0044	2.86 × 10⁻³	3	150	30	29.00	5.2	11.08	5.0	0.00393	155	21.5	H = 8	8	2	50

[b]Armature resistance value *plus* terminal resistance.

[c]Number of windings.

[d]1, Electrocraft; 2, Torque Systems.

719

1 to 5 V) reduction in the voltage *actually reaching the motor*. Consequently, it is necessary to take this into account when sizing the supply. The effect of ohmic loss can also be minimized by paralleling two or more wires, thereby reducing the current that each has to carry and hence the voltage drop. A sample list of motors is shown in Table B.3.1.

B.4 FLOWCHART OF MOTOR SELECTION PROCEDURE

A flowchart of the 13-step procedure described in Section B.2 is shown in Figure B.4.1. It is also necessary to have compiled a data base which lists a variety of servomotors that are currently available from different manufacturers. Called the *list of motors*, this data base must be updated periodically for the procedure to be most effective.

B.5 EXAMPLE

To illustrate the use of the procedure described in Section B.2, let us find a motor that will move a weight a short distance in a linear fashion very rapidly. Note that such a situation is analogous to a robotic prismatic axis. It is assumed that a lead screw is employed as the mechanical drive between the motor and load. The specifications for this problem are given below.

Problem Specifications

W_L = weight of load to be moved = 188 oz
P = lead screw pitch = 5 turns/in.
η = lead screw efficiency = 1 (i.e., 100%)
L_{LS} = lead screw length = 6.15 in.
r_{LS} = lead screw radius = 5/16 in.
ΔX = linear displacement of load = 0.060 in.
$t_a = t_s$ = 0.015 s
t_{cv} = 0
T_{L_f} = total lead screw friction = 10 oz-in.
B_L' = 0

In addition, it is assumed that the motion duty cycle is 100%, so that the dwell time $t_d = 0$.

B.5.1 Preliminary Calculations

1. *Lead screw moment of inertia, J_{LS}*

$$\text{Weight} = \text{volume} \times \text{density} = r_{LS}^2 L_{LS} \rho_{LS}$$

$$\rho_{LS} = 0.28 \text{ lb/in.}^3 = 4.48 \text{ oz/in.}^3$$

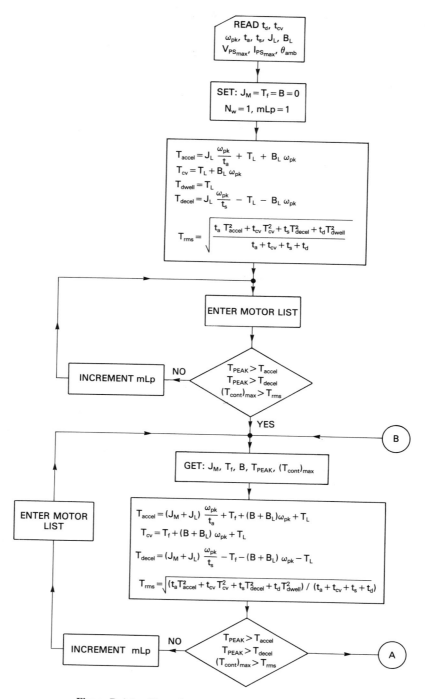

Figure B.4.1. Flow chart for motor selection procedure.

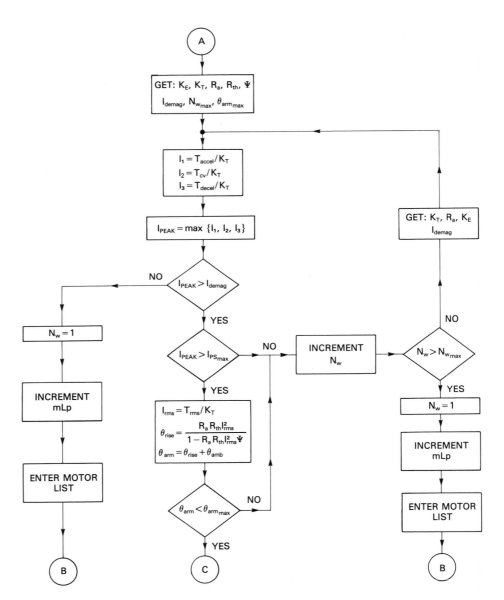

Figure B.4.1. *(continued)*.

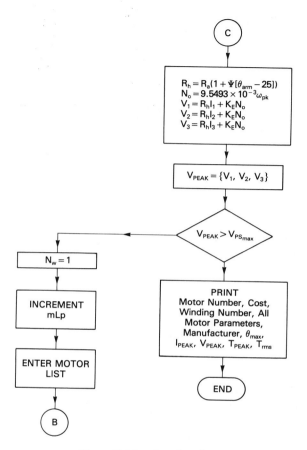

C

$R_h = R_a(1 + \Psi[\theta_{arm} - 25])$
$N_o = 9.5493 \times 10^{-3}\omega_{pk}$
$V_1 = R_h I_1 + K_E N_o$
$V_2 = R_h I_2 + K_E N_o$
$V_3 = R_h I_3 + K_E N_o$

$V_{PEAK} = \{V_1, V_2, V_3\}$

$V_{PEAK} > V_{PS_{max}}$

$N_w = 1$

INCREMENT mLp

ENTER MOTOR LIST

B

PRINT
Motor Number, Cost,
Winding Number, All
Motor Parameters,
Manufacturer, θ_{max},
I_{PEAK}, V_{PEAK}, T_{PEAK}, T_{rms}

END

Figure B.4.1.   (*continued*).

$$\text{Weight}_{LS} = \pi \times (5/16)^2 \times 6.15 \times 4.48 = 8.45 \text{ oz}$$

$$m_{LS} = \frac{\text{weight}_{LS}}{g} = \frac{8.45}{386.09} = 0.0219 \text{ oz-s}^2/\text{in.}$$

$$J_{LS} = 0.5\, m_{LS}r_{LS}^2 = 0.001069 \text{ oz-in.-s}^2$$

2. *Load equivalent moment of inertia, J'_L*

$$J'_L = \frac{m_L}{\eta(2\pi P)^2}$$

$$= \frac{188/386.09}{1[2\pi(5)]^2}$$

$$= 0.000493 \text{ oz-in.-s}^2$$

3. *Peak motor shaft angular velocity*, ω_{pk}

$$\omega_{pk} = \frac{2 \times (2\pi P) \times \Delta X}{t_a + 2t_{cv} + t_s}$$

$$= \frac{2 \times 2\pi(5) \times 0.06}{0.015 + 0 + 0.015}$$

$$= 125.66 \text{ rad/s}$$

B.5.2 Program Inputs

$$\omega_{pk} = 125.66$$

$$J_L = J_{LS} + J'_L = 0.001562$$

$$B_L = 0$$

$$T_L = 10$$

$$t_a = t_s = 0.015$$

$$t_{cv} = t_d = 0$$

$$I_{PS_{max}} = V_{PS_{max}} = \infty \text{ (no limit on power supply given)}$$

$$\theta_{amb} = 25°C$$

B.5.3 Complete Example

With the values in Section B.5.2 as inputs, it is possible to utilize the procedure from Section B.3. The pertinent calculations and comments about each program step are summarized in Table B.5.1. It is noted that the initial guess concerning a likely motor candidate (i.e., an Electrocraft E530) is proven to be incorrect. As a consequence, steps 3 and 4 must be repeated to find a motor that meets the specifications.

B.6 UNITS AND CONVERSION FACTORS

In the following list of quantities used in the motor selection procedure presented in Section B.2, the preferred unit (i.e., the unit to be used in the procedure) is given first. When a quantity is specified using another unit, the conversion to the preferred unit is accomplished by multiplying by the factor shown. Defining equations are given, where appropriate.

TABLE B.5.1 SUMMARY OF CALCULATIONS FOR EXAMPLE

Program step	Calculations	Comments
1	$$T_{accel} = \frac{J_L \times \omega_{pk}}{t_a} + T_L$$ $$= \frac{0.0015162 \times 125.664}{0.015}$$ $$= 23.09 \text{ oz-in.}$$ $$T_{decel} = \frac{-J_L \times \omega_{pk}}{t_s} + T_L$$ $$= -3.09 \text{ oz-in.}$$ $$T_{rms} = 16.47 \text{ oz-in.}$$	There is no "run" period since t_{cv} is zero. Therefore, $T_{cv} = 0$.
2	$T_{PEAK} = 240 > T_{accel}$ and also $|T_{decel}|$ $(T_{CONT})_{max} = 29 > T_{rms}$	Using an Electrocraft E530/110 motor, inequality (B.2.1) is satisfied. Continue.
3	$J_M = 0.004$ oz-in.-s^2 $T_f = 3$ oz-in. $B = 9.55 \times 10^{-4}$ oz-in.-s/rad $$T_{accel} = \frac{(0.004 + 0.001562) \times 125.66}{0.015} + 3 + 10 + (9.55 \times 10^{-4}) \times 125.66$$ $$= 59.72 \text{ oz-in.}$$ $T_{decel} = 33.48$ oz-in. $T_{rms} = 48.41$ oz-in.	J_M is moment of inertia of the motor, 110 tach, and bearings.
4	$T_{PEAK} = 240 > T_{accel}$ and also $|T_{decel}|$ $(T_{CONT})_{max} = 29 < T_{rms}$	Inequality (B.2.1) is violated. Go to next motor on list of motors—assume M1030 and go back to step 3.
3	$J_M = 0.00071$ oz-in.-s^2 $B = 9.55 \times 10^{-3}$ oz-in.-s/rad $T_f = 2.5$ oz-in. $T_{accel} = 32.73$ oz-in. $T_{decel} = 5.33$ oz-in. $T_{rms} = 23.45$ oz-in.	

TABLE B.5.1 *(continued)*

Program step	Calculations	Comments
4	$T_{\text{PEAK}} = 348 > T_{\text{accel}}$ and also $|T_{\text{decel}}|$ $(T_{\text{CONT}})_{\text{max}} = 35 > T_{\text{rms}}$	If T_{PEAK} is not given, it is calculated as $K_T \times I_{\text{demag}}$. Inequality (B.2.1) is satisfied. Continue.
5	$K_T = 5.8$ oz.in./A $I_{\text{demag}} = 60$ A $I_1 = \dfrac{32.73}{5.8} = 5.64$ A $I_3 = \dfrac{5.33}{5.8} = 0.92$ A $I_{\text{PEAK}} = 5.64$ A	
6	$I_{\text{PEAK}} = 5.64 < I_{\text{demag}} = 60$ A	Magnet not demagnetized. Continue.
7	$I_{\text{PEAK}} < \infty$	Power supply maximum current not specified. Continue.
8	$R_a = 0.7 + 0.15 = 0.85\ \Omega$ $\psi = 0.00393$ $R_{\text{th}} = 2.8°\text{C/W}$ $K_E = 4.3$ V/1000 rpm $\theta_{\text{arm}_{\text{max}}} = 155°\text{C}$ $I_{\text{rms}} = 23.45/5.8 = 4.04$ A $\theta_{\text{rise}} = \dfrac{0.85 \times 2.8 \times (4.04)^2}{1 - 0.85 \times 2.8 \times (4.04)^2 \times 0.00393}$ $= 45.9°\text{C}$ $\theta_{\text{arm}} = 45.9 + 25 = 70.9°\text{C}$	
9	$\theta_{\text{arm}} = 70.9 < \theta_{\text{arm}_{\text{max}}} = 155°\text{C}$	Motor runs "cool" in continuous operation. Continue.

TABLE B.5.1   (*continued*)

Program step	Calculations	Comments
10	$R_h = 0.85[1 + 0.00393 \times (70.9 - 25)] = 1.003 \ \Omega$ $N_o = 9.5493 \times 10^{-3} \times 125.66 = 1.2 \times 10^3$ rpm $V_1 = 1.003 \times 5.64 + 4.3 \times 1.2 = 10.8$ V $V_2 = 1.003 \times 0.92 + 4.3 \times 1.2 = 6.1$ V $V_{\text{PEAK}} = 10.8$ V	
11	$V_{\text{PEAK}} = 10.8 < \infty$	Power supply voltage not specified. Continue.

Conclusion: An Electrocraft M1030 motor with 110 tachometer will perform well with a power supply capable of supplying 6 A at 11 V.

The table format is as follows:

Quantity (symbol, if any) = preferred unit
$\qquad\qquad\qquad\qquad$ = secondary unit × factor needed to convert to preferred unit
$\qquad\qquad\qquad$ (Defining equation, if any)

1. Mass (m) = ounce-second^2/inch = weight in ounces/386.09
$\qquad\qquad\qquad$ = weight in pounds × 16/386.09
2. Length = inch = feet × 12 = meter × 39.37 = centimeter × 0.3937
3. Linear velocity (v) = inch/second = feet × 12/second
$\qquad\qquad\qquad$ = meter × 39.37/second
$\qquad\qquad\qquad$ = centimeter × 0.3937/second
4. Angular velocity $\omega(t)$ = radians/second = rpm × 0.10472
$\qquad\qquad\qquad$ = 1000 rpm × 104.72
5. Force (F) = ounces = pound × 16 = newton × 3.5969
6. Torque (T) = ounce-inch = pound-feet × 192
$\qquad\qquad\qquad$ = newton-meter × 141.612
7. Moment of inertia (J) = ounce-inch-second^2
$\qquad\qquad\qquad$ = kilogram-meter^2 × 141.612
$\qquad\qquad\qquad$ = kilogram-centimeter^2 × 0.014162
$\qquad\qquad\qquad$ = gram-centimeter^2 × 1.41612 × 10^{-5}
8. Viscous damping (B) = ounce-inch-second/radian
$\qquad\qquad\qquad$ = ounce-inch/1000 rpm × 9.5493 × 10^{-3}
$\qquad\qquad\qquad$ = newton-meter-second/radian × 141.612
$\qquad\qquad\qquad$ = newton-meter/1000 rpm × 1.352

9. Temperature (θ) = degrees Celsius

$$\theta = \frac{\text{degrees Fahrenheit} - 32}{1.8}$$

10. Torque constant (K_T) = ounce-inch/ampere
    = 1000 rpm/ampere $\times$ 1.3887 $\times$ 10^3
    = newton-meter/ampere $\times$ 141.612

11. Voltage (back EMF) constant (K_E) = volts/1000 rpm
    = volt-second/radian $\times$ 104.72

APPENDIX C

Digital Control of a Single Axis

C.0 INTRODUCTION

This appendix discusses the implementation of a digital control for a single axis. From a top-down approach, we discuss the various concepts that must be considered when implementing a computer controlled servo loop used in a robot.

C.1 SYSTEM DESCRIPTION

The block diagram of a single-axis servo is shown in Figure C.1.1. Essentially, it consists of two components:

- Digital servo loop
- Profile generator

The digital servo loop contains the motor, its load, and a power amplifier and the electronics and software necessary to implement some type of closed-loop control. The profile generator computes set points (over some time interval) and sends them to the digital servo loop so that the ultimate position of the robot axis driven by the motor's shaft follows some prescribed function of time.

The remainder of our discussion will concern ultimate control we will have over the motor's shaft position. Thus our control strategy assumes that the profile generator provides discrete position versus time set points, $\theta_s(nT)$. Over a given time interval, these set points define a specified function which the motor's shaft

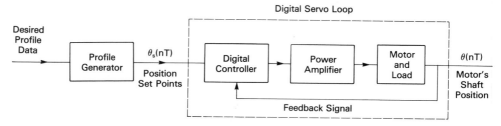

Figure C.1.1. Profile generator and digital servo loop.

position, $\theta(nT)$, should follow. If we wish the axis to remain fixed at some position, the same set point must be sent at each update to the digital servo loop. It is extremely important to understand that set points must always be sent to the input of the digital servo loop even if no motion is to occur.

C.2 THE PROFILE GENERATOR

Figure C.2.1 shows a detailed description of the profile generator. It should be apparent that the implementation of this block requires an output stream of data in real time—that is, the output corresponds to a desired position as a function of time. This discrete data signal also carries implicit information about the desired velocity and acceleration states of the motor's shaft.

A possible scenario of the operation of the profile generator is as follows:

1. The displacement and desired maximum time for a motion are sent to the profile generator.
2. Based on the constraint data, information is sent to the displacement algorithm so that it can comply with the request.
3. When a "start" signal is received, the appropriate number of set points are output spaced by the clock's period.
4. After all the set points have been generated, a "profile completed" signal is generated.

As can be seen from Figure C.2.1, there are two major components: the displacement algorithm and the control logic. The displacement algorithm is the rule by which the set points are generated. The control logic accepts commands, reports status, and checks the validity of the data presented to the displacement algorithm.

The profile generator is usually implemented in software. Since the data sequence $[\theta_s(nT)]$ must be output in real time, it is necessary to provide a clock for synchronization. Later we will see that this clock is also needed to synchronize the digital servo loop.

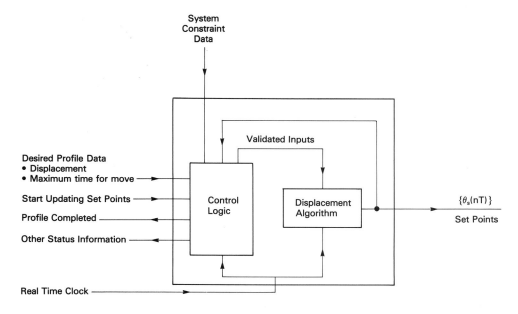

Figure C.2.1. Details of the profile generator.

Certain constraints defined by the physical system (i.e., motor, power supplies, and so on) are either resident in the algorithm or supplied as input data to the control logic. These constraints are used to prevent this block from producing profiles that are beyond the physical capabilities of the system. Thus if the time required for a motion would result in a value of acceleration that could demagnetize a motor or could stress a transmission, either an error signal could be sent or the system would report the exception but compensate and possibly use the maximum value that was permissible. This latter condition may actually produce a profile that does not comply with the initial request.

The desired profile data constitute the minimum set of information needed to tell the displacement algorithm the information it needs to generate set points. For instance, if the total time for the move and the displacement are specified, then based on known relationships, the algorithm can determine the values of acceleration, constant velocity, and deceleration that satisfy the request without violating the constraint data.

Figure C.2.2 shows the position and acceleration waveforms for a trapezoidal velocity profile. While this profile is common practice throughout industry and quite simple to analyze, it contains impulses in the jerk function (i.e., the third derivative of position) which can excite various modes of the system, which may result in undesirable operation. To reduce or eliminate this problem, higher order polynomials can be used whose characteristics are to constrain the magnitude of the jerk; however, we will use the trapezoidal velocity profile in our discussion due to its simplicity.

It can be shown, using calculus, that the corresponding position waveform for the trapezoidal velocity profile can be described by the following set of equa-

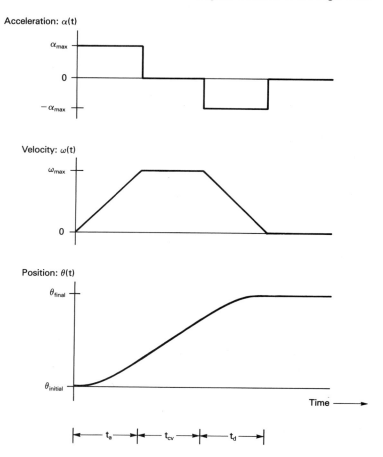

Figure C.2.2. Trapezoidal velocity profile and corresponding acceleration and position profiles.

tions. These equations are solely functions of the three time intervals, (t_a, t_{cv}, t_d) corresponding to the acceleration, the constant velocity, and the deceleration times respectively and the maximum velocity, ω_{max}.

$$0 \le t \le t_a$$

$$\theta(t) = \tfrac{1}{2} \left[\omega_{max}/t_a\right] t^2 \tag{C.2.1}$$

$$t_a \le t \le (t_a + t_{cv})$$

$$\theta(t) = \left[\tfrac{1}{2} \omega_{max} t_a\right] + (t - t_a)\omega_{max} \tag{C.2.2}$$

$$(t_a + t_{cv}) \le t \le (t_a + t_{cv} + t_d)$$

$$\theta(t) = \omega_{max}\left[\tfrac{1}{2}t_a + t_{cv}\right] + \tag{C.2.3}$$

$$\omega_{max}\left[(t - (t_a + t_{cv}))\left[1 - (t - (t_a + t_{cv}))/2t_d\right]\right]$$

By substituting nT for t (where $n = 0, 1, \ldots k$) we obtain the position for any discrete time. The maximum value, k, must be an integer. By design choice, the total motion time $(t_a + t_{cv} + t_d)$ can be forced to be an integer multiple of T. Note that T is the sample period for the waveform generator and that this set of equations gives an exact value for the desired profile, $\theta(t)$, at any discrete time nT.

While Eqs. (C.2.1) through (C.2.3) give an exact solution, they involve quite a bit of computation especially if a higher order polynomial was used to define the velocity profile. Since it was stated previously that the stream of data was required to be in real time, this solution may not be practical for the computational power of the hardware chosen. Of course, we could always get a more powerful computer, but this may not be cost effective.

One possibility is to precompute the information and store it in a lookup table. This table can then provide the data from memory at the correct times. At face value, this seems to be the solution, and in fact it is for certain cases, specifically when there are a finite number of profiles and sufficient memory. However, from a practical point of view, we probably only want to store final position points and a descriptor of the desired profile as defined by parameters, such as the desired velocity and desired acceleration for a particular motion. This set of information is a better model for the concepts discussed in Chapter 2.

An interesting alternative to either table lookup or exact equations is to do the integration digitally in real time. By starting with a description of the acceleration profile, in terms of the magnitude of the maximum acceleration and the three time intervals needed to obtain the desired displacement, one could implement the following set of equations, which are quite simple.

$$\omega(nT) = \omega([n - 1]T) + T\,\alpha(nT) \qquad\qquad (C.2.4)$$

$$\theta(nT) = \theta([n - 1]T) + T\,\omega(nT) \qquad\qquad (C.2.5)$$

These equations are based on a simple forward rectangular integration scheme. Eq. (C.2.4) is an exact solution, while Eq. (C.2.5) will have an error associated with it. Even though the possibility of an error exists, if the integration rate (defined by T) is fast enough, it may be quite small or can be removed by modifying several positions of the profile by adding part of the error term. If the error in Eq. (C.2.5) is unacceptable, a bilinear integration scheme can be used which yields the exact result for integration of a trapezoid. If higher order polynomials are used to describe the displacement curve, the error due to simple forward integration will be more pronounced and either the integration period will have to be reduced or more accurate integration schemes chosen.

Further simplification is possible by eliminating the constant coefficient T. This can be accomplished by normalizing the output by T. Additionally, α_{max} may be scaled so that the output data stream is in the same granularity as defined by the encoding device. That is, properly scaling Eqs. (C.2.4) and (C.2.5) will give a data stream whose units are encoder pulses per update and encoder pulses respectively.

As mentioned previously, when the profile generator is implemented, it must recognize certain physical constraints about the actual system and produce data that are consistent with the ratings of the components. For example, if the motor and amplifier are sized for the worst-case conditions (using a scheme such as outlined in Appendix B), the profile generator must have knowledge of the limitations based on parameters it understands, such as absolute maximum allowable acceleration and absolute maximum allowable velocity. With this information and a desired displacement, an algorithm can find values such as t_a, t_{cv}, t_d, and ω_{max} which will generate a position data stream that can be tracked by the digital servo loop.

In the actual implementation of a complete robotic system, the calculations required to generate the set points for all of the servos may take a considerable amount of time. For example, consider the requirements of the back solution as discussed in Chapter 8. If the updates to the digital servo loop are too slow, the performance may be unaceptable or instability may result. A method of alleviating the lack of data in the servo is to use an interpolator. Essentially, the interpolator takes the difference between two successive profile generator set points (which may occur every 16 servo updates), divides the difference by the number of digital servo loop updates, and feeds the set point data at the servo rate. This allows the profile generator sufficient time to do its computations yet provides the servo with set point data at the desired rate.

C.3 THE DIGITAL SERVO LOOP

The digital servo loop is depicted in Figure C.3.1. Although we are calling it digital, it does contain analog components. However, since the control algorithm is done by a digital process, we are justified in using the digital terminology. As shown, the analog components consist of the dc motor and its associated load, a power amplifier, and a D/A converter. The controller and summing junction are implemented by difference equations within some computational element (e.g., a microprocessor). In practice, the feedback transducer could be an optical encoder (as discussed in Chapter 5) whose output pulses would cause an up/down counter to be updated whenever motion occurs. We have modeled the position feedback transducer by an integrator (which converts the motor's shaft velocity to position) immediately followed by a uniform rate sampler to convert the continuous position signal into the digital domain so that it can be applied directly to the summing junction. To implement uniform sampling, the register connected to the counter would only be read every T seconds.

To summarize the operation of this block, at a given rate (every T seconds) a digital value is placed on the input of the D/A converter which causes the motor's

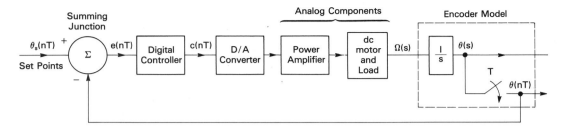

Figure C.3.1.   The digital servo loop.

shaft to move.   The shaft's position is monitored by an encoder which updates a counter.   The value of the counter, $\theta(nT)$, is read by the computer at the same rate the D/A updates and is compared with the set point $\theta_s(nT)$ generating an error signal, $e(nT)$.   The error signal is operated on by the difference equation of the digital controller to produce the control signal driving the D/A converter, $c(nT)$.

For purposes of our discussion, we let the digital controller be a PID algorithm (see Chapter 4).   Figure C.3.2 illustrates the block diagram of the digital controller with difference equations performing the differentiation and integration operations.

Figure C.3.3 shows the corresponding z domain representation of the controller.

At this point, we will mention a very important concept: the actual implementation of the integrator.   Figures C.3.2 and C.3.3 both utilize a bilinear transformation for the integrator.   It can be shown that this particular implementation preserves both regions of stability and instability when mapping from the s to the z planes.   Both of the other common forms of integration (i.e., forward rectangular

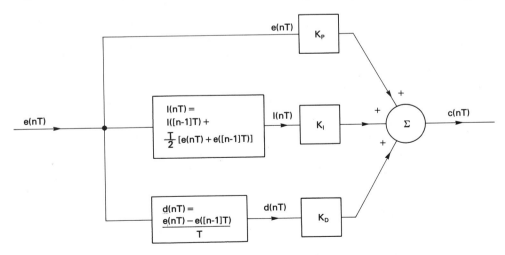

Figure C.3.2.   Digital implementation of PID controller.

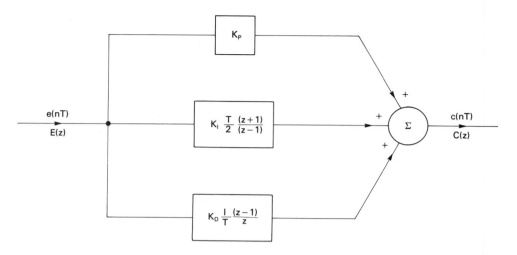

Figure C.3.3.   z-transform model of PID controller.

and backward rectangular) do not preserve both the stable and unstable regions simultaneously.   Specifically, forward rectangular preserves stability but under certain conditions maps unstable regions into the unit circle, while backward rectangular preserves the unstable regions but may map a stable region outside the unit circle.   These concepts become important if one chooses to model a system with a digital controller as a continuous system and are dependent on the sample rate chosen.   If sampling is fast enough, even without the bilinear implementation, no problems should occur; however, in cases of undersampling, the results are instability and inaccuracy.   Modeling of a system such as shown in Figure C.3.1 in the continuous domain also requires the addition of a phase shift of T/2 radians due to the D/A converter.   However, if one models the system in the z domain, regardless of the form of integrator chosen, and ensures stability by the unit circle criteria, no problem will exist by design.

Another point to consider when choosing an integration scheme is the accuracy that the algorithm will provide.   Although stability may be designed in, the actual value of the integrator may be far from the desired result.   Of course, the accuracy is dependent on the sample period.   It can be shown that bilinear integration running at an update rate of T seconds provides about the same accuracy as a forward rectangular scheme running with a T/10 second update rate.

Examination of Figure C.3.2 will show that under certain conditions, the output from the integrator may grow very large.   If fixed point arithmetic is used to implement the PID control algorithm, it is entirely possible that the value of the integral term may exceed the length of the word chosen as memory for the integrator.   Unless something is done, a wraparound may occur and an unpredictable operation will result.   One practical way to circumvent this situation is to place a saturation nonlinearity after the integrator. Effectively, this will let the integrator reach a defined magnitude, say the maximum word length chosen, after which its value stays fixed until an input causes it to be reduced (i.e., move in the opposite direction).

When implementing a saturation after an integrator, it is important to actually hold the integrator at its maximum (or minimum) value once the limit has been reached.   That is, the integrator must stop increasing for inputs that would make it grow, but should respond immediately to inputs that would make it decrease. It should be clear that this is very different from letting the integrator always run but forcing the output to a constant if the value of the integrator is greater than some limit.

Saturation is extremely important when interfacing the control signal $c(nT)$ to the D/A convertor.   Since D/A convertors have a finite word length, typically 12 to 16 bits, it is necessary to ensure that the magnitude of $c(nT)$ does not exceed the capability of the D/A converter. This is accomplished easily by adding a saturation nonlinearity before the converter so that the signal applied is in the converter's range.

Saturation may also be introduced to prevent some conditions from occurring. For example, if a constant torque is applied to a joint driven by a motor that is attempting to maintain a constant set point, it may be desirable to limit the input to the D/A converter so as not to burn out the motor.   For the case of a new equilibrium position (different from the set point), the contribution from the derivative term would be zero, the contribution from the proportional term would be a constant, but the contribution from the integral term would grow since the error term was constant.   The saturation after the integral term may not be sufficient to prevent damage and the value of the saturation nonlinearity preceding the D/A input could be reduced to limit the magnitude of the signal applied to the D/A converter in order to ensure that the motor current is kept below some maximum magnitude.

At this point it is interesting to note that whereas the implementation shown in Figure C.3.1 includes the controller and a physical system, one could conceivably replace the physical components (D/A, power amplifier, motor, and load) with a linear model to simulate the digital servo loop.   In fact, the blocks starting with the D/A converter and ending with the ideal sampler can be modeled in the z-domain and ultimately replaced with a difference equation. Thus by running exactly the same algorithms for the summing junction and controller in the model and using the difference equation to replace the physical components, the system may be tuned by modifying the PID parameters to obtain a desired performance response.   These tuned parameters could then be placed in the actual controller. Although the model and the physical system may not be exactly the same, simulation will provide a reasonable starting point for the controller parameters which can then be fine-tuned in the actual system.

C.4 AN IMPLEMENTATION

To give a perspective on the interaction of the profile generator and the digital control loop, the following example may be helpful: A system consisting of both the profile generator and digital control loop will be implemented using a single microprocessor, which has a real-time interrupt occurring every T milliseconds. To simplify our example, all the parameters to describe the desired trapezoidal velocity profile will be stored in memory (maximum acceleration and the times t_a, t_{cv}, and t_d) as well as the PID gains. Additionally, we will assume that initially, the motor is stationary (i.e., maintaining position) at some position θ_1, until a certain signal, *enable_profile_gen*, is asserted. At this point, the profile generator will begin operation and generate set points. The control loop will use the set points and the actual position of the motor's shaft (available from a register called *encoder_counter_register*) to generate a value for the control which it will ultimately put on the D/A converter. After the profile generator has completed its sequence, it will assert the *profile_complete* flag and clear the *enable_profile_gen* flag, and the set points will remain fixed at the value of the final position, θ_2. Remember that θ_2 is the sum of the initial position, θ_1 and the change in displacement caused by running the profile generator.

Table C.4.1 defines one possible algorithm. The variables used are consistent with the previous figures, equations, and discussions in this appendix. Note that in this implementation, the control algorithm executes first, using the value in the set point register. When motion is in progress, the profile generator computes the set point for the next update immediately following the current update. This approach guarantees that the data are placed on the D/A converter at a fixed rate, since the code executing from the beginning of the interrupt to the D/A output are straight line and deterministic in nature.

Due to contributions from friction, or the actual system's response (as defined by the PID parameters), it is possible that the motor's shaft position is not at the final position defined by the profile generator after the *profile_complete* flag is asserted. Consider for instance, the case where the system is somewhat underdamped. Even though the profile generator is finished and the set point is constant, the system must respond as defined by its difference equation. Therefore, it may be necessary to test the value of the error signal, $e(nT)$, and set another flag to indicate when some settling criteria has been attained.

C.4.1 The Model Paradox

A careful examination of the algorithm in Table C.4.1 and the block diagram of Figure C.3.1 will show a subtle difference. Specifically, Figure C.3.1, which is the typical topology shown in most textbooks, shows the system output $\theta(nT)$ being combined with the set point information and the control algorithm to immediately

TABLE C.4.1 PSEUDOCODE IMPLEMENTATION FOR A PROFILE GENERATOR AND DIGITAL CONTROL LOOP

```
    ON INTERRUPT
{
        θ = encoder_counter_register; /* current position   */
        θs = setpoint_register;          /* set point θs   */
/*   summing junction   */
        e = θs − θ;
/*   PID controller */
        i = i_old + (e + e_old)/2;   /* integrator */
        i_old = i;
        d = e − e_old;              /* derivative term */
        e_old = e;
        c = (kp * e) + (ki * i) + (kd * d);
/*   put control signal c on D/A converter       */
        d_a = c;
/*   Profile generator */
        if (enable_profile_gen = = TRUE && profile_complete = = TRUE){
            n = 1;
            profile_complete = FALSE;
            α = max_acceleration;
        }
        while (profile_complete = = FALSE)
        {
            if (n > ta)
                α = 0;
            else if (n > (ta + tcv))
                α = − max_acceleration;
            else if (n > (ta + tcv + td)){
                α = 0;
                profile_complete = TRUE;
                enable_profile_gen = FALSE;
            }
            ω = ω_old + α;     /* compute new velocity */
            set point_register = old_set point + ω;
            old_set point = set point_register;
            ω_old = ω;
            n = n + 1;
        }
}   /*   end of interrupt routine */
```

generate $\theta(nT)$. This is mathematically correct; however, this particular topology cannot be implemented in the real world because of communication and response delays. Examination of the pseudocode will show that while the current value of the system output, θ, and the set point information, θ_s are also combined to feed the control algorithm, the result of this control is not seen until the encoder is read again at the next update.   A little thought should convince the reader that in order to make Figure C.3.1 model the actual system under computer control requires the addition of a unit delay (z^{-1}) in the feedback path along with the difference equation (D/A through sampler) used to model the physical components.   This technique will make the block diagram in the z domain model the physical system and therefore allow analysis in terms of stability and transient performance.

Index

744